READY-TO-EAT SNACKS

Emerging Technologies for Production and Safety

Postharvest Biology and Technology Series

READY-TO-EAT SNACKS

Emerging Technologies for Production and Safety

Ravi Pandiselvam, PhD
Gülşah Çalışkan Koç, PhD
Editors

APPLE ACADEMIC PRESS

First edition published 2025

Apple Academic Press Inc.
1265 Goldenrod Circle, NE,
Palm Bay, FL 32905 USA

760 Laurentian Drive, Unit 19,
Burlington, ON L7N 0A4, CANADA

CRC Press
2385 NW Executive Center Drive,
Suite 320, Boca Raton FL 33431

4 Park Square, Milton Park,
Abingdon, Oxon, OX14 4RN UK

Library and Archives Canada Cataloguing in Publication

CIP data on file with Canada Library and Archives

Library of Congress Cataloging-in-Publication Data

CIP data on file with US Library of Congress

ISBN: 978-1-77491-864-7 (hbk)
ISBN: 978-1-77491-865-4 (pbk)
ISBN: 978-1-00357-012-7 (ebk)
DOI: 10.1201/9781003570127

DEDICATION

Dr. Ravi Pandiselvam would like to dedicate this book to his mother, Anbu Mani.

ABOUT THE BOOK SERIES: POSTHARVEST BIOLOGY AND TECHNOLOGY

Editor-in-Chief

Mohammed Wasim Siddiqui, PhD

Scientist-cum-Assistant Professor, Department of Food Science and Technology, Bihar Agricultural University, Sabour, Bhagalpur, Bihar, India

AAP Acquisitions Editor | Horticultural Science

Founding/Managing Editor | *Journal of Postharvest Technology*

Email: wasim_serene@yahoo.com

As we know, preserving the quality of fresh produce has long been a challenging task. In the past, several approaches were in use for the postharvest management of fresh produce, but due to continuous advancement in technology, the increased health consciousness of consumers, and environmental concerns, these approaches have been modified and enhanced to address these issues and concerns.

The Postharvest Biology and Technology series features edited books that address several aspects related to postharvest technology of fresh produce. The series presents existing and novel management systems that are in use today or that have great potential to maintain the postharvest quality of fresh produce in terms of microbiological safety, nutrition, and sensory quality.

The books are aimed at professionals, postharvest scientists, academicians researching postharvest problems, and graduate-level students. This series is intended to be a comprehensive venture that provides up-to-date scientific and technical information focusing on postharvest management for fresh produce.

Books in the series address the following themes

- Nutritional composition and antioxidant properties of fresh produce
- Postharvest physiology and biochemistry
- Biotic and abiotic factors affecting maturity and quality
- Preharvest treatments affecting postharvest quality

- Maturity and harvesting issues
- Nondestructive quality assessment
- Physiological and biochemical changes during ripening
- Postharvest treatments and their effects on shelf life and quality
- Postharvest operations such as sorting, grading, ripening, de-greening, curing etc.
- Storage and shelf-life studies
- Packaging, transportation, and marketingVase life improvement of flowers and foliage
- Postharvest management of spice, medicinal, and plantation crops
- Fruit and vegetable processing waste/byproducts: management and utilization
- Postharvest diseases and physiological disorders
- Minimal processing of fruits and vegetables
- Quarantine and phytosanitory treatments for fresh produce
- Conventional and modern breeding approaches to improve the postharvest quality
- Biotechnological approaches to improve postharvest quality of horticultural crops

BOOKS IN THE SERIES

Postharvest Biology and Technology of Horticultural Crops: Principles and Practices for Quality Maintenance
Editor: Mohammed Wasim Siddiqui, PhD

Insect Pests of Stored Grain: Biology, Behavior, and Management Strategies
Ranjeet Kumar, PhD

Postharvest Management of Horticultural Crops: Practices for Quality Preservation
Editors: Mohammed Wasim Siddiqui, PhD, and Asgar Ali, PhD

Innovative Packaging of Fruits and Vegetables: Strategies for Safety and Quality Maintenance
Editors: Mohammed Wasim Siddiqui, PhD,
Mohammad Shafiur Rahman, PhD, and Ali Abas Wani, PhD

Advances in Postharvest Technologies of Vegetable Crops

Editors: Bijendra Singh, PhD, Sudhir Singh, PhD, and
Tanmay K. Koley, PhD

Plant Food By-Products: Industrial Relevance for Food Additives and Nutraceuticals
Editors: J. Fernando Ayala-Zavala, PhD, Gustavo González-Aguilar, PhD, and Mohammed Wasim Siddiqui, PhD

Emerging Technologies for Shelf-Life Enhancement of Fruits
Editors: Basharat Nabi Dar, PhD, and Shabir Ahmad Mir, PhD

Sensor-Based Quality Assessment Systems for Fruits and Vegetables
Editors: Bambang Kuswandi, PhD, and Mohammed Wasim Siddiqui, PhD

Emerging Postharvest Treatment of Fruits and Vegetables
Editors: Kalyan Barman, PhD, Swati Sharma, PhD, and
Mohammed Wasim Siddiqui, PhD

Packaging and Storage of Fruits and Vegetables: Emerging Trends
Editor: Tanweer Alam, PhD

Postharvest Physiological Disorders of Fruits
Editors: Ghulam Khaliq, PhD, and Sajid Ali, PhD

Bioengineered Fruit and Vegetables: Trends in Postharvest Quality Improvement
Editors: Mohammad Irfan, PhD, and Pankaj Kumar, PhD

Novel Packaging Systems for Fruits and Vegetables
Editors: Kirtiraj K. Gaikwad, PhD, and Suman Singh, PhD

Ozone Applications in the Food Industry: Trends and Future Prospects
Editors: Ravi Pandiselvam, PhD, Anjineyulu Kothakota, PhD, and Kaavya Rathnakumar, PhD

Nonthermal Food Processing Technologies: Impact on Color Profile
Editors: Kaavya Rathnakumar, PhD, and R. Pandiselvam, PhD

Ready-to-Eat Snacks: Emerging Technologies for Production and Safety
Editors: R. Pandiselvam, PhD and Gülşah Çalışkan Koç, PhD

ABOUT THE SERIES EDITOR

Mohammed Wasim Siddiqui, PhD
Scientist-cum-Assistant Professor, Department of Food Science and Technology, Bihar Agricultural University, Sabour, Bhagalpur, Bihar, India

Mohammed Wasim Siddiqui, PhD, is an Assistant Professor and Scientist in the Department of Food Science and Post-Harvest Technology, Bihar Agricultural University, Sabour, India, and author or co-author of peer-reviewed research articles, book chapters, two manuals, and conference papers. He has over 20 edited and an authored books to his credit, published by Elsevier, CRC Press, Springer, and Apple Academic Press. Dr. Siddiqui has established the international peer-reviewed *Journal of Postharvest Technology.* He is Editor-in-Chief of two book series (*Postharvest Biology and Technology and Innovations in Horticultural Science*), published by Apple Academic Press. Dr. Siddiqui is also a Senior Acquisitions Editor for Apple Academic Press, USA, for Horticultural Science. He has been serving as an editorial board member and active reviewer of several international journals.

Dr. Siddiqui has received several grants and respected awards for his research work by a number of organizations. He is an active member of the organizing committees of several national and international seminars, conferences, and summits. He is one of key members in establishing the World Food Preservation Center (WFPC), LLC, USA, and is currently an active associate and supporter.

Dr. Siddiqui acquired a BSc (Agriculture) degree from Jawaharlal Nehru Krishi Vishwa Vidyalaya, Jabalpur, India. He received his MSc (Horticulture) and PhD (Horticulture) degrees from Bidhan Chandra Krishi Viswavidyalaya, Mohanpur, Nadia, India, with specialization in Postharvest Technology. He is a member of Core Research Group at the Bihar Agricultural University (BAU), which provides appropriate direction and assistance in prioritizing research.

ABOUT THE EDITORS

Ravi Pandiselvam, PhD
Scientist, ICAR–Central Plantation Crops Research Institute (CPCRI), Kasaragod, India

Ravi Pandiselvam, PhD, is actively engaged in food engineering research. After completion of doctorate studies at Tamil Nadu Agricultural University (2015), he joined ICAR as a scientist (Agricultural Process Engineering) at the Central Plantation Crops Research Institute (CPCRI), Kasaragod, India, in 2016. Dr. Pandiselvam has contributed in the design and development of a minimal processing machine for tender coconut, a continuous type coconut Testa removing machine, a tender coconut punching and cutting machine, and preservation protocol for trimmed tender coconut. He is also a developer/co-developer of many value-added food products (Kalpa Krunch, coconut flakes, foam-mat dried coconut milk powder, frozen coconut delicacy, and dark chocolate and matured coconut water based beverages), which are commercialized for many entrepreneurs. Dr. Pandiselvam has authored more than 180 articles published in national and international journals of repute. He has been author or co-author of six books and 67 book chapters. He was conferred with a National Academy of Agricultural Sciences (NAAS) Young Scientist Award 2020, NAAS Associate 2021, and Indian Society of Agricultural Engineers (ISAE) Distinguished Service Award 2020. He received his BTech in Agricultural Engineering in 2010 and MTech in Agricultural Processing & Food Engineering in 2012 from the Tamil Nadu Agricultural University, India.

Gülşah Çalışkan Koç, PhD
Associate Professor, Food Technology Program, Esme Vocational High School, Usak University, Turkey

Gülşah Çalışkan Koç, PhD, is an Associate Professor in the Food Technology Program at Esme Vocational High School, Usak University, Turkey. She teaches courses on unit operations along with other food science courses. Her research focuses on drying, freezing, baking, extrusion, enrichment of bakery products, traditional fermented products, etc. She worked as a Research Assistant in the Department of Food Engineering at Ege University, Turkey, between 2010 and 2018. She also worked as an Assistant Professor in the Department of Gastronomy and Culinary Arts, Alanya Hamdullah Emin Paşa University from April 2018 to September 2019. She is a member of the International Food Technologists and Chamber of Food Engineers. She holds a Best Paper Award from the 1st International Conference of Food Properties, Kuala Lumpur, Malaysia. In addition, she has received Publons Peer Review Awards Top 1% of Reviewers in Agricultural Science, 2017–2018 and 2018–2019 and Publons Peer Review Awards Top 1% of Reviewers in Cross-Field, 2018–2019, from Clarivate Analytics. Dr. Koç has authored 33 articles published in international journals and 16 articles published in national journals. She has been the editor of a book and author or co-author of seven book chapters.

CONTENTS

CONTRIBUTORS

Nancy Aswati
Lactalis USA, Buffalo, New York, USA

A. Burcu Atalay
Department of Food Engineering, Faculty of Engineering, Iğdır University, Iğdır, Turkey

Soumitra Banerjee
Associate Professor, Center for Incubation, Innovation, Research, and Consultancy (CIIRC), Department of Food Technology, Tataguni, Bangalore, Karnataka, India

Shalini Billa
Assistant Professor, Department of Biotechnology, Gowdara Mallikarjunappa Institute of Technology, Davangere, Karnataka, India

Dibya Ranjan Dash
Department of Food Process Engineering, National Institute of Technology, Rourkela, Odisha, India

Mathala Juliet Gupta
Horticulture Division, ICAR–Central Coastal Agricultural Research Institute, Goa, India

Rubeka Idrishi
School of Agro and Rural Technology, Indian Institute of Technology, Guwahati, Assam, India

Hilal Isleroglu
Tokat Gaziosmanpasa University, Faculty of Engineering and Architecture, Department of Food Engineering, Tokat, Turkey

Anandu Chandra Khanashyam
Horticulture Division, ICAR–Central Coastal Agricultural Research Institute, Goa, India

Gülşah Çalişkan Koç
Food Technology Program, Food Processing Department, Eşme Vocational School, Uşak University, Uşak, Turkey

V. Ajesh Kumar
Center of Excellence for Soybean Processing and Utilization, ICAR–Central Institute of Agricultural Engineering, Nabibagh, Bhopal, Madhya Pradesh, India

Krishna H. Menon
Department of Food Process Engineering, School of Bioengineering, College of Engineering and Technology, SRM IST, Kattankulathur, Tamil Nadu, India

G. Nagamaniammai
Department of Food Process Engineering, School of Bioengineering, College of Engineering and Technology, SRM IST, Kattankulathur, Tamil Nadu, India

Venkata Sarath Pamu
Oklahoma State University, Stillwater, Oklahoma, USA

Ravi Pandiselvam
Physiology, Biochemistry, and Post-Harvest Technology Division, ICAR–Central Plantation Crops Research Institute, Kasaragod, Kerala, India

Anirudh Gururaj Patil
Assistant Professor, Department of Food Technology, Annasaheb Dange College of Engineering and Technology, Ashta, Sangli, Maharashtra, India

Gouri Patil
Department of Polymer Science and Technology, Sri Jayachamarajendra College of Engineering (SJCE), Jagadguru Sri Shivarathreeshwara Science and Technology University, Mysore, Karnataka, India

B. Pooja
Department of Food Process Engineering, School of Bioengineering, College of Engineering and Technology, SRM IST, Kattankulathur, Tamil Nadu, India

M. Pravitha
Agro-Produce Processing Division, ICAR–Central Institute of Agricultural Engineering, Nabibagh, Bhopal, Madhya Pradesh, India

Latha Rangan
Department of Biosciences and Bioengineering, Indian Institute of Technology, Guwahati, Assam, India

Rahul Rathee
Department of Food Process Engineering, National Institute of Technology, Rourkela, Odisha, India

Kartik Shah
Sargento Foods, Elkhart Lake, Wisconsin, USA

Shubhra Shekhar
Department of Food Process Engineering, National Institute of Technology, Rourkela, Odisha, India

M. Selim Şılbır
Department of Food Engineering, Faculty of Engineering, Iğdır University, Iğdır, Turkey

Sushil Kumar Singh
Department of Food Process Engineering, National Institute of Technology, Rourkela, Odisha, India

Poonam Singha
Department of Food Process Engineering, National Institute of Technology, Rourkela, Odisha, India

Siddhartha Singha
School of Agro and Rural Technology, Indian Institute of Technology, Guwahati, Assam, India

Özge Süfer
Osmaniye Korkut Ata University, Food Engineering Department, Osmaniye, Turkey

Yeliz Tekgül
Food Processing Department, Köşk Vocational School, Aydın Adnan Menderes University, Aydın, Turkey

Taruna Varghese
PhD Scholar, Food Process Engineering, Department of Food Engineering, National Institute of Food Technology, Entrepreneurship, and Management, Thanjavur, Tamil Nadu, India

N. Venkatachalapathy
Professor and Head, Department of Food Engineering, National Institute of Food Technology, Entrepreneurship, and Management, Thanjavur, Tamil Nadu, India

Ajay Yadav
Agro-Produce Processing Division, ICAR–Central Institute of Agricultural Engineering, Nabibagh, Bhopal, Madhya Pradesh, India

Snehal T. Yadav
Assistant Professor, Department of Food Technology, Annasaheb Dange College of Engineering and Technology, Ashta, Sangli, Maharashtra, India

Ayşe Nur Yüksel
Department of Gastronomy and Culinary Arts, Faculty of Engineering, Architecture, and Design, Kahramanmaraş Istiklal University, Kahramanmaraş, Turkey

PREFACE

The intense pace of modern life has raised a demand for quick, prepared foods like ready-to-eat snacks. This demand is growing as a result of several reasons, including changing lifestyles, an increase in the participation of women in the workforce, a lack of time for meals due to hectic work schedules, and a rising preference among health-conscious people for wholesome and useful snacks over processed food. Snacks or snack foods can be defined as light-content foods suitable for snacking between regular meals (breakfast-lunch-dinner). Ready-to-eat snacks, including fruits, vegetables, meat, cheese, cookies, granola, breakfast cereals, protein bars, chips, popcorn, nuts, candies, bread, rolls, pizzas, and more, have become increasingly significant in the food industry. These widely available and convenient snacks are not only well-liked but also highly regarded for their nutritive content and ease of use.

Extrusion, drying, frying, roasting, and baking processes are the most used methods in the production of snacks. In addition to traditional methods, a range of emerging technologies, including 3D printing, microwave cooking, vacuum impregnation, osmotic dehydration, puffing drying, air frying, and more, have been employed to create ready-to-eat snacks. These novel approaches provide different ways to create practical and delicious snack options. The food sector has been able to improve the manufacturing process and meet the rising demand for ready-to-eat snacks in several innovative and effective ways by using these technologies.

This book provides an overview of today's science on ready-to-eat snacks, traditional production methods of ready-to-eat snacks and their disadvantages, the possibilities of using emerging technologies in the production of ready-to-eat snacks, the impact of these methods on the quality of ready-to-eat snacks, and packaging of ready-to-eat snacks, and covers all aspects of this multidisciplinary field. This book provides an in-depth analysis of how emerging technologies such as 3D printing, extrusion, high pressure, vacuum impregnation, etc., are used in ready-to-eat snacks.

Starting with the significance of ready-to-eat snacks and the benefits and drawbacks of conventional production techniques, this book covers the uses of developing technologies for the manufacture of ready-to-eat snacks as well as the difficulties and future opportunities. It also investigates how the bioactive components of the ready-to-eat snacks change during processing and production. For food technologists, food scientists, the food industry, and nutritionists who want to comprehend the effects of cutting-edge food processing technology on the creation of ready-to-eat snacks, this book is a helpful resource.

We anticipate that this book will be a valuable resource for all food scientists, food technologists, processors, and businesses who believe that ready-to-eat snacks have a bright future. We owe thanks to the most knowledgeable authors who are subject-matter experts for their invaluable contributions.

We would like to thank Almighty God for His strength and blessings, as well as our family, friends, mentors, colleagues, and supervisors for their unwavering support. Last but not least, we would like to thank all the contributors for their valuable book chapters. We also would like to express our gratitude to Apple Academic Press, Inc.'s editorial staff for their assistance and support during the assignment.

—Editors

Ravi Pandiselvam, PhD

Gülşah Çalışkan Koç, PhD

CHAPTER 1

READY-TO-EAT SNACKS: IMPORTANCE, NUTRITIONAL VALUE, PROCESSING METHODS, AND NOVEL TRENDS

AYSE NUR YÜKSEL,[1] GÜLŞAH ÇALIŞKAN KOÇ,[2] YELIZ TEKGÜL,[3] and RAVI PANDISELVAM[4]

[1]*Department of Gastronomy and Culinary Arts, Faculty of Engineering, Architecture, and Design, Kahramanmaraş Istiklal University, Kahramanmaraş, Turkey*

[2]*Food Technology Program, Food Processing Department, Eşme Vocational School, Uşak University, Uşak, Turkey*

[3]*Food Processing Department, Köşk Vocational School, Aydın Adnan Menderes University, Aydın, Turkey*

[4]*Physiology, Biochemistry, and Post-Harvest Technology Division, ICAR–Central Plantation Crops Research Institute, Kasaragod, Kerala, India*

ABSTRACT

Factors such as the modern age's change in lifestyle, people's lack of time or unwillingness to spare time for eating out of work, and the fact that women are more active in the business world have led to a change in eating habits today. Snacks or snack foods can be defined as light-content foods suitable for snacking between regular meals (breakfast-lunch-dinner). Various products such as chewing gums, chocolates, cookies, granolas, breakfast cereals, protein bars, biscuits, chips, popcorn, nuts, candies, bread, rolls, and pizzas, are included in the scope of snacks. In the past, when snacks were

Ready-to-Eat Snacks: Emerging Technologies for Production and Safety.
Ravi Pandiselvam and Gülşah Çalışkan Koç

mentioned, oily chips, and unhealthy high-sugar biscuit derivative products came to mind. Today, this perception has begun to change with the production of healthy snacks. Within the scope of healthy snacks, products such as dried fruits, and bars with high protein and fiber contents have started to be produced. In fact, healthy snacks have started to be produced by changing both product compositions and preparation and production methods. Extrusion technology is one of the most used methods in the production of snacks. In addition, drying, frying, roasting, and baking are also frequently used. New developments have begun to add fiber, protein, and food by-products or wastes that contain valuable phytochemicals to the extruded product composition. In addition, cold extrusion technology is one of the new production methods in order to produce snacks. Instead of hot air-drying methods, freeze-drying methods have started to be used for drying fruits and vegetables, so that the nutritional value, color, and sensory properties of the products are preserved. Instead of deep fat frying, techniques like vacuum frying and microwave frying techniques have begun to be used. In these ways, the oil absorption of the products has been reduced and the formation of acrylamide has been prevented. Furthermore, microwave roasting has to be started to be used as an alternative to oven and sand roasting due to prevent heat-induced contaminants (HMF, furfural, furan, and acrylamide) and losses of phenolic compounds and antioxidant activity, improve functional properties, etc. Besides traditional methods, and new approaches, emerging technologies such as 3D printing, microwave cooking, vacuum impregnation, osmotic dehydration, puffing drying, air frying, etc., have been applied to generate ready-to-eat snacks. Within the scope of this book chapter, the concepts of snacks and healthy snacks are discussed, and the traditional production methods (drying, extrusion, frying, roasting, and baking), nutritional content, and economic value of the snacks are examined in detail.

1.1 INTRODUCTION

Modern life has created a high demand for ready-to-eat foods or easy-to-prepare foods due to change in lifestyles, participation of women in working life, decreased time for meals due to intense work schedules, and conscious consumers want to consume healthy, and functional snacks instead of processed foods. Little is known about how and why individuals choose snacks, and there is no static definition for "snack" [1]. Snacks or snack foods can be defined as light-content foods suitable for snacking between regular meals (breakfast-lunch-dinner). Snacks are foods that suppress hunger, are ready-to-eat or easy to process, are small in size, and can be solid or liquid or semiliquid. Although snacks differ from country to country, they are usually consumed between meals or for entertainment purposes [2, 3]. Snacks are also defined as "all foods and beverages that contain calories and can be consumed between or outside the three main meals" [4]. There are several types of snacks such as savory snacks, corn snacks, meat snacks, snacks nuts, cheese snacks, snack bars, popcorns, pretzels, potato chips, tortilla chips, crackers, cookies, dried fruits, candies, frozen appetizers, etc. (Figure 1.1) [2].

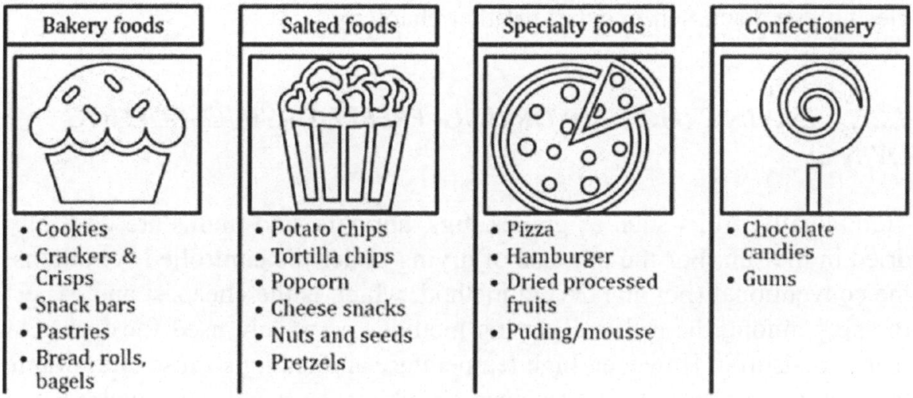

FIGURE 1.1 Some examples of ready-to-eat snacks.

In the past, when it comes to snacks, harmful products that trigger obesity come to mind due to the high fat, sugar, and/or salt contents of snacks. However, the results of epidemiological studies do not support this claim [5]. The concept of healthy snacks, which can be considered as a new concept, includes dried fruits and vegetables, and products such as bars, cereal products, etc., made from them that do not contain glucose or fructose syrup. Freeze-dried fruit and vegetable slices are among the healthy snacks that are gaining popularity day by day [6]. In addition to fruit and vegetable snacks which are rich in vitamins, minerals, and fiber; protein-rich meat products are among the healthy snacks whose demand is increasing day by day. For the production of meat snacks, extrusion, frying, roasting, baking, and drying technologies are generally used [7, 8].

1.2 PRODUCTION METHODS

Different production technologies are used according to the types of snacks. For example, frying for potato chips, drying for fruit and vegetable slices, roasting for seeds such as walnuts and nuts, and extrusion technology for snacks such as cereal-based breakfast cereals are used. Production methods, process parameters, and key findings for different types of snack foods are given in Tables 1.1 and 1.2. In this chapter, the traditional methods of producing snacks are briefly mentioned. In addition to traditional methods, emerging technologies such as 3D printing, microwave cooking, vacuum impregnation, osmotic dehydration, puffing drying, etc., have been applied to generate ready-to-eat snacks. The above-mentioned emerging technologies will be discussed in detail in other chapters.

1.2.1 DRYING (HOT AIR DRYING–FREEZE DRYING–PUFFING DYING)

Traditionally, fruits such as grapes, figs, apricots, and plums are generally dried in the sun. For the purpose of drying under the controlled conditions, the conventional (hot air) drying method, which is the cheapest and easiest to apply among the artificial drying methods, is widely used for drying of many foodstuffs. However, high temperature applications cause undesirable taste, aroma, texture (dense structure, shrinkage, etc.), color (browning), and vitamin and mineral losses, etc., in the product, and also lead to low quality product in terms of textural and sensory [6, 9]. To overcome these problems, the freeze-drying method is generally preferred to produce healthy snacks instead of traditional drying methods. Although this method is an expensive method (4–8 times higher) compared to hot air drying, the obtained final product is superior both in terms of sensory and nutrition [6, 9, 10]. Since the product is not exposed to heat in this method, heat-sensitive vitamins, phenolic compounds, color pigments, etc., are protected, and the color and texture of the product are sensory appealing. In addition, this product, which has a porous structure, is preferred in terms of its rehydration properties (4–6 times higher than traditional hot air dried). This property also makes products which are freeze dried excellent for ready-to-eat soups or instant meals [9]. In addition, it can be very advantageous when used in breakfast cereals. Guava type pears were dried with three different methods in order to produce snacks: hot air convection (at 60±0.5°C in tray dryer and velocity of air 0.6±0.02 m/s), freeze dryer (–35°C and vacuum pressure 8 kPa, drying times 1, 2, 3, 4, and 5 hours), and reflectance window (tray speed 4.16 x 10^{-5} m/s, temperature of water 90°C, drying time 10, 20, 30, 60, and 120 min). The colors of the products dried with the reflectance window showed little changes compared to other methods, and the retention of physical properties and nutritional quality was better. In addition, the loss of vitamin C and carotenoids was greater as a result of convection drying [11].

Another method that can produce snacks by protecting the product from the effects of heat is the puff drying method. In this method, foods with reduced moisture content are dried using a system working with the principle of pressure difference. Puffing drying is a fast and inexpensive method and the final product has desired color, flavor, and rehydration capacity [12] as well as textural (porous, crispy, etc.,) structure and high nutritional value. Microwave puffing was used to produce raspberry chips by Ma et al. [13]. Dehydration rate and heating rate were strongly affecting the quality of the snack. They observed that raspberry snacks produced using high microwave density had a more porous structure, low firmness but high crispness.

Based on the principle of immersing the food to be dried in a hypertonic solution containing osmotic agents such as glucose, corn syrup, sodium chloride, so on and replacing the moisture in the food with the solute in the solution, osmotic dehydration is a mild technology and is thought to have a positive impact on innovative fruit snacks and consumer health. The advantages of the osmotic dehydration can be listed as; higher color and flavor retention, softer textures to vegetable and fruit, so on. Osmotic dehydration processes are applied before drying and shorten the drying time, ensuring the preservation of heat-sensitive components such as color and vitamins that affect the quality of the product [14–16].

1.2.2 EXTRUSION

Although not a new technology, one of the most common techniques used in the production of snack foods is extrusion technology. Snacks of different shapes, textures, and colors can be easily produced by an extruder using various components. In addition, extruders are widely used in the production of foods such as breakfast cereal, baby foods, noodles, snacks, animal feed, pasta, etc. The extrusion process is a continuous process in which ingredients are mixed, cooked, and shaped. The food ingredients are mixed at high temperatures (100–180°C) in a short time (30–120 s), cooked, and shaped by mixing under pressure in a screw barrel. When the food mixture in molten form, which is cooked under pressure and at high temperature, falls to atmospheric pressure, the melt expands, and a fluffy structure is formed. The final product (extrudate) is usually ready for consumption, but in some cases, final drying may be required. During this process, many structural and chemical changes, such as starch gelatinization, inactivation of enzymes and microorganisms, depigmentation, and complex formation between amylose and lipids, occur. Therefore, determining the parameters of the extrusion process is extremely important for the quality of the final product.

Extruded products can be dried and seasoned directly (2nd generation or expanded snack) for consumption or puffed (3rd generation snacks or pellets) using frying in deep fat, puffing with hot air, deep fat frying, baking, or heating with a microwave, etc., as a final process. Third generation snacks are gaining popularity due to their high bulk density, ease of packaging and storage, shelf life, taste, etc. [19, 20]. The extrusion process can be applied to both cold (40–70°C, 60–90 bar) and hot (130–180°C, 120–250 bar) conditions [19]. In the extrusion process, there are three important parameters that determine the sensory properties of the end product: moisture content, feed rate, and screw speed [21]. Betaine-added whole grain snack samples were produced with a co-rotating twin screw extruder, and it was understood that the feed moisture had an effect on the lightness (L*) value, while the screw speed had an effect on redness (a*) and yellowness (b*) values [21]. Furthermore, extrusion temperature was also found to be effective on functional properties like antioxidant activity, and total phenolic and anthocyanin contents of snacks produced with purple sweet potato and butterfly pea flower [20]. In another study, Delic et al. [22] produced snacks containing cornmeal, Brewer's spent grain, and mechanically deboned poultry meat using a co-rotating twin-screw extruder (screw speed at 800 rpm and feed rate at 35 kg/h). Nutritional contents were found to be better in functional snacks. Moreover, the addition of Brewer's spent grain lowered lateral expansion as well as firmness and hardness of the extruded snack. Extruded snacks obtained by mixing rice and chickpea flour at different rates were produced with a twin-screw extruder. With the response surface and regression model evaluation, optimum conditions for barrel temperature at 102°C and screw speed at 281 rpm were determined [23].

In the study of Awolu et al. [24], the optimum snack formulation was optimized using response surface methodology, with rice and peanut flour and lemon pomace as independent variables, and functional and proximal properties as dependent variables. It was observed that the nutritional quality of the snacks produced with a single screw extruder using 75 g of rice flour, 20 g of peanut flour, and 5 g of lemon pomace was better than the control product (100% wheat flour). In addition, no significant difference was observed among all the snacks produced in terms of taste according to sensory evaluations, and they were found to be generally acceptable to consumers. Moreover, snacks that were chickpea-sorghum extruded (50:50, 60:40, and 70:30 chickpea: sorghum, w/w) were produced after mixing chickpeas and sorghum with a co-rotating twin screw extruder by Bekele et al. [25]. Changes in sensory properties and oxidative stability were observed at 55°C accelerated and 25°C ambient temperature storage conditions. Since chickpeas have more fat content than other pulses, it has been observed that the product mixed with a ratio of 50:50 has a lower peroxide value and lower rancid aroma and off-flavor compared to the others.

1.2.3 FRYING (DEEP–FAT DRYING–VACUUM FRYING–MICROWAVE FRYING)

Frying is one of the oldest cooking methods and is a widely used cooking method based on immersing food in oil at a temperature of 150–190°C (deep frying) or adding oil between the heated surface and the food (shallow frying) The deep-fried snack foods such as potato chips, fish snacks, etc., are desired due to their unique flavor, color (golden brown), and crispy texture. As a result of the high temperature in the oil being transmitted to the food during the frying process, the water in the food is removed and oil penetration into the spaces in the tissues occurs. Due to high temperature and oxygen, decomposition reactions occur in the frying oils, the nutritional value of the fried foods decreases, and some undesirable compounds, especially acrylamides, are released. As consumers want to consume foods with low calorie and low-fat content, the vacuum frying technique, in which the nutritional value of foods is preserved without the occurrence of burning and darkening reactions; oil absorption is low in an oxygen-free environment, has been developed. This method is generally used for frying fruits and vegetables. Under vacuum (negative pressure, <60 Torr, 8 kPa), the boiling points of oil is lowered as well as moisture and oil absorption is reduced, the lipid oxidation and browning reactions are inhibited [26–29]. In addition, the microwave heating technique creates volumetric heating in the food, thus reducing the heating time significantly. In combined systems, where microwave heating and vacuum application are used together, rapid mass and energy transfer are provided at low temperatures.

The most common ready-to-eat snack which is produced by frying is potato chips. In the study of Granda et al. [30], seven different potato types were fried using two different techniques (vacuum and deep frying) and some quality characteristics of the chips and acrylamide contents were compared. Frying under atmospheric conditions was carried out at different temperatures (150, 165, and 180°C). In the vacuum frying system, chips were produced at 118, 125, and 140°C under constant pressure (10 Torr). It has been determined that vacuum frying reduces acrylamide formation by 94%. It was determined that reducing the temperature from 180°C to 165°C in the frying process under atmospheric conditions reduced the acrylamide formation in potato chips by 51%. It has been shown that the decrease in temperature (from 140°C to 125°C) in the vacuum frying system reduces acrylamide formation by 63%. It was determined that the difference between potato types had an effect on the color parameters and textural properties of the chips, especially on the formation of acrylamide. In another study made by Song et al. [29], vacuum frying process conditions were optimized after vacuum microwave pre-drying. Vacuum microwave drying time (3, 6, and 9 min) and frying time (5, 10, 15, 20, 25, and 30 min), and temperature (90, 100, and 110°C) were determined as independent variables. The amount of oil and moisture, and the breaking strength were chosen as the responses in the optimization. It was found that the increase in the vacuum microwave pre-drying time caused a shortening of the oil absorption and cooking times of the samples. After 8–9 minutes of vacuum microwave drying pre-treatment, frying at 108–110°C for 20–21 minutes was revealed as the optimum processing conditions.

Moreover, Quan et al. [31] fried potato samples in soybean oil for varying times at three different power levels (600, 800, and 1,000 W) using a vacuum microwave frying system, and chips were produced. The effects of different microwave power levels on the texture properties (breaking strength), color parameters (brightness, redness, and yellowness), moisture loss, and oil absorption level in the samples were investigated. Microwave pretreatment increased the rate of drying of the potato chips and increased the oil absorption rate in the samples. While the increase in microwave power caused an increase in the brightness and yellowness values of the chip samples, it led to a decrease in the redness values. It was found that the oil absorption rates of the samples that were only fried under vacuum were higher and the cooking times were longer compared to the samples that were subjected to microwave pretreatment. In addition, the effects of vacuum frying and vacuum microwave frying systems on the quality of potato chips were also investigated [32]. Potato chips were produced using palm oil at different power levels of 12, 16, and 20 W/g, different temperature values of 100, 110, and 120°C and varying vacuum levels of 0.065, 0.075, and 0.085 MPa in a vacuum microwave frying system. It was determined that the vacuum microwave frying system preserved the textural properties and color of the chips better than the vacuum frying system. It was found that the oil content of the samples fried in the vacuum microwave frying system was lower and the difference between the two methods was significant. While the increase in vacuum level increased the drying rate, it caused a decrease in the oil content of the chips. In another study of comparing vacuum frying and vacuum microwave frying systems, potato chips were produced. Production was made in vacuum microwave frying systems at varying power levels (600, 800, and 1,000 W), constant temperature (100°C), and constant vacuum (15 kPa). Vacuum frying was carried out at 100°C, under a vacuum of 15 kPa, and palm oil was used in both systems. The frying process in the systems was modeled mathematically and found to be compatible with the logarithmic model. The increase in power level in the vacuum microwave frying system caused an increase in the effective moisture diffusivity and drying speed. It has been found that the total oil content and surface oil content of the products obtained from the vacuum microwave frying system are lower than those fried in the vacuum system [33]. The investigation on the effects of drying in a vacuum infrared dryer on the amount of oil and some quality characteristics of potatoes fried in a microwave-assisted vacuum infrared frying system was made by Su et al. [34]. Potato samples were dried in a vacuum infrared dryer at 120°C, constant pressure of 0.085 MPa, for different times (20, 25, 30, and 35 minutes). The dried samples were fried under a vacuum of 0.085 MPa at 100°C for varying times (2, 4, 6, 8, and 10 min). Palm oil is preferred as frying oil. It was determined that the chips obtained after the vacuum infrared drying process for 30 minutes and the frying process for 2 minutes contained less oil compared to the other chips' samples. The increase in vacuum infrared drying time increased the drying rate. When the chips samples were evaluated in terms of sensory, the samples fried in a vacuum microwave were liked the most in terms of color, while the samples in the group fried at atmospheric pressure were least appreciated. The products fried in a vacuum infrared dryer after drying for 35 minutes were the most liked group in terms of texture. Furthermore, Elfaitouri et al. [35] investigated the effects of frying with microwave on acrylamide formation in chips. Potato chips were obtained by frying in palm and corn oil at different power levels (200, 400, 800 W), 3 different temperatures (160, 170, and 180°C), and varying times (30, 60, 90, 120, and 150 minutes). It was found that the increase in temperature, power of microwave and cooking time caused an increase in acrylamide formation, and there was a significant difference among oil types in terms of acrylamide content of potato chips fried in palm and corn oil.

Different from potato chips, apple slices were fried in vacuum microwave and ultrasound-assisted vacuum microwave frying systems. Drying kinetics of apple chips were modeled, and oil absorption, color parameters, and textural properties were investigated. Trials were carried out under constant pressure (0.01 MPa), 800 and 1,000 W, at 2 different power levels, at constant temperature (90°C), at varying times (6, 8, 10, 12, 14, and 16 minutes). The frying processes were found to be mathematically compatible with the logarithmic model. The frying process in the ultrasound assisted vacuum microwave system took a shorter time compared to the frying process in the vacuum microwave, it was revealed that the chips obtained from the system absorb less oil and are more texturally crispy. In both systems, it was determined that the increase in power level caused an increase in the brightness value and oil content of the products. Prolongation of the frying process increased the oil content of the samples [36]. In a study where mushroom chips were produced in a vacuum microwave frying system, the effects of ultrasound treatment on the browning rate and product quality were investigated. It has been determined that ultrasound pretreatment accelerates moisture loss in the frying process performed at different power (800, 900, and 1,000 W) and temperatures (80, 85, and 90°C). By applying ultrasound pretreatment, it was determined that the oil absorption levels of the vacuum microwave fried mushroom chips were lower than the vacuum fried and vacuum microwave fried samples, while the color and textural properties were superior [37]. In another study by the same researchers, the effects of ultrasound and microwave pretreatment on frying kinetics and nutritional values of products were investigated. Mushroom samples were turned into chips at two different temperatures (80 and 90°C), and under three different frying processes (vacuum frying, vacuum microwave frying, and vacuum microwave frying with ultrasound pretreatment). The pressure was kept constant at 12 kPa in vacuum frying and vacuum microwave frying processes. It has been determined that the oil absorption levels and moisture losses of the products are compatible with the logarithmic model and the Pedrischi model. It has been determined that the nutritive values of mushroom chips produced with ultrasound pretreatment vacuum microwave frying systems are higher and the products absorb less oil [38].

Apple chips were obtained by using a vacuum microwave frying (at two different power levels of 800 and 1,000 W) system after ultrasound pretreatment by Faruq et al. [39]. The fried samples were compared in terms of color, odor, texture, and oil content. It was determined that the Maillard reaction products were released in lesser amounts in the ultrasound-applied samples. It was found that ultrasound application caused a shortening of the browning times of the samples. It was determined that there was no difference between the oil absorption levels of the chips obtained from the vacuum microwave frying system and the samples fried in the vacuum microwave using ultrasound. Furthermore, Islam et al. [40] obtained chips by frying green soybeans with ultrasound pretreatment in a vacuum microwave frying system. In the study, the frying process was carried out at different temperatures (80, 90, and 100°C) by keeping the microwave power and vacuum constant (1,000 W and 10±2 kPa). It was determined that ultrasound pretreatment caused less oil shrinkage in the samples. It was found that the water activities of the samples that were fried after the ultrasound was found to be lower, while the chlorophyll level and vitamin C amount in the samples were higher than the ones that were not pre-treated. It has been revealed that the frying process after ultrasound application preserves the color and texture of green soybeans better. According to Fan et al. [41] revealed the effects of osmotic ultrasound dehydration pretreatment on moisture absorption isotherms of sweet purple potatoes fried in a vacuum microwave oven. Different salt solutions (CH_3COOK, $MgCl_2$, K_2CO_3, NaBr, KI, NaCl, and KCl) were prepared to be used in osmotic dehydration. The samples were fried for 15 minutes at a power of 1,000 W, at 90°C, at a vacuum value of 0.090 MPa. Soybean oil was used as frying material.

The air frying method can also be used to produce ready-to-eat snacks. In the study of Liu et al. [42], 3D printed potato snacks were air fried. First, potato flakes and wheat flour were mixed at a ratio of 2:1, and a slurry was obtained with bovine fat and water. Then, the products were printed with a 3D printer with a diameter of 40 mm, and a height of 6 mm, with 30%, 50%, and 70% infill. The products coming out of the printer were immediately frozen at –18°C for 12 hours. The air-frying process was carried out at 140°C, but for different times depending on the infill ratio (21, 16, and 12 minutes for 70%, 50%, and 30% infill, respectively). Infill ratio of 100% cast and printed products were prepared as control products. The hardness and fracturability of the products decreased as the amount of infill decreased. Besides, cast and printed products containing 100% infills showed large puffing during the air-frying process.

1.2.4 ROASTING (SAND ROASTING–PAN ROASTING–OVEN ROASTING–MICROWAVE ROASTING–INFRARED ROASTING)

Roasting (sand, pan, oven, microwave, etc.) is a common food processing method in order to improve the digestibility by starch gelatinization and protein denaturation, color, shelf-life palatability, texture (crispiness), palatability, etc., of the foods such as nuts, seeds, coffee beans, pulses and legumes such as beans, chickpea, wheat, etc. Roasting of the foods performed at high temperatures such as at 150–400°C and in a short time where resulted in higher rate of water evaporation, lower value of water activity, higher antioxidant activity, longer shelf life and crispy structure, etc. [43–46]. The sensory quality parameters such as texture, color, taste, and aroma of foods are affected by the degree of roasting. Nuts to be used as snacks or added to food by roasting are often roasted with hot air. New roasting methods and roasters are being designed to minimize the energy cost of the hot air roasting system. Infrared roasting and microwave roasting recently gained more attention as alternative methods of roasting for nuts. Infrared roasting and microwave roasting methods are more advantageous in terms of roasting time, heat efficiency, and diffusion coefficient compared to traditional roasting methods [47]. It has been revealed that roasting using the infrared roasting method of pistachios, one of the popular nuts rich in antioxidant compounds, shortens the time of roasting by 50% on the contrary to other methods [48]. In a study by Mosayebi et al. [49], sunflower seeds were roasted with 492.5 W IR power for 9.1 min.

Sunflower seeds roasted with the infrared roasting method were more appreciated than those roasted with the conventional method (hot air). Mokhtari and Ziaiifar [50] roasted wild almond kernels using the infrared roasting method. It has been determined that the increase in power and time causes a decrease in pH and moisture content in the processes performed at different infrared powers and roasting times. Roasting at 200 W infrared power for 15 minutes was found to be the optimum condition considering optimum color, sensory and chemical properties.

1.2.5 BAKING

Baking is another method used in ready-to-eat snack production. It can be used after the extrusion process or alone. It provides an appealing brown color with the Maillard browning reaction. In addition, it contributes to the formation of aroma, taste, and flavor components specific to the product.

Baking is used in the production of granolas, snack bars, crackers, biscuits, cakes, and bread types [51–53]. Although the cooking process is generally done in a convection oven (with or without a fan), studies on microwave cooking have increased considerably in recent years. However, although the cooking time is very short in microwave baking the desired sensory properties do not develop on the surface of the product and the product does not swell. Rahmi et al. [54] produced snack bars containing soybeans and purple sweet potatoes. After making snack dough, it was spread on a tray and baked in a gas oven at 160°C for 30 min. The produced bars showed significant differences in terms of nutrition, antioxidant activity, color, and sensory acceptability. Snacks of cakes and biscuits containing yellow fleshed sweet potato and wheat flour were baked in the oven at 180°C for 20 min and 160°C for 30 min and, respectively. Chin-chin samples were fried at 160°C for 3 min in vegetable oil. Yellow fleshed sweet potato substitution provided better nutritional properties in terms of ash, dietary fiber, and carbohydrate contents. Besides, up to 40% of substitution was found acceptable according to sensory evaluations [55].

Gluten-free corn snacks were produced by adding yerba mate (*Ilex paraguariensis*) extract and flaxseed (*Linum usitatissimum* L.) flour in different proportions by baking at 160°C for 25 minutes. The addition of flaxseed flour caused high moisture content and water activity, resulting in a decrease in crispness and stiffness. Yerba mate extract, on the other hand, positively affected the textural properties and made it brittle. Snacks with added flaxseed flour and yerba mate extract were acceptable in the sensory evaluation, although they were darker than the control (corn snack) product. In addition, since the antioxidant activity of the products is high, a product with improved functional properties was obtained [56]. In addition, baking can be applied after the expansion process with an extruder. In the snack formulation, the mixture was prepared by replacing the corn grits with 10% and 15% of rice bran (the control product was produced with 100% corn grits.) Then the mixture was hydrated to 16% moisture content and fed into a co-rotating twin screw extruder. After extrusion, expanded products in the form of balls were dried at 50°C for 3 hours and analyzed. The dietary fiber, protein, fat, and mineral contents of the snacks produced by adding rice bran were increased. Rice bran added products decreased their hardness while their crispness increased. In addition, rheological solid-like behavior and increased elastic interactions were observed. It has been seen that with the substitution of rice bran, snacks with higher nutritional value and good textural and rheological properties can be produced [57].

For the production of functional snacks, freeze-dried algae powders of *Chlorella vulgaris* (Chlorella) and *Arthrospira platensis* (Spirulina) were replaced with pastry wheat flour in different proportions and the batter was prepared by adding egg powder, sugar, water, sunflower oil, baking powder, and milk powder. Later, cylindrical snacks (diameter: 35 mm and height: 8 mm) were produced with a 3D printer. In order to prevent collapse during baking, the snacks were first frozen at −18°C for 90 minutes and were cooked in the convection oven at 105°C for 30 minutes. The printability of the algae added batters in the extrusion-based 3D printer was found to be better than the control product due to the increase in extrusion force and shear modulus. On the contrary, snacks with a high amount of algae were not well received by consumers due to their dark and green color after baking [58].

1.3 NUTRITIONAL VALUE

Ready-to-eat snacks (fruits, vegetables, meat, cheese, seafood, poultry, etc.,) are important, healthy, and popular food products and gaining importance day by day due to their high nutritional content, low calorie value, and ease of consumption [78]. A wide range of raw materials can be used to produce snacks. Nowadays natural, minimally processed, and organic foods are preferred due to health concerns of the consumers [20].

When it comes to ready-to-eat snacks, high-calorie foods such as chips, chocolate, and candies come to mind. These snacks are calorie-dense foods. It is known that consuming high amounts of calorie-dense foods can cause obesity, type 2 diabetes, stroke, hypertension, heart diseases and cancer [79]. Today, people give importance to a healthy, active life and a balanced diet. For this reason, the concept of snacks has also changed direction. High-fat and sugary snacks have been replaced by dried fruits, nuts, granola bars, and oatmeal breakfast cereals.

In order to increase the nutrient content, food wastes such as malt, fruit, and vegetable industry wastes, etc., that include high nutritional content can be used as binders in cereal bars. In addition, the use of dietary fiber will both increase the nutrient content of cereal bars and also eliminate the need for the use of additional binding agents [80].

In the extrusion process, flour (corn, wheat, potato, rice, oat, etc.), starch (corn, tapioca, wheat and potato), liquid shortenings, and monoglyceride are the main products for the production of snacks [81]. Food by-products such as pomaces, bagasse, cereals, oil-seed, whey, cocoa, etc., are generally

TABLE 1.1 Production Methods (Extrusion, Drying, and Frying), Process Parameters, and Key Findings for Different Types of Snack Foods

	Snack Type	Ingredients	Production Method	Process Parameters	Key Findings	References
Extrusion	Purpled third-generation snack	Purple sweet potato powder, butterfly pea flower, potato flake, corn flour, salt, sugar, monoglyceride and baking soda	Extrusion (co-rotating, twin-screw extruder) + Puffing (Deep fat frying + microwave puffing)	Extrusion: Screw speed: 200 rpm, Zone temperatures: 100, 105, and 110°C. Deep fat frying: Palm oil: Frying temperature: 170 and 190°C Frying time: 10 and 8 secs. Microwave puffing: Microwave power (MP): 210, 420, and 560 W Microwave puffing time: 35, 25 and 15 secs,	❖ The best functional properties were obtained at 110°C, ❖ Hardness decreased and antioxidant activity increased with increasing MP and frying temperature ❖ Better color values, sensory scores and antioxidant activity values were obtained from microwave puffed snacks	[20]
	Spirulina enriched extruded snack	*Spirulina*, organic corn and rice fours	Extrusion (co-rotating, twin-screw extruder)	Screw speed: 250 rpm, Zone temperatures: 75, 100, 120, and 120–160°C. Feed moisture content: 15, 16.2, 18, 19.8, and 21% *Spirulina sp.* LEB 18 Concentration: 0.4, 1.0, 1.8, 2.6, and 3.2% Drying temperature: 85°C	❖ Optimum conditions: ❖ 143°C zone temperature, 2.6% *Spirulina sp.* LEB 18, and 16.2% feed moisture content ❖ The protein content, hardness, and ΔE of samples increased and water absorption and color values decreased with *Spirulina sp.* LEB 18	[59]
	Whey protein concentrate enriched extruded corn snack	Whey protein concentrate and corn starch	Extrusion (twin-screw extruder) + Hot air drying	Zone temperatures: 70/100/130/130/130°C and 70/100/150/150/150°C Drying temperature: 40°C	❖ An increase in the zone temperature (130 to 150°C) resulted in lower specific mechanical energy, expansion ratio, water and protein solubility in phosphate buffer, -mercaptoethanol and sodium dodecyl sulfate, however, higher hardness values were observed. ❖ Higher whey protein concentrate concentrations resulted in lower specific mechanical energy, water and protein solubility index, sulfhydryl groups content, gelatinization degree in phosphate buffer, 2-mercaptoethanol and sodium dodecyl sulfate.	[60]

TABLE 1.1 *(Continued)*

	Snack Type	Ingredients	Production Method	Process Parameters	Key Findings	References
	Ready-to-eat breakfast cereal	Low-amylose rice flour (LARC), seeded banana (SB), and carambola pomaces (CP)	Extrusion (single-screw extruder)	Screw speed: at 200–400 rpm Barrel temperature: at 90–130°C LARC:SB:CP ratio: 80:10:10 and 60:30:10 Feed moisture content: 9–21% (wet basis)	❖ Optimum conditions (high water absorption and solubility indexes, dietary fiber, overall acceptability): 350 rpm screw speed of 120°C barrel temperature, 65:25:10 LARC:SB:CP ratio and 12% feed moisture content	[61]
	β-carotene enriched snack type extrudate	Wheat Flour, α-lactose monohydrate, maltodextrin, sunflower oil+β-carotene (lipid-phase), whey protein isolate (emulsifier), gum arabic (polyelectrolyte)	Extrusion (twin-screw extruder)	Screw speed: 300 rpm Zone temperatures:105, 120, 145, and 155°C.	❖ Higher β-carotene loss in extrudates with LBL emulsion ❖ Lower L* and b* values were obtained for extrudate with SL emulsion ❖ Temperature and storage time increased the β-carotene loss	[62]
	Pearl millet-based extruded snacks with whey protein concentrate (WPC)	Pearl millet grains, whey protein concentrate	Extrusion (co-rotating twin-screw extruder) + Hot air drying	Screw speed: 350 rpm Zone temperatures:30, 60, and 130°C Drying temperature: 40°C WPC: 0%, 2.5%, 5.0% and 7.5%	❖ Acceptable extrudate snack was obtained from 5% WPC ❖ Higher WPC amounts caused lower expansion ratio and sensory score and higher bulk density, hardness, breaking strength and nutritional value	[63]
Drying	Turkey breast meat	-	Puffing drying	Pre-drying methods: Freeze drying (FD): Pressure: 15, 20, 25 Pa Hot Air Drying (HAD): Temperature: 65, 75, and 90°C Microwave drying (MD): Power: 180, 360, and 540 W Puffing drying (PD): 10 min at 100°C, 190 kPa pressure + 80°C 25 Pa vacuum pressure	❖ Higher fat and salt contents, TBARS value, overall acceptability score and porous structure and lower moisture content, water activity, and bulk density values were obtained from FD+PD samples	[7]

TABLE 1.1 *(Continued)*

Snack Type	Ingredients	Production Method	Process Parameters	Key Findings	References
Peach chip	Peaches	Puffing drying	Pretreatment: Impregnation liquids: distiled water, 2% of citric acid solution, a mix of 2% citric acid and 24% sucrose, a mix of 2% citric acid, 24% sucrose and 2% sodium dihydrogen phosphate. Pre-drying: Temperature: at 70°C with electrical thermostatic drying oven Puffing drying: Temperature: 105°C Pressure: 0.2 MPa	❖ Pretreatments affected the brittleness of chips with improvement of firmness (crispy texture) ❖ Acid provides better swelling with higher porosity, while sugars result in smaller pores. Therefore, optimization is required for pretreatments	[64]
Apple chip snack	–	Hot air drying, freeze drying, hot air+ puffing drying and hot air drying+ CO_2 puffing drying	Hot air drying: Temperature: 60°C Air flow rate:2.4 m/s Freeze drying: Pressure: 100 Pa pressure, Condenser temperature: –50°C Heating plate temperature: 60°C Hot air drying + puffing drying: 60°C, 2.0 m/s+ 60°C, 150 Pa Hot air drying + CO_2 puffing drying: 60°C, 2.0 m/s+ puffing tank until reach 4 MPa pressure+ 60°C, 150 Pa	❖ Highest overall acceptance score and crispiness, better textural properties and low shrinkage rate and rehydration ratio was obtained for hot air drying combined with CO_2 puffing drying ❖ Freeze dried apple snacks have advantages such as high rehydration ratio, porosity, total color change, taste and smell scores	[65]

TABLE 1.1 *(Continued)*

Snack Type	Ingredients	Production Method	Process Parameters	Key Findings	References
Vegetable snack bar	Broccoli, cauliflower, green pepper, yellow and green beans, potato, carrot, corn grain, salt basil, dill, chive, sodium alginate, calcium lactate, locust bean gum and xanthan gum	Freeze drying	Pressure: 63 Pa Safety pressure:–137 Pa Shelf temperature: –30°C Time: 72 h	❖ Higher porosity and hardness were observed from snacks with sodium alginate ❖ No clear relationship detected between the type of hydrocolloid and dry matter content and water activity ❖ Dry matter content was found higher than 96%	[6]
Carrot-potato chips	-	Combined freeze-drying and instant controlled pressure drop process (DIC)	Freeze drying: Pressure: 0.37 mbar Condensate temperature: −50°C DIC Drying: Pressure: 3–5 kPa vacuum Temperature: 60°C	❖ Optimum conditions and numerical optimization suggestions were found as 46–54% w/w and 47.43% carrot, 0.20–0.35 g/g and 0.29 g/g moisture content of partially dried samples, and 85–95°C and 90.57°C equilibrium temperature of DIC, respectively. ❖ Carrot-potato chips produced by the combined drying method have higher quality than the freeze-dried chips.	[66]
Apple and carrot chips	-	Microwave vacuum drying and freeze drying	Microwave vacuum drying: Microwave power density: 2 W/g Freeze drying: Heating plate temperature: 30°C Pressure 0.2 kPa Conventional hot air drying: Temperature: 60–65°C Air Flow Rate: 0.5 m/s	❖ Microwave-vacuum dried carrot and apple slices have higher carotene and vitamin C retentions compared to hot air–dried samples. ❖ Microwave-vacuum dried carrot and apple slices have similar carotene, vitamin C and color retentions, rehydration capacity, and texture with those of the freeze-dried ones.	[67]

TABLE 1.1 *(Continued)*

	Snack Type	Ingredients	Production Method	Process Parameters	Key Findings	References
Frying	Potato chips	Potato	Deep fat frying	Oils: Refined coconut and soybean oils, pure olive oil, and vegetable shortening Repeated use of oils: 1, 20, 40, 60, and 80 times Frying: Temperature:180±5°C Time: 4 min.	❖ Repeated use of oils did not affect the moisture and fat contents of potatoes ❖ Highest aldehydes were observed from potato chips that were fried in pure olive oil ❖ Refined coconut oil preserves the quality of the potato better and is more suitable for reuse.	[68]
	Sweet potato, green beans, tommy Atkins mango, and blue potato	-	Vacuum frying	Vacuum frying: Pressure: P<10 Torr=1.33 kPa Temperature: 120–130°C Frying time: 120 s for sweet potato, 330 s for green beans, 180 s for mango, and 420 s for blue potato Atmospheric frying: Temperature: 160–165°C Frying Time: 240 s for sweet potato, 300 s for green beans, 240 s for mango, and 180 s for blue potato Oil: Fresh canola oil	❖ Vacuum frying resulted in lower oil content and higher anthocyanin and total carotenoid contents ❖ Higher sensory scores were observed for vacuum fried samples ❖ Oxidation was observed less and color and flavor were better in vacuum fried products.	[27]

TABLE 1.1 *(Continued)*

Snack Type	Ingredients	Production Method	Process Parameters	Key Findings	References
Potato chips	-	Vacuum microwave pre-drying vacuum frying	Vacuum microwave pre-drying (VMPD): Microwave power: 146.5 W Vacuum pressure: –0.06 MPa Vacuum frying (VF): Temperature: 90, 100, and 110°C Vacuum pressure: –0.09 MPa Frying time: 5, 10, 15, 20, 25, and 30 min Oil: Soybean oil	❖ Optimum conditions: VMPD: 8–9 min, VF temperature and time: 108–110°C and 20–21 min. ❖ Higher VMPD resulted in lower moisture and oil contents ❖ Higher frying temperature and time resulted in lower moisture content	[29]

TABLE 1.2 Production Methods (Roasting and Baking), Process Parameters, and Key Findings for Different Types of Snack Foods

	Snack Type	Ingredients	Production Method	Process Parameters	Key Findings	References
Roasting	Chickpea	-	Malting and sand roasting	Temperature: 180°C Roasting time: 28 s	❖ Malting and roasting processes reduced the tannin content and trypsin inhibitor activity. ❖ Malting and roasting processes improved the protein efficiency ratio, biological value, nutritional index, and protein digestibility.	[69]
	Pistachio	-	Hot-air, infrared, and microwave roasting	Hot-air temperature: 100, 125, and 150°C Hot-air roasting time: 5, 12.5, and 20 min Infrared power:75, 85, and 95 V Infrared roasting time: 10, 20, and 30 min Microwave power:180, 270, and 360 W Microwave roasting time: 12, 14, and 16 min	❖ The acrylamide amount of raw and roasted samples ranged between 57 ± 0.86 and 851 ± 2.8 μg/kg. ❖ Infrared roasting caused the highest acrylamide amount in unsalted raw pistachio, whereas microwave roasting resulted in the lowest acrylamide formation. ❖ Salting resulted in lower acrylamide formation. ❖ An increase in the temperature, time, voltage, and power caused higher acrylamide formation.	[70]
	Nut (Almond, peanut, cashew nuts, hazelnut, and sachainchi)	-	Sand roasting	Temperature: 150–200°C Roasting Time: 5–25 min	❖ High temperature and roasting time caused higher furan levels due to decomposition of linolenic and linoleic acids ❖ 0-order reaction kinetic described the furan formation ❖ The highest and lowest furan levels were observed from cashew nuts and peanuts, respectively.	[71]

TABLE 1.2 *(Continued)*

	Snack Type	Ingredients	Production Method	Process Parameters	Key Findings	References
	Wild almond	-	Superheated steam, hot-air, frying, and infrared roasting	Superheated steam:160 and 180°C Hot air temperature:160 and 180°C Oil temperatures: at 160 and 180°C Infrared power: at 200 W and 300 W Roasting time: 15, 25, and 35 min	❖ Optimum conditions; 200 W, 15 min for infrared, 160°C, 15 min for hot air, and 160°C, 25 min for frying roasted wild almond samples. ❖ An increase in the hot air temperature, infrared power and roasting time resulted in lower pH and the moisture content. ❖ The highest overall acceptance score observed from frying roasted samples, whereas the lowest score observed from infrared roasted sample.	[50]
	Peanut kernel	-	Infrared roasting	Infrared power: 250, 350, and 450 W Roasting time: 10, 20, and 30 min	❖ Optimum roasting conditions: 370 W for 20 min. ❖ Higher infrared power and time resulted in higher energy consumption, a*, b* and ΔE values, whereas, lower L* value, moisture content, hardness and compressive energy values were observed at high infrared power and long roasting time.	[72]
	Chickpea	-	Microwave roasting	Microwave power: 450, 600, and 900 W Roasting time: 0, 5, 10, and 15 min	❖ Roasting increased the antioxidant activity, total phenolic content, amylose-lipid complex, and water absorption capacity, whereas L* value, water solubility index, oil absorption capacity, and total quercetin content decreased ❖ An increase in the roasting time and microwave power increased the a*, b*, water absorption properties, amylose-lipid complex onset temperature and decreased L*, bulk density, scratch gelatinization onset temperature, foaming properties, oil absorption capacity	[73]
Baking	Bread	Whole wheat flour, Siyez wheat flour, yeast, salt, olive oil, brown sugar, water or Kombucha (from black tea and green tea)	Baking	Baking in an electric oven at 180°C for 45 min. Steam spraying every 5 min	• Brown color, sour taste and higher hardness were obtained for breads with kombucha • Bread with green tea kombucha have higher total phenolic content and antioxidant activity	[74]

TABLE 1.2 *(Continued)*

Snack Type	Ingredients	Production Method	Process Parameters	Key Findings	References
Sunroot snack bar	Sunroot paste, potato paste, oat powder, creamy cheese, roasted nuts, and olive or cheese	Baking	Baking in an electric oven at 160°C for 1 hour followed by 140°C for 30 min	• No microbial growth was observed for all sunroot snack bars which were stored at 4°C for 90 days. • Firmness and sensory scores were improved with the higher amount of sunroot.	[51]
Snack crackers	Wheat flour, dried green curd of pea peels (DGCPp), shortening, sugar, salt, and water	Baking	Baking at 150°C for 25 min	• Nutritional value of crackers containing 15% dried green curd pea peels was the highest. • Pea peels, which are rich in protein and insoluble dietary fiber, were successfully used in the production of crackers.	[53]
Kiwi, pumpkin, quince powder fortified cake	Wheat flour, sugar, egg, oil, baking powder, milk, freeze-dried kiwi powder, freeze-dried pumpkin powder, freeze-dried quince powder (5, 10, and 20% by weight)	Baking	Baking at 175°C for 20 min in a conventional oven	• Higher vitamin C content and lower moisture content, water activity, L*, volume and symmetry index values were obtained for fruit powders fortified cakes • Higher overall acceptability score for cake with pumpkin powder	[75]
Cookie	Flour, sucrose, nonfat dry milk, NaCl, Sodium bicarbonate, Shortening, Ammonium bicarbonate, high fructose corn syrup, water	Baking	Baking at 205°C for 9 min in a conventional oven	• Gelatinization temperature of wheat starch, pasting peak viscosity, and wheat starch granule average particle size increased by sweetener amount. • Higher solution concentration resulted in the larger degree of polymerization, lower paste viscosity, granule swelling.	[76]
Salty cereal bar	Peanut, rice flakes, oats, sesame, cashew nut, brazil nuts, chia, flax seed, salt, collagen, guar gum, xanthan gum, psyllium	Baking	Baking at 240°C for 45 min	• Higher appearance, texture, flavor and color scores and lower L* values were obtained from psyllium containing bars • Highest and lowest hardness values were obtained from collagen and guar gum containing bars	[77]

used for the enrichment of extruded snacks due to their high fiber, protein, and antioxidant contents [82]. In order to improve the nutrition value of the extruded snack several ingredients such as fish [83], spirulina [59], whey [63, 84], fruit pomaces [61, 85–88], brans [89–94], oilseed by-products [95] were added to the formulation. According to Lucas et al. [59], it was determined that the addition of spirulina at the rate of 2.6% to snack products increased the protein, oil, and mineral contents of the products. In the study by Chen et al. [96], specific mechanical energy, water solubility index, the content of sulfhydryl (SH) groups, gelatinization degree (GD), solubility of protein in Pb + 2-Me + SDS, and darker color of the extruded samples were found to be lower in extruded corn starches with added whey protein concentrate. During extrusion cooking, the increase in the amount of banana flour and pomegranate pomace in the content caused a decrease in expansion by diluting the starch content and increasing the fiber content [61]. The composition of food products influences nutritional quality. Valuable bioactive components such as β-carotene can be encapsulated and used for enrichment in extruded snacks. The interface structures of emulsions may affect the protection of bioactive compounds. In the production of snack type extracts in a study in which β-carotene was encapsulated using single-layer and layer-by-layer emulsions as a component, it was found that the β-carotene content was better preserved in layer-by-layer extrudates [62].

Rajagukguk et al. [97] evaluated the antioxidant activities and probiotic viability of the chickpea and green lentil snack bars as well as their sensory properties. Chickpea snack bars had higher antioxidant activity and probiotic viability even after two months of storage at 20°C when compared with green lentil ones. Besides, according to sensory evaluation, chickpea snack bars were found to be more preferable than green lentil one. In addition, innovative formulations were also studied by Singh et al. [98]. They produced a functional snack bar with popped amaranth grain, oat, and banana peel powder. The produced bar was found rich in essential amino acid composition, total phenolic content, dietary fiber, and mineral content in which these contents boost immunity and are beneficial for all ages.

Other than the conventional ingredients, edible insect components have gained importance as environmentally friendly energy sources due to food shortages that may occur in the future. In the study by Cho et al. [99], the nutrients such as protein, fatty acid, fats, and amino acids of the snacks produced using honey bee (*Apis mellifera* L.) drone pupae powder were found to be higher than the control product (puffed rice snack). In the study of Maya et al. [100], alternative snacks (muffins, cookies, and puddings) were produced for diabetics using sago starch (*Metroxylon* sp.) and tempeh. The ratios of sago starch to tempeh were 2:1 and 1:1. The sensory evaluations showed that for puddings and muffins 1:1 ratio, and cookies 2:1 ratio were liked. Moreover, produced snacks contain high levels of dietary fiber.

1.4 ECONOMIC VALUE

Ready-to-eat snacks that contain biscuits, crackers, ice cream, cakes, pizza, hamburgers, hot dogs, dried fruits, etc., are a billion-dollar industry. In 2020, the total revenue of the confectionery and snacks market was approximately 300 billion dollars in the United States, 150 billion dollars in Japan, and 121 billion dollars in China [101]. On the other hand, the imported quantity of prepared foods obtained by swelling or roasting cereals or cereal products, e.g., corn flakes were 1,788,704 tons in 2021 worldwide. The imported values of these snacks were changed between 4,238,996 and 4,903,334 thousand U.S. dollars between the years 2017 to 2021. The exported value of these snacks was 5,117,173 thousand U.S. dollars in the year 2021 worldwide. According to the amount of exports, Germany is followed by the United States, while according to the amount of imports, Canada ranks first [102].

Other snack food like the meat of bovine animals, salted, in brine, dried or smoked has 301,992 thousand U.S. dollars' worth exportation value, where the import value is 289,589 thousand U.S. dollars in 2021 worldwide. Italy (88,488 thousand U.S. dollars) ranks first followed by Switzerland (66,384 thousand U.S. dollars) in the exporters list in terms of value in 2021 while, France ranks first in the importers list based on value (60,839 to 81,909 thousand U.S. dollars from 2017 to 2021). Moreover, for mixtures of nuts or dried fruits, the most importer countries are France, United Kingdom and Germany (43,522, 39,492, and 24,441 thousand U.S. dollars, respectively) where, Germany ranks first by far as an exporter country with 116,197 thousand U.S. dollars in 2021 [102].

It can be clearly understood that snacks made by cereals and cereal products are the most imported and exported ones in terms of value (U.S. dollars) worldwide (Figures 1.2 and 1.3). The trade of these products is beneficial and important for the economies of the countries.

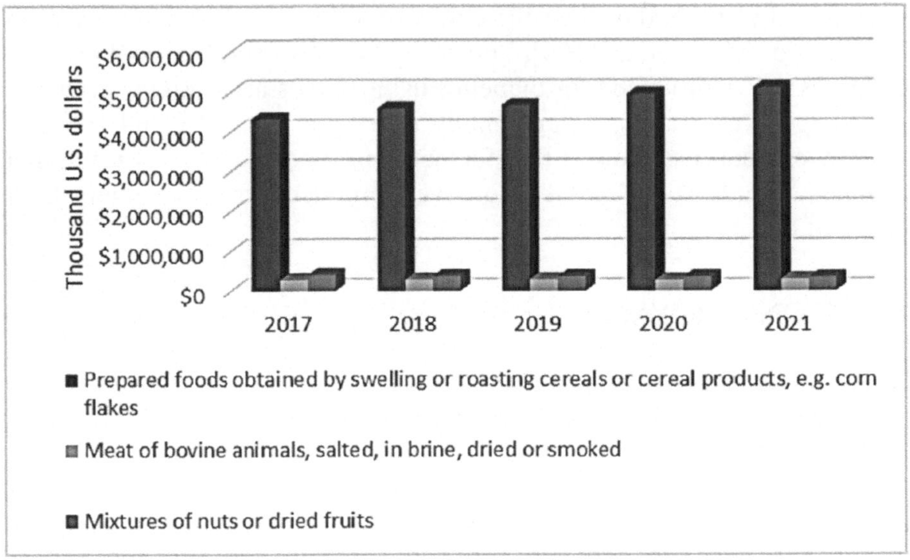

FIGURE 1.2 Worldwide export values of some snacks.

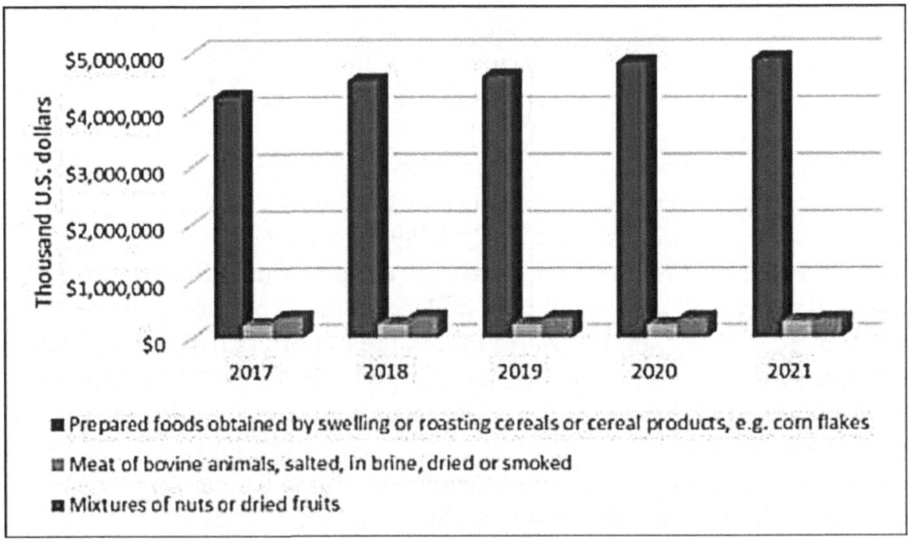

FIGURE 1.3 Worldwide import values of some snacks.

1.5 CONCLUSION

Extrusion, drying and frying technologies are generally used in the production processes of food snacks. While these technologies can be used separately, there are studies where they are used together such as extrusion microwave puffing, extrusion deep fat frying, freeze drying puffing drying, etc. In order to eliminate the disadvantages of these methods, emerging technologies like air frying, vacuum microwave, and vacuum frying have started to be used in the production of snacks. Crisp snack products are obtained by extrusion, puff drying, and deep-frying methods. The crispy structure gives consumers a pleasant feeling as they perceive the sound when biting, and has a positive effect on snack products. Recent studies on snacks have focused on developing innovative products that can be considered healthy snacks, such as freeze-dried fruit or vegetable gels, fruit peels, and fruit or vegetable slices, which can be considered healthy with high nutritional value [10]. Another trend in snacks is the production of calorie-reduced snacks with increased nutritional value. In this context, the production of snacks enriched with valuable components such as phenols, antioxidants, dietary fiber, and fatty acids, especially for use in food waste, has become very important [18]. Studies on the production of gluten-free and vegetarian cereal bars that contain high nutritional value ingredients could be increased [80].

KEYWORDS

- **drying**
- **extrusion**
- **infrared roasting**
- **nutritional value**
- **puffing**
- **snack**

REFERENCES

1. Hess, J. M., Jonnalagadda, S. S., & Slavin, J. L. (2016). What is a snack, why do we snack, and how can we choose better snacks? A review of the definitions of snacking, motivations to snack, contributions to dietary intake, and recommendations for improvement. *Advances in Nutrition, 7*(3), 466–475. https://doi.org/10.3945/an.115.009571.

2. Lusas, E. W., & Rooney, L. W. (2001). *Snack Foods Processing*. CRC Press.
3. Pérez-Navarrete, C., González, R., Chel-Guerrero, L., & Betancur-Ancona, D. (2006). Effect of extrusion on nutritional quality of maize and lima bean flour blends. *Journal of the Science of Food and Agriculture, 86*(14), 2477–2484. https://doi.org/10.1002/jsfa.2661.
4. Chaplin, K., & Smith, A. P. (2011). Breakfast and snacks: Associations with cognitive failures, minor injuries, accidents and stress. *Nutrients, 3*(5), 515–528. https://doi.org/10.3390/nu3050515.
5. Chapelot, D. (2011). The role of snacking in energy balance: A biobehavioral approach. *The Journal of Nutrition, 141*(1), 158–162. https://doi.org/10.3945/jn.109.114330.
6. Ciurzyńska, A., Marczak, W., Lenart, A., & Janowicz, M. (2020). Production of innovative freeze-dried vegetable snack with hydrocolloids in terms of technological process and carbon footprint calculation. *Food Hydrocolloids, 108*, 105993. https://doi.org/10.1016/j.foodhyd.2020.105993.
7. Elmas, F., Bodruk, A., Köprüalan, Ö., Arıkaya, Ş., Koca, N., Serdaroğlu, F. M., Kaymak-Ertekin, F., & Koç, M. (2021). The effect of pre-drying methods on physicochemical, textural and sensory characteristics of puff dried turkey breast meat. *LWT, 145*, 111350. https://doi.org/10.1016/j.lwt.2021.111350.
8. Karmas, E. (1976). *Processed Meat Technology*. Noyes Data Corp.
9. Ratti, C. (2001). Hot air and freeze-drying of high-value foods: A review. *Journal of Food Engineering, 49*(4), 311–319.
10. Ciurzyńska, A., Cieśluk, P., Barwińska, M., Marczak, W., Ordyniak, A., Lenart, A., & Janowicz, M. (2019). Eating habits and sustainable food production in the development of innovative "healthy" snacks. *Sustainability, 11*(10), 2800.
11. Leiton-Ramirez, Y. M., Ayala-Aponte, A., & Ochoa-Martinez, C. I. (2020). Physicochemical properties of guava snacks as affected by drying technology. *Processes, 8*(1), 106. https://doi.org/10.3390/pr8010106.
12. Du, L. J., Gao, Q. H., Ji, X. L., Ma, Y. J., Xu, F. Y., & Wang, M. (2013). Comparison of flavonoids, phenolic acids, and antioxidant activity of explosion-puffed and sun-dried jujubes (Ziziphus jujuba Mill.). *Journal of Agricultural and Food Chemistry, 61*(48), 11840–11847.
13. Ma, W., Feng, S., Zhao, W., Xue, L., Shen, L., & Zheng, X. (2022). Formation of texture quality of raspberry snack under microwave puffing. *Journal of Texture Studies, 53*(2), 242–254. https://doi.org/10.1111/jtxs.12646.
14. Ahmed, I., Qazi, I. M., & Jamal, S. (2016). Developments in osmotic dehydration technique for the preservation of fruits and vegetables. *Innovative Food Science & Emerging Technologies, 34*, 29–43. https://doi.org/10.1016/j.ifset.2016.01.003.
15. Kowalska, J., Kowalska, H., Marzec, A., Brzeziński, T., Samborska, K., & Lenart, A. (2018). Dried strawberries as a high nutritional value fruit snack. *Food Science and Biotechnology, 27*, 799–807. https://doi.org/10.1007/s10068-018-0304-6.
16. Kowalska, H., Marzec, A., Domian, E., Kowalska, J., Ciurzyńska, A., & Galus, S. (2021). Edible coatings as osmotic dehydration pretreatment in nutrient-enhanced fruit or vegetable snacks development: A review. *Comprehensive Reviews in Food Science and Food Safety, 20*(6), 5641–5674. https://doi.org/10.1111/1541-4337.12837.
17. Obadina, A. O., Oyewole, O. B., & Williams, O. E. (2013). Improvement in the traditional processing method and nutritional quality of traditional extruded cassava-based snack

(modified Ajogun). *Food Science & Nutrition, 1*(4), 350–356. https://doi.org/10.1002/fsn3.43.
18. Maskan, M., & Altan, A. (2012). *Advances in Food Extrusion Technology*. CRC Press.
19. Huber, G. (2001). Snack foods from cooking extruders. In E. W. Lusas & L. W. Rooney (Eds.), *Snack Foods Processing* (pp. 315–367). CRC Press.
20. Kantrong, H., Klongdee, S., Jantapirak, S., Limsangouan, N., & Pengpinit, W. (2022). Effects of extrusion temperature and puffing technique on physical and functional properties of purpled third-generation snack after heat treatment. *Journal of Food Science and Technology, 59*(6), 2209–2219. https://doi.org/10.1007/s13197-021-05234-x.
21. Kojic, J., Belovic, M., Krulj, J., Pezo, L., Teslic, N., Kojic, P., Tukuljac, L. P., Seregelj, V., & Ilic, N. (2022). Textural, color and sensory features of spelt wholegrain snack enriched with betaine. *Foods, 11*(3), 475. https://doi.org/10.3390/foods11030475.
22. Delic, J., Ikonic, P., Colovic, R., Peulic, T., Jokanovic, M., Banjac, V., & Rakita, S. (2020). Mechanically deboned poultry meat and brewer's processing by-product as promising ingredients for nutritionally valuable extruded snacks. *Emirates Journal of Food and Agriculture, 32*(6), 453–460. https://doi.org/10.9755/ejfa.2020.v32.i6.2115.
23. Altaf, U., Hussain, S. Z., Qadri, T., Iftikhar, F., Naseer, B., & Rather, A. H. (2021). Investigation on mild extrusion cooking for development of snacks using rice and chickpea flour blends. *Journal of Food Science and Technology-Mysore, 58*(3), 1143–1155. https://doi.org/10.1007/s13197-020-04628-7.
24. Awolu, O. O., Magoh, A. O., & Ojewumi, M. E. (2020). Development and evaluation of extruded ready-to-eat snack from optimized rice, Kersting's groundnut and lemon pomace composite flours. *Journal of Food Science and Technology-Mysore, 57*(1), 86–95. https://doi.org/10.1007/s13197-019-04033-9.
25. Bekele, E. K., Nosworthy, M. G., Henry, C. J., Shand, P. J., & Tyler, R. T. (2020). Oxidative stability of direct-expanded chickpea-sorghum snacks. *Food Science & Nutrition, 8*(8), 4340–4351. https://doi.org/10.1002/fsn3.1731.
26. Choe, E., & Min, D. B. (2007). Chemistry of deep-fat frying oils. *Journal of Food Science, 72*(5), R77–R86. https://doi.org/10.1111/j.1750-3841.2007.00352.x.
27. da Silva, P. F., & Moreira, R. G. (2008). Vacuum frying of high-quality fruit and vegetable-based snacks. *LWT–Food Science and Technology, 41*(10), 1758–1767. https://doi.org/10.1016/j.lwt.2008.01.016.
28. Fan, L., Zhang, M., & Mujumdar, A. S. (2005). Vacuum frying of carrot chips. *Drying Technology, 23*(3), 645–656. https://doi.org/10.1081/DRT-200054159.
29. Song, X., Zhang, M., & Mujumdar, A. S. (2007). Optimization of vacuum microwave predrying and vacuum frying conditions to produce fried potato chips. *Drying Technology, 25*(12), 2027–2034. https://doi.org/10.1080/07373930701728638.
30. Granda, C., Moreira, R., & Tichy, S. (2004). Reduction of acrylamide formation in potato chips by low-temperature vacuum frying. *Journal of Food Science, 69*, E405–E411. https://doi.org/10.1111/j.1365-2621.2004.tb09903.x.
31. Quan, X., Zhang, M., Zhang, W., & Adhikari, B. (2014). Effect of microwave-assisted vacuum frying on the quality of potato chips. *Drying Technology, 32*(15), 1812–1819.
32. Su, Y., Zhang, M., Zhang, W., & Adhikari, B. (2016). Application of novel microwave-assisted vacuum frying to reduce the oil uptake and improve the quality of potato chips. *LWT, 73*, 490–497.
33. Su, Y., Zhang, M., Fang, Z., & Zhang, W. (2017). Analysis of dehydration kinetics, status of water and oil distribution of microwave-assisted vacuum frying potato

chips combined with NMR and confocal laser scanning microscopy. *Food Research International, 101*, 188–197.

34. Su, Y., Zhang, M., Zhang, W., Liu, C., & Bhandari, B. (2018). Low oil content potato chips produced by infrared vacuum pre-drying and microwave-assisted vacuum frying. *Drying Technology, 36*(3), 294–306. https://doi.org/10.1080/07373937.2017.1326500.
35. Elfaitouri, T. A., Ghazali, H. M., Sumnu, G., Ariffin, A. A., & Tan, C. P. (2018). Effect of microwave frying on acrylamide formation in potato chips. *World, 2*(2), 33–37.
36. Faruq, A. A., Zhang, M., & Fan, D. (2019). Modeling the dehydration and analysis of dielectric properties of ultrasound and microwave combined vacuum frying apple slices. *Drying Technology, 37*(3), 409–423. https://doi.org/10.1080/07373937.2018.1465433.
37. Devi, S., Zhang, M., & Law, C. L. (2018). Effect of ultrasound and microwave assisted vacuum frying on mushroom (*Agaricus bisporus*) chips quality. *Food Bioscience, 25*, 111–117. https://doi.org/10.1016/j.fbio.2018.08.004.
38. Devi, S., Zhang, M., Ju, R., & Mujumdar, A. S. (2020). Co-influence of ultrasound and microwave in vacuum frying on the frying kinetics and nutrient retention properties of mushroom chips. *Drying Technology, 38*(15), 2102–2113. https://doi.org/10.1080/07373937.2019.1604542.
39. Faruq, A. A., Zhang, M., & Adhikari, B. (2019). A novel vacuum frying technology of apple slices combined with ultrasound and microwave. *Ultrasonics Sonochemistry, 52*, 522–529. https://doi.org/10.1016/j.ultsonch.2018.12.033.
40. Islam, M. M., Zhang, M., Bhandari, B., & Guo, Z. (2019). A hybrid vacuum frying process assisted by ultrasound and microwave to enhance the kinetics of moisture loss and quality of fried edamame. *Food and Bioproducts Processing, 118*, 326–335. https://doi.org/10.1016/j.fbp.2019.10.004.
41. Fan, K., Zhang, M., & Bhandari, B. (2019). Osmotic-ultrasound dehydration pretreatment improves moisture adsorption isotherms and water state of microwave-assisted vacuum fried purple-fleshed sweet potato slices. *Food and Bioproducts Processing, 115*, 154–164. https://doi.org/10.1016/j.fbp.2019.03.011.
42. Liu, Z. B., Dick, A., Prakash, S., Bhandari, B., & Zhang, M. (2020). Texture modification of 3D printed air-fried potato snack by varying its internal structure with the potential to reduce oil content. *Food and Bioprocess Technology, 13*(3), 564–576. https://doi.org/10.1007/s11947-020-02408-x.
43. Gahlawat, P., & Sehgal, S. (1992). Phytic acid, saponins, and polyphenols in weaning foods prepared from oven-heated green gram and cereals. *Cereal Chemistry, 69*(4), 463–464.
44. Gunel, Z., Tontul, İ., Dincer, C., Topuz, A., & Sahin-Nadeem, H. (2018). Influence of microwave, the combined microwave/hot air and only hot air roasting on the formation of heat-induced contaminants of carob powders. *Food Additives & Contaminants: Part A, 35*(12), 2332–2339.
45. Sharma, P., Gujral, H. S., & Rosell, C. M. (2011). Effects of roasting on barley β-glucan, thermal, textural and pasting properties. *Journal of Cereal Science, 53*(1), 25–30. https://doi.org/10.1016/j.jcs.2010.08.005.
46. Sharma, P., & Gujral, H. S. (2011). Effect of sand roasting and microwave cooking on antioxidant activity of barley. *Food Research International, 44*(1), 235–240.
47. Ashtiani, S. H. M., Salarikia, A., & Golzarian, M. R. (2017). Analyzing drying characteristics and modeling of thin layers of peppermint leaves under hot-air and infrared treatments. *Information Processing in Agriculture, 4*(2), 128–139.

48. Bagheri, H., Mosayebi, M., Kashaninejad, M., & Najafian, L. (2020). Application of infrared heating for roasting nuts. *Journal of Food Quality*, *2020*, 1–10. https://doi.org/10.1155/2020/8810143.
49. Mokhtari, Z., & Ziaiifar, A. M. (2018). The effect of different methods of roasting on the physico-chemical properties of wild almond. *Innovative Food Technologies*, *6*(1), 55–73.
50. Abedelmaksoud, T. G., Smuda, S. S., Altemimi, A. B., Mohamed, R. M., Pratap-Singh, A., & Ali, M. R. (2021). Sunroot snack bar: Optimization, characterization, consumer perception, and storage stability assessment. *Food Science & Nutrition*, *9*(8), 4394–4407. https://doi.org/10.1002/fsn3.2412.
51. Iqbal, S., Thanushree, M. P., Sudha, M. L., & Crassina, K. (2021). Quality characteristics of buckwheat (*Fagopyrum esculentum*) based nutritious ready-to-eat extruded baked snack. *Journal of Food Science and Technology*, *58*(5), 2034–2040. https://doi.org/10.1007/s13197-020-04940-2.
52. Mousa, M. M. H., El-Magd, M. A., Ghamry, H. I., Alshahrani, M. Y., El-Wakeil, N. H. M., Hammad, E. M., & Asker, G. A. H. (2021). Pea peels as a value-added food ingredient for snack crackers and dry soup. *Scientific Reports*, *11*(1), 22747. https://doi.org/10.1038/s41598-021-02202-5.
53. Rahmi, Y., Kurniawati, A. D., Widyanto, R. M., Ariestiningsih, A. D., Al Farahi, A. Z., Ruchaina, A. N., Sihombing, E. V., Istira, F. B., Nafsiyah, I., Permatasari, K. D., Anjani, R. D., Simanjuntak, S. A. Y. M., & Rahma, Y. A. (2021). The sensory, physical and nutritional quality profiles of purple sweet potato and soy-based snack bars for pregnant women. *Journal of Public Health Research*, *10*(2). https://doi.org/10.4081/jphr.2021.2241.
54. Agu, H. O., Yunana, R., & Jideani, A. I. O. (2021). Development and acceptability of snacks from yellow-fleshed sweet potato and wheat flours. *Journal of Food Processing and Preservation*, *45*(1). https://doi.org/10.1111/jfpp.14851.
55. Pineda, C. G., Yamul, D. K., & Navarro, A. S. (2021). Effect of flaxseed (*Linum usitatissimum* L.) flour and yerba mate (*Ilex paraguariensis*) extract on physicochemical and sensory properties of a gluten-free corn-based snack. *Journal of Food Science and Technology-Mysore*, *58*(10), 3890–3901. https://doi.org/10.1007/s13197-020-04850-3.
56. Renoldi, N., Peighambardoust, S. H., & Peressini, D. (2021). The effect of rice bran on physicochemical, textural and glycemic properties of ready-to-eat extruded corn snacks. *International Journal of Food Science and Technology*, *56*(7), 3235–3244. https://doi.org/10.1111/ijfs.14939.
57. Uribe-Wandurraga, Z. N., Zhang, L., Noort, M. W. J., Schutyser, M. A. I., Garcia-Segovia, P., & Martinez-Monzo, J. (2020). Printability and physicochemical properties of microalgae-enriched 3D-printed snacks. *Food and Bioprocess Technology*, *13*(11), 2029–2042. https://doi.org/10.1007/s11947-020-02544-4.
58. Lucas, B. F., Morais, M. G., Santos, T. D., & Costa, J. A. V. (2017). Effect of spirulina addition on the physicochemical and structural properties of extruded snacks. *Food Science and Technology*, *37*, 16–23. https://doi.org/10.1590/1678-457x.06217.
59. Yu, C., Liu, J., Tang, X., Shen, X., & Liu, S. (2017). Correlations between the physical properties and chemical bonds of extruded corn starch enriched with whey protein concentrate. *RSC Advances*, *7*, 11979–11986. https://doi.org/10.1039/C6RA26764E.
60. Borah, A., Lata Mahanta, C., & Kalita, D. (2016). Optimization of process parameters for extrusion cooking of low amylose rice flour blended with seeded banana and carambola

pomace for development of minerals and fiber rich breakfast cereal. *Journal of Food Science and Technology*, *53*(1), 221–232. https://doi.org/10.1007/s13197-015-1772-9.

61. Caliskan, G., Lim, A. S. L., & Roos, Y. H. (2015). Beta-carotene stability in extruded snacks produced using interface-engineered emulsions. *International Journal of Food Properties, 18(10),* 2256–2267. https://doi.org/10.1080/10942912.2014.973963.
62. Yadav, D. N., Anand, T., Navnidhi, & Singh, A. K. (2014). Co-extrusion of pearl millet-whey protein concentrate for expanded snacks. *International Journal of Food Science & Technology, 49(3),* 840–846. https://doi.org/10.1111/ijfs.12373.
63. Liu, C., Xue, Y., Guo, J., Ren, H., Jiang, S., Li, D., Song, J., & Zhang, Z. (2021). Citric acid and sucrose pretreatment improves the crispness of puffed peach chips by regulating cell structure and mechanical properties. *LWT, 142,* 111036. https://doi.org/10.1016/j.lwt.2021.111036.
64. Zhu, J., Liu, Y., Zhu, C., & Wei, M. (2022). Effects of different drying methods on the physical properties and sensory characteristics of apple chip snacks. *LWT, 154,* 112829. https://doi.org/10.1016/j.lwt.2021.112829.
65. Yi, J., Hou, C., Bi, J., Zhao, Y., Peng, J., & Liu, C. (2018). Novel combined freeze-drying and instant controlled pressure drop drying for restructured carrot-potato chips: Optimized by response surface method. *Journal of Food Quality.* Article ID 6157697. https://doi.org/10.1155/2018/6157697.
66. Cui, Z. W., Li, C. Y., Song, C. F., & Song, Y. (2008). Combined microwave-vacuum and freeze drying of carrot and apple chips. *Drying Technology, 26(12),* 1517–1523. https://doi.org/10.1080/07373930802463960.
67. Kim, N., Yu, K. S., Kim, J., Lim, T., & Hwang, K. T. (2018). Chemical characteristics of potato chips fried in repeatedly used oils. *Food Measurement and Characterization, 12,* 1863–1871. https://doi.org/10.1007/s11694-018-9800-x.
68. Kaur, R., & Prasad, K. (2022). Effect of malting and roasting of chickpea on functional and nutritional qualities of its protein fractions. *International Journal of Food Science & Technology, 57,* 3990–4000. https://doi.org/10.1111/ijfs.15769.
69. Asadi, S., Aalami, M., Shoeibi, S., Kashaninejad, M., Ghorbani, M., & Delavar, M. (2020). Effects of different roasting methods on formation of acrylamide in pistachio. *Food Science & Nutrition, 8(6),* 2875–2881.
70. Cha, C. Y., & Lee, K. G. (2020). Effect of roasting conditions on the formation and kinetics of furan in various nuts. *Food Chemistry, 331,* 127338.
71. Bagheri, H., Kashaninejad, M., Ziaiifar, A. M., & Aalami, M. (2019). Textural, color and sensory attributes of peanut kernels as affected by infrared roasting method. *Information Processing in Agriculture, 6(2),* 255–264. https://doi.org/10.1016/j.inpa.2018.11.001.
72. Jogihalli, P., Singh, L., & Sharanagat, V. S. (2017). Effect of microwave roasting parameters on functional and antioxidant properties of chickpea (*Cicer arietinum*). *LWT–Food Science and Technology, 79,* 223–233. https://doi.org/10.1016/j.lwt.2017.01.047.
73. Kaya, Z., & Asir, Y. (2022). Assessment of instrumental and sensory quality characteristics of the bread products enriched with Kombucha tea. *International Journal of Gastronomy and Food Science, 29,* 100562.
74. Ergün Altay, K., Çalışkan Koç, G., & Dirim, S. N. (2021). The effect of freeze-dried fruit powders on some of the physicochemical properties of the layer cake. *Pamukkale Üniversitesi Mühendislik Bilimleri Dergisi (PAJES), 27(7),* 813–819.

75. Woodbury, T. J., Lust, A. L., & Mauer, L. J. (2021). The effects of commercially available sweeteners (sucrose and sucrose replacers) on wheat starch gelatinization and pasting, and cookie baking. *Journal of Food Science, 86(3),* 687–698.
76. Melati, J., Lucchetta, L., Prado, N. V. D., Oliveira, D. F. D., & Tonial, I. B. (2021). Physical and sensory characteristics of salty cereal bar with different binding agents. *Food Science and Technology, 41,* 150–154.
77. Ochirov, V. D., Altukhov, I. V., Bykova, S. M., & Tsuglenok, N. V. (2021). Investigation of infrared drying of carrot chips. *IOP Conference Series: Earth and Environmental Science, 659(1),* 012037. https://doi.org/10.1088/1755-1315/659/1/012037.
78. Asbridge, S. C. M., Pechey, E., Marteau, T. M., & Hollands, G. J. (2021). Effects of pairing health warning labels with energy-dense snack foods on food choice and attitudes: Online experimental study. *Appetite, 160,* 105090. https://doi.org/10.1016/j.appet.2020.105090.
79. Kowalska, H., Masiarz, E., Ignaczak, A., Marzec, A., Hać-Szymańczuk, E., Salamon, A., Cegiełka, A., Żbikowska, A., & Galus, S. (2022). Advances in multigrain snack bar technology and consumer expectations: A review. *Food Reviews International,* 1–26. https://doi.org/10.1080/87559129.2022.2094402.
80. Panak Balentić, J., Babić, J., Jozinović, A., Ačkar, D., Miličević, B., Muhamedbegović, B., & Šubarić, D. (2018). Production of third-generation snacks. *Croatian Journal of Food Science and Technology, 10(1),* 98–105.
81. Grasso, S. (2020). Extruded snacks from industrial by-products: A review. *Trends in Food Science & Technology, 99,* 284–294. https://doi.org/10.1016/j.tifs.2020.03.012.
82. Goes, E., Souza, M., Campelo, D., Yoshida, G., Xavier, T., de Moura, L., & Monteiro, A. (2015). Extruded snacks with the addition of different fish meals. *Food Science and Technology (Campinas), 35(4),* 683–689. https://doi.org/10.1590/1678-457X.6818.
83. Fernandes, A., Madeira, R., PilerCarvalho, C., & Pereira, J. (2016). Physical and sensory characteristics of pellets elaborated with different levels of corn grits and whey protein concentrate. *Ciência e Agrotecnologia, 40,* 235–243. https://doi.org/10.1590/1413-70542016402031515.
84. Alam, M. S., Pathania, S., & Sharma, A. (2016). Optimization of the extrusion process for development of high fiber soybean-rice ready-to-eat snacks using carrot pomace and cauliflower trimmings. *LWT, 74,* 135–144. https://doi.org/10.1016/j.lwt.2016.07.031.
85. O'Shea, N., Arendt, E., & Gallagher, E. (2014). Enhancing an extruded puffed snack by optimizing die head temperature, screw speed and apple pomace inclusion. *Food and Bioprocess Technology, 7(6),* 1767–1782. https://doi.org/10.1007/s11947-013-1181-x.
86. Selani, M. M., Brazaca, S. G. C., dos Santos Dias, C. T., Ratnayake, W. S., Flores, R. A., & Bianchini, A. (2014). Characterization and potential application of pineapple pomace in an extruded product for fiber enhancement. *Food Chemistry, 163,* 23–30. https://doi.org/10.1016/j.foodchem.2014.04.076.
87. Wang, S., Kowalski, R. J., Kang, Y., Kiszonas, A. M., Zhu, M. J., & Ganjyal, G. M. (2017). Impacts of the particle sizes and levels of inclusions of cherry pomace on the physical and structural properties of direct expanded corn starch. *Food and Bioprocess Technology, 10(2),* 394–406. https://doi.org/10.1007/s11947-016-1824-9.
88. Fleischman, E. F., Kowalski, R. J., Morris, C. F., Nguyen, T., Li, C., Ganjyal, G., & Ross, C. F. (2016). Physical, textural, and antioxidant properties of extruded waxy wheat flour snack supplemented with several varieties of bran. *Journal of Food Science, 81(11),* E2726–E2733. https://doi.org/10.1111/1750-3841.13511.

89. Ogunmuyiwa, O. H., Adebowale, A. A., Sobukola, O. P., Onabanjo, O. O., Obadina, A. O., Adegunwa, M. O., Kajihausa, O. E., Sanni, L. O., & Keith, T. (2017). Production and quality evaluation of extruded snack from blends of Bambara groundnut flour, cassava starch, and corn bran flour. *Journal of Food Processing and Preservation, 41(5),* e13183.
90. Sharma, R., Srivastava, T., & Saxena, D. C. (2016). Development of nutritious snack from rice industry waste using twin screw extrusion. *MATEC Web of Conferences, 57,* 04006.
91. Sharma, R., Srivastava, T., & Saxena, D. C. (2018). Physico-chemical and functional properties of deoiled rice bran and its utilization in the development of extruded product. *The Pharma Innovation, 7(5, Part B),* 109.
92. Srivastava, T., Saxena, C. D., & Sharma, R. (2016). Effects of twin screw parameters upon the mechanical hardness of ready-to-eat extrudates enriched with deoiled rice bran. *Food and Environment Safety Journal, 13(4),* 299–308.
93. Srivastava, T., Saxena, C. D. (2017). Effects of different process and machine parameters on physical properties of rice flour, corn flour and deoiled rice bran extruded product. *Current Nutrition & Food Science, 13(3),* 219–226.
94. Bhise, S. R., Kaur, A., Manikantan, M. R., & Singh, B. (2015). Development of textured defatted sunflower meal by extrusion using response surface methodology. *Acta Alimentaria, 44(2),* 251–258.
95. Chen, B., Yu, C., Liu, J., Yang, Y., Shen, X., Liu, S., & Tang, X. (2017). Physical properties and chemical forces of extruded corn starch fortified with soy protein isolate. *International Journal of Food Science & Technology, 52(12),* 2604–2613. https://doi.org/10.1111/ijfs.13547.
96. Rajagukguk, Y. V., Arnold, M., Sidor, A., Kulczynski, B., Brzozowska, A., Schmidt, M., & Gramza-Michalowska, A. (2022). Antioxidant activity, probiotic survivability, and sensory properties of a phenolic-rich pulse snack bar enriched with *Lactiplantibacillus plantarum*. *Foods, 11(3),* 309. https://doi.org/10.3390/foods11030309.
97. Singh, A., Kumari, A., & Chauhan, A. K. (2022). Formulation and evaluation of novel functional snack bar with amaranth, rolled oat, and unripened banana peel powder. *Journal of Food Science and Technology,* 1–11. https://doi.org/10.1007/s13197-021-05344-6.
98. Cho, W. H., Park, J. M., Kim, E. J., Mohibbullah, M., Choi, J. S. (2022). Evaluation of the quality characteristics and development of a puffed-rice snack enriched with honeybee (*Apis mellifera* L.) drone pupae powder. *Foods, 11(11),* 1599. https://doi.org/10.3390/foods11111599.
99. Maya, S., Sulaeman, A., & Sinaga, T. (2020). Alternative snack for diabetic patients from sago (*Metroxylon* sp.) starch and tempeh. *Journal Gizi Dan Pangan, 15(1),* 27–36. https://doi.org/10.25182/jgp.2020.15.1.27-36.
100. Statista. (2021). *Revenue of the Confectionery & Snacks Market Worldwide by Country 2020.* Published by statista research department. Retrieved from: https://www.statista.com/forecasts/758648/revenue-of-the-confectionery-market-worldwide-by-country (accessed on 9 August 2024).
101. Trade Map, (2022). *Import and Export Information for Different Snacks*. Retrieved from: http://trademap.org/Index.aspx (accessed on 9 August 2024).

CHAPTER 2

ROASTING TECHNOLOGY FOR PRODUCTION OF READY-TO-EAT SNACKS

ÖZGE SÜFER

Osmaniye Korkut Ata University, Food Engineering Department, Osmaniye, Türkiye

ABSTRACT

This comprehensive chapter reviews the main titles; principles of roasting system, roasted ready to eat (RTE) snack products and the influences of roasting process on the individual's health. It begins with the guidelines of roasting technology which explains the concept of related operation by some mathematical equations, and then the roasting techniques are introduced. Sand roasting, oven roasting and pan roasting are categorized as classical procedures, even though microwave roasting, infrared roasting and superheated steam roasting are mentioned as novel/emerging methods. Roasted RTE snack foods are divided into three groups – (i) those with cereal or legume; (ii) products containing nut or edible seed; and (iii) the others. Popular members of first subtitle are chickpea, leblebi, and flaked rice, and peanut and sunflower seed are the most favorite ones in the second. Bakery products, fruit snacks and sweet potato are evaluated under the last class. While sand roasting is usually preferred for roasting step of RTE snacks with cereal or legume, oven roasting is the most common technique. At the end of the chapter, positive (total phenolics and antioxidant activity) and negative (aflatoxin, acrylamide, 5-hydroxymethylfurfural and ochratoxin) health effects of roasted food materials are presented.

Ready-to-Eat Snacks: Emerging Technologies for Production and Safety.
Ravi Pandiselvam and Gülşah Çalışkan Koç

2.1 INTRODUCTION

One of the fundamental food processing techniques is roasting, and this operation is significantly related to the sensory properties of the food material [1]. Generally, heat is applied in a temperature range of 150–300ºC, and the processing time is dependent on the type of the product. Roasting contains several chemical reactions like hydrolysis, oxidation, polymerization etc., and as such in other heat treatment procedures, it is able to contribute both digestibility and palatability [2]. However, many bioactive compounds occur at this high temperature levels in the medium, some carcinogenic and mutagenic components (heterocyclic amines, acrylamide and so on) may also be formed in the food matrix [2, 3]. There are a lot of methods used for roasting of foods, and they can be classified under two main groups; traditional and novel methods. Sand roasting [4], hot air roasting [5], and open pan roasting [6] are conventional ones, and microwave [7], infrared [7], superheated steam [5], and roasting procedures are accepted as new. Although both sand and open pan roasting seem unhygienic, most of novel roasting processes have disadvantages as well. These are high-energy consumption, non-homogenous properties and direct exposure of the food to flame and/or smoke [6]. In order to overcome the related limitations, some combined roasting methodologies (hybridized infrared-hot air, microwave-hot air) have been developed by researchers [8, 9].

Ready to eat (RTE) foods need no further heating or processing and can be consumed hot or cold. A wide range of products including sandwiches, biscuits, salads and crisps are considered in RTE category [10], and the roasted members of these foods are comprised of especially nuts and legumes (almond, groundnut, chickpea, pistachio, sunflower seed, pumpkin seed etc.,) and cereals (popped rice, puffed rice, popcorn [4], popped buckwheat etc.). On the other hand, recent advances in RTE roasted snacks are dominated by coated products. Chocolate and honey are used for coating applications, because of distinctive flavor and naturally sweet taste, respectively [12, 13].

This chapter has the following objectives:

1. To explain the theory and mechanism of roasting technology and equipment/systems utilized in RTE snack food production,
2. To sum up traditional, mostly consumed cereal and legume based roasted RTE foods, as well as emerging trends,
3. To focus on the effects of roasting process on individual's health considering bioactive compounds and toxic substances.

2.2 ROASTING TECHNOLOGY

2.2.1 PRINCIPLES OF ROASTING

The definition of "roasting" includes the application of dry heat, and is valid for generally meat and snacks. If a dough is available, "baking" term is more suitable. The main goal of roasting process is to ensure short term stability of foods by heating and drying. The heat is simultaneously supplied by three transfer mechanisms which may vary at different stages of the process: conduction, convection and radiation. When hot medium (gas, fire, etc.,) causes convection, conduction occurs between hot surfaces (pan, tray, and so on) and food. Radiation can be originated from the walls of oven [13] or only electromagnetic waves depending on the type of roasting method.

During roasting, rapid starch gelatinization and protein denaturation happen which help the conversion of many nutritional elements into more digestible form. As a result of these complex reactions, volume and texture improve, and an elevated puffing index is obtained. In addition, aroma, color and shelf-life of the material advance, the amount of antinutrient molecules and bulk density decrease [4]. A heterogeneous roasting can be occurred in classical methods, because the external surface of the food is over roasted, however the inside of the product is not sufficiently processed. Furthermore, bitter burned taste and unpleasant aroma are able to arise [14].

Both temperature and time have great importance on the quality of final product as well as other factors (e.g., size and composition of the material, pressure, airflow rate). Hence, mathematical models should be concentrated on understanding and simulating heat and mass transfer phenomena occurred in multicomponent biomass during roasting, in order to present relationships between process conditions and product characteristics [2]. Generally, the experimental moisture/mass data are fitted to the following equation and solved by using appropriate "n" values (sign reaction degree) which corresponds commonly utilized mathematical expressions in literature [15].

$$\frac{dx}{dt} = k\,t^{n} \tag{1}$$

where; x is the dimensionless moisture ratio; t is the time (s); and k is the velocity constant.

The coefficient of moisture is estimated by solving Fick's second law equation analytically [15];

$$\frac{\partial x}{\partial t} = D_e \nabla^2 x \tag{2}$$

where; D_e is the effective moisture diffusivity (m^2/s). The product geometry plays an important role in calculation.

Rojas et al. [16] studied the roasting process of cocoa particle in μ-reactor, and the heat transfer mechanism was accepted as conduction. Both conduction and radiation were regarded in boundary conditions, and some part of heat was used in chemical reactions. The energy balance was expressed as;

$$\rho_s Cp_s \frac{\partial(T)}{\partial t} + div(\rho_v Cp_v v_v T) = div(\lambda_{eff} \nabla T) + Q_{reactions} \quad 3)$$

where; ρ_s is the density of solid phase (kg/m^3); Cp_s and Cp_v is the heat capacities of solid and gas (J/kgK); T is the absolute temperature (K); t is the time (s); v_v is the velocity vector of volatiles; λ_{eff} is the effective thermal conductivity (W/mK); and $Q_{reactions}$: amount of energy released or required through volatiles (W).

In a spouted bed coffee roaster, between coffee beans and hot air, an unsteady state heat transfer was investigated by Eqn. (4) [17].

$$T = \frac{(T_a - T_i)}{(T_a - T_p)} = exp\left(\frac{UAt}{m_p c_p}\right) \quad (4)$$

where; T_a is the air temperature (K); T_i is the initial temperature of product (K); T_p is final temperature of product (K); U is overall heat transfer coefficient (W/m^2K); A is total surface area of product (m^2); t is time (s); m_p is mass of product (kg); and C_p is specific heat (J/kgK).

Besides that, the degradation of natural compounds and kinetic analysis of formed volatiles can be useful for the better interpretation of roasting extensively. These also will contribute to optimize the process in a successful way [18]. In a very recent study, during oven roasting of pumpkin seed, peanut, flaxseed, almond, and sunflower seed at 160 and 180°C, both Maillard and caramelization reactions were monitored, and multi response kinetic models were proposed by Şen and Gökmen [19]. In fluidized bed roasting of cocoa nibs, the formation of nitrogen-heterocycles as a function of temperature was described as follows [20];

$$C_i = C_o + k_{ref} exp\left(\frac{E_a}{RT_f}\left(1 - \frac{T_{ref}}{T}\right)\right) t \quad (5)$$

where; C_i is the relative concentration of a volatile at time t; C_o is the initial concentration of a volatile; E_a is the activation energy (J/mol); R is the universal gas constant (8.314 J/mol K); T is the absolute temperature (K); T_{ref} is the reference temperature (K); and k_{ref} is the reference rate of volatile formation.

Nowadays, all industries have started to regard process optimization as significant, because of saving time and production costs, establishing the best operating conditions and gaining more acceptable results. Before nearly 70 years ago, only one independent variable was considered, and therefore the number of trials increased and inaccurate findings were observed. In order to make better predictions, evaluate several parameters at the same time and improve the technique, new methods like response surface methodology (RSM) have been developed [21]. Recent papers have begun to concentrate on modeling some responses using RSM tool during roasting. Peanut kernels [22], *Opuntia fiscus-Indica* (prickly pear) seeds [23], sunflower seed kernels [24] and *Argania spinosa* L. (argan) kernels [25] are the some encountered examples in literature. Moreover, Artificial Neural Network and Genetic Algorithm are robust optimization procedures, and the combination of both methods were tested in hot air roasting of Chironji kernels at 25 combinations of two process parameters (time and temperature) [26].

2.2.2 TYPES OF ROASTING EQUIPMENT/SYSTEMS

Conventional and commonly used roasting methods are known as sand roasting, oven roasting and pan roasting (Figure 2.1), and some of the novel ones are microwave roasting, infrared roasting and superheated steam roasting (Figure 2.2) [2].

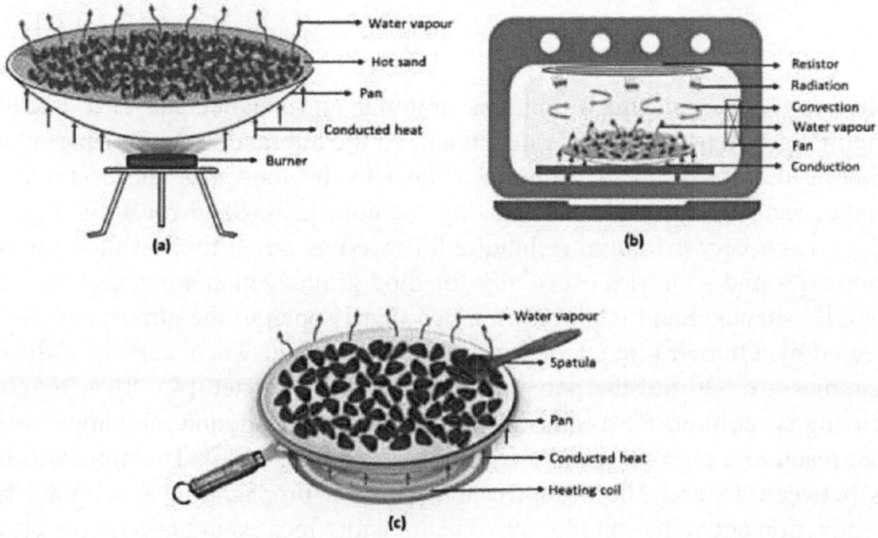

FIGURE 2.1 Traditional roasting methods a) Sand roasting, b) Oven roasting, c) Pan roasting.
Source: Adapted with permission from Ref. [2]. © 2021 Elsevier.

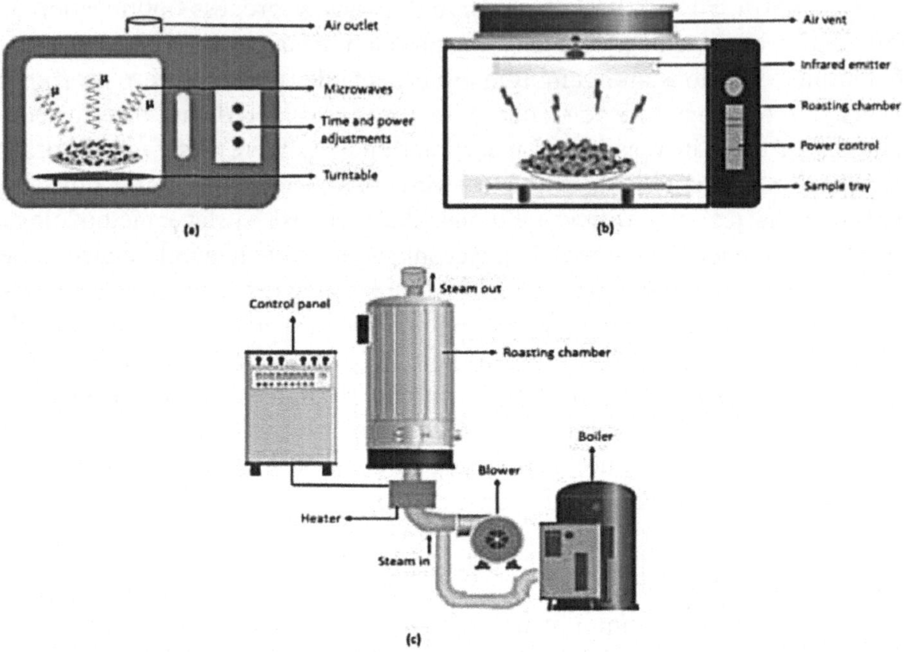

FIGURE 2.2 Newly Developed roasting techniques a) Microwave roasting, b) Infrared roasting, c) Superheated steam roasting.

Source: Adapted with permission from Ref. [2]. © 2021 Elsevier.

2.2.2.1 SAND ROASTING

Sand is a material which contains granular and smaller particles, and its thermal properties are strictly dependent on the interactions among particles. Having the ability of distribution of heat to the food in a successful way makes sand being used as a roasting medium [27]. Sand roasting (Figure 2.1(a)) is a very primitive technique followed by street food vendors, some industries and countries especially for food grains which are generally sold as RTE snacks. Sand is placed in a pan that is open to the atmosphere, and heated by a burner (coal, wood, etc.,) in this method. Then, cereals, mills or legumes are put into the pan, and the roasting is started [4, 28]. A regular stirring is required for a uniform heating and roasting, and the temperature can reach to a high level (350°C) during the process [28]. The process time is between 15 and 100 seconds, and the heat propagation is supplied by conduction and radiation [27, 29]. The moisture located in grain is converted into the superheated steam vapor which causes puffing, and unsaturated fats and dextrins become appear [28].

Sari et al. [27] simulated the heat transfer of peanut sand roasting process in one dimension by using finite difference technique, and benefitted from the following mathematical equations;

$$T_{wok}(t + \Delta t) = T_{wok}(t) + \frac{P_{coil}\Delta t}{m_{wok} cp_{wok}} \quad (6)$$

where; T_{wok} is the temperature of wok (K); P_{coil} is the power produced from coil (W); m_{wok} is the mass of wok (kg); Cp_{wok} is the specific heat of wok (J/kgK); and t is the time (s).

$$P_{coil} = \frac{V_{in}^2}{R_{coil}} \quad (7)$$

where; V_{in} is the Given voltage (V); and R_{coil}; and Resistance of coil (ohm).

Sand roasting technique have some disadvantages which limit the usage of this method. First of all, temperature control is not easy, and a non-uniform temperature distribution can be observed, if an effective stirring is not performed. Due to having no enclosed medium, roasted products are open to physical and microbial contamination. Also, both personnel and food are prone to smoke, flame, heat, gas and charcoal which are potential damages. It needs labor force, has low productivity, and is an energy inefficient procedure [2].

2.2.2.2 OVEN ROASTING

Baking/roasting ovens (Figure 2.1(b)) are the most popular equipment in traditional roasting process, and the convectional movement of heat is generally preferred [30]. Thanks to the circulation of air and steam, temperature and density differences arise, and hence convectional (natural or forced) heat transfer occurs. Vertical circulation of air is stemmed from rising of hot gas and falling of cool gas [31]. Convection ovens enable good product quality, and their designs are very easy. But longer roasting times are required, and they have less energy efficient [32]. The temperature range of 120–180°C is normally used, and elevated temperature levels are able to speed up some undesired reactions in proteins and lipids [30].

In several ovens, conduction and convection are used together. If a hot surface contacts directly to the food which will be roasted, and there is an important temperature difference between surfaces, conduction results. But conduction mechanism is not very efficient, and the heat transfer is heterogeneous. Fan-forced ovens circulate hot gases very rapidly when compared to the others, and not only vertical motion but also horizontal movement can be created. Furthermore, impingement ovens jet hot gas through nozzles, provide uniform heat, and thus roasting time and temperature reduce [31].

Computational fluid dynamics technique is capable of observing a 3D flow and temperature distribution inside roasting ovens, and it is based on the following conservation equations in cartesian coordinates [33];

Conservation of mass:

$$\frac{\delta u}{\delta x}+\frac{\delta v}{\delta y}+\frac{\delta w}{\delta z}=0 \tag{8}$$

Conservation of momentum:

$$\rho\left(u\frac{\delta u}{\delta x}+v\frac{\delta u}{\delta y}+w\frac{\delta u}{\delta z}\right)=-\frac{\delta P}{\delta x}+\mu\left(\frac{\delta^2 u}{\delta x^2}+\frac{\delta^2 u}{\delta y^2}+\frac{\delta^2 u}{\delta z^2}\right)+f_x \tag{9}$$

$$\rho\left(u\frac{\delta v}{\delta x}+v\frac{\delta v}{\delta y}+w\frac{\delta v}{\delta z}\right)=-\frac{\delta P}{\delta y}+\mu\left(\frac{\delta^2 u}{\delta x^2}+\frac{\delta^2 v}{\delta y^2}+\frac{\delta^2 v}{\delta z^2}\right)+f_y \tag{10}$$

$$\rho\left(u\frac{\delta w}{\delta x}+v\frac{\delta w}{\delta y}+w\frac{\delta w}{\delta z}\right)=-\frac{\delta P}{\delta z}+\mu\left(\frac{\delta^2 w}{\delta x^2}+\frac{\delta^2 w}{\delta y^2}+\frac{\delta^2 w}{\delta z^2}\right)+f_z \tag{11}$$

where; P is the pressure (Pa); v is the velocity (m/s); ρ is the density (kg/m^3); μ is the dynamic viscosity (Pa. s); f is the volumetric mass generation rate.

2.2.2.3 PAN ROASTING

Pan roasting (Figure 2.1(c)) is so simple that it only requires a pan as an equipment and heating medium, except the food that will be roasted. Pans are usually made of stainless steel or cast iron [2], which accelerates heat propagation, because of its high thermal conductivity. As such in sand roasting, a continuous and proper stirring is essential in order to avoid over-heating and/or burning. In cashew processing, the dimensions of pans change between 600 and 675 mm, and roasting is conducted by an open fire in India [34].

2.2.2.4 MICROWAVE ROASTING

In microwave roasting (Figure 2.2(a)), energy transfer occurs with the aid of electromagnetic waves, unlike thermal radiation. The wavelength ranges of microwaves that are partly absorbed, reflected and transmitted, are between 0.1 and 1 m in air, which correspond 0.3–3 GHz frequency range. Microwaves are able to penetrate deep into the food, and heat is produced inside the material (dielectric heating) independently from surface resistance or thermal conductivity. The heat production rate (W/m^3) can be calculated by Eqn. (12) per unit volume [13];

$$W_v = 2\pi\varepsilon_0 ffE^2\varepsilon'' \quad (12)$$

where; ε_0 is the vacuum permittivity (8.86×10^{-12} F/m); ε″ is the loss factor; ff is the field frequency (1/s); and E is the potential gradient (V/m).

Loss factor is the ratio of microwave energy which is converted to heat, and determined by loss tangent (Eqn. 13) as in the following [13];

$$\tan\delta = \frac{\varepsilon''}{\varepsilon'} \quad (13)$$

where ε′ is the dielectric permittivity.

On the other hand, the temperature gradient in microwave processing is specified by Eqn. (14) [13, 35]:

$$\frac{dT}{dt} = \frac{2\pi f\varepsilon_0\varepsilon'' E^2}{Cp\rho} \quad (14)$$

where; dT: T is the temperature (K); dt: t is the time (s); Cp is the heat capacity (J/kg K); and ρ is the density (kg/m^3).

It should not be forgotten that, chemical composition and accordingly polarity of the food material plays an important role in roasting process, because microwaves interact with water molecules [13].

Microwaves bring a new perspective into non-conventional roasting process in RTE production, and provide several advantages. They eliminate pre-heating, and permit both rapid and selective heating. It is an energy efficient procedure, and instant start-up and space savings are the other major benefits. In spite of being successful in food processing operations besides roasting, some problems related to the product quality can be seen. Starch gelatinization may be insufficient, color development might not be desired or gummy or dense structure can appear [32]. In order to compensate these deficiencies, more optimization studies should be conducted and/or the researches on hybrid roasting techniques are ought to be developed.

2.2.2.5 INFRARED ROASTING

The wavelength range of infrared radiation is 0.78–1,000 μm, and this gap can be categorized into three groups named as near infrared, middle infrared and far infrared which correspond the wavelength range between 0.78 and 1.4 μm, 1.4 and 3 μm and 3 and 1,000 μm, respectively. In infrared roasting of food products (Figure 2.2(b)), some parts of infrared radiation is absorbed, reflected and transmitted like microwaves [36]. During roasting, absorbed and penetrated infrared radiation by food causes strong rotations in water molecules, and the kinetic energies of these molecules enhance because of oscillations and vibrations. Then, the elevated kinetic energy is converted to friction work and heat [13, 36]. Heat contributes to rise both the water vapor pressure and temperature inside the product, and therefore moisture transfer starts as a result of fast heating [36]. In grains, high pressure triggers cotyledon in order to collapse or explode, and so biomolecules of food such as starch and protein modify their microstructure [37].

Pawar and Pratape [38] proposed a mathematical expression for the designation of infrared heat flux which is given by Eqn. (15);

$$Qr = \frac{(\sigma A_p(T_r^4 - T_s^4)}{\frac{1-\varepsilon_p}{\varepsilon_p} + \frac{1}{F_{pr}} + \frac{1-\varepsilon_r}{\varepsilon_r(\frac{A_r}{A_p})}} \tag{15}$$

where; Q_r is the radiative flux (W/m^2); σ is the Stefan–Boltzmann constant (5.67×10^{-8} W/m^{2}K^4); A_p is the surface area of product (m^2); A_r is the radiator surface area (m^2); T_r is the radiator temperature (K); T_s is the surface temperature (K); F_{pr} is the view factor; ε_r is the radiator emissivity; and ε_p is the product emissivity.

Infrared heaters/radiators can be grouped under two categories; gas-fired heaters and electrical heaters [37]. Electrical heaters are also classified considering the studied wavelengths, as quartz, ceramic and halogen ones. The maximum temperature levels and efficiencies reached by these heaters, are between 600 and 2,500 K, and 78 and 85%, respectively. The efficiency values decrease to the stage of 40 and 46% in gas-fired ovens [36]. Infrared equipment have gained popularity because of less processing time, energy and infrared lamb costs, giving an efficient heating, providing a homogenous surface temperature and good sensory and nutritive attributes especially when compared to microwave. Nevertheless, some disadvantages like low penetration and being not susceptible to reflective characteristics of coatings can put limits for the industrial utilization of this method [32].

2.2.2.5 SUPERHEATED STEAM ROASTING

Although the superheated steam procedure is generally used for roasting coffee and cocoa beans [39–42], this technology is promising in order to be utilized in the manufacturing process of RTE snacks. Because low energy consumption and environmental impact, and high heat transfer rates [43] make superheated steam technology be a candidate for an emerging phenomenon enhancing RTE snack quality.

Superheated steam has a temperature above the boiling point, and a typical system of related engineering application (Figure 2.2(c)) contains five units: heating device, a generator which the steam is produced, processing chamber, blower/fan and a heat exchanger. The heat from the heating unit is transferred to the generator for the production of superheated steam, and then this steam is flowed through the closed chamber. After that, the steam condenses on the material which will be roasted, and huge amount of heat is released. At the same time, the product temperature rises, and the diffused moisture is also recycled for being as a part of heating medium. A wide range of temperature (373–1,273 K) and pressure can be applied (low or high) in superheated steam technology [43].

2.2.3 ROASTED RTE SNACKS

When a deep literature survey is performed, several types of RTE snacks processed by roasting are available. In order to catch the current trends, studies only carried out in last five years were regarded, and three basic categories were created; RTE snacks including cereal or legume, RTE snacks with nuts or edible seeds and the others.

2.2.3.1 ROASTED RTE SNACKS WITH CEREAL AND LEGUME

Since the beginning of humanity, cereals have been the most basic food source due to being cultivated all over the world. In addition, cereals are strategic food materials, their producers are supported by many countries, and their markets are regulated by the states [44]. Ones of the member of Gramineae family are cereals, and this group involves wheat, rice, corn, oat, rye and barley [4]. Gluten-free cereals are rice and corn, and also pseudo-cereals containing quinoa, buckwheat and amaranth are gluten-free. There are lots of biologically active materials such as polyphenols, minerals, soluble and insoluble dietary fibers and starch in cereals [45].

Legumes are the good sources of proteins, and they comes after grains in the ranking, among the most consumed foods [45]. Nowadays, lots of nutrition styles such as “Mediterranean diet,” “Karatay diet,” “Vegetarian diet," and "Stone Age diet" suggest the consumption of legumes [46]. Bean, chickpea, broad bean, kidney bean, soybean, pea, black-eyed pea, red lentil, green lentil and yellow lentil are in the group of legumes, and all of them are gluten-free. Two globulin proteins named as legumin and vicilin are in the structure of pea, and chickpea proteins have high bioavailability levels [45]. Moreover, daily consumption of chickpea in small amounts can be helpful for decreasing the acidity level of stomach [46].

Table 2.1 summarizes RTE snack products which include cereal and/or legume in roasted form. Bean, chickpea, rice, maize, corn, green lentil and soybean are used as ingredient in roasted RTE snacks, and the most favorable roasting techniques are oven roasting, hot-air roasting, microwave roasting, microwave-hot air combination roasting, roasting by an agitated dry roaster, pan roasting and sand roasting. Also, a new method has been introduced by Jogihalli et al. for roasting of chickpea [6]. In this system, there are a heating element, temperature controller, a screw, electrical DC motor and a feed hopper. Feed hopper helps to feed chickpea into the cylinder, and the screw transfers chickpea. A proportional integral derivative (PID) controller has been installed into the electrical heaters, which provide energy into the surface of the cylinder, and an electric motor controls the speed of this screw. Screw conveyor enables both a homogenous mixing and carrying. Conduction and convection heat transfer mechanisms are valid for propagation of heat from the cylinder and hot air to the chickpea. At the same time, superheated steam has been generated during roasting that increase the rate of heating. The authors declared the benefits of this novel technology as low cost, simple design, continuous stirring, and uniform heating. Preliminary trials have created source for determining process conditions, and selected temperature levels are 250, 270, 290, 310, 330, and 350°C for roasting time of 6 minutes. Burnings are observed above the temperatures of 350°C, while unroasted chickpeas are produced below 250°C [6].

On the other hand, one of the traditional RTE snack of Ottoman-Turkish cuisine is roasted chickpea, in other words leblebi. Different names of leblebi are known, and it is called Garbanzo (in Latin America), Shimbra (in Ethiophia), and Bengal gram (in India) in the world. Even though, chickpea contains a significant amount of starch (nearly 30%), the glycemic index, which is the indicator of blood glucose level, of leblebi is not high (about 20–30) [46]. Yellow leblebi is the most favorite roasted chickpea by Turkish people, and it is sold in in sealed packages in almost every supermarket [60]. In previous papers, hot-air roasting, microwave roasting and microwave-hot air combination roasting techniques were used in leblebi production, and

TABLE 2.1 RTE Snacks Containing Roasted Cereal or Legume

RTE Snack Product	Type of Roasting	Process Conditions	Key Findings	References
Biscuit with adzuki bean	Oven roasting	155°C, 30 min	Resulted in pleasant flavor.	[47]
Chickpea	Novel continuous roasting	Temperature: 250, 270, 290, 310, 330, 350°C; Time: 6 min	Decrease in total flavonoid content, Increase in total phenolics and antioxidant activity, Maximum puffing and expansion index, minimum length/width ratio at the highest roasting temperature.	[6]
Chickpea (Leblebi)	ND	Temperature: 150, 180, 200°C; Time: 4, 6, 8 min	Decrease in L*, increase in a* and b*, L* followed 0th order, a* and b* followed 1st order reaction kinetics, The lowest speckling was found at 150°C–4 min combination.	[48]
Chickpea (Leblebi)	Hot-air roasting, Microwave roasting, microwave-hot air combination	Hot air roasting: 250°C, 20 min; Microwave roasting: 600 W, 20 min Microwave-hot air roasting: 360 W, 10 min + 150°C, 5 min	Roasting process did not cause an increase in acrylamide content. The concentration of heat-induced contaminants was minimum in microwave method. The maximum HMF content was belonged to hot air roasting.	[49]
Chocolate coated roasted flaked rice	Agitated dry roaster	Temperature: 260–340°C; Time: 20–60 s	Decrease in bulk density.	[11]

TABLE 2.1 *(Continued)*

RTE Snack Product	Type of Roasting	Process Conditions	Key Findings	References
Donkua (a local snack with maize and groundnut seed)	Pan roasting	Temperature: 180°C; Time: 20, 30, 40, 50, 60 min	Reduced carbohydrate, ash and crude fiber content, Increased fat and moisture levels.	[50]
Flaked rice	Sand roasting	Temperature: 178°C; Time: ND	Some part of native starch converted into resistant one. The physicochemical structure of grain changed. Increase in b* value.	[51]
Flaked rice	Agitated dry roaster, sand roasting	Agitated dry roaster: Temperature: 260–340°C; Time: 20–60 s; Sand roasting: Temperature: 178–180°C; Time: 28 s	Important effect on whiteness and yellowness indexes, HMF, expansion ratio, hardness and overall acceptability. Residual sand in sand roasted samples.	[52]
Half-popped corn	Oven roasting	Salt rotary oven: 175–185°C, 1.5–2.5 min; Electrical oven: 160–180°C, 4–6 min; Convection oven: 175–200°C, 4–9 min, 3.4 m/s	The convection oven method was chosen in cannon puffing technology.	[53]
Mixed extrudates (rice, chickpea, green lentil and tomato powder)	Microwave roasting	Power: 25%, 50%, 75%, 100%; Time: 20, 30, 40, 50, 60, 70, 80, 90 s	More porous structure and higher expansion ratio and puffiness.	[54]

TABLE 2.1 *(Continued)*

RTE Snack Product	Type of Roasting	Process Conditions	Key Findings	References
Orange maize	Local roasting (Hot-charcoal burning on wire gauze)	Temperature: ND Time: 10, 12, 15 min for dehusked cobs; 10, 15, 20 min for undehusked cobs	Roasting techniques influenced color properties of products significantly.	[55]
Pregnant women's snack bar (with soybean and purple sweet potato)	ND	Temperature: ND; Time: 25 min	Brownish yellow color and savory flavor were detected. Increased sugar content.	[56]
Puffed rice	Sand roasting (puffing)	260–280°C, 29 s	Development in dimensional parameters and surface area, Decrease in bulk density, Affected water solubility and water absorption index.	[57]
Snack with maize flour and groundnut paste	ND	For groundnut: Temperature: 300°C; Time: 60 min	Improved taste and aroma.	[58]
Yellow and black soybean	Oven roasting	Yellow soybean: 230°C, 25 min and 210°C, 35 min; Black soybean: 230°C, 25 min and 210°C, 30 min	Increased isoflavones, Decreased trypsin inhibitor.	[59]

ND: Not determined.

microwave method was the most efficient procedure in obtaining the lowest amount of heat-induced contaminants [49].

2.2.3.2 ROASTED RTE SNACKS WITH NUT AND EDIBLE SEED

Almond, cashew, Brazilian nut, pecan, pistachio and hazelnuts are tree nuts that are widely consumed all over the world. Their annual global trade is over 32 billion American dollars [61, 62]. Despite the fact that peanuts are legumes, they have nearly same nutritional compounds with tree nuts. Reported phenolic substances in nuts are anthocyanins, flavanols, flavones, dihydrochalcones, flavanonols, phenolic acids, flavanones, tyrosols, hydrolysable tannins, flavan-3-ols and stilbenes which contribute to health-promoting activities in body. The net positive effects of nuts are – (i) improving glycemic control and vasomotion; (ii) lowering blood pressure and cholesterol; and (iii) antioxidant potential [63].

Edible seeds are rich in healthy fats, proteins, micronutrients like minerals (potassium, magnesium, iron, and zinc) and vitamins such as vitamin E, B1 and B3. Sunflower seeds, sesame seeds, opium seeds, mustard seeds and flax seeds are considered as edible seeds, and they have the good potential of being substituted with fish, egg or meat. It is recommended that, seeds should be consumed with an antioxidant rich food, so that the absorption of iron can boost up. Flax seeds are the good source of both fibers and omega 3 fatty acids which promote to produce energy and strengthen the immune system [64].

Table 2.2 depicts RTE snacks which contain roasted edible seed or nut. According to the results of reviewing actual articles, almond, cashew, chironji kernels, fenugreek seeds, fox nut, hazelnut, pistachio, sunflower seeds, watermelon seeds and tidbit from cassava and benniseed are popular RTE snacks, and the most favorable roasting techniques are pan roasting, hot-air roasting, oven roasting, microwave roasting, roasting by vertical spouted bed roaster, ultra superheated steam roasting, forced blast roasting, electric muffle furnace roasting and infrared roasting in this category.

As in roasted RTE snacks with cereal and legume, newly developed technologies and systems have adapted into nut-based snacks, and the outstanding one is roasting by a vertical spouted bed [70]. The equipment of spouted bed roaster contains a dryer bed, centrifugal fan, frequency inverter, cyclone, container, programmable logic controller and a resistor. For setting up a spouted bed, the velocity of air which comes from fan is set carefully

TABLE 2.2 RTE Snacks Containing Roasted Nut or Edible Seed

RTE Snack Product	Type of Roasting	Process Conditions	Key Findings	References
Almond	ND	1st roasting: 140°C, 2nd roasting: 145°C, 3rd roasting: 150°C Time: 23 min	The highest acrylamide content was determined at 150°C.	[3]
Cashew	Oven roasting	Temperature: 135–145°C, Time: 20–30 min	Honey coated cashews had higher protein, ash and crude fiber levels than roasted ones. Mineral amounts of roasted samples were greater than honey coated specimens.	[65]
Chironji kernels	Hot-air roasting	Temperature: 80, 100, 120, 140, 160°C, Time: 10, 20, 30, 40, 50 min	Decrease in hardness, water activity and bulk density, Increase in browning index, bioactives and water absorption capacity.	[26]
Dried nut (cashew, almond, peanut and hazelnut)	Oven roasting	Temperature: 200°C, Time: 4 min (rare roasted), 6 min (medium roasted), 8 min (well roasted)	Saturated and monounsaturated fatty acid levels were not influenced by roasting degree, although polyunsaturated ones had significantly affected. Heavy polycyclic aromatik hydrocarbons were not determined.	[66]
Fenugreek seeds	Pan roasting	130°C, 7 min	Decrease in total dietary fiber content, Increase in vitro digestibility, Reduction in phytates.	[67]
Fox nut	Oven roasting	130°C, 10 min	Include high amount of bioactives and have low glycemic index value	[68]
Fox nut (Makhana seed)	Pan roasting and popping machine	1st roasting: 270–280°C Time: ND, 2nd roasting: 260–270°C, Time: 3–5 min	First roasting caused the gelatinization and swelling of starch, however aforementioned process did not decrease moisture content.	[69]

TABLE 2.2 *(Continued)*

RTE Snack Product	Type of Roasting	Process Conditions	Key Findings	References
Peanut	ND	1st roasting: 150°C, 2nd roasting: 155°C, 3rd roasting: 160°C, Time: 40 min	The highest acrylamide content was determined at 160°C.	[3]
Peanut	Vertical spouted bed roaster	Aflatoxin amount: 35, 332, 695 μg/kg 180°C, up to 20 min, Aflatoxin amount: 85 μg/kg, Temperature: 160, 180, 200°C, Time: 5, 10, 15, 20, 25 min	Second conditions gave better results. Only roasting is not enough in order to decrease aflatoxins.	[70]
Peanut kernels	Infrared roasting	Power: 250–450 W, Time: 10–30 min	Decrease in L*, compressive energy, moisture and hardness.	[71]
Peanut	Ultra superheated steam roasting	Temperature: 300°C, Time: between 40 and 80 s Temperature: 350 and 400°C, Time: between 10 and 40 s, Speed: 60–70 rpm	Roasting degree was directly correlated with final surface temperature. Ultra superheated steam application detoxified aflatoxin B1 contamination successfully.	[72]
Pistachio	Hot-air roasting, infrared roasting, microwave roasting,	Hot air: Temperature: 100, 125, 150°C, Time: 5, 12.5, 20 min, Infrared: Source voltage: 75, 85, and 95 V, Time: 10, 20, 30 min, Microwave: Power: 180, 270, and 360 W, Time: 12, 14, and 16 min	Acrylamide was produced in both salted and roasted specimens in all techniques. Increased roasting grade caused also an increment in acrylamide concentration. Maximum and minimum acrylamide amounts were detected in pictachios roasted by microwave (180 W–12 min) and infrared (95 V–30 min), respectively.	[7]

TABLE 2.2 *(Continued)*

RTE Snack Product	Type of Roasting	Process Conditions	Key Findings	References
Sacha inchi	Forced blast roasting	Temperature: 150, 160, 170, 180, 190, and 200°C, Time: 5, 10, 15, 20 and 25 min	The conditions of roasting process had a positive impact on the concentration of furans.	[73]
Sunflower seeds with shell	ND	1^{st}roasting: 160°C, 2^{nd}roasting: 165°C, 3^{rd}roasting: 170°C, Time: 23 min	The highest acrylamide content was determined at 170°C.	[3]
Sunflower seeds	Oven roasting	200°C, 10 min	Polycyclic aromatic hydrocarbons (anthracene, fluoranthene, benzo[a] pyrene, and phenanthrene) were detected.	[74]
Sunflower seeds	Oven roasting	Temperature: 125, 135, 145°C Time: 15, 30, 45, 60 min	Increased content of volatiles.	[75]
Tidbit from cassava with benniseed	Electric muffle furnace roasting	200°C, 60 min	Reduced protein content.	[76]
Watermelon seeds	Oven roasting	Temperature: 180–260°C, Time: 5–15 min	Decrease in moisture, hardness and total color change, Increase in antioxidant activity and total polyphenol content.	[77]

ND: Not determined.

in order not to drag off the food to the cyclone and to permit the entrance the material into the container [70].

2.2.3.3 OTHER RTE SNACKS

Feasibility of usage in several disaster times and feeding programs have been increased the offers and demands to the baked food products. Among them, biscuits have a notable position because of being carried easily and also their shelf lifes, sweety tastes and textures. Regarding Britain, crackers, cookies and wafers are also seemed as biscuit [78]. Cookies are a type of biscuits comprised of flour, butter, sugar, hazelnuts, peanuts, almonds, chocolate, cocoa, etc., and have a high fat content. Depending on sort, they could be made of sugar or salt, and are easily dispersed in the mouth. But it is very difficult to define the cake, because of the many cake types and formulations are available in the food industry. In general terms, cake is a soft flour product with higher moisture content than biscuits which is obtained by cooking the dough that is prepared using flour, sugar, butter, egg, baking powder, water (sometimes milk) and sweetener. Different creams and sauces are able to be utilized in cake manufacturing, as well [44].

Sweet potato is an ancient crop, and considered in a main food category together with cereals, legumes and tubers. Its glycemic index value is not as high as potato (*Solanum tuberosum* L.), and dietary fiber content is between 1.3 and 3.8% [79]. Japanese, Korean, Chinese and Asian people love roasted RTE sweet potatoes in virtue of tender texture, delicious and sweet taste and attractive flavor. Orange, yellow and purple-fleshed varieties are utilized for producing this famous street RTE snack. Though, acylated anthocyanins are available in purple-fleshed ones, there are phenolic acids and carotenoids in orange and yellow varieties. Since the shelf life is short, and the storage is a bit difficult, some improvements like freezing have been developed [80]. Also, edible insects may be good sources of environmentally friendly proteins, and high and regular breeding, low greenhouse gas emissions and feed demands make them superior, when compared to the other animal origin proteins [81].

Table 2.3 shows RTE snacks involving a roasted fruit, vegetable or ingredient, and beetroot chips, cake with clove bud, cake with terebinth, chocolate, food bar with salak fruit, biscuit with longan stone, biscuit with *Tamarindus indica* seed, biscuit with sweet potato, bakery with carob flour, larvae of insect species, plantain fruit, sweet potato, terebinth fruit and cookie with terebinth are placed in other roasted RTE snacks group. Most common

TABLE 2.3 Other RTE Snacks with Any Roasted Ingredient

RTE Snack Product	Type of Roasting	Process Conditions	Key Findings	References
Beetroot chips	Oven roasting	105°C, 60 min	High market or economical value.	[82]
Cake with clove bud	Pan roasting, microwave roasting	Pan roasting: 180°C, 10 min; Microwave roasting: 450 W, 30 s	Positive effects on antimicrobial characteristics, reducing power potential, total phenolic content	[83]
Chocolate	Oven roasting	24°C, 0 min 64°C, 55 min 84°C, 80 min 105°C, 11 min 114°C, 40 min 135°C, 80 min 151°C, 54 min	Inverse proportion between epicatechin and procyanidin B2 concentrations and roasting, Increase in theobromine and caffeine amounts.	[84]
Food bar with salak (Sidimpuan cultivar) fruit	Oven roasting	Temperature: 80, 90, 100°C; Time: 40, 30, 30 min	Maximum fat and carbohydrate amounts and calorie at 80°C, High mineral, fiber and protein content at 100°C.	[85]
Functional biscuit with longan stone	Fluidized bed roaster	Temperature: 170°C; Time: 5, 15, 25 min	Increase in bioactivity and acceptability of longan stone powder.	[86]
Functional biscuit with Tamarindus indica seed	Microwave roasting	Power: 600 W; Time: 8, 16, 24 min	Increased a* value and antioxidative properties, Decreased L*, b*, water absorption capacity, hardness, fracturability	[87]
Gluten-free bakery with carob flour	ND	Temperature: 100°C; Time: 60, 75, 90 min	Low degrees of roasting did not change hardness and specific volume significantly, Reduced cake staling and dimension, higher acceptability at lower times of roasting.	[88]
Gluten-free biscuit with sweet potato	Oven roasting	160°C, 30 min	Reduced the hardness levels.	[89]

TABLE 2.3 *(Continued)*

RTE Snack Product	Type of Roasting	Process Conditions	Key Findings	References
Larvae of insect species (*Tenebrio molitor* and Zophobas morio)	Oven roasting	Temperature: 160, 180, 200°C; Time: 10, 15, 20 min	Pleasant smell was recognized in mealworms roasted at 180°C, Undesirable burnt smell was detected in samples roasted at 200°C.	[81]
Plantain (*Musa paradisiaca*) fruit	Smoking methods: Roasting with charcoal, firewood, charcoal with polyethylene; Non-smoking methods: Frying and oven	Smoking methods: Temperature: Above 250°C Time: ND; Non-smoking methods: 170°C, 10 min for frying, and 220°C, 5 min for oven drying	Smoking techniques do increase polycylic aromatic hydrocarbon amounts of food rather than the other. Oven method is suggested.	[90]
Sponge cake with terebinth	Oven roasting	200°C, 20 min	Developed mouthfeel.	[91]
Sweet potato	ND	Temperature: 200, 240, 280°C; Time: 30, 40, 45 min; Speed: 30, 50, 70 rad/min	Decrease in protein, starch and vitamin C, Increase in antioxidant activity and total polyphenol content.	[92]
Sweet potato	ND	Temperature: 220°C; Time: 0, 0.5, 1, 1.5, 2, 2.5, 3 h	Increase in browning degree, maltose amount, a* value and total acidity, Maltol, 2–furanmethanol and furfural were the main aroma compounds.	[93]
Terebinth fruit	Pan roasting	200°C, 20 min	5–methylfurfural and pyrazines were observed.	[94]
Wire-cut cookie with terebinth	Oven roasting	Temperature: Group 1: 100°C, Group 2: 125°C, Group 3: 150°C, Time: 20 min	Increase in ash, protein, fat, and a* value, Decrease in average diameter of cookie, phytic acid level, and L* and b* values	[95]

ND: Not determined

roasting methods are oven roasting, pan roasting, microwave roasting, roasting by a fluidized bed and roasting with charcoal, firewood, charcoal containing polyethylene. Also, frying, grilling and oven drying are accepted by Nworah et al. [90], as roasting procedures for smoking unripe plantain.

Although a fluidized bed roaster is commonly used in roasting of coffee beans, fluidized bed roasting is faster and more successful than oven roasting, and it is mostly particular to convective heat propagation mechanism. Balances between upward forces of hot air flux and downward gravity forces of roasted food material employ fluidized bed [20], and the temperatures of particles are uniform inside of the chamber, due to their high mobile activity. Hot spots, dead zones and quality deficiencies are dramatically reduced, and sustainable energy conversion or production occur as well. However, this technique is very old, industrial practice are not prevalent [20, 96]. Furthermore, particles which will be roasted should not be too heavy for being suspended in the air, accordingly the efficiency of the process principally [20].

2.3 THE EFFECTS OF ROASTING PROCESS ON HEALTH AND BIOAVAILABILITY

The amount of bioactive compounds and toxic substances in a food material is strongly related to processing steps, and the roasting is the most important one. The roasting conditions, especially low water content and high temperature induce the formation of new molecules and thermal degradation. Both Maillard reaction (between reducing sugars and amino acids) and degradation operations cause some alterations in physical and chemical attributes of the product [97]. Therefore, critical positive and negative points of roasting procedure in mankind's health are discussed in this section.

2.3.1 TOTAL PHENOLIC CONTENT AND ANTIOXIDANT ACTIVITY

Plant-derived phenolics are phenolic acids, tannins, lignans, coumarins, flavonoids and stilbenes, and their antimicrobial, anti-enterotoxin, antioxidant and anti-biofilm potentials have been gaining deep attention by several researchers [98]. Phenolic acids can be found in foods in the form of esterified, free, insoluble bound or glycosidic, and not bounded phenolic acids contribute to taste. Tannins are present either hydrolysable or condensed.

Flavonoids are the most crowded group and contain flavononols, isoflavones, anthocyanidins, flavanones, flavonols, flavanols and flavones [99]. Sung et al. [100] claimed that, heating can provide flavonoids to convert into aglycones which have the capability of destruction of lipid accumulation in fibroblast cells. Also Pelvan et al. [101] declared that, roasting process raised protocatechuic acid, salicylic acid, syringic acid, vanillic acid, 4-hydroxybenzoic acid, caffeic acid, ferulic acid and o-coumaric acid levels of hazelnut samples. But, antioxidant activity by DPPH method gave the highest IC_{50} value (350 mg sample/ml), and the lowest total phenolic content (17 mg gallic acid equivalent, GAE/100 g sample), ABTS (0.08 μg trolox equivalent, TE/100 g sample) and ORAC (946 μmol TE/100 g sample) results [101]. IC_{50} value is defined as the required concentration in order to scavenge 50% of DPPH or any antioxidant solution [102], and the minimum levels are the indicator of maximum antioxidant behavior.

Chocolate is a RTE snack product which is made of cocoa seeds, and flavonoids of cocoa seeds give quite bitter taste to the chocolate. It is known that, methylxanthines (caffeine and theobromine) and flavan-3-ols (procyanidin B2, epicatechin, and catechin) are related to the bitterness [84]. A study conducted by McClure et al. [84] investigated eight different conditions in the roasting of three cocoa seeds (from Peru, Madagascar, and Ghana), and showed that, both procyanidin B2 and epicatechin levels decreased, however cyclo (proline-valine) and catechin increased during roasting. Furthermore, caffeine and theobromine amounts enhanced, probably due to water removal.

Pumpkin seeds are expressed as excellent sources of protein, carotenoids, phytosterols, oil and vitamins, and they are used for protein supplement or RTE snack in some vicinities. Nevertheless, after processing, the seeds are usually thrown away. They contain p-hydroxybenzoic acid and some phenolic acids in desirable amounts [103]. Peng et al. roasted pumpkin seeds in oven at 120, 160 and 200°C for 10 minutes, and total phenolic and flavonoid contents and antioxidant activity enhanced with increased roasted temperature. Probably reasons were associated to being released bound bioactives because of cell destruction, and new compounds occurring from Maillard reactions which may be able to react Folin-Ciocalteu solvent. Gallic acid, caffeic acid, *p*-coumaric acid, epicatechin, rutin, ferulic acid and *p*-hydroxybenzoic acid were detected individual phenolic substances [103].

The cotyledons of *Gevuina avellana* (Chilean hazelnut) is roasted for consuming as RTE snack [104], and some researchers from Chile, Spain and France investigated interesting roasting procedures in the heat treatment of this product. Tostador and Callana techniques take the advantages of metals during roasting, but Tostador uses a metal plate, although Callana is performed in a container. Hydroxycinnamic acid and hydroxybenzoic acid were the dominant individual phenolics, and roasting increased DPPH radical scavenging activity. Total phenolic contents of Chilean hazelnuts were evaluated as low [104].

2.3.2 AFLATOXIN, ACRYLAMIDES AND THE OTHER DELETERIOUS SUBSTANCES

Aflatoxin B1, B2, G1 and G2 are the most hazardous chemical materials produced by *Aspergillus flavus* and *Aspergillus parasiticus* which can cause cancer [105, 106]. Aflatoxin contamination may occur during growing of food in the field, harvesting, storage or turning into final product. Aflatoxins are especially found in hard-shelled fruits such as pistachios, peanuts, almonds, dried fruits like figs, and can also be detected in the milk of animals that contaminated by feed, and their products [106]. Salve et al. declared that, roasted peanuts (temperature: 120, 150, 180°C, time: 5, 10, 15 minutes) had 97.7% less aflatoxin concentration [105]. In conjunction with the study of Salve et al. roasting operation of pistachios with 30 ml of water, 2.25 g citric acid and 15 ml of lemon juice, and roasting with 30 ml of water, 6 g citric acid and 30 ml of lemon juice for 60 minutes at 120°C dramatically decreased the aflatoxin levels 49.2% and 93.1%, respectively [107].

One of the products of Maillard reaction is acrylamide, and soaking and bleaching pretreatments as well as time and temperature control can be a useful way for reducing acrylamide formation in foods [108]. In addition, decarboxylated Amadori product and acrolein can form acrylamide, different from the mechanism of Maillard reaction [3]. Acrylamide is known neurotoxic, hepatoxic, carcinogenic, genotoxic, and immunotoxic effects [108]. The research conducted by Süvari, Süvari et al. [3] proved that, acrylamide occurrence increased with promoted roasting temperature in sunflower seed, almond and peanut, and raw seeds and nuts contained no acrylamide. An interesting finding was published by Chinese scientists, and drum-roasted adzuki beans had the lower acrylamide content than microwave baked beans thank to volumetric heating of microwaves [109].

Thermal processing of sugars procures to arise 5-hydroxymethylfurfural (HMF), and the surface area of the product and operation conditions are contributed to the concentration of HMF in the medium. This complex molecule has antioxidant, antihypoxic, anticarcinogenic, anti-apoptotic and anti-sickling behaviors, but also HMF has negative effects such as carcinogenic action, genotoxicity, organotoxic potential and mutagenicity [110]. Per kg body weight, 80–100 mg HMF do not cause a harmful effect [111], even though the highest HMF limit must be 40 mg/kg in honey according to Codex Alimentarius. Above this level, it is understood that, heating has been applied to the honey sample [110]. Göncüoğlu Taş and Gökmen [112] roasted hazelnuts at 150°C for half an hour, and said that, HMF contents of Tombul and Levant varieties increased importantly, when compared to unroasted specimens. Moreover, for storage, HMF were not accepted as a good indicator/marker.

Ochratoxin is another toxic material which can be detectable in nuts and cereals and produced by *Aspergillus* and *Penicillium* fungi. Up to 125°C, it is a heat stable carcinogen [113]. Capei, Pettini and Tacconi conducted a study about the ochratoxin A levels of 12 different wheat based RTE sweet snacks determined by ELISA technique, and concluded that, 51% of RTE snacks contaminated with ochratoxin A in a range of 0.5 to 2.1 ng/g in Italy. Adults seemed more lucky than both teenagers and children [114]. To the best of my knowledge, experiments that determine roasting impact on ochratoxin A levels in RTE snacks are scarce, although it is a very dangerous compound for human health.

2.4 CONCLUSION

This chapter highlighting a common technology applied in food industry widely, roasting, has been subjected for screening its potential in producing RTE snacks. A changing world, lack of time and heavy working conditions have caused people to need something to eat without wasting time and effort, and at this point of view, RTE snacks become important in daily life. Roasting is a mixture of several chemical reactions induced by heat transfer, and the conditions (temperature/power and time, sometimes air flow rate and rotating speed) of this process are directly related to the quality of RTE snack. Generally, conduction and convection heat propagation mechanisms are dominant during roasting, and some traditional and novel methods have been using successfully. Modeling researches focused on heat transfer and degradation kinetics of some noteworthy substances, will be helpful for understanding the underlying phenomena of roasting correctly. In order to obtain a well-qualified and hygienic RTE snack, some methods like sand roasting should be leaved, and novel methods such as superheated steam roasting are ought to be preferred. Also, hybrid techniques may be useful for compensating undesirable changes in food. Oven roasting is the most used procedure all over the world. While RTE snacks contained cereal and legume are almost rich in carbohydrate, nuts and edible seeds have large amounts of healthy fats. In Turkey, leblebi (chickpea) and globally, roasted rice may have a strong potential for being favorite RTE snacks. Functional RTE snacks including any roasted ingredient have been started to manufacture, especially in bakery and chocolate industry. More researches should be performed about toxic compounds occurred during roasting.

ACKNOWLEDGMENT

I would like to present my special thanks to Food Engineer Nesibe Onar for valuable helps.

KEYWORDS

- **electromagnetic waves**
- **heat transfer**
- **ready-to-eat**
- **roasting**
- **superheated steam roasting**

REFERENCES

1. Lee, J., Cho, J. J., Hong, S. J., Kim, D. S., Boo, C. G., & Shin, E. C. (2020). *Platycodon grandiflorum* roots: A comprehensive study on odor/aroma and chemical properties during roasting. *Journal of Food Biochemistry, 44*(9), 1–14. https://doi.org/10.1111/jfbc.13344.
2. Sruthi, N. U., Premjit, Y., Pandiselvam, R., Kothakota, A., & Ramesh, S. V. (2021). An overview of conventional and emerging techniques of roasting: Effect on food bioactive signatures. *Food Chemistry, 348,* 129088. https://doi.org/10.1016/j.foodchem.2021.129088.

3. Süvari, M., Sivri, G. T., & Öksüz, Ö. (2017). Effect of different roasting temperatures on acrylamide formation of some different nuts. *IOSR Journal of Environmental Science, Toxicology and Food Technology, 11*(4), 38–43. https://doi.org/10.9790/2402-1104013843.
4. Kora, A. J. (2019). Applications of sand roasting and baking in the preparation of traditional Indian snacks: Nutritional and antioxidant status. *Bulletin of the National Research Centre, 43,* 158. https://doi.org/10.1186/s42269-019-0199-2.
5. Chindapan, N., Soydok, S., & Devahastin, S. (2019). Roasting kinetics and chemical composition changes of Robusta coffee beans during hot air and superheated steam roasting. *Journal of Food Science, 84*(2), 292–302. https://doi.org/10.1111/1750-3841.14422.
6. Jogihalli, P., Singh, L., Kumar, K., & Sharanagat, V. S. (2017). Novel continuous roasting of chickpea (*Cicer arietinum*): Study on physico-functional, antioxidant and roasting characteristics. *LWT–Food Science and Technology, 86,* 456–464. https://doi.org/10.1016/j.lwt.2017.08.029.
7. Asadi, S., Aalami, M., Shoeibi, S., Kashaninejad, M., Ghorbani, M., & Delavar, M. (2020). Effects of different roasting methods on formation of acrylamide in pistachio. *Food Science and Nutrition, 8*(6), 2875–2881. https://doi.org/10.1002/fsn3.1588.
8. Bagheri, H., Kashaninejad, M., Ziaiifar, A. M., & Aalami, M. (2016). Novel hybridized infrared-hot air method for roasting of peanut kernels. *Innovative Food Science & Emerging Technologies, 37,* 106–114. https://doi.org/10.1016/j.ifset.2016.08.014.
9. Akkarachaneeyakorn, S., Laguerre, J. C., Tattiyakul, J., Neugnot, B., Boivin, P., Morales, F. J., & Birlouez-Aragon, I. (2010). Optimization of combined microwave-hot air roasting of malt based on energy consumption and neo-formed contaminants content. *Journal of Food Science, 75*(4), E201–E207. https://doi.org/10.1111/j.1750-3841.2010.01567.x.
10. Brennan, M. A., Derbyshire, E., Tiwari, B. K., & Brennan, C. S. (2013). Ready-to-eat snack products: The role of extrusion technology in developing consumer acceptable and nutritious snacks. *International Journal of Food Science & Technology, 48*(5), 893–902. https://doi.org/10.1111/ijfs.12055.
11. Kumar, S., Baniwal, P., Nayik, G. A., Prasad, K., Khan, K. A., Ghramh, H. A., Kumar, H., & Karabagias, I. K. (2021). Optimization and development of ready to eat chocolate coated roasted flaked rice as instant breakfast food. *Foods, 10*(7), 1658. https://doi.org/10.3390/foods10071658.
12. Olalekan-Adeniran, M. A., & Ogunwolu, S. O. (2018). Comparative quality evaluation of oven-roasted and honey-coated cashew (*Anarcadium occidentale* L.) nut produced using locally fabricated cashew nut processing machine in Nigeria. *International Journal of Agriculture and Environmental Biotechnology, 3*(5), 1796–1803. https://doi.org/10.22161/ijeab/3.5.31.
13. Berk, Z. (2018). Frying, baking, and roasting. *Food Process Engineering and Technology,* 583–590. Academic Press, 3rd Edition. https://doi.org/10.1016/b978-0-12-812018-7.00024-5.
14. Bagheri, H. (2020). Application of infrared heating for roasting nuts. *Journal of Food Quality,* Article ID 8813047. https://doi.org/10.1155/2020/8813047.
15. Domínguez-Pérez, L. A., Concepción-Brindis, I., Lagunes-Gálvez, L. M., Barajas-Fernández, J., Márquez-Rocha, F. J., & García-Alamilla, P. (2019). Kinetic studies and moisture diffusivity during cocoa bean roasting. *Processes, 7,* 770. https://doi.org/10.3390/pr7100770.

16. Rojas, M., Granados, D., Osorio, J., & Chejne, F. (2022). Analysis of cocoa particle roasting process in a μ-reactor. *Journal of Food Engineering, 330,* 111102. https://doi.org/10.1016/j.jfoodeng.2022.111102.
17. Heryoki, Yohanes., Sang, Kompiang, W., & Eko Agus, S. (2022). Study of heat transfer and product characterization in spouted bed coffee roaster. *Journal of Advanced Research in Fluid Mechanics and Thermal Sciences, 92*(2), 182–190. https://doi.org/10.37934/arfmts.92.2.182190.
18. Rojas, M., Hommes, A., Heeres, H. J., & Chejne, F. (2022). Physicochemical phenomena in the roasting of cocoa (*Theobroma cacao* L.). *Food Engineering Reviews, 14,* 509–533. https://doi.org/10.1007/s12393-021-09301-z.
19. Şen, D., & Gökmen, V. (2022). Kinetic modeling of Maillard and caramelization reactions in sucrose-rich and low moisture foods applied for roasted nuts and seeds. *Food Chemistry, 395,* 133583. https://doi.org/10.1016/j.foodchem.2022.133583.
20. Peña-Correa, R. F., Ataç-Mogol, B., Van Boekel, M. A. J. S., & Fogliano, V. (2022). Fluidized bed roasting of cocoa nibs speeds up processing and favors the formation of pyrazines. *Innovative Food Science & Emerging Technologies, 79,* 103062. https://doi.org/10.1016/j.ifset.2022.103062.
21. Tirado-Kulieva, V. A., Sánchez-Chero, M., Villegasyarlequé, M., Aguilar, G. F. V., Carrión-Barco, G., Cruz, A. G. Y. S., & Sànchez-Chero, J. (2021). An overview on the use of response surface methodology to model and optimize extraction processes in the food industry. *Current Research in Nutrition and Food Science, 9*(3), 745–754. https://doi.org/10.12944/CRNFSJ.9.3.03.
22. Bagheri, H., Kashaninejad, M., & Aalami, M. (2019). Optimization of hot air roasting of peanut kernels using response surface methodology. *Journal of Food Technology and Nutrition, 16*(2), 33–44.
23. Guezzane, C. E., Moudden, H. E., Harhar, H., Warad, I., Bellaouchou, A., Guenbour, A., Zarrouk, A., & Tabyaoui, M. (2021). Optimization of roasting conditions on the bioactive compounds and their antioxidant power from *Opuntia fiscus-indica* seeds using response surface methodology (RSM). *Bio Interface Research in Applied Chemistry, 11*(3), 10510–10532. https://doi.org/10.33263/BRIAC113.1051010532.
24. Mahdis, M., Mahdi, K., & Leila, N. (2019). Evaluation of quality characteristics and engineering of infrared-roasted sunflower seeds kernels by RSM. *Journal of Food Science and Technology (Iran), 16*(88), 271–287.
25. Demnati, D., Pacheco, R., Martínez, L., & Sánchez, S. (2020). Optimum roasting conditions of argan kernels (*Argania spinosa* L.) for the production of high-quality edible argan oil. *Journal of Food Science and Technology, 57*(3), 840–847. https://doi.org/10.1007/s13197-019-04115-8.
26. Pradhan, D., & Pradhan, R. C. (2020). Application of a neural network mathematical model in the development of hot air roasting process technology for chironji (*Buchanania lanzan*) kernels. *Journal of Food Processing and Preservation, 44*(12), 1–12. https://doi.org/10.1111/jfpp.14907.
27. Sari, R., Lubis, N. F., Fitriyani, A., & Mayadi. (2020). The simulation of heat propagation rate in smart roasting process using finite difference method. *Piksel, 8*(1), 59–66.
28. Jogihalli, P., Singh, L., Kumar, K., & Sharanagat, V. S. (2017). Physico-functional and antioxidant properties of sand-roasted chickpea (*Cicer arietinum*). *Food Chemistry, 237,* 1124–1132. https://doi.org/10.1016/j.foodchem.2017.06.069.

29. Van Rooyen, J., Simsek, S., Oyeyinka, S. A., & Manley, M. (2022). Holistic view of starch chemistry, structure, and functionality in dry heat-treated whole wheat kernels and flour. *Foods, 11*(2), 1–19. https://doi.org/10.3390/foods11020207.
30. Devu, S. S., Dileepmon, R., Kothakota, A., Venkatesh, T., Pandiselvam, R., Garg, R., Jambrak, A., Mediboyina, M. K., Kumar, M., Rajkumar, R., & Khaneghah, A. M. (2022). Recent advancements in baking technologies to mitigate formation of toxic compounds: A comprehensive review. *Food Control, 135,* 108707. https://doi.org/10.1016/j.foodcont.2021.108707.
31. Cappelli, A., Lupori, L., & Cini, E. (2021). Baking technology: A systematic review of machines and plants and their effect on final products, including improvement strategies. *Trends in Food Science & Technology, 115,* 275–284. https://doi.org/10.1016/j.tifs.2021.06.048.
32. Chhanwal, N., Bhushette, P. R., & Anandharamakrishnan, C. (2019). Current perspectives on non-conventional heating ovens for baking process—A review. *Food Bioprocess Technology, 12*(1), 1–15. https://doi.org/10.1007/s11947-018-2198-y.
33. Piaia, J. C. Z., Claumann, C. A., Quadri, M. B., & Bolzan, A. (2018). Air flow CFD modeling in an industrial convection oven. In Driss, Z., Necib, B., & Zhang, H. C. (Eds.), *CFD Techniques and Thermo-Mechanics Applications*. Springer, Cham, 208 pages. https://doi.org/10.1007/978-3-319-70945-1_1.
34. Banana, K. (2017). Traditional methods of cashew processing. *ZENITH International Journal of Business Economics & Management Research, 7*(12), 77–96.
35. Orfeuil, (1987). *Electrical Process Heating*. Battelle Press, Columbus, Ohio.
36. Delfiya, D. S. A., Prashob, K., Murali, S., Alfiya, P. V., Samuel, M. P., & Pandiselvam, R. (2022). Drying kinetics of food materials in infrared radiation drying: A review. *Journal of Food Processing Engineering, 45*(6), 1–19. https://doi.org/10.1111/jfpe.13810.
37. Berk, O. M., & Kayitesi, E. (2019). Influence of infrared heating processing technology on the cooking characteristics and functionality of African legumes: A review. *Journal of Food Science and Technology, 56*(4), 1669–1682. https://doi.org/10.1007/s13197-019-03661-5.
38. Pawar, S. B., & Pratape, V. M. (2017). Fundamentals of infrared heating and its application in drying of food materials: A review. *Journal of Food Processing Engineering, 40,* e12308. https://doi.org/10.1111/jfpe.12308.
39. Chindapan, N., Chaninkun, N., & Devahastin, S. (2022). Comparative evaluation of phenolics and antioxidant activities of hot air and superheated steam roasted coffee beans (*Coffea canephora*). *International Journal of Food Science & Technology, 57*(1), 342–350. https://doi.org/10.1111/ijfs.15393.
40. Rattanarat, P., Chindapan, N., & Devahastin, S. (2021). Comparative evaluation of acrylamide and polycyclic aromatic hydrocarbons contents in Robusta coffee beans roasted by hot air and superheated steam. *Food Chemistry, 341,* 128266. https://doi.org/10.1016/j.foodchem.2020.128266.
41. Zzaman, W., Bhat, R., Yang, T. A., & Easa, A. M. (2017). Influences of superheated steam roasting on changes in sugar, amino acid, and flavor active components of cocoa bean (*Theobroma cacao*). *Journal of the Science of Food and Agriculture, 97*(13), 4429–4437. https://doi.org/10.1002/jsfa.8302.
42. Agus, B. A. P., Mohamad, N. N., & Hussain, N. (2018). Composition of unfermented, unroasted, roasted cocoa beans and cocoa shells from Peninsular Malaysia. *Journal of*

Food Measurement and Characterization, 12(4), 2581–2589. https://doi.org/10.1007/s11694-018-9875-4.
43. Fang, J., Liu, C., Law, C. L., Mujumdar, A. S., Xiao, H. W., & Zhang, C. (2022). Superheated steam processing: An emerging technology to improve food quality and safety. *Critical Reviews in Food Science and Nutrition, 63(27), 8720–8736.* https://doi.org/10.1080/10408398.2022.2059440.
44. Köksel, H. (2021). *Hububat Bilimi ve Teknolojisi* (1st ed.). Sidas Medya, İzmir, Türkiye. (In Turkish).
45. Qamar, S., Manrique, Y. J., Parekh, H., & Falconer, J. R. (2020). Nuts, cereals, seeds, and legumes proteins derived emulsifiers as a source of plant protein beverages: A review. *Critical Reviews in Food Science and Nutrition, 60*(16), 2742–2762. https://doi.org/10.1080/10408398.2019.1657062.
46. Özbey, F. (2022). The non-dairy probiotic potential of the prebiotic Turkish snack "Leblebi." *Food Science and Technology (Campinas), 42*, e38321. https://doi.org/10.1590/fst.38321.
47. Bi, S., Wang, A., Lao, F., Shen, Q., Liao, X., Zhang, P., & Wu, J. (2021). Effects of frying, roasting, and boiling on aroma profiles of adzuki beans (*Vigna angularis*) and potential of adzuki bean and millet flours to improve flavor and sensory characteristics of biscuits. *Food Chemistry, 339*(17), 127878. https://doi.org/10.1016/j.foodchem.2020.127878.
48. Sağlam, H., & Seydim, A. C. (2020). Effect of roasting temperature and time on some physical and sensory properties of roasted chickpea (*Leblebi*), and kinetic studies of the changes in *Leblebi* color. *Harran University Journal of the Faculty of Agriculture, 24*(2), 140–150. https://doi.org/10.29050/harranziraat.647181.
49. Günel, Z. (2022). The effect of different roasting methods on the formation of heat-induced contaminants in roasted chickpea. *Journal of Food, 47*(2), 296–309. https://doi.org/10.15237/gida.GD21151.
50. Moses, M. O., & Olanrewaju, M. J. (2018). Effects of roasting time of maize (*Zea Mays*) and groundnuts (*Arachis hypogeae*). In *11th International Science, Technology, Arts, Education, Management & the Social Sciences Conference*, Lagos, Nigeria, *13*, 57–62.
51. Kumar, S., Haq, R., & Prasad, K. (2018). Studies on physico-chemical, functional, pasting, and morphological characteristics of developed extra thin flaked rice. *Journal of the Saudi Society of Agricultural Sciences, 17*(3), 259–267. https://doi.org/10.1016/j.jssas.2016.05.004.
52. Kumar, S., & Prasad, K. (2017). Optimization of flaked rice dry roasting in common salt and studies on associated changes in chemical, nutritional, optical, physical, rheological, and textural attributes. *Asian Journal of Chemistry, 29*(6), 1380–1392. https://doi.org/10.14233/ajchem.2017.20563.
53. Rajha, H. N., Debs, E., Abi, R., Rachelle, M. A., Nicolas, M., Khoury, K., & El. (2021). Innovation in cannon puffing technology for the homogenization of bulk treatment: Half-popped purple corn, a new healthy snack. *Journal of Food Process Engineering, 44*(6), e13695. https://doi.org/10.1111/jfpe.13695.
54. Singh, P., Vatankhah, H., & Ramaswamy, H. S. (2022). Extruded snacks from rice, green lentil, chickpea, and tomato powder finished with frying/microwave roasting. *Revista Facultad de Ingeniería–Universidad de Antofagasta.* https://doi.org/10.17533/udea.redin.20220269.
55. Alamu, E. O., Dixon, B. M., Menkir, A., & Ogunlade, A. O. (2020). Harvesting time and roasting effects on color properties, xanthophylls, phytates, tannins, and vitamin C

contents of orange maize hybrid. *Scientific Reports, 10*, Article number: 21327. https://doi.org/10.1038/s41598-020-78433-9.

56. Rahmi, Y., Kurniawati, A. D., Widyanto, R. M., Ariestiningsih, A. D., Zahratul, A., Farahi, A. A., Ruchaina, A. N., Sihombing, E. V., Istira, F. B., Nafsiyah, I., Permatasari, K. D., Anjani, R. D., Simanyuntak, S. A. Y. M., & Rahma, Y. A. (2021). The sensory, physical, and nutritional quality profiles of purple sweet potato and soy-based snack bars for pregnant women. *Journal of Public Health Research, 10*, 390–398. https://doi.org/10.4081/jphr.2021.2241.
57. Kumar, S., & Prasad, K. (2018). Effect of parboiling and puffing processes on the physicochemical, functional, optical, pasting, thermal, textural, and structural properties of selected indica rice. *Journal of Food Measurement and Characterization, 12*(3), 1707–1722. https://doi.org/10.1007/s11694-018-9786-4.
58. Dauda, A. O., Kayode, R. M. O., & Salami, K. O. (2020). Quality attributes of snack made from maize substituted with groundnut. *Ceylon Journal of Science, 49*(1), 21–27. https://doi.org/10.4038/cjs.v49i1.7702.
59. Zhou, R., Cai, W., & Xu, B. (2017). Phytochemical profiles of black and yellow soybeans as affected by roasting. *International Journal of Food Properties, 20*(12), 3179–3190. https://doi.org/10.1080/10942912.2017.1280678.
60. Müjdeci, G. N., Kayalar, F., Mehder, G., & Kılıç, F. (2021). Investigation of chemical and microbiological properties of *Leblebi* (roasted chickpeas) sold unpackaged in Corum Province. *Turkish Bulletin of Hygiene and Experimental Biology, 78*(4), 535–544.
61. International Nut and Dried Fruit Council (INC). (2017). *World Nut and Dried Fruit Trade Map*. Web address: https://inc.nutfruit.org/world-nut-and-dried-fruit-trade-map-now-available-online/ Access date: July, 2022.
62. Gama, T., Wallace, H. M., Trueman, S. J., & Hosseini-bai, S. (2018). Quality and shelf life of tree nuts: A review. *Scientia Horticulture, 242*, 116–126. https://doi.org/10.1016/j.scienta.2018.07.036.
63. Alasalvar, C., Salvadó, J., & Ros, E. (2020). Bioactive and health benefits of nuts and dried fruits. *Food Chemistry, 314*, 126192. https://doi.org/10.1016/j.foodchem.2020.126192.
64. De, L. C. (2020). Edible seeds and nuts in human diet for immunity development. *International Journal of Recent Scientific Research, 11*(06 (B)), 38877–38881. https://doi.org/10.24327/ijrsr.2020.1106.5395.
65. Olalekan-Adeniran, M. A., & Ogunwolu, S. O. (2018). Comparative quality evaluation of oven-roasted and honey-coated cashew (*Anacardium occidentale* L.) nut produced using locally fabricated cashew nut processing machine in Nigeria. *International Journal of Environment Agriculture and Biotechnology, 3*(5), 1796–1803. https://doi.org/10.22161/ijeab/3.5.31.
66. Öz, E., Ekiz, E., Savaş, A., Aoudeh, E., El-Aty, A. M., & Öz, F. (2021). Impact of roasting level on fatty acid composition, oil, and polycyclic aromatic hydrocarbon contents of various dried nuts. *Turkish Journal of Agriculture and Forestry, 45*(2), 213–221. https://doi.org/10.3906/tar-2006-43.
67. Wani, S. A., Alwahibi, M. S., Elshikh, M. S., Abdel Gawwad, M. R., Ali, M. A., Alhaji, J. H., Naik, H. R., & Kumar, P. (2021). Sensory, functional characteristics and *in vitro* digestibility of snacks supplemented with non-traditional ingredient raw and processed fenugreek. *International Journal of Food Science, 57*(8), 4716–4725. https://doi.org/10.1111/ijfs.15441.

68. Liaquat, M., Pasha, I., Ahsin, M., & Salik, A. (2022). Roasted fox nuts (*Euryale ferox* L.) contain higher concentration of phenolics, flavonoids, minerals, and antioxidants, and exhibit lower glycemic index (GI) in human subjects. *Food Production, Processing and Nutrition, 4*(1), 1–12. https://doi.org/10.1186/s43014-021-00081-x.
69. Devi, M., Sharma, K., Narayan, S., Simran, J., Shadanan, A., Kumar, Y., & Kumar, R. (2020). Effect of popping on physicochemical, technological, antioxidant, and microstructural properties of makhana seed. *Journal of Food Processing and Preservation, 44*, e14787. https://doi.org/10.1111/jfpp.14787.
70. Martinsi, L. M., Sant'Ana, A. S., Iamanaka, B. T., Berto, M. I., Pitt, J. I., & Taniwaki, M. H. (2017). Kinetics of aflatoxin degradation during peanut roasting. *Food Research International, 97*, 178–183. https://doi.org/10.1016/j.foodres.2017.03.052.
71. Bagheri, H., Kashaninejad, M., Ziaiifar, A. M., & Aalami, M. (2019). Textural, color, and sensory attributes of peanut kernels as affected by infrared roasting method. *Information Processing in Agriculture, 6*(2), 255–264. https://doi.org/10.1016/j.inpa.2018.11.001.
72. Pukkasorn, P., Ratphitagsanti, W., & Haruthaitanasan, V. (2018). Effect of ultra-superheated steam on aflatoxin reduction and roasted peanut properties. *Journal of the Science of Food and Agriculture, 98*(8), 2935–2941. https://doi.org/10.1002/jsfa.8788.
73. Cha, C., & Lee, K. G. (2020). Effect of roasting conditions on the formation and kinetics of furan in various nuts. *Food Chemistry, 331*, 127338. https://doi.org/10.1016/j.foodchem.2020.127338.
74. Akbari-Adergani, B., Khaki, T., Jahed, G., Ahmadkhaniha, R., Rezapour, S., & Sadighara, P. (2017). Polycyclic aromatic hydrocarbons in shell and kernel of roasted sunflower seeds. *Journal of Pharmaceutical Sciences and Research, 9*(6), 994–996.
75. Shuangshuang, G., Kriskamol, N. J., & Yan, G. (2019). Influence of roasting condition on flavor profile of sunflower seeds: A flavoromics approach. *Scientific Reports, 9*, 11295. https://doi.org/10.1038/s41598-019-47811-3.
76. Arise, A. K., Malomo, S. A., Owolabi, O., & Arise, R. O. (2021). Proximate, antioxidant, and sensory properties of tidbit snacks from cassava enriched with processed benniseeds. *ACS Food Science and Technology, 1*, 268–274. https://doi.org/10.1021/acsfoodscitech.0c00054.
77. Gholami, Z., & Ansari, S. (2021). Effects of roasting conditions on physicochemical properties of the watermelon seed. *Iranian Journal of Chemistry and Chemical Engineering, 40*(2), 615–626. https://doi.org/10.30492/ijcce.2020.38167.
78. Arepally, D., Reddy, R. S., Goswami, T. K., & Datta, A. K. (2020). Biscuit baking: A review. *LWT–Food Science and Technology, 131*, 109726. https://doi.org/10.1016/j.lwt.2020.109726.
79. Bach, D., Bedin, A. C., Lacerda, L. G., Nogueira, A., & Demiate, I. M. (2021). Sweet potato (*Ipomoea batatas* L.): A versatile raw material for the food industry. *Brazilian Archives of Biology and Technology, 64*, 1–14. https://doi.org/10.1590/1678-4324-2021200568.
80. Mu, T. H., Sun, H. N., & Ma, M. M. (2019). Sweet potato snack foods (Chapter 11). In *Sweet Potato, Chemistry, Processing and Nutrition*, edited by Tai-Hua Mu and Jaspreet Singh, Elsevier Inc., 303–324. https://doi.org/10.1016/B978-0-12-813637-9.00011-9.
81. Aydın, M., Arslan, D., & Türker, S. (2021). Gıda teknolojisinde yenilikçi yaklaşımlar. *Journal of Halal Ethics and Research, 3*(1), 19–36. (In Turkish). https://doi.org/10.51973/head.925628.

82. Hossain, B., Kamrul, N., & Biswas, B. (2019). Studies of the compositional characteristics of commercial roasted beet root chips snacks. *Journal of Engineering Research and Reports, 4*(1), 1–8. https://doi.org/10.9734/jerr/2019/v4i116894.
83. Dev, M., Ghosh, M., & Bhattacharyya, D. Kumar. (2021). Physico-chemical, antimicrobial, and organoleptic properties of roasted aromatic spice (clove bud) in baked product. *Applied Biochemistry and Biotechnology, 193*(6), 181–1835. https://doi.org/10.1007/s12010-021-03504-0.
84. McClure, A. P., Spinka, C. M., & Grün, I. U. (2021). Quantitative analysis and response surface modeling of important bitter compounds in chocolate made from cocoa beans with eight roast profiles across three origins. *Journal of Food Science, 86*(11), 4901–4913. https://doi.org/10.1111/1750-3841.15924.
85. Siregar, N. S., Julianti, E., & Silalahi, J. (2021). The effect of roasting temperature on proximate and dietary fiber of food bar salak (Sidimpuan cultivar) fruit. *E3S Web of Conferences, 332*, 03001. https://doi.org/10.1051/e3sconf/202133203001.
86. Bölek, S. (2022). Valorization of roasted longan stone in production of functional biscuits with high antioxidant activity and dietary fiber. *Food Science and Technology (Brazil), 42*, 1–6. https://doi.org/10.1590/fst.69820.
87. Bölek, S. (2022). Valorization of microwave roasted *Tamarindus indica* seed in functional biscuit production and effects on rheological and textural properties of biscuit dough. *Journal of Food Science and Technology, 59*(5), 2080–2086. https://doi.org/10.1007/s13197-021-05224-z.
88. Román, L., González, A., Espina, T., & Gómez, M. (2017). Degree of roasting of carob flour affecting the properties of gluten-free cakes and cookies. *Journal of Food Science and Technology, 54*(7), 2094–2103. https://doi.org/10.1007/s13197-017-2649-x.
89. Limam, S. (2020). Physical and nutritional properties of gluten-free biscuits formulated with multi-grains and roasted sweet potato. *Journal of Food and Dairy Sciences, 11*(1), 39–44. https://doi.org/10.21608/jfds.2020.77977.
90. Nworah, F. N., Nkwocha, C. C., Nwachukwu, J. N., & Ezeako, E. C. (2019). Comparative analysis of the polycyclic aromatic hydrocarbon (PAH) content and proximate composition of unripe *Musa paradisiaca* (plantain) fruit exposed to varying methods of roasting. *Journal of Environmental Health Science and Engineering, 17*(1), 105–113. https://doi.org/10.1007/s40201-018-00331-0.
91. Köten, M. (2021). Influence of raw/roasted terebinth (*Pistacia terebinthus* L.) on the selected quality characteristics of sponge cakes. *International Journal of Gastronomy and Food Science, 24*, 100342. https://doi.org/10.1016/j.ijgfs.2021.100342.
92. Hou, F., Mu, T., Ma, M., & Blecker, C. (2019). Optimization of processing technology using response surface methodology and physicochemical properties of roasted sweet potato. *Food Chemistry, 278*, 136–143. https://doi.org/10.1016/j.foodchem.2018.11.034.
93. Tsai, Y. J., Lin, L. Y., Yang, K. M., Chiang, Y. C., Chen, M. H., & Chiang, P. Y. (2021). Effects of roasting sweet potato (*Ipomoea batatas* L. Lam.) on quality, volatile compound composition, and sensory evaluation. *Foods, 10*, 2602. https://doi.org/10.3390/foods10112602.
94. Amanpour, A., Guclu, G., Kelebek, H., & Selli, S. (2019). Characterization of key aroma compounds in fresh and roasted terebinth fruits using aroma extract dilution analysis and GC-MS-olfactometry. *Microchemical Journal, 145*, 96–104. https://doi.org/10.1016/j.microc.2018.10.024.

95. Köten, M. (2021). Influence of roasted and unroasted terebinth (*Pistacia terebinthus*) on the functional, chemical, and textural properties of wire-cut cookies. *Food Science and Technology (Brazil), 41*(1), 245–253. https://doi.org/10.1590/fst.17020.
96. Di Renzo, A., Scala, F., & Heinrich, S. (2021). Recent advances in fluidized bed hydrodynamics and transport phenomena—Progress and understanding. *Processes, 9*(4), 639. https://doi.org/10.3390/pr9040639.
97. Herawati, D., Giriwono, P. E., Dewi, F. N. A., Kashiwagi, T., & Andarwulan, N. (2019). Critical roasting level determines bioactive content and antioxidant activity of robusta coffee beans. *Food Science and Biotechnology, 28*(1), 7–14. https://doi.org/10.1007/s10068-018-0442-x.
98. Takó, M., Beáta, Kerekes, E., Zambrano, C., Kotogán, A., Papp, T., Krisch, J., & Vágvölgyi, C. (2020). Plant phenolics and phenolic-enriched extracts as antimicrobial agents against food-contaminating microorganisms. *Antioxidants, 9*(2), 1–21. https://doi.org/10.3390/antiox9020165.
99. Chang, S. K., Alasalvar, C., Bolling, B., & Shahidi, F. (2016). Nuts and their co-products: The impact of processing (roasting) on phenolics, bioavailability, and health benefits—A comprehensive review. *Journal of Functional Foods, 26*, 88–122. https://doi.org/10.1016/j.jff.2016.06.029.
100. Sung, J., Suh, J. H., & Wang, Y. (2019). Effects of heat treatment of mandarin peel on flavonoid profiles and lipid accumulation in 3T3-L1 adipocytes. *Journal of Food and Drug Analysis, 27*(3), 729–735. https://doi.org/10.1016/j.jfda.2019.05.002.
101. Pelvan, E., Olgun, E., Karadağ, A., & Alasalvar, C. (2018). Phenolic profiles and antioxidant activity of Turkish tombul hazelnut samples (natural, roasted, and roasted hazelnut skin). *Food Chemistry, 244*, 102–108. https://doi.org/10.1016/j.foodchem.2017.10.011.
102. Süfer, Ö., & Bozok, F. (2020). Characterization of essential oil from *Matricaria sevanensis* by microwave-assisted distillation. *Journal of Thermal Analysis and Calorimetry, 140*(1), 253–261. https://doi.org/10.1007/s10973-019-08829-x.
103. Peng, M., Lu, D., Liu, J., Jiang, B., & Chen, J. (2021). Effect of roasting on the antioxidant activity, phenolic composition, and nutritional quality of pumpkin (*Cucurbita pepo* L.) seeds. *Frontiers in Nutrition, 8*, 1–11. https://doi.org/10.3389/fnut.2021.647354.
104. Pino Ramos, L. L., Jiménez-Aspee, F., Theoduloz, C., Burgos-Edwards, A., Domínguez-Perles, R., Oger, C., Durand, T., Gil-Izquierdo, À., Bustamante, L., Mardones, C., Márquez, K., Contreras, D., & Schmeda-Hirschmann, G. (2019). Phenolic, oxylipin, and fatty acid profiles of the Chilean hazelnut (*Gevuina avellana*): Antioxidant activity and inhibition of pro-inflammatory and metabolic syndrome-associated enzymes. *Food Chemistry, 298*, 125026. https://doi.org/10.1016/j.foodchem.2019.125026.
105. Salve, A. R., LeBlanc, J. G., & Arya, S. S. (2021). Effect of processing on polyphenol profile, aflatoxin concentration, and allergenicity of peanuts. *Journal of Food Science and Technology, 58*(7), 2714–2724. https://doi.org/10.1007/s13197-020-04779-7.
106. Türkiye Cumhuriyeti Gıda Tarım ve Hayvancılık Bakanlığı, Gıda ve Kontrol Genel Müdürlüğü. (2014). Aflatoksin nedir? E-bülten (In Turkish). https://www.tarimorman.gov.tr/GKGM/Belgeler/Tuketici_Bilgi_Kosesi/E-Bultenler/04.pdf (accessed on 9 August 2024).
107. Rastegar, H., Shoeibi, S., Yazdanpanah, H., Amirahmadi, M., Khaneghah, A. M., Campagnollo, F. B., & Sant'Ana, A. S. (2017). Removal of aflatoxin B1 by roasting

with lemon juice and/or citric acid in contaminated pistachio nuts. *Food Control, 71*, 279–284. https://doi.org/10.1016/j.foodcont.2016.06.045.

108. Rifai, L., & Saleh, F. A. (2020). A review on acrylamide in food: Occurrence, toxicity, and mitigation strategies. *International Journal of Toxicology, 39*(2), 93–102. https://doi.org/10.1177/1091581820902405.

109. Yao, X., Zheng, X., Zhao, R., Li, Z., Shen, H., Li, T., Gu, Z., Zhou, Y., Xu, N., Shi, A., Wang, Q., & Lu, S. (2021). Quality formation of adzuki bean baked: From acrylamide to volatiles under microwave heating and drum roasting. *Foods, 10*(11), 3762. https://doi.org/10.3390/foods10112762.

110. Choudhary, A., Kumar, V., Kumar, S., Majid, I., & Suri, S. (2021). 5-Hydroxymethylfurfural (HMF) formation, occurrence and potential health concerns: Recent developments. *Toxin Reviews, 40*(4), 545–561. https://doi.org/10.1080/15569543.2020.1756857.

111. Shapla, U. M., Alam, N., Khalil, I., & Gan, S. H. (2018). 5-Hydroxymethylfurfural (HMF) levels in honey and other food products: Effects on bees and human health. *Chemical Central Journal, 1–18*. https://doi.org/10.1186/s13065-018-0408-3.

112. Göncüoğlu Taş, N., & Gökmen, V. (2019). Effect of roasting and storage on the formation of Maillard reaction and sugar degradation products in hazelnuts (*Corylus avellana* L.). *Journal of Agricultural and Food Chemistry, 67*(1), 415–424. https://doi.org/10.1021/acs.jafc.8b05048.

113. Gu, K., Ryu, D., & Lee, H. J. (2021). Ochratoxin A and its reaction products affected by sugars during heat processing. *Food Chemistry, 348*, 129038. https://doi.org/10.1016/j.foodchem.2021.129038.

114. Capei, R., Pettini, L., & Tacconi, F. M. (2019). Occurrence of ochratoxin A in breakfast cereals and sweet snacks in Italy: Dietary exposure assessment. *Annali di Igiene, 31*, 130–139. https://doi.org/10.7416/ai.2019.2265.

CHAPTER 3

OSMOTIC DEHYDRATION COMBINED WITH DRYING TECHNOLOGY FOR PRODUCTION OF READY-TO-EAT SNACKS

M. PRAVITHA,[1] V. AJESH KUMAR,[2] and AJAY YADAV[1]

[1]*Agro-Produce Processing Division, ICAR–Central Institute of Agricultural Engineering, Nabibagh, Bhopal, Madhya Pradesh, India*

[2]*Center of Excellence for Soybean Processing and Utilization, ICAR–Central Institute of Agricultural Engineering, Nabibagh, Bhopal, Madhya Pradesh, India*

ABSTRACT

Osmotic dehydration is a conventional food processing method widely used on fruits and vegetables to partially remove water, which facilitates the production of high-quality dried food products. With the advancement of the processing sector, osmotic dehydration is also witnessing tremendous changes in its approach. Presently, osmotic dehydration is coupled with other drying treatments to produce a high-quality final product. The combination of osmotic dehydration with convective drying is one of the largely exploited areas. Other advanced drying methods like infrared drying, freeze-drying, and vacuum drying in combination with osmotic dehydration have also been experimented with in various fruits. This chapter comprehensively reviews the available combination treatment of osmotic dehydration with other standard drying methods for producing high-value dried snacks. Specific advantages of this combination method are also discussed elaborately by taking some examples. Furthermore, future perspectives of osmotic

Ready-to-Eat Snacks: Emerging Technologies for Production and Safety.
Ravi Pandiselvam and Gülşah Çalışkan Koç

dehydration-based combination treatments were also discussed to guide upcoming researchers in this particular area.

3.1 INTRODUCTION

Fruits and vegetables are essential items in the human diet because of their nutritive value and health-promoting effects. In developing countries, up to 40% of agricultural produce is wasted, owing to a lack of storage and processing facilities as well as a lack of knowledge of processing technologies. According to the USDA fruits and vegetable losses in the retail and consumption stages in the United States are estimated to be 8.3 million tons (28%) and 11.4 million tons (30%), respectively [1]. To minimize these losses, preservation methods must be used on fruits and vegetables. Drying is one of the most popular and straightforward preservation strategies employed on fruits and vegetables, which removes the water from the food tissue, thus lowering the risk of microbial spoilage in food by reducing its water activity [2]. Dried fruits and vegetables are commonly used as an ingredient in various food preparations like soups, confectioneries, ice cream, and other snacks. Moreover, efficient drying in combination with various food processing techniques can directly yield a variety of dried snacks.

Traditionally, sun drying is used to dry fruits and vegetables in developing and underdeveloped countries. The use of sun drying results in producing a final product with lower quality and nutritional value due to more cell damage caused by the prolonged drying. Hot air drying is another widely used drying method used industrially for producing dried fruits and vegetables. Unlike sun drying, control of the drying rate is possible here to some extent, but the case hardening effect and product shrinkage act as a challenge. In most of the drying processes, water removal with the use of high temperatures for a long time imparts a loss of nutritive components, flavor, and color to the final product. In order to increase the product quality and reduce the drying time, some pretreatment can be introduced before proceeding to the major drying operations. Osmotic dehydration is one such treatment widely exploited as an excellent pretreatment in fruits and vegetables for the partial removal of water from the food tissues. The process of osmotic dehydration involves the immersion of food products in a hypertonic solution made by dissolving one or more solutes, which forces the diffusion of moisture from the food tissue to the solvent due to the osmotic pressure gradient. Various combinations of osmotic dehydration with different drying methods were experimented with by different researchers on different products [3]. This chapter discusses the overview and some of the popular combination drying methods with osmotic dehydration, challenges, and ways forward to understanding and popularizing these kinds of combination drying methods.

3.2 OVERVIEW OF OSMOTIC DEHYDRATION

Water activity plays a major role in maintaining the perishability and overall acceptability of food products. Generally, the lowering of water activity is correlated with the enhanced shelf stability of food products in terms of chemical and microbial stability [4]. The water activity is lowered by different techniques of drying. The use of osmotic dehydration is one of the conventionally known dewatering processes employed in food products [5]. Osmotic dehydration is a process that involves the partial removal of moisture from food sample tissue by immersing it in a hypertonic solution. The water that would have been drawn from the food tissue is replaced, and a hypertonic solution fills the space between the cell wall and the cell membrane [6].

Osmotic dehydration works on the principle of osmosis, involving multicomponent diffusion. In the osmotic dehydration process, the whole or cut solid food sample is immersed in a solution of a higher concentration known as the osmotic solution (hypertonic), such as sucrose or brine solutions, which results in the migration of water molecules from the solid sample (hypotonic solution) into the osmotic solution through a semi-permeable membrane till equilibrium is achieved [7, 8]. In osmotic dehydration, along with the water molecules, there is leaching of solutes (water soluble vitamins and minerals) in the osmotic solution and the removal of water from the cells of the food material is dependent on the concentration gradient established between the intracellular fluid and the hypertonic solution [6, 9]. A schematic diagram representing the osmotic dehydration is depicted in Figure 3.1.

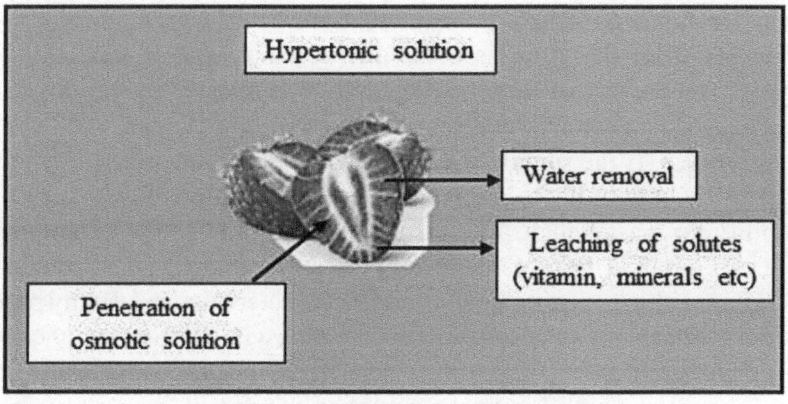

FIGURE 3.1 Schematic diagram representing the osmotic dehydration.

During the treatment of osmotic dehydration of food products, parameters such as solute gain (SG), water loss (WL), and moisture reduction are calculated for estimating the efficacy of the osmotic dehydration process. The value of water loss and solute gain can be determined by the Eqns. (1) and (2).

$$WL = \frac{M_0 X_{mo} - M_t X_{mt}}{M_0} \tag{1}$$

$$SG = \frac{M_t X_{st} - M_0 X_{so}}{M_0} \tag{2}$$

Where; M_0 and M_t are the weight of the sample prior and after osmotic dehydration (g), respectively; X_{m0} X_{mt} are moisture content of the fresh and osmotically dehydrated samples (g water/100 g), respectively; and X_{s0} and X_{st} are total solid content in the sample before and after osmotic dehydration (g solid/100 g). Like other processing methods, use of osmotic dehydration is also associated with advantages and disadvantages. Major advantages include:

- The process is a low-temperature dehydration method, so the thermal degradation of the sample will be less.
- Considerable flavor and nutrient enrichment is possible in the Osmo dried product due to the infusion of solutes.
- The chances of enzymatic and oxidative browning in the product are minimized due to the presence of sugar.
- Less energy is required for the process, thus it can be used as a pre-treatment before proceeding to a further energy intensive drying operation.
- The process is simple and does not require expensive equipment.
- Products have more quality in terms of color, flavor, texture, and aroma.

The process also has some drawbacks, such as loss of original taste, nutrient leeching, particularly the mass transfer of low molecular weight components from the fruit tissue to the solution, and it takes longer to complete. The process of osmotic dehydration is affected by various factors. In which, product geometry and size can have a significant effect on the mass transfer process. If the sample is big or thickly sliced, then it will take more time for dehydration due to the lengthy diffusion path and less surface area availability for mass transfer [11]. Hence, it is recommended to optimize the slice thickness while doing osmotic dehydration to attain the desired degree of dehydration within a specific time. Product characteristics, especially porosity, play an important role in mass transfer. Samples with more porosity can have more dehydration ability and the associated shrinkage in the final product will also be minimized [7]. Osmotic solution dependent factors like solution concentration, temperature, molecular weight, ionic strength, solubility, and viscosity are other important factors influencing the osmotic dehydration process [12]. Those solutes with low molecular weight and viscosity at higher concentrations and temperature conditions can increase the water loss ability of the process. In addition to these process-dependent parameters, sample to solution ratio, agitation or stirring, and pre-treatment giving to sample can all have an impact on mass transfer. The solution to sample ratio should be carefully chosen to ensure that the driving force for moisture removal continues through the entire process. The driving force to release water is lessened when osmotic solutions are diluted. This limitation can be reduced to some extent by using an adequate solution to sample ratio [13]. At static conditions, a concentrated solution usually creates a resistance to mass transfer. Application of agitation or stirring could be effective to reduce this mass transfer resistance. Use of other processing methods in combination with osmotic dehydration can also enhance the efficacy of osmotic dehydration.

3.3 OSMOTIC DEHYDRATION IN COMBINATION WITH OTHER DRYING METHODS

Osmotic dehydration is one of the most basic pretreatment methods for reducing the moisture content of food by immersing it in a hypertonic solution. It offers considerable savings in drying time and an enhancement in product quality for most of the drying processes. Various studies have been reported to prove the benefits of osmotic dehydration in combination with common drying methods like hot air drying, infrared drying, freeze-drying, etc. A pictorial representation of various processes and treatments with product quality analysis is given in Figure 3.2 and details of some of the combination methods of osmotic dehydration are elaborated below.

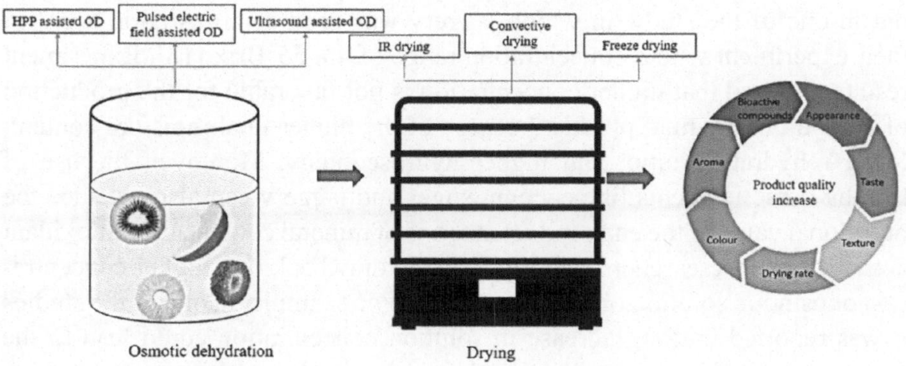

FIGURE 3.2 Pictorial representation of osmotic dehydration in combination with various other processing method and advantage.

3.3.1 OSMOTIC DEHYDRATION IN COMBINATION WITH CONVECTIVE DRYING

Osmotic dehydration is widely considered as a better pretreatment prior to convective drying in the production of dried snacks from various fruits and vegetables. During the pretreatment process, cell membrane changes can occur. These changes may have a positive influence on the further air-drying step. It has been reported that the process of osmo-convective drying results in the production of crispier products with favorable color, texture, and flavor [14, 15]. Some examples are given in Figure 3.3.

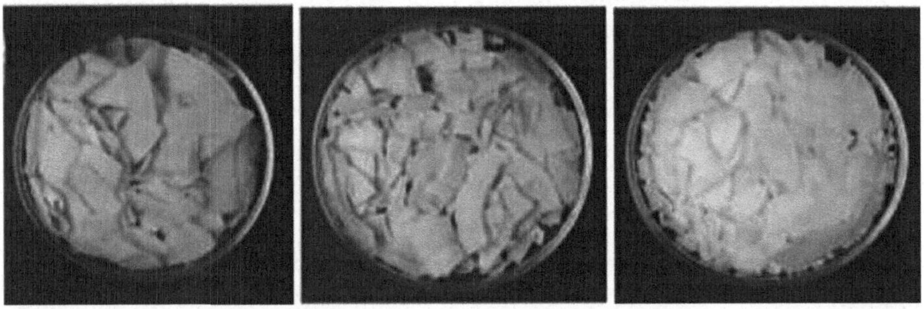

FIGURE 3.3 Coconut chips obtained after osmo-convective drying with solute: a. Jaggery; b. Coconut sugar; and c. Sucrose.

Source: Reprinted with permission from Ref. [20]. Copyright © 2022 The Authors. Published by Elsevier Ltd.

Osmo-convective drying in coconut slices for the production of coconut chips was reported by various authors [16–18]. Optimization of the concentration of initial osmotic dehydration and further drying temperature was carried out in one of the study, in which jaggery was used as an osmotic agent in their experiment with a concentration range of 45–55°Brix [18]. Experiment results revealed that higher concentration is not favorable for the production of a good quality final product because of the higher final moisture content, lower rehydration ratio, and higher hygroscopicity. Moreover, the use of healthy osmotic agents like coconut sugar and jaggery can also enhance the nutritional value of the end product in terms of mineral content and antioxidant activity. Other researchers also reported the drawback of a higher concentration of osmotic solution on product quality. For example, in one of the studies it was reported that an increase in solution concentration could lead to the formation of an impervious layer at the product surface, which further causes a reduction in the drying rate [19]. The solute layer could restrict the water movement during the air-drying process, which causes a case hardening effect in the product. Hence, the final dried product could have a higher moisture content and lower rehydration ability. Nevertheless, conducting the process in precisely optimized conditions could positively influence the product quality by combining osmotic dehydration with convective drying. Initial dewatering with an osmotic solution also significantly lowers drying time due to water removal and solute uptake during the pretreatment. In the case of coconut slices, it was shown that initial osmotic dehydration could reduce the moisture content of coconut slices from 52.43% to a range of 28.50–35.26 (%w.b) [20]. This reduction is further reflected in the drying process, with higher exposure times corresponding to untreated samples. Drying time reduction due to osmotic dehydration has also been reported [21]. In their experiment, osmotic dehydration was conducted by using salt solutions with different concentrations at different temperatures (30–50°C) conditions with a solution to sample ratio of 3:1–5:1. Their experiment revealed that processing of jackfruit slices at an optimum condition of 10% salt concentration at a temperature of 40°C in a solution to sample ratio of 4:1 for 2 hours could reduce the processing time by 4 hours during air drying at 60°C. A number of researchers claimed that further enhancement in the quality of the product and process efficacy for convective drying could be achieved by using ultrasonication along with osmotic dehydration [22–24]. Since the novel processing treatments can considerably enhance cell permeability, which helps for easy moisture removal during the air-drying operation [25].

3.3.2 OSMOTIC DEHYDRATION IN COMBINATION WITH INFRARED DRYING

Conventional hot air drying is mostly associated with the loss of color and nutritive value, with a lower rehydration ratio and porosity. The use of advanced drying techniques can minimize this product degradation. Infrared drying is gaining popularity nowadays due to its increased drying rate, better energy savings, and more uniform product drying. Osmotic dehydration combined with infrared (IR) drying has been proven to be an efficient way to produce better quality dried products. The influence of osmotic dehydration treatment on the infrared drying of peach slices to produce peach chips has been evaluated [26]. Their studies revealed that using osmotic dehydration at a lower concentration of sucrose solution would increase the drying rate during final infrared drying. At the same time, the higher concentration negatively affected the water movement and thus reduced the drying rate. They concluded that osmotic dehydration at an appropriate concentration improves the textural and drying characteristics of peach. The effect of osmotic dehydration in microwave-assisted infrared drying was also evaluated [25]. They found that the process of osmotic dehydration reduces drying time with an average effective moisture diffusivity value of 7.100×10^{-9} -1.445×10^{-8} m^2/s. However, a higher concentration of osmotic solution imparts a detrimental effect on effective moisture diffusivity even at a higher power level of IR-microwave treatment. The author observed that at lower IR power and osmotic solution concentration, there was a better improvement in color parameters in the final product in the eggplant.

Ultrasound-assisted osmotic dehydration (USOD) has been identified as a better method to enhance the drying characteristics and product quality before proceeding to IR treatment. For example, researchers studied the effect of USOD prior to IR drying in the case of Pakchoi stems [27]. They showed that the application of USOD can induce more cell structure damage and microchannel formation. This eventually contributes to an enhancement in the drying rate during the final IR drying process. Better retention in color and rehydration ratio with a slight reduction in ascorbic acid content and chlorophyll was observed in the sample treated with USOD. Conversely, in the case of kiwifruit drying, a considerable increase in ascorbic acid, total phenolic, antioxidant activity, and flavonoid content was observed in the IR dried sample pretreated with an osmotic solution [28]. Similarly, increased drying rate with product quality improvement by applying osmotic dehydration prior to IR drying has also been reported in other food products [29–32].

3.3.3 OSMOTIC DEHYDRATION IN COMBINATION WITH COMMON DRYING METHODS

Osmotic dehydration is also used as a pretreatment for other common drying methods like freeze-drying, vacuum drying, microwave drying, etc. Freeze-drying is commonly considered a better drying technique to maintain the product quality after drying. During freeze-drying, moisture in the food product is first converted into a solid-state, which is then directly converted into a gaseous state. The temperature involvement in this drying is less, so the minimum product degradation occurs. However, this technique comes with the major drawback of higher drying time requirements to reach the desired moisture content. Hence, the process becomes time-consuming with the high-cost involvement [33]. So, introducing osmotic dehydration before proceeding to freeze-drying could be beneficial in increasing the drying rate. It has been reported that the use of osmotic dehydration could increase the drying rate, but more shrinkage was observed in the osmotically dehydrated sample, which further lowers the rehydration ratio of the product. However, this combination method can be utilized for the production of dried products where the rehydration ratio of the product is not much considered [34]. Effects of sucrose pretreatment on product nature before proceeding with freeze-drying of raspberries was evaluated [2]. Osmotic dehydration treatment resulted in a reduction in Tg (glass transition temperature) of the fruit compared to that of the control. The final product was characterized by low rehydration ability, which was considered favorable in the case of the dried fruits, which are mixed with cereal mix, where faster rehydration of the product is not desirable. Another popular combination method of drying is microwave freeze-drying (MWFD). MWFD is a freeze-drying method that uses microwaves as an energy source for the water sublimation. Osmotic dehydration is reported as an important auxiliary treatment for MWFD to ensure increased process efficiency. The efficacy of this combination method has been demonstrated in various food products [35, 36]. Osmotic dehydration is considered as a successful method for enhancing the drying rate in other hybrid drying methods like microwave hot air drying [37–40], microwave vacuum drying [41, 42] and pulsed fluidized microwave drying [36, 43].

3.4 APPLICATION OF NOVEL PROCESSING TECHNIQUES IN OSMOTIC DEHYDRATION

Osmotic dehydration has the potential to improve the product's shelf life considerably by lowering its moisture. However, under normal conditions, it takes a long time to achieve sufficient water removal from the product. Therefore, heating is widely used in industrial processing to increase the transfer process during osmotic dehydration, although it results in a loss of quality and bioactive compounds. This quality degradation induced by heat treatment can be overcome by the use of novel processing treatments along with osmotic dehydration [44].

The use of high-pressure processing (HPP) for increasing mass transfer during osmotic dehydration was demonstrated as a successful process in various fruits and vegetables. It has been reported that HPP could be able to increase the effective moisture diffusivity in Wumei fruit during their osmotic dehydration in comparison with the untreated samples [45]. Application pressure of less than 200 MPa was recommended as more suitable because of the increased cell rupture effect in fruit under the application of high pressure. In strawberries, osmotic dehydration combined with high pressure processing exhibited a beneficial effect on the retention of their bioactive components like antioxidant activity, phenolic compounds, and vitamin C during the refrigerated storage condition. As per their findings, a condition of 400 MPa for 10 min using commercial sugar syrup with a 40°Brix concentration is the ideal processing condition to achieve better retention in the properties [46].

In one of the studies it has been showed that pretreatment of banana slices with HPP before proceeding to osmotic dehydration and final finishing drying was beneficial for increasing water loss and solute gain [47]. According to their findings, this beneficial effect is contributed by the cell wall rupture during the pressurization of the sample, making the cell wall more permeable for the mass transfer. They also confirm morphological changes induced by the HPP with the scanning electron microscope (SEM) images. Ideal processing conditions for banana slices are reported as 200 MPa pressure for a dwell time of 5 min at ambient conditions with a sucrose concentration of 60°Brix. Pretreatment at this condition, along with the final drying at 55°C, could be able to produce an excellent quality dried product from the banana.

Another interesting way to enhance osmotic dehydration ability in fruits and vegetables is pretreatment with ohmic heating. This novel processing method cause's cell damage by combining electrical and thermal effects produced during the process. The efficacy of ohmic heating in strawberries during osmotic dehydration was studied in one of the experiments [48]. According to them, a drastic increase in mass transfer and diffusion rate could be possible by applying ohmic heating during blanching treatment. A significant increase in dry matter content of 68% was observed after osmotic dehydration for the samples pretreated with ohmic heating at 85°C for 3 min compared with control which records 20.3% of dry matter after osmotic dehydration. Similar results were also obtained for apples blanched by ohmic heating followed by osmotic dehydration [49–51]. Further quality improvement in the final product of osmotic dehydration in combination with ohmic heating can be achieved by performing the process under vacuum conditions. Many works have been done combining ohmic heating and vacuum osmotic dehydration [52, 53]. The efficiency of osmotic dehydration could be significantly improved by performing osmotic dehydration at a vacuum pressure rather than at atmospheric pressure with a lower requirement for process temperature [50].

The use of pulsed electric fields (PEF) is also recognized as an effective pretreatment for osmotic dehydration for enhancing water loss and solute gain through tissue permeabilization. In drying goji berries, the introduction of PEF and osmotic dehydration was found to be effective in the retention of color, phenolic compounds, and antioxidant activity during convective drying at 60°C. Better retention ability was obtained during osmotic dehydration at 55°C temperature for 60 minutes of immersion and a PEF treatment of 2.8 kV/cm [54]. Effect of PEF application on the mass transfer and product quality of kiwifruit has been investigated [55]. Their findings also confirm that applying PEF could enhance mass transfer in fruits with a higher value mass transfer results at higher field strength and temperature conditions. The use of PEF treatment along with osmotic dehydration in various agricultural commodities like apples, bananas, and carrots was investigated [56]. During their study at both static and dynamic conditions of osmotic dehydration in sucrose solution, a considerable increment in water loss and solid gain was observed with the application of PEF. Similar findings in the case of kiwi fruit [57], apple slices [58–60], red bell pepper [61] and carrot [23] were also reported in the literature.

Low heat and mass transfer efficiency in the drying process for producing fruit and vegetable-based dried snacks can also be overcome by using ultrasonication in combination with conventional osmotic dehydration pretreatment. By accelerating the mass transfer rates of solids and water to and from a drying material, ultrasound-assisted osmotic dehydration (UAOD) improves osmotic dehydration (OD) by increasing the mass transfer rates of water to and from a drying material. Application of USOD to strawberries was used as a pretreatment for freeze drying [36]. The experiment was conducted at different power levels (180–300 W) by using different osmotic agents. Analysis of various parameters revealed that the introduction of USOD reduced the drying time by nearly 10% for better moisture diffusivity. Moreover, pretreated samples were also characterized by better color, texture, and nutritional quality than untreated dried samples. USOD was also found to have its application as a pretreatment for the production of fried potatoes. USOD for 30 min in a binary solution of sodium chloride (15%) and sucrose (50%) was shown to reduce the oil uptake by 12.5% during the frying process at 170°C. Moreover, better color in the final potato chips was also observed due to the pretreatment. The use of ultrasound reduced the time taken for osmotic dehydration by 67% [63]. The use of USOD in pomegranate aril before air drying was also studied [64]. They conducted the experiment at different frequencies of 25 and 40 kHz. From the moisture removal rate calculation, it was found that a 2 times and 2.7 times increase in water loss was achieved by using US at a frequency of 25 and 40 kHz, respectively. It was inferred that US treatment causes more detachment of water from the cell wall and causes changes in the microstructure of the food product. These structural alterations at the microscopic level ultimately resulted in more water diffusivity during the final hot air-drying process also. A better result of coupling the US with osmotic dehydration was also reported for other fruits like cranberries [65], kiwifruit [66, 67], and sapota [68], etc. Details of reported studies were also given in the Table 3.1.

TABLE 3.1 Application of Novel Processing Method in Osmotic Dehydration

Commodity	Treatment and Condition	Inference	References
High Pressure Processing (HPP)			
Strawberry	OD concentration: 40°Brix Pressure: 100–500 MPa Time: 10 min	• HPP helps in retaining Vitamin C • Processing at 400 MPa for 10 min is optimum condition for better product quality.	[46]
Red abalone	Drying temperature: 40 to 60°C Pressure: 350–550 MPa Dwell time:5 and 10 min OD concentration: 10 and 15% NaCl	• Increased moisture diffusivity • Better energy efficiency during drying	[69]
Pineapple	Pressure: 100–700 MPa Time: 5 min OD concentration: 50°Brix sugar OD temperature: 40°C	• Higher water transport due to cell breakage • Diffusivity of water and solute increased till 400 MPa • Lower dehydration in treated sample due to cell rupture	[70]
Banana	Pressure:100–500 MPa Sucrose concentration:30–70°Brix OD time: 5–9 h OD temperature: 30–70°C	• Treated samples exhibitedmore WL and SG during OD • Optimized operating conditions: Pressure 200 MPa, Time–5 min at room temperature (26°C), sucrose concentration–60°Brix, OD time – 5 h and OD temperature –40°C.	[47]
• Ohmic heating			
Pears	Voltage: 100 v Electric field intensity: 13 V/cm Time:5 h Sucrose concentration: 65°Brix	• Greater water loss, Solute uptake and structural changes • Less firmness loss • 40% reduction in processing time	[50]
Strawberry	Ohmic heating temperature: 65 and 85°C OD temperature: 26 and 37°C Sucrose concentration: 30–70°Brix	• Dry matter content in treated sample increased up to 68%.	[48]

TABLE 3.1 *(Continued)*

Commodity	Treatment and Condition	Inference	References
Apple	Ohmic heating temperature: 60–95°C Time:0–6 min Sucrose: 70°Brix	• Short duration treatment enhance mass transfer without causing structural damage	[49]
Quince	Electrical field strengths: 20, 30 and 40 V/cm Holding time: 10, 20 and 30 minutes Microwave power: 90, 180 and 270 WSucrose concentration: 0%, 25% and 50%	• Optimum conditions: 40 V/cm electrical field strength, 30 min holding time, 16.67% sucrose concentration and 270 W microwave power • Total phenolic content, rehydration ratio, color differences and dielectric properties higher than that of control.	[71]
• Pulsed electric field			
Kiwifruit	Field strengths:0.7, 1.1 and 1.8 kV/cm OD solution: consisting of glycerol, trehalose, maltodextrin, ascorbic acid, sodium chloride, calcium chloride and citric acid (wfood/wsolution = 1/5) Immersion time: 40 min Immersion temperature:25, 35 and 45°C	• Product retained optimum quality characteristics and adequate decrease of aw. • Acceptable level of color change, increased firmness, high vitamin C content	[55]
Apple	Sucrose concentrations: 44.5–6°Brix Field strength: 0.90 kV/cm Time of PEF: 0.75 s Syrup to sample ratio: 3:1 (w/w)	• Both WL and SG increased with more effect on WL.	[59]
Strawberries	Pulse width:100 μs Repetition time: 10 ms Electric fields strength: 100, 200 and 400 V cm^{-1}. OD concentration: 40% w/w sucrose and trehalose	• Positive effect on mass transfer even at low field strength. • Maintain the quality of final product.	[72]
Goji berry	Field strength: 0.9–2.8 kV/cm Number of pulses: 7,500 pulses	• Processing at PEF of 2.8 kV/cm for 750 pulses at an osmotic dehydration temperature of 55°C for 60 min with an air dying at 60°C could decrease processing time by 33%. • Better retentionin color, antioxidant capacity and total phenolic content.	[54]

TABLE 3.1 *(Continued)*

Commodity	Treatment and Condition	Inference	References
• Ultrasonication			
Kiwifruit	Frequency: 35 kHz US time:10, 20 and 30 min. Sucrose concentration: 61.5% Immersion time: 0, 10, 20, 30, 60, and 120 min.	• US alone positively affected the chlorophyll content, during 20 min of application with a negative effect on texture. • Combined OD, US able to maintain and improve the kiwifruit characteristics.	[73]
Apricot	OD concentration: 35°Brixsorbitol OD temperature: 55°C Immersion time: 30 and 45 min US frequency: 25 and 35 kHz Air drying temperature: 60°C	• US OD reduces total phenolic compounds (TPC), β-carotene contents especially at higher frequency. • Antioxidant capacity and β-carotene content higher than control without US	[74]
Plum	Ultrasound frequency:40 kHz Intensity treatments of ultrasound: 0.45, 0.90, and 1.35 W/g. US time: 15 min at 30°C. OD concentration: 50°Brix sucrose	• Treatment reduced the drying time, enhanced drying rate, and water diffusivity. • US treatment decreases the hardness and chewiness	[75]

3.5 CHALLENGES IN CONVENTIONAL AND NOVEL OSMOTIC DEHYDRATION TECHNIQUES

The management of syrup resulting from osmotic dehydration creates a greater difficulty in practical situations [76]. The constituent of the syrup obtained after dewatering from the food tissue greatly varies because of the various mass transfer processes such as solute gain and water loss. Hence, it could not be utilized for the treatment of repeated mode. Because of the dilution of the solution during the process, it is difficult to maintain the concentration of the solution during the process, especially at low solution-to-sample ratios. It is advisable to keep a solution to sample ratio of 2:1. Using a large amount of solute is also not economically feasible if we are not reusing the solution. Therefore, it is always advisable to reuse the solution after mixing the required quantity of solute to achieve the desired concentration. Several studies have reported the feasibility of using osmotic solutions repeated times [8, 77]. In one of the studies conducted in grapes it has been reported that it is possible to reuse osmotic solutions without re-concentration up to five times [77]. However, it is recommended to pasteurize the solution before processing for the subsequent treatment to obtain a higher shelf life in the final product. In another study for apple cube osmotic dehydration, it has been reported that after re-concentration to 60°Brix, it is possible to use the sample OD solution 20 times without compromising the mass transfer ability and final physical properties. Nevertheless, they have found that syrup is highly loaded with yeast and mold after 15th use [78]. Consequently, proper solution treatment or solution replacement should be followed regularly to ensure the final product's microbiological safety. During the reuse, it is also taken care to do the filtration treatment, especially when the treatment is given for minimally processed fruits and vegetables. Since there is a chance of the presence of seed and pulp portion in the osmotic solution, this has been removed for efficient operation. Treatment of leftover solution is also challenging since the osmotic solution is considered heavily polluted sewage due to the rich presence of suspended solids. Hence, the cost involved in treating osmotic solutions is higher, so some effective solution needs to be implemented. Even though the reuse of solutions is recommended in most cases, it may not be viable in certain circumstances. For example, in the case of osmotic dehydration of cherry, the solution may become dark due to the leaching of anthocyanin content into the syrup. As a result, it could not be used multiple times [76].

In the reuse of osmotic solutions, restoring solution concentration is the primary step, which can be achieved by processes like evaporation, solute addition, and membrane concentration. Due to the high temperature required to achieve the required degree of concentration while using evaporation processes, significant loss in nutritive value and organoleptic properties of the solution is possible. The use of vacuum evaporation can minimize this, but the high cost of this process restricts its use in industrial applications. Electro dialysis is another option that is widely used in the Japanese chemical industry as the most effective way to concentrate brines. The main advantage of electro dialysis is its extremely low energy consumption, especially when combined with the membranes with high durability.

Despite the various quality improvements and time-saving techniques in the process, the use of osmotic dehydration pretreatment can cause substantial changes in the food's nature compared to the original one, with a loss of nutritional and sensory values. In some cases, the solute may act as a layer on the surface of the fruits or vegetables and cause resistance to water removal during further drying processes. Sugar uptake can also change the flavor and rehydration ability of the dried product due to its lower rehydration capacity [10, 62]. In the osmotic dehydration process, mechanical agitation is usually given during the immersion treatment to increase the mass transfer. This mechanical agitation sometimes causes the breakage of fruit or vegetable tissue and disrupts the final product shape and quality.

As mentioned earlier, mass transfer in the osmotic dehydration process could be attained using novel processing methods. This fact was confirmed on a laboratory scale by conducting experiments on various fruits and vegetables. However, up-scaling all novel combination methods is still challenging for their commercial application.

3.6 CONCLUSION

In combination with various drying methods, osmotic dehydration is a promising strategy to reduce post-harvest losses in fruits and vegetables, which also offers the production of dried snack food with versatile taste and quality. Using an appropriate osmotic agent for the treatment can offer excellent organoleptic properties and nutritional value in the end product. Utmost care should be taken to decide the optimum concentration and immersion condition to yield a product with desirable properties. Otherwise, considerable nutritive loss and cell damage can occur during the process.

Further enhancement in the osmotic dehydration-assisted mass transfer can be obtained by the use of novel processing methods. In that aspect, the application of ultrasonication, high-pressure processing, and pulsed electric field treatment in combination with osmotic dehydration are gaining popularity among food researchers. Food material that has been processed by osmotic dehydration further undergoes drying operation to attain shelf-stable snacks. The standard drying methods like convective drying, freeze-drying, microwave drying, and infrared drying are commonly used for the final drying processes. Application of lower temperature, altered pressure, and innovative methods can result in the reduction of heat-sensitive compounds with a simultaneous improvement in the quality and sensory attributes of the product.

Future research should focus on up-scaling the novel processing strategy for industrial applications. Process optimization specific for different food material with different osmotic agent need to be carried out to minimize the adverse effects. The introduction of healthy novel osmotic agents must be promoted for the nutritional enrichment of the final product. Research on scientific reusing and recycling of osmotic solutions should be promoted to reduce economic losses and environmental pollution.

KEYWORDS

- **combination treatment**
- **hypertonic solution**
- **microwave drying**
- **novel processing**
- **osmotic dehydration**

REFERENCES

1. Porat, R., Lichter, A., Harker, R., & Buzby, J. (2018). Postharvest losses of fruit and vegetables during retail and in consumers' homes: Quantifications, causes, and means of prevention. *Postharvest Biology and Technology, 139*, 135–149. https://doi.org/10.1016/j.postharvbio.2018.02.003.
2. Sette, P., Salvatori, D., & Schebor, C. (2016). Physical and mechanical properties of raspberries subjected to osmotic dehydration and further dehydration by

air- and freeze-drying. *Food and Bioproducts Processing, 100*, 156–171. https://doi.org/10.1016/j.fbp.2016.01.003.

3. Pandiselvam, R., Tak, Y., Olum, E., Sujayasree, O. J., Tekgül, Y., Çalışkan Koç, G., Kaur, M., Nayi, P., Kothakota, A., & Kumar, M. (2021). Advanced osmotic dehydration techniques combined with emerging drying methods for sustainable food production: Impact on bioactive components, texture, color, and sensory properties of food. *Journal of Texture Studies, 53*(6), 737–762. https://doi.org/10.1111/jtxs.12622.
4. Fontana, A. J. (2000). Understanding the importance of water activity in food. *Cereal Foods World, 45*, 7–10.
5. Yadav, A. K., & Singh, S. V. (2014). Osmotic dehydration of fruits and vegetables: A review. *Journal of Food Science and Technology, 51*, 1654–1673. https://doi.org/10.1007/s11483-013-9367-3.
6. Chandra, S., & Kumari, D. (2015). Recent development in osmotic dehydration of fruit and vegetables: A review. *Critical Reviews in Food Science and Nutrition, 55*(4), 552–561. https://doi.org/10.1080/10408398.2012.664830.
7. Ramya, V., & Jain, N. K. (2017). A review on osmotic dehydration of fruits and vegetables: An integrated approach. *Journal of Food Process Engineering, 40*(3), e12440. https://doi.org/10.1111/jfpe.12440.
8. Ciurzyńska, A., Kowalska, K., Czajkowska, A., & Lenart, A. (2016). Osmotic dehydration in production of sustainable and healthy food. *Trends in Food Science & Technology, 50*, 186–192. https://doi.org/10.1016/j.tifs.2016.01.015.
9. Rahman, M. S. (2007). Osmotic dehydration of foods. In *Handbook of Food Preservation* (pp. 451–464). CRC Press.
10. Tortoe, C. (2010). A review of Osmo dehydration for food industry. *African Journal of Food Science, 4*(6), 303–324. https://doi.org/10.5897/AJFS10.010.
11. Kadam, D. M., & Dhingra, D. (2011). Mass transfer kinetics of banana slices during osmo-convective drying. *Journal of Food Process Engineering, 34*(2), 511–532. https://doi.org/10.1111/j.1745-4530.2008.00376.x.
12. Bekele, Y., & Ramaswamy, H. (2010). Going beyond conventional osmotic dehydration for quality advantage and energy savings. *Ethiopian Journal of Applied Science and Technology, 1*(1), 1–15. https://doi.org/10.4314/ejast.v1i1.59089.
13. Singh, B., Kumar, A., & Gupta, A. K. (2007). Study of mass transfer kinetics and effective diffusivity during osmotic dehydration of carrot cubes. *Journal of Food Engineering, 79*, 471–480. https://doi.org/10.1016/j.jfoodeng.2006.01.015.
14. Pisalkar, P. S., Jain, N. K., & Jain, S. K. (2011). Osmo-air drying of aloe vera gel cubes. *Journal of Food Science and Technology, 48*, 183–189. https://doi.org/10.1007/s11483-010-0276-2.
15. Velickova, E., Winkelhausen, E., & Kuzmanova, S. (2014). Physical and sensory properties of ready-to-eat apple chips produced by osmo-convective drying. *Journal of Food Science and Technology, 51*(12), 3691–3701. https://doi.org/10.1007/s11483-014-0823-3.
16. da Silva, W. P., do Amaral, D. S., Duarte, M. E. M. C., Mata, M. E. R. M. C., e Silva, C. M. D. P. S., & Pinheiro, R. M. M. (2013). Description of the osmotic dehydration and convective drying of coconut (*Cocos nucifera* L.) pieces: A three-dimensional approach. *Journal of Food Engineering, 115*(1), 121–131. https://doi.org/10.1016/j.jfoodeng.2012.09.017.

17. da Silva, W. P., da Silva, C. M. D. P., de Farias Aires, J. E., & da Silva Junior, A. F. (2014). Osmotic dehydration and convective drying of coconut slices: Experimental determination and description using one-dimensional diffusion model. *Journal of the Saudi Society of Agricultural Sciences, 13*(2), 162–168. https://doi.org/10.1016/j.jssas.2013.06.002.
18. Pravitha, M., Manikantan, M. R., Ajesh Kumar, V., Beegum, S., & Pandiselvam, R. (2021). Optimization of process parameters for the production of jaggery infused osmo-dehydrated coconut chips. *LWT, 146*, 111441. https://doi.org/10.1016/j.lwt.2021.111441.
19. Kaur, P., Zalpouri, R., Singh, M., & Verma, S. (2021). Process optimization for dehydration of shelled peas by osmosis and three-stage convective drying for enhanced quality. *Journal of Food Processing and Preservation, 44*(12), e14983. https://doi.org/10.1111/jfpp.14983.
20. Pravitha, M., Manikantan, M. R., Ajesh Kumar, V., Shameena Beegum, P. P., & Pandiselvam, R. (2022). Comparison of drying behavior and product quality of coconut chips treated with different osmotic agents. *LWT, 162*, 113432. https://doi.org/10.1016/j.lwt.2022.113432.
21. Bakhara, C. K., Pal, U. S., & Bal, L. M. (2018). Drying characteristics and physico-chemical evaluation of tender jackfruit slices during osmo-convective drying. *Journal of Food Measurement and Characterization, 12*, 564–572. https://doi.org/10.1007/s11483-018-9581-2.
22. Mierzwa, D., & Kowalski, S. J. (2016). Ultrasound-assisted osmotic dehydration and convective drying of apples: Process kinetics and quality issues. *Chemical and Process Engineering, 37*(3), 383–391. https://doi.org/10.1515/cpe-2016-0035.
23. Amami, E., Fersi, A., Khezami, L., Vorobiev, E., & Kechaou, N. (2007). Centrifugal osmotic dehydration and rehydration of carrot tissue pre-treated by pulsed electric field. *LWT–Food Science and Technology, 40*, 1156–1166. https://doi.org/10.1016/j.lwt.2006.06.014.
24. Kroehnke, J., Szadzińska, J., Radziejewska-Kubzdela, E., Biegańska-Marecik, R., Musielak, G., & Mierzwa, D. (2021). Osmotic dehydration and convective drying of kiwifruit (*Actinidia deliciosa*): The influence of ultrasound on process kinetics and product quality. *Ultrasonics Sonochemistry, 71*, 105377. https://doi.org/10.1016/j.ultsonch.2021.105377.
25. Aydogdu, A., Sumnu, G., & Sahin, S. (2015). Effects of microwave-infrared combination drying on quality of eggplants. *Food and Bioprocess Technology, 8*(6), 1198–1210. https://doi.org/10.1007/s11947-015-1568-5.
26. Lyu, J., Yi, J., Bi, J., Chen, Q., Zhou, L., & Liu, X. (2017). Effect of sucrose concentration of osmotic dehydration pretreatment on drying characteristics and texture of peach chips dried by infrared drying coupled with explosion puffing drying. *Drying Technology, 35*(15), 1887–1896. https://doi.org/10.1080/07373937.2017.1353474.
27. Wu, X. F., Zhang, M., Mujumdar, A. S., & Yang, C. H. (2019). Effect of ultrasound-assisted osmotic dehydration pretreatment on the infrared drying of pakchoi stems. *Drying Technology, 38*(15), 2015–2026. https://doi.org/10.1080/07373937.2019.1602704.
28. Chakraborty, R., & Samanta, R. (2017). Concurrent osmotic dehydration and vacuum drying of kiwi fruit (*Actinidia deliciosa* cv. Hayward) under far infrared radiation: Process optimization, kinetics and quality assessment. *Journal of Food Process Engineering, 40*, e12391. https://doi.org/10.1111/jfpe.12391.

29. Tan, M., Chua, K. J., Mujumdar, A. S., & Chou, S. K. (2001). Effect of osmotic pre-treatment and infrared radiation on drying rate and color changes during drying of potato and pineapple. *Drying Technology, 19*(9), 2193–2207. https://doi.org/10.1081/DRT-100107339.
30. Deepika, S., Sutar, P. P., & Pandiselvam, R. (2018). Combining osmotic-steam blanching with infrared-microwave-hot air drying: Production of dried lemon (*Citrus limon* L.) slices and enzyme inactivation. *Drying Technology, 36*(14), 1719–1737. https://doi.org/10.1080/07373937.2018.1489312.
31. Rongtong, B., Suwonsichon, T., Ritthiruangdej, P., & Kasemsumran, S. (2018). Determination of water activity, total soluble solids and moisture, sucrose, glucose and fructose contents in osmotically dehydrated papaya using near-infrared spectroscopy. *Agriculture and Natural Resources, 52*(6), 557–564. https://doi.org/10.1016/j.anres.2018.04.003.
32. Zhou, C., Wang, Z., Wang, X., Yagoub, A. E., Ma, H., Sun, Y., & Yu, X. (2021). Effects of tri-frequency ultrasound-ethanol pretreatment combined with infrared convection drying on the quality properties and drying characteristics of scallion stalk. *Journal of the Science of Food and Agriculture, 101*, 2809–2817. https://doi.org/10.1002/jsfa.11043.
33. Karam, M. C., Petit, J., Zimmer, D., Djantou, E. B., & Scher, J. (2016). Effects of drying and grinding in production of fruit and vegetable powders: A review. *Journal of Food Engineering, 188*, 32–49. https://doi.org/10.1016/j.jfoodeng.2016.01.004.
34. Prosapio, V., & Norton, I. (2017). Influence of osmotic dehydration pre-treatment on oven drying and freeze-drying performance. *LWT, 80*, 401–408. https://doi.org/10.1016/j.lwt.2017.03.046.
35. Wang, R., Zhang, M., & Mujumdar, A. S. (2010). Effect of osmotic dehydration on microwave freeze-drying characteristics and quality of potato chips. *Drying Technology, 28*(6), 798–806. https://doi.org/10.1080/07373931003740571.
36. Jiang, J., Zhang, M., Devahastin, S., & Yu, D. (2021). Effect of ultrasound-assisted osmotic dehydration pretreatments on drying and quality characteristics of pulsed fluidized bed microwave freeze-dried strawberries. *LWT, 145*, 111300. https://doi.org/10.1016/j.lwt.2021.111300.
37. Pereira, N. R., Marsaioli, A., & Ahrné, L. M. (2007). Effect of microwave power, air velocity and temperature on the final drying of osmotically dehydrated bananas. *Journal of Food Engineering, 81*(1), 79–87. https://doi.org/10.1016/j.jfoodeng.2006.12.025.
38. Botha, G. E., Oliveira, J. C., & Ahrné, L. (2012). Quality optimization of combined osmotic dehydration and microwave assisted air drying of pineapple using constant power emission. *Food and Bioproducts Processing, 90*(2), 171–179. https://doi.org/10.1016/j.fbp.2011.07.002.
39. Assis, F. R., de Morais, R. M. S. C., & de Morais, A. M. M. B. (2018). Osmotic dehydration combined with freeze-drying of apple cubes and comparison with microwave drying and hot air drying. *Advances in Food Science and Engineering, 21*(1), 38–47.
40. Dehghannya, J., Hosseinlar, S. H., & Heshmati, M. K. (2018). Multi-stage continuous and intermittent microwave drying of quince fruit coupled with osmotic dehydration and low temperature hot air drying. *Innovative Food Science & Emerging Technologies, 45*, 132–151. https://doi.org/10.1016/j.ifset.2017.10.007.
41. Changrue, V., Orsat, V., & Raghavan, S. V. (2008). Osmotically dehydrated microwave-vacuum drying of strawberries. *Journal of Food Processing and Preservation, 32*(5), 798–816. https://doi.org/10.1111/j.1745-4549.2008.00202.x.

42. Bórquez, R. M., Canales, E. R., & Redon, J. P. (2010). Osmotic dehydration of raspberries with vacuum pretreatment followed by microwave-vacuum drying. *Journal of Food Engineering, 99*(2), 121–127. https://doi.org/10.1016/j.jfoodeng.2010.01.022.
43. Li, L., Zhang, M., & Wang, W. (2020). Ultrasound-assisted osmotic dehydration pretreatment before pulsed fluidized bed microwave freeze-drying (PFBMFD) of Chinese yam. *Food Bioscience, 35*, 100548. https://doi.org/10.1016/j.fbio.2020.100548.
44. Luo, W., Tappi, S., Wang, C., Yu, Y., Zhu, S., Dalla Rosa, M., & Rocculi, P. (2019). Effect of high hydrostatic pressure (HHP) on the antioxidant and volatile properties of candied wumei fruit (*Prunus mume*) during osmotic dehydration. *Food Bioprocess Technology, 12*, 98–109. https://doi.org/10.1007/s11947-018-2173-4.
45. Luo, W., Tappi, S., Wang, C., Yu, Y., Zhu, S., & Rocculi, P. (2018). Study of the effect of high hydrostatic pressure (HHP) on the osmotic dehydration mechanism and kinetics of wumei fruit (*Prunus mume*). *Food Bioprocess Technology, 11*, 2044–2054. https://doi.org/10.1007/s11947-018-2197-8.
46. Nuñez-Mancilla, Y., Pérez-Won, E. M., Uribe, A., Vega-Gálvez, A., & Di Scala, K. (2013). Osmotic dehydration under high hydrostatic pressure: Effects on antioxidant activity, total phenolics compounds, vitamin C, and color of strawberry (*Fragaria vesca*). *LWT – Food Science and Technology, 52*(2), 151–156. https://doi.org/10.1016/j.lwt.2012.09.016.
47. Verma, D., Kaushik, N., & Rao, P. S. (2014). Application of high hydrostatic pressure as a pretreatment for osmotic dehydration of banana slices (*Musa cavendishii*) finished-dried by dehumidified air drying. *Food Bioprocess Technology, 7*(5), 1281–1297. https://doi.org/10.1007/s11947-013-1130-3.
48. Allali, H., Marchal, L., & Vorobiev, E. (2010). Blanching of strawberries by ohmic heating: Effects on the kinetics of mass transfer during osmotic dehydration. *Food Bioprocess Technology, 3*, 406–414. https://doi.org/10.1007/s11947-009-0201-2.
49. Allali, H., Marchal, L., & Vorobiev, E. (2009). Effect of blanching by ohmic heating on the osmotic dehydration behavior of apple cubes. *Drying Technology, 27*(6), 739–746. https://doi.org/10.1080/07373930902991804.
50. Moreno, J., Simpson, R., Estrada, D., Lorenzen, S., Moraga, D., & Almonacid, S. (2011). Effect of pulsed-vacuum and ohmic heating on the Osmo dehydration kinetics, physical properties, and microstructure of apples (*cv. Granny Smith*). *Innovative Food Science & Emerging Technologies, 12*(4), 562–568. https://doi.org/10.1016/j.ifset.2011.01.006.
51. Moreno, J., Zúñiga, P., Dorvil, F., Petzold, G., Mella, K., & Bugueño, G. (2017). Osmo dehydration assisted by ohmic heating/pulse vacuum in apples (*cv. Fuji*): Retention of polyphenols during refrigerated storage. *International Journal of Food Science & Technology, 52*(5), 1203–1210. https://doi.org/10.1111/ijfs.13445.
52. Moreno, J., Simpson, R., Pizarro, N., Parada, K., Pinilla, N., Reyes, J. E., & Almonacid, S. (2012). Effect of ohmic heating and vacuum impregnation on the quality and microbial stability of osmotically dehydrated strawberries (*cv. Camarosa*). *Journal of Food Engineering, 110*, 310–316. https://doi.org/10.1016/j.jfoodeng.2011.12.035.
53. Moreno, J., Espinoza, C., Simpson, R., Petzold, G., Nuñez, H., & Gianelli, M. P. (2016). Application of ohmic heating/vacuum impregnation treatments and air drying to develop an apple snack enriched in folic acid. *Innovative Food Science & Emerging Technologies, 33*, 381–386. https://doi.org/10.1016/j.ifset.2015.11.007.
54. Dermesonlouoglou, E., Chalkia, A., Dimopoulos, G., & Taoukis, P. (2018). Combined effect of pulsed electric field and osmotic dehydration pre-treatments on mass transfer

and quality of air-dried goji berry. *Innovative Food Science & Emerging Technologies, 49*, 106–115. https://doi.org/10.1016/j.ifset.2018.06.013.
55. Dermesonlouoglou, E., Zachariou, I., Andreou, V., & Taoukis, P. S. (2016). Effect of pulsed electric fields on mass transfer and quality of osmotically dehydrated kiwifruit. *Food and Bioproducts Processing, 100*, 535–544. https://doi.org/10.1016/j.fbp.2016.03.002.
56. Amami, E., Khezami, L., Jemai, A. B., & Vorobiev, E. (2014). Osmotic dehydration of some agro-food tissue pre-treated by pulsed electric field: Impact of impeller's Reynolds number on mass transfer and color. *Journal of King Saud University-Engineering Sciences, 26*(1), 93–102. https://doi.org/10.1016/j.jksues.2012.06.005.
57. Traffano-Schiffo, M. V., Laghi, L., Castro-Giraldez, M., Tylewicz, U., Romani, S., Ragni, L., Dalla Rosa, M., & Fito, P. J. (2017). Osmotic dehydration of organic kiwifruit pre-treated by pulsed electric fields: Internal transport and transformations analyzed by NMR. *Innovative Food Science & Emerging Technologies, 41*, 259–266. https://doi.org/10.1016/j.ifset.2017.01.002.
58. Taiwo, K. A., Angersbach, A., & Knorr, D. (2003). Effects of pulsed electric field on quality factors and mass transfer during osmotic dehydration of apples. *Journal of Food Process Engineering, 26*(1), 31–48. https://doi.org/10.1111/j.1745-4530.2003.tb00385.x.
59. Amami, E., Vorobiev, E., & Kechaou, N. (2006). Modelling of mass transfer during osmotic dehydration of apple tissue pre-treated by pulsed electric field. *LWT–Food Science and Technology, 39*(9), 1014–1021. https://doi.org/10.1016/j.lwt.2005.02.007.
60. Parniakov, O., Bals, O., Lebovka, N., & Vorobiev, E. (2016). Effects of pulsed electric fields assisted osmotic dehydration on freezing-thawing and texture of apple tissue. *Journal of Food Engineering, 183*, 32–38. https://doi.org/10.1016/j.jfoodeng.2016.03.009.
61. Ade-Omowaye, B. I. O., Rastogi, N. K., Angersbach, A., & Knorr, D. (2003). Combined effects of pulsed electric field pre-treatment and partial osmotic dehydration on air drying behavior of red bell pepper. *Journal of Food Engineering, 60*(1), 89–98. https://doi.org/10.1016/S0260-8774(03)00176-1.
62. Ghosh, P. K., Agrawal, Y. C., Jayas, D. S., & Kumbhar, B. K. (2004). Mass transfer kinetics model of osmotic dehydration of carrots. *Transactions of the ASAE, 47*(4), 1179–1185. https://doi.org/10.13031/2013.16613.
63. Karizaki, V. M., Sahin, S., Sumnu, G., Mosavian, M. T. H., & Luca, A. (2013). Effect of ultrasound-assisted osmotic dehydration as a pretreatment on deep fat frying of potatoes. *Food Bioprocess Technology, 6*(12), 3554–3563. https://doi.org/10.1007/s11947-013-1085-2.
64. Allahdad, Z., Nasiri, M., Varidi, M., & Varidi, M. J. (2019). Effect of sonication on osmotic dehydration and subsequent air-drying of pomegranate arils. *Journal of Food Engineering, 244*, 202–211. https://doi.org/10.1016/j.jfoodeng.2018.09.014.
65. Nowacka, M., Fijalkowska, A., Dadan, M., Rybak, K., Wiktor, A., & Witrowa-Rajchert, D. (2018). Effect of ultrasound treatment during osmotic dehydration on bioactive compounds of cranberries. *Ultrasonics, 83*, 18–25. https://doi.org/10.1016/j.ultras.2017.10.004.
66. Nowacka, M., Tylewicz, U., Laghi, L., Dalla Rosa, M., & Witrowa-Rajchert, D. (2014). Effect of ultrasound treatment on the water state in kiwifruit during osmotic dehydration. *Food Chemistry, 144*, 18–25. https://doi.org/10.1016/j.foodchem.2013.07.125.

67. Prithani, R., & Dash, K. K. (2020). Mass transfer modelling in ultrasound-assisted osmotic dehydration of kiwi fruit. *Innovative Food Science & Emerging Technologies, 64*, 102407. https://doi.org/10.1016/j.ifset.2020.102407.
68. Rodrigues, S., Gomes, M. C., Gallão, M. I., & Fernandes, F. A. (2009). Effect of ultrasound-assisted osmotic dehydration on cell structure of sapotas. *Journal of the Science of Food and Agriculture, 89*(4), 665–670. https://doi.org/10.1002/jsfa.3564.
69. Pérez-Won, M. (2016). Modelling of red abalone (*Haliotis rufescens*) slices drying process: Effect of osmotic dehydration under high pressure as a pretreatment. *Innovative Food Science & Emerging Technologies, 34*, 127–134. https://doi.org/10.1016/j.ifset.2016.02.001.
70. Rastogi, N. K., & Niranjan, K. (1998). Enhanced mass transfer during osmotic dehydration of high pressure treated pineapple. *Journal of Food Science, 63*(3), 508–511. https://doi.org/10.1111/j.1365-2621.1998.tb15695.x
71. Kutlu, N. (2022). Optimization of ohmic heating-assisted osmotic dehydration as a pretreatment for microwave drying of quince. *Food Science & Technology International, 28*, 60–71. https://doi.org/10.1177/10820132211029279.
72. Tylewicz, U. (2017). Effect of pulsed electric field (PEF) pre-treatment coupled with osmotic dehydration on physico-chemical characteristics of organic strawberries. *Journal of Food Engineering, 213*, 2–9. https://doi.org/10.1016/j.jfoodeng.2017.04.016.
73. Nowacka, M., Tylewicz, U., Romani, S., Dalla Rosa, M., & Witrowa-Rajchert, D. (2017). Influence of ultrasound-assisted osmotic dehydration on the main quality parameters of kiwifruit. *Innovative Food Science & Emerging Technologies, 41*, 71–78. https://doi.org/10.1016/j.ifset.2017.01.003.
74. Sakooei-Vayghan, R., Peighambardoust, S. H., Hesari, J., & Peressini, D. (2019). Effects of osmotic dehydration (with and without sonication) and pectin-based coating pretreatments on functional properties and color of hot-air dried apricot cubes. *Food Chemistry, 311*, 125978. https://doi.org/10.1016/j.foodchem.2019.125978.
75. Li, L. (2021). Effect of ultrasound-assisted osmotic dehydration pretreatment on the drying characteristics and quality properties of Sanhua plum (*Prunus salicina L.*). *LWT, 138*, 110653. https://doi.org/10.1016/j.lwt.2021.110653.
76. Dalla Rosa, M. (2001). Osmotic treatments (OT) and problems related to the solution management. *Journal of Food Engineering, 49*, 223–236. https://doi.org/10.1016/S0260-8774(00)00125-6
77. Moraga, M. J., Moraga, G., & Martínez-Navarrete, N. (2011). Effect of the re-use of the osmotic solution on the stability of osmodehydro-refrigerated grapefruit. *LWT, 44*, 35–41. https://doi.org/10.1016/j.lwt.2010.08.005.
78. Valdez-Fragoso, A., Mujica-Paz, H., Giroux, F., & Welti-Chanes, J. (2002). Reuse of sucrose syrup in pilot-scale osmotic dehydration of apple cubes. *Journal of Food Processing and Engineering, 25*(2), 125–139. https://doi.org/10.1111/j.1745-4530.2002.tb00414.x.

CHAPTER 4

HOT EXTRUSION TECHNOLOGY FOR THE PRODUCTION OF LEGUME-BASED READY-TO-EAT SNACKS

ANIRUDH GURURAJ PATIL,[1] SNEHAL T. YADAV,[1] SHALINI BILLA,[2] GOURI PATIL,[3] and SOUMITRA BANERJEE[4]

[1]*Assistant Professor, Department of Food Technology, Annasaheb Dange College of Engineering and Technology, Ashta, Sangli, Maharashtra, India*

[2]*Assistant Professor, Department of Biotechnology, Gowdara Mallikarjunappa Institute of Technology, Davangere, Karnataka, India*

[3]*Department of Polymer Science and Technology, Sri Jayachamarajendra College of Engineering (SJCE), Jagadguru Sri Shivarathreeshwara Science and Technology University, Mysore, Karnataka, India*

[4]*Associate Professor, Center for Incubation, Innovation, Research, and Consultancy (CIIRC), Department of Food Technology, Tataguni, Bangalore, Karnataka, India*

ABSTRACT

Snacks have always been the savior of our modern life. They are diverse, always expanding, and ever-changing group of foods. Customers seek convenient snacks with fascinating sensory and textural essentials and therefore ready-to-eat (RTE) foods are very likely to grow rapidly in popularity over the next coming years. The behavior of various legume type flours in extrusion cooking has been studied in recent years, demonstrating that the

Ready-to-Eat Snacks: Emerging Technologies for Production and Safety.
Ravi Pandiselvam and Gülşah Çalışkan Koç

legumes have an unmatched potential for producing extrusion-based RTE foods, that are made by partially or entirely replacing cereals. Extrusion cooking has a very strong influence on the nutritional features as it leads to significant changes in starch and proteins improving their digestibility and lowering the level of trypsin inhibitors. This chapter focuses on the various types of legumes that are utilized, and a comparison of the nutritional benefits of legumes and cereals, with a broad knowledge of the ideal processing conditions for both extruded snacks that are legume-based and those that also contain legumes is presented. This chapter also focuses on the important properties of extrudates, packaging materials and food safety regulations for the legume-based RTE snacks discussing the future aspects of the same.

4.1 INTRODUCTION

The increasing demand for convenient and nutritious food options has spurred innovation in the food processing industry, leading to the exploration of novel technologies for snack production. One such technology that has gained prominence is hot extrusion, a versatile and efficient method that facilitates the transformation of raw ingredients into ready-to-eat (RTE) products. When applied to legume-based snacks, hot extrusion offers an exciting opportunity to enhance the nutritional profile of conventional snack foods, capitalizing on the rich protein, fiber, and essential micronutrients that legumes provide.

Hot extrusion technology involves the application of heat and mechanical shear under controlled conditions, allowing for the transformation of raw legume flour into expanded, crispy snack forms. The process not only enhances the digestibility and palatability of legumes but also plays a vital role in developing unique textures and flavors that appeal to a broad consumer base. By utilizing legumes, manufacturers can cater to the growing consumer preference for healthy, plant-based snacks while simultaneously addressing issues related to sustainability and environmental impact, given the lower resource requirements associated with legume cultivation compared to traditional snack ingredients.

Furthermore, the global shift toward healthier eating habits and the rising awareness of the importance of dietary diversity align perfectly with the potential of hot extrusion technology. This method not only addresses the challenges of conventional snack production but also opens new avenues for product innovation. As the industry strives to meet consumer expectations for taste, texture, and nutrition, legume-based RTE snacks produced through hot extrusion stand out as a viable solution that combines health benefits with convenience, paving the way for a new era in snacking.

4.1.1 LEGUMES FOR EXTRUDED FOOD PRODUCTS

Modern lifestyles have increased customer demand for RTE foods since they save time. Numerous RTE products like collects, runs, puffs, instant soups, breakfast cereals, meat alternatives and foods for sports are produced using extrusion cooking [1, 2]. Because of their ease in preparation, appealing look, and pleasing texture, extruded foods have the potential to attract consumers [3, 4]. Because of their superior expansion qualities, cereals are widely used in extrusion cooking. Having said that, extruded foods, in addition to delivering energy from starch, also serve as transporters of additional nutrients when enriched with other components. Protein, carbohydrates (starch and dietary fiber), vitamins, and minerals are abundant in legumes. When animal protein consumption is restricted due to cost or dietary choices, plant demand for plant-based protein sources increases. Additionally, legumes are considered sustainable crops due to their adaptability to marginal terrain [5–7].

Soybean has been used in the development of extruded food products in the preceding 10 years. Yet, several studies have been undertaken in recent years to incorporate various legumes such as bean, lentil, pea, chickpea, and faba bean into extruded meals to boost the nutritional value. Extruded multi-legume bars enriched with palm oil, honey, and whey protein concentrate have been proposed as a technique of addressing malnutrition in developing countries [8, 9]. Extrusion cooking reduces anti-nutritional factors present in legumes such as tannin, trypsin inhibitors, lectins, and phytic acid. Moreover, extrusion cooking improves the digestion of proteins and carbohydrates [10].

4.1.2 BASICS OF EXTRUSION COOKING

The single-screw extruder was invented in 1946 in the United States for quick cooking and expanding maize and rice-based snacks, and it was followed by double (twin) screw extruders in the mid-1970s [1, 2]. Further during extrusion cooking, the usual conditions are pressures of up to 20 Megapascals (MPa) and temperature of 200°C, respectively. Extruder consists of one or two rotating screws, before being fed to the extruder the raw ingredients should be thoroughly grounded and conditioned to certain level of moisture. With a spinning screw, the raw material is transported and combined. The first section of the barrel, also known as the feeding zone. In the kneading zone, the material is further kneaded and crushed using shear energy. Due to friction and extra heating of the barrel, the contents in the last component of the machine reach melting point and plasticize. This area is also known as the cooking zone [11, 12].

As the extruded material is heated over 100°C and the plasticized starchy material exits from the extruder's die, air bubbles become caught within the matrix and expand due to pressure drop immediately [13]. When cooked, the plasticized food matrix shrinks and becomes crispy and glassy in texture. As a result of the aerating effect, extrusion cooking is an excellent method of transforming thick and hard materials into more pleasing and indeed lighter goods [14, 15].

4.1.3 OPTIMUM EXTRUSION COOKING PARAMETERS FOR LEGUME FLOURS

Temperature, screw speed, and moisture content of feed components must all be controlled during the extrusion cooking process since these critical factors have a substantial influence on extruded product qualities. As a result, several research has been done to examine the various combinations of the aforementioned characteristics in order to reach the ideal circumstances for each variety of legume. The experiments began with legume-based foods. Second, there are legume-added extruded foods, which include legume flour into a cereal-based extruded product. Apart from just the processing parameters the factors like content of legume flour, refinement degree, type of legume, particle size and effect of legume type on expansion efficiency in the extrudates are been studied. Legumes are ground into flour using entire cotyledons or after split flours have been removed (hulls). The removal of the hull decreases the mineral and fiber content while increasing extrusion expansion [16, 17]. The interactions of swelling starch with proteins and lipids limit the puffing effect, while insoluble fibers can enclose air bubbles and impede their maximal expansion [18]. There is a direct relationship between the protein and fiber content of starch-based extruded goods and their bulk density [19]. Additionally, the findings reveal that the degree of extrusion of different bean types varies according to their composition. Spilled yellow pea flour has the same expansion ratio and bulk density as maize meal because to its higher starch content and lower protein content when compared to lentils and chickpeas. Chickpea ejected snacks tend to have the most reduced extension capacities when compared to other beats because to their lower starch and higher fat content. Sodium bicarbonate, which emits CO_2, can be used to increase the amount of extension in ejected items. In any case, it should be noted that the addition of sodium bicarbonate leads in an undesirable alkalization of the final product [18, 19].

4.1.4 CURRENT SCENARIO OF LEGUME – EXTRUDED SNACKS

Working hours, long commutes, and unexpected workloads have all contributed to a radical change in our eating habits in recent years. It is estimated that around 22% of the overall UK population consumes snacks throughout the day as opposed to three regular meals [20, 21].

Just 39% of snackers sought out healthy options all or most of the time, whereas 10–15% sought out legume-based extruded snacks. The firms which create legume-based snacks typically claim the snacks to be "high fiber and protein." Because of the low fat, calories, and salt content, 42–47% of UK customers polled preferred legume -based extruded snacks. Because to their high protein and dietary fiber content, they were favored by around 21–23% of consumers [20, 22]. Between 2013 and 2017, there was a 39% increase in goods manufactured from legumes. Legume based snacks are also viewed as alternative to meat extruded items and has experienced rise of 128%. Gluten-free product advertising has climbed by 73%, while vegetarian product promotion has increased by 196%. This can be ascribed to health claims made with legume-based food items and a growth in the number of health-conscious people [23].

4.2 LEGUMES

4.2.1 LEGUME CLASSIFICATION

The term legume is a Latin origin word "Legere" that meaning "to collect." As a result, it refers to plant species that grow fruits contained in structures known as pods [24, 25]. FAO says that pulses are the leguminous crop that is harvested in form of dry seedpod except to those of green pods of green peas [26]. For more than thousands of years, legumes have been a part of the human diet. Legumes belong to the family *Leguminosae* which is third largest family in the plantae kingdom [7]. This family has more than 750 genera and nearly 2 lakh species among which around 60 species have been domesticated which include soybeans, mung beans, lentils, and chickpeas. These legumes are economically significant [24].

4.2.2 BENEFITS OF USING LEGUMES

FAO recognized 2016 as the World Year of Pulses to increase awareness of the nutritional benefits of plants and to highlight how they may help with economic farming. Pulses include vitamins and bioactive substances that promote health and have a part in nitrogen fixation, which has a monetary contribution through agriculture. Pulses are a rich source of plant-based proteins that can help reduce malnutrition throughout the world since they have a good amino acid profile and greater quantities of the important amino acid lysine [28].

They also provide an excellent source of energy-giving carbs as well as other essentials such as water-soluble vitamins and minerals. Thiamine, folic acid, and Vitamin C are among the most important vitamins [29]. Genistein, daidzein, and glycitein are also abundant in legumes. They have minimal levels of saturated fats and cholesterol. Foods with a glycemic index of 16 to 30 are considered low glycemic [28].

The legume cultivation has a great contribution towards sustainable agriculture system as they have symbiotic relationship with root-soil-microbiome. As discussed, they are involved in nitrogen fixation from atmosphere and they by reduce the need of using nitrogen rich fertilizers [30]. They also aid in minimizing environmental and climatic change by lowering water and carbon footprints. As a result, they are said to be a completely renewable source of energy for both human and animal diet. They also improve soil fertility by promoting the growth of microorganisms [24, 31]. European Union has come to awareness about unsustainable demand for meat-based protein worldwide. The EU has funded a project called Protein 2 Food, which aims to develop sustainable and innovative processing technology for producing high-quality, cost-effective, resource-efficient, and cost-effective protein from legumes in order to provide protein-rich functional meals. The initiative helps to support feed conversion while also increasing the quality and quantity of protein.

The report of this project throws light on the arable land devoted for pulse production in EU. The report says around 25% increase in production of protein and 10% increase in the land share devoted for pulses cultivation [32].

4.3 LEGUMES AND CEREALS' NUTRITIONAL PROPERTIES COMPARISON

Nutrition and health will likely remain the most discussed and demanding topics of food science forever. Some traditional cooking methods include exposing the food material to greater temperatures for longer periods of time, which drastically alters the nutritional properties of the dish. High temperature and short time processing technologies, such as extrusion cooking, might be explored from the standpoint of high productivity and nutritional value retention [32, 33]. Extrusion is a hot cooking process that involves combining raw ingredients and forcing them through a screw die under varying pressure, moisture, heat, and mechanical shear conditions. The functional characteristics of food alter throughout the extrusion process due to the breakage of covalent and non-covalent bonds in the food matrix. The functional and nutritional benefits of extrusion cooking are gaining popularity. It is utilized in the manufacture of RTE cereals, sweet and pleasant snack meals, and many other products. Extrusion cooking has both beneficial and bad consequences on the nutritional content of meals, according to research [34].

4.3.1 PROTEIN

Protein is the second most prevalent and vital component in legumes and grains; proteins have a significant impact on the nutritional and physical properties of the extruded food product. The major impacts of extrusion on plant-based protein include changes in protein quality, such as amino acid composition and digestibility, as well as protein functionality, such as texture and solubility. This section focuses on the nutritional and functional aspects of legume and grain protein as a result of diverse extrusion settings [34].

Protein quality is defined by total protein, necessary amino acid concentration, and protein digestibility. Protein quality is influenced by several parameters, including bioavailability, source, processing conditions, and dietary matrix. Extrusion has little impact on the overall protein content of fed mixed grains. The most prevalent effect seen is in terms of protein availability and essential amino acid digestibility [35]. Extruding a high protein biscuit mix with flour has been observed to result in a loss of up to 40% of lysine accessible at low feed rates and low humidity, regardless of the starting lysine concentration of the feed mix. Grains often have lower lysine content, which can lead to a drop in protein quality. According to research, the loss of accessible lysine during grain extrusion is inversely related to extruder temperature and proportionate to feed water content. By lowering the heat load inside the barrel, high screw speeds and smaller die sizes can considerably enhance lysine retention during extrusion [36].

The free amino group of lysine and reducing sugar found naturally in grain or added to feed mixture. Millard-type reactions may be the primary cause of the loss of accessible lysine. As a result, the blackness of the extruded grain product can be utilized as an indirect measure of lysine loss. When lysine is attached to the Millard-type reaction product, the extruded product has less protein than the beginning feed. Lysine can form iso-peptide connections with its free amine group and the amide group of asparagine or glutamine in specific cereal mixes. It is a rare occurrence in grain proteins; however, it has been identified in several studies. Apart from lysine, extrusion affects other important amino acids such as threonine, valine, isoleucine, and leucine. Thermal denaturation occurs as a result of extrusion circumstances such as greater barrel temperature. These decreases in accessible lysine amino acid have been widely reported in the literature [37].

According to recent studies, essential amino acid digestibility and bioavailability should be enhanced. Lentils, for example, have a higher total amino acid content, a higher digestibility score for key amino acids, and a higher protein efficiency ratio (freshly baked or cooked). Extruded maize meal and soya protein concentrates have been shown to greatly increase protein biological value while retaining key amino acids such as accessible lysine [36, 37]. To address the loss of available lysine and enhance the protein digestibility of different cereals, the extrusion situation is changed. Low dietary moisture improves in vitro and in vivo protein digestibility in legumes, dry beans, maize, rice, quinoa, carob, pea, and sorghum, according to research. When wheat protein is extruded, its digestibility improves, whereas others report no noticeable difference when raw wheat flour is used [38, 39]. Digestibility improves because of anti-nutrient molecule inactivation and heat denaturation of proteins during the extrusion process. For example, some whole cereals and beans contain highly condensed tannins and other polyphenols, which have anti-oxidant qualities but bond to protein in indigestible complexes. Protease inhibitors found in cereals and legumes attach to enzymes like trypsin and disrupt protein digestion. Several efforts are made to decrease the activity of protease inhibitors that interfere with protein digestion by mixing high extrusion temperatures and shear [40, 41].

Extrusion process promotes denaturation due to mechanical shear and thermal effect lead to protein unfolding which get hydrolyzed by proteolytic enzymes. Compared to boiling, extrusion enhances the sorghum protein degradation resulting in decrease in digestibility [42]. This impact is caused by the cleavage of di-sulfide bridges at high screw velocities, which results in reduced digestibility of sorghum flour. Higher temperatures during extrusion, on the other hand, have been shown to decrease the digestibility of proso-millet protein by nearly 50% because they promote the creation of hydrophobic aggregates that are difficult to break by shear [34]. As a result, the impact of extrusion on the nutrition element of grain and legume protein is difficult and specific to the protein type, but in most instances, extrusion has increased protein digestibility. Upon legumes addition to grain-based extrusion improves the final protein content of the extruded product because of higher protein content of legumes.

For example, adding 30% green beans and red beans to corn flour increases the crude protein level by 12 times. The protein content of the final extrudates increases from 6.59% to 10.04% when common beans (*Phaseolus vulgaris L.*) are added to maize flour in percentages ranging from 17 to 40% [4]. Extruded puffed items made from Vicia faba beans increased protein content by three times when compared to corn-based alternatives. The addition of 20–40% soy protein to wheat flour boosted the protein level from 20.75% to 27.5%. Furthermore, fortifying rice flour with bean flour doubles the protein level [43, 44]. Extrusion improves the in vitro protein digestibility of plant protein-based meals. Sumargo et al. [45] discovered that adding 15–45% pinto bean flour to brown rice flour improves protein digestion [45]. Similarly, Patil et al. [12] discovered that supplementing wheat-based snacks with legume flours such as chickpea, yellow pea, and lentils in levels ranging from 5 to 15% enhanced protein digestibility [12].

4.3.2 CARBOHYDRATE

As we know, carbohydrates are the primary, important energy source. Normally legumes have lower energy content of around 301 to 359 kilocalories per 100 grams compared to that of cereals that range between 316 and 397 kcal per 100 grams. Pinto beans have the lowest energy content of 301 kcal/100 g among legumes, whereas chick peas have the most at 359 kcal/100 g [24]. Having said that, the energy density for cereals particularly for sorghum is 316 kcal/100 g and 397 kcal/100 g for maize [46]. As we all know, starch is the major storage carbohydrate in plants, as well as the second most common polysaccharide after cellulose. The starch is composed of two glucan units, amylose and amylopectin, which are granularly organized in plant substances. During the extrusion process these granules melt, the combination of gelatinized and melted starch changes the structural and functional properties. Further, in few certain conditions of extrusion dextrinization and depolymerization of amylopectin can take place [47]. The properties like pasting, technological and physiochemical differ in legumes as their starch content is different. The temperature of gelatinization for legumes is higher, therefore starch present in legumes do not swell during instead gets easily decomposed. For example, the gelatinization temperature of kidney bean starch is 78.46°C, whereas it is 67.25°C for wheat starch [48, 49].

The starch of legumes contains significant amount of esterified phosphate that increases the strength and viscosity of the gel. For that reason, the starch of legumes has better shear stability compared to that of cereals. Due to excellent stability against heat and mechanical shear stress legume starch is an ideal raw material for extruded snack products [48]. Because of the high retrogradation propensity of legume starch paste, legume starches are more resistant to the action of digestive enzymes, resulting in low GI values. For example, whole meal wheat bread has a GI value of roughly 74, while lupin bread has a GI value of 34 [20]. In this respect, legume starch is an ideal ingredient for the development of diabetic foods. The foods that release reducing sugars slowly are called low glycemic foods. The glycemic response of corn and wheat based extruded foods is achieved by addition of various levels (5, 10, 15%) of yellow peas, green peas or chick peas [12].

Starch is classified into three major groups based on the time necessary for digestion in a well-defined in vitro experimental setting. Under the prescribed circumstances, the RDS (rapidly digestible starch) digests in 20 minutes. SDS (slowly digestible starch) is digested in 20–120 minutes, but RS (resistant starch) is not digested after 120 minutes [20, 50].

The presence or uptake of legumes as ingredient lowers the RDS content, increases SDS and RS content and thereby lowers the glycemic levels. For example, adding 15–24% pinto bean flour to brown rice flour decreases the SDS while increasing the RS levels in the final extrudates. Replacing potato flour with red lentil flour and chick pea flour in the proportion of 10% and 80%, respectively decreased starch digestibility in the extrudate and this effected was observed to increase at higher substitution [51, 52]. Based on enzyme resistance of resistant starch (RS), they are classified into five groups; RS1 starch granules that are physically inaccessible to enzyme as they are contained in the refractory matrix and cannot be digested without gelatinization. Chemically modified starches such as RS2 crystal structure, RS3 aged starch, RS4 starch ester / ether crosslinked starch and RS5 amylose lipid complex. Legumes are a possible source of RS1 and are found in intact plant cells where the plant cell wall (cellulose fiber) wraps starch and severely restricts amylase enzyme access in the gastrointestinal system. The treatment methods currently used for the legume flour production like dry milling, further processing like par-boiling, prolonged hydrothermal treatment led to reactions and disruption of cell structures, reduce RS1 levels leading to increase in blood sugar [53, 54].

Therefore, in recent times the studies are conducted in development of alternative legume flour which has got higher proportions of intact cells. In the recent studies, a new ingredient called PulseON® was developed and checked for preparation of delicious biscuits. PulseON® is new resistant starch rich chickpea powder derived from whole Kabuli chickpea through a new specific process. The treatment had hydro-thermal treatment that runs for 90 minutes at 100°C. The ingredients were precooked and PulseON® was added to replace 0.25, 50, 75, 100% of the flour in a cookie recipe and RS1 appeared structurally unaffected during the process. The study found that upon replacing 50% of flour with PulseON® reduced the starch hydrolysis index of biscuits by nearly 60% [55, 56]. In a subsequent study, Edwards et al. [56] compared the properties of pulses from seven different botanical sources, including green lentils, butter beans, green split peas, yellow peas, kidney peas, and chickpeas, to flours obtained from the same source. All PulseON® flours have lower starch digestibility (lesser than 40% starch digestion in 90 minutes) and better water retention capacity than cooked bean flour (80% or more starch digestion in 30 minutes) and has a larger swelling potential and water retaining capacity than their counterpart [56]. As a result, the use of this component in extruded snacks will be investigated further in the future.

4.3.3 LIPIDS

The lipids are present in trace amounts in most of the extruded products. Lower concentrations of lipids (less than 5%) have tendency to stabilize the vitrification of extruded products. The higher lipid concentration provides additional lubricity within the extruder barrel reducing the expansion [32]. In terms of nutrition, lipids tend to bond with proteins and carbohydrates during the extrusion process. Extrusion cooking causes interactions between lipids and starch grains. Lipids bond to starch, making extraction using non-polar solvents problematic. Because nut flours naturally include lipids, the extrusion feed combination typically comprises 6.5% lipids. According to the research, the development of lipid-starch complexes rose as the barrel temperature and feed moisture levels increased. However, under these conditions, increased extruder density, stiffness and oil loss were found in the extruder barrel [57]. This research focused on the synthesis of starch-fatty acid complexes using natural lipids from natural sources. These investigations also concluded that saturated fatty acids interact more strongly with starches than unsaturated fatty acids, and that amylopectin-rich starches form more complexes with fatty acids [58]. In general, legumes possess minimal lipid content; predominantly legumes have a lipid content of 1–3%, except for lupines (5 to 15%) and chickpeas (2 to 7%). Cereal grains have a lipid content ranging from 7 to 12%, with some maize varieties having a lipid content of 17%. The reduced lipid content levels of legumes when included into flour/grain-based forms is observed, for example, the final extrudate lipid content was reduced when the bean (*Phaseolus vulgaris*) flour was included in the corn four by 48 to 50% [4, 59].

4.3.4 MICRONUTRIENTS

The legumes are rich source of vitamin B complexes that includes thiamine, folates, and niacin. They also have crucial minerals essential for metabolism and maintenance like magnesium, calcium, zinc, and iron [60]. Vitamin retention can be observed in extruded cooked products due to high temperature, mechanical shear, and high specific energy. Vitamin retention decreases with decrease in humidity, feed rate and die diameter. Several options are available to mitigate the vitamin loss. Few to mention here are use of thermal stable vitamins, adding supplementary vitamins to compensate for loss, addition of vitamins in post-extrusion process to the extrudate. For an illustration, the vitamins that are most sensitive Vitamins D and K, as well as the water-soluble vitamins C, B1, folic acid and the fat-soluble vitamins A and E are generally stable [32]. Foods with high fiber content may grind on the extruder barrel and screws, increasing the mineral content. The absorption of most of the minerals enhance during extrusion cooking which is elucidated by the reduction of anti-nutritional features like condensed tannins and phytates. Thermal effects bring chemical changes in fiber structure [61]. Gulati and Rose [62] reported that after extrusion there is substantial increase in dialyzability of K, P, Mg and Fe by 30%, 50%, 79%, respectively in great northern beans when parameters maintained were a temperature of 90–140°C, moisture content of 17–25% and speed of screw 156 to 250 rpm.

With a decrease in moisture content, there was a considerable fall in phytate (46%), and folate (24%). Lower barrel temperature, intermittent moisture content and higher screw speeds were identified as the optimum conditions for maximum retention of vitamins, dialyzability of minerals and folate and keeping check on phytates (anti-nutritional entities) [62].

4.3.5 MICRONUTRIENTS

Most bioactive compounds in legumes include; phenolic acids, proanthocyanidins, saponins, flavanols, small proteins and peptides [20, 60]. Pigmented and dark colored pulses contain increased number of phenolic compounds when compared to light colored varieties of pulses [32]. These above said compounds present in legume/pulses relate to wide range of bio-activities that are anti-inflammatory, anti-oxidant, anti-microbial and anti-diabetic both *in vitro* and *in vivo* and also in combatting obesity cardiovascular diseases. The final contents of bio-active chemicals in extruded product are determined by the kind of pulse, variety, food matrix structure, chemical nature of its ingredients, and extrusion process parameters [63].

Many studies have been undertaken to confirm the negative influence of extrusion-cooking on the bioactive component levels in foods [64]. In distinction the research conducted by Arribas et al. [65] inferred that higher amount of total phenolic compounds having anti-oxidant property was found significantly in all the extruded samples depended upon the blend of pea, carob fruit, and rice in comparison to the commercially available products [65]. Ciudad-Mulero et al. [66] conducted study and stated that, the increase in beneficial alpha-galactosidase that have pre-biotic characteristics in unique formulation of high protein nutritional yeast powders and lentil. The optimized extrusion cooking conditions result in higher phenolic compounds and anti-oxidants content in extrudates made containing flour mixture of corn/common bean [66]. F'elix–Medina et al. [4] discovered a high concentration of key bioactive substances such as polyphenols, flavonoids, and polyunsaturated fatty acids (PUFA) in expanded snacks manufactured from a 70:30 blend of maize and beans [4].

4.3.6 ANTI-NUTRITIONAL FACTORS

Enzyme inhibitors such as trypsin inhibitors, chymotrypsin inhibitors and alpha-amylase inhibitors as well as phenolic substances such as tannins, phytates, saponins, and lectins are the anti-nutritional factors found in legumes. These non-nutritional substances can limit nutrient absorption, decrease digestibility, and simultaneously decrease bioavailability. They are known to provide an off-flavor, bitter, and unpalatable taste to food [24, 67, 68]. Nevertheless, these anti-nutritional components could be deactivated or decreased with the help of extrusion cooking. After the extrusion procedure of the peas the antinutrient factors significantly reduced, trypsin inhibitor by 94.6%, tannins by 91.7%, phytic acid by 5.9% and chymotrypsin inhibitor by 65.4% [69]. Another study found that unique formulations developed with lentil and nutrient-rich yeast flours had a considerable reduction in undesired inositol phosphates, lectins, and trypsin inhibitor activity [66].

4.4 EXTRUSION TECHNOLOGY

Extrusion is a hot cooking technique where raw ingredients are assorted and pressed into a screw die under various combinations of moisture, pressure, heat, and mechanical shear. The extrusion method has the capacity to disrupt covalent and non-covalent connections between the constituents of a complex food matrix. As a result, the functional characteristics are changed. The extrusion method has several uses, including the creation of ready-to-eat cereal snacks, confectionary, and sweet and savory snack items. In the past few years, a surge in understanding functional and nutritional components towards the extrusion process has been observed [34].

4.4.1 TECHNOLOGY INVOLVED IN EXTRUSION-COOKING

Extrusion cooking is a highly heat-efficient technique utilized in the food industry and is considered as a multifunctional technology. It is thought to be perfect for creating a variety of nutrient-rich food products, reasonably priced and has longer shelf-life. Products like pasta, traditional snack foods, pet foods, crackers, RTE foods, breakfast cereals, textured vegetable protein, legume-based puff snacks and many more can be manufactured using extrusion cooking. Extrusion is considered as continuous efficient procedure having amalgamation of operations like shearing, mixing, heating, molding and sizing [70].

The extruders contain one or more screws that push the pre-mixed ingredients into barrel. This technology makes use of two equipment configuration or designs. One is single screw and another is twin-screw extruder. The twin-screw extruder has got many sub-components that resembles single screw extruder that include, flow hopper, feed screw, pre-conditioning barrel, extruder barrel, rotary screw and jacketed head. A twin-screw extruder is much complex compared to single-screw extruder; nonetheless simultaneously it suggests greater adaptability and suitable control over procedure. Both the single and intermeshing co-rotating twin extruder are most frequently used in feed and food industry [71, 72]. The food ingredients are heated at temperatures between 100 and 180°C, subjected to high shear pressure, and kept at comparatively low moisture levels between 10 and 40% wet basis in an extruder [73].

The raw material is entirely transformed into a functional shape within 30 to 120 seconds thanks to the high temperature, which additionally speeds up the processing time. The food material is then forced into a small opening known as a die and a cut to the suitable size [74]. The formulation of feed, the feed moisture content, the feed rate, the temperature of the barrel, and the screw-die configuration are the primary operational parameters of the extrusion process. Die pressure, die temperature, and motor torque/current are used for indirect output measurement for assessing the quality of the produced material in a majority of typical extruder usages. High temperature, high shear, and high pressure combined trigger several reactions that change the nutrient content and functional qualities of food.

The steps involved, which can be basic or complex, include protein breakdown, polymerization cross-linking, texturization, dextrinization, gelatinization, and solubilization. Enzymes like chymotrypsin, trypsin, disaccharides, and alpha-amylases are either entirely or partially inactivated. In addition, the elimination of microbial load, decrease of lipid oxidation and rancidity, and annihilation of antinutritional factors that lower food nutritional values [74, 75].

4.4.2 OPTIMIZATION OF PROCESS PARAMETERS FOR LEGUME BASED EXTRUDED FOODS

The experiments related to extrusion was carried out by utilizing 100% board bean meal with process parameters, screw speed between 200 and 300 rpm, and varying the meal particle size lesser than 0.5 mm, 1 to 2 mm, and 2.5 mm while keeping extruder temperature constant. It was inferred that, 140°C extruder temperatures, screw speed of 300 and particle size of less than 0.5 mm was optimized and that yielded the firm, inflated, and crispy snacks with similar sensory characteristics that of commercially available corn snack [76]. Temperature, content of moisture and screw speed was investigated throughout manufacturing a 100% cow pea extruded snack. The RSM Technique was utilized to analyze and assess the effects of the moisture content from 16–24% range, temperature of the die ranging from 160–180°C and speed of the screw ranging between 160 and 200 rpm (RSM) [77].

The studies showed that low content of moisture, fast screw speed and high die pressure resulted in fewer dense and rigid extrudates with higher water absorption index (WAI), water solubility index (WSI) and sensory ratings [78]. With a moisture level of 16%, die temperature of 180°C and screw speed of 200 rpm, the optimum products could be produced. Like that, a twin screw extruder's favorable conditions for extruding chick peas included comparatively low water content, a higher die temperature and a high speed of screw [33]. The impact of water content was confirmed once more throughout the production of 100% puffed yellow pea munchies using a cp-rotating twin extruder running at 150°C, 200 rpm. The feed moisture content values (14%, 16%, and 18%) were examined. Lowest moisture levels produced low density extrusion, which was the ideal and finest outcome [79]. Screw speeds of 150 rpm, 200 rpm and 250 rpm along with feed moisture levels of 14, 18, and 22% were kept while producing lentil extrudates at different die temperatures of 140, 160, and 160°C. In twin screw extruder 18% moisture, temperature of the die being 160°C and a screw speed of 200 rpm were found to produce the finest sensory and nutritional properties [64].

4.4.3 OPTIMIZATION OF PROCESS PARAMETERS FOR LEGUME ADDED EXTRUDED FOODS

In some studies, as an alternative of making 100% bean/legume extrudates, the grains are partially replaced with legume flour/meal to supplement the deficiency of essential amino acids and to enhance the overall nutritional value of grain based extruded products. These studies also placed a strong emphasis on identifying the ideal working conditions. As anticipated, typically established the results of extrusion trial where legumes showed 100% [80, 81].

Similar studies were carried out to assess the impacts of feed water levels between 15 and 25% and screw speeds between 50 and 240 rpm on extruded corn-based snacks foods containing 30% black soya bean with particle size of 0.4 mm, 15% humidity and speed of screw 238 rpm were the processing circumstances that guaranteed a high expansion coefficient. It utilized a uniaxial screw extruder. In the twin screw extruder utilized to produce extruded rice-based munchies with 30% mung bean flour, the impacts of the identical parameters, such as moisture content of feed and screw speed range of 14 to 18% and 400 to 550 rpm, were also investigated. Again, at a lower feed moisture level of 14% and a high screw speed of 550 rpm, the best and optimal outcomes were attained. The lowest density and highest water absorption index (WAI), which are related to the water solubility index (WSI), were also ensured by these conditions [82]. Experiments designed to create healthy puffed snacks using fermented chickpea flour of 50%, yogurt and potato starch extruded using a twin screw extruder produced similar outcomes. In order to get rid of the undesirable legume flavors, chickpeas are fermented. These experiments used process parameters, varying the die temperature between 130 and 150°C and the speed of screw between 266 and 434 rpm. The experiment suggested that temperature of 140°C and a speed of screw 434 rpm were ideal conditions. The attractive extruded products met the necessary low-density standards [83].

The extrusion cooking influences the color of the product. This is a crucial aspect in acceptability of the product by consumers. Studies were conducted which focused on highlighting the outcome of these cooking techniques on appearance (cooler). Millet based extruded product was produced from a blend containing 12 to 28% pigeon pea flour using single screw extruder. The parameters were, temperature ranging between 160 and 200°C and screw speeds ranging between 100 and 400 rpm. The overall acceptance was extruded products with higher lightness values which resulted in selection of optimal process parameters. Further, for the number of legumes the highly affecting parameter was die head temperature. The optimal conditions in a given experiment were found to be 24% of moisture content, speed of screw 104 rpm and temperature of die 171°C with legume flour constituting to 19% [84]. In the preparation of cow pea based extruded product, the redness and yellowness found to increase and the brightness faded as the temperature inclined. With a rise in moisture content of the feed, the redness, yellowness intensified while the lightness reduced. This study indicates that if the consumers dark colors, then the optimal conditions required for bright colors are contradictory to those required to ensure higher expansion rates. Therefore, equilibrium is necessary to achieve both product sensory and physical property of the product [19, 64].

4.4.4 DIVERSITY OF SNACKS

Snacks are largely divided into 1st generation, 2nd generation and 3rd generation. Snacks from the first generation include simple snacks like popcorn and potato crisps [20, 85]. Second Generation snacks also known as direct extended snacks, ready to eat snacks or collets.

These require higher shear stress for proper expansion of collets. When the moisture content is maintained between 5 and 8% the product swells on an extrusion die and normally do not require any post-treatment other than drying. Drying facilitates further reduction in moisture content to 1 to 2%. Examples include corn tortilla snacks, onion rings, and puffed corn curls, among many others. Due to the use of corn starch in their preparation, they have comparatively lower nutritional values [4]. Since the products are dried to reach moisture content of around 10% after extrusion, snacks are also referred to as indirectly expanded snacks, pellets, or half products. The desiccated pellets are then fried, heated in hot air, or even heated in a microwave. If packed correctly to prevent moisture from entering, these pellets can last up to a year without refrigeration [85].

4.5 PRODUCT PROPERTIES OF EXTRUDATES WITH LEGUMES

Pre-conditioning parameters for the extruder, such as mixing and pre-conditioning, feed formulation, die design for the extruder screw, operating conditions (feed rate, water injection rate, screw speed), and post-extrusion operations, such as drying, roasting or frying, and flavoring addition, all affect the quality of the extruded product [86]. All the steps are essential and must be properly controlled in order to produce the proper product quality. Figure 4.1 shows an outline of important control points in the form of a pictorial representation. The formulation, extrusion pre-conditioning (preconditioning and mixing), screw arrangement of extruder, design of die, operating condition of extruder and post-operation-extruder (roasting, frying, drying, microwaving, and adding flavorings) all affect the quality of the extruded product.

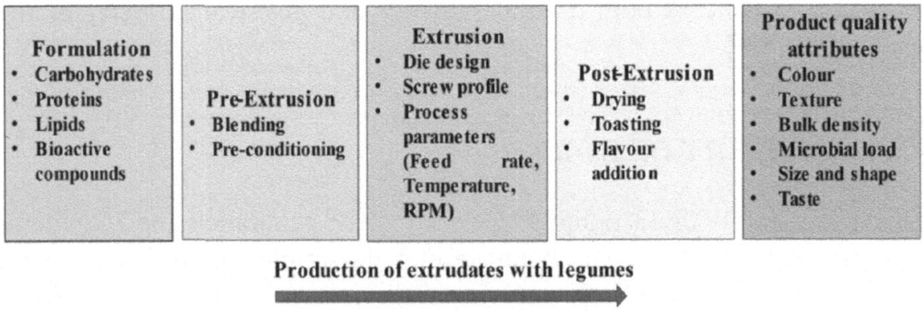

FIGURE 4.1 Control points of importance in the manufacturing of legume-based extrudates.

4.5.1 EXPANSION RATIO (ER) AND BULK DENSITY (BD)

Degree of cooking stays closely correlated with the extensibility of the product, which is a very fundamental characteristic that describes the quality of the product. The melt's expansion process profoundly impacts the textural and structural characteristics of extruded snacks [86]. As soon as the melt leaves the die, there is a sudden drop in pressure, leading to massive internal moisture evaporation. Foaming and expansion of product is the result of extrusion melting that is caused due to difference between water vapor pressure and atmospheric pressure [87]. As we know, the extruder characteristic is strongly influenced by extruder operation conditions like feed rate, barrel temperature, rate of water injection and screw speed. Additionally, the expansion efficacy is influenced by the type, composition, and particle size of the legume flour [1]. Extrudate expansion is quantified by ratio of the extrudate's cross-sectional diameter to that of the die this is known as the expansion ratio. In general, speed of screw, moisture content of feed, and temperature of barrel all have an impact on expansion ratio. It increases with reduction in feed moisture content and increase in barrel temperature and screw speed [1, 86].

Expansion rate (ER) and bulk density (BD) are physical properties that are inter-dependent. With increase in ER there is a decrease in BD and vice-versa. From the view point of commercial production, BD is a very important parameter as most extrusion is occupied with weight and not by volume. During production if bulk density changes, then packaging will not be sufficient or will overflow [4]. The expansion ratio is significantly connected to the comparative composition and the interactions of starch, fiber and protein. Increased levels of ER are associated with greater levels of starch, lower levels of protein, and lower levels of fiber [88]. Numerous studies have found that increased supplementation with substances strong in fiber and protein like bean flour is beneficial. As an example, when the extrusion conditions are set at 16% feed moisture, 180°C and speed of screw 250 rpm, addition of chickpea (30%) to corn extrudates reduces the expansion rate [89]. Further, the addition of soya bean flour and chick peas significantly reduced expansion ration value of corn extrudate snacks for the process parameters of feed rate at 30 Kgs per hour, barrel temperature 150°C and speed of screw 110 rpm [90].

Addition of lentil powder in the proportion of 10–15% to the extruded corn snacks reduced ER when the processing parameters were the following; feed rate of 2.5 to 6.8 Kgs per hour, extruder temperature ranging between 170 and 230°C and feed moisture between 13 and 19% wet basis (wb) were kept. The expansion rate was reported to increase with the increment in feeding rate and decrease in humidity, temperature, and corn-lentil ratio. In a different study conducted, extrusions produced from corn meal and carioca bean in proportion (4.8 to 55.2 grams per 100 grams of corn meal) expanded less when measure using cross sectional expansion index. Higher levels of moisture content are reported to show lubricative effects within extruder, lowering both shear rate and temperature inside the equipment. Extruded pinto bean and brown rice composite flours exhibit increases in bean flour content ranging from 0 to 45% and moisture content ranging from 17.2 to 20.1% during processing, as well as increases in density and hardness or firmness and decreases in expansion when the following processing parameters are used: speed of screw 300 rpm and barrel temperature from inlet to die of 80 to 100°C [45].

In research using pea-based expanded snack foods made with pea flour/fiber and pea starch, the ER substantially reduced as the temperature was raised from 110°C to 135°C. Furthermore, it was observed that there were additional variables to consider, including the extrusion conditions and blend formulations. Similarly, F'elix-Medina et al. [4] found that beam composition, barrel temperature ranging from 159 to 164°C, and screw speed ranging from 165 to 214 rpm all influenced expansion ration, with the latter two increasing it specifically at lower bean flour contents in corn /common bean flour mixtures [4]. Further, Rangira et al. [92] described a thought-provoking result that using pea starch lowers the barrel temperatures and increased the expansion which is contradictory to additional cereal starches like corn and rice with process parameters; moisture content of 15%, 17.5 and 20% wet basis, respectively, temperature of 120, 135 and 150°C, respectively and speed of screw of 150 rpm 200 rpm and 250 rpm, respectively [92]. Low temperature with greater moisture content increased expansion ratio values in formulations of 60:40 chickpea and barley combination with process parameters of 150 to 170°C and 16 to 20% wet basis [93].

Philipp et al. [94] examined the result of pea protein fortification, utilizing 50% weight/weight pea protein isolate (PPI) on micro-structural and physical features of puffed (expanded) rice snacks. The expansion rates of the extrudates were increased by raising s speed of the screw from 400 rpm to 600 rpm. Increase in die temperature from 130°C–150°C and content of moisture from 23% to 26% did not have noticeable results on the ER. Extrudates containing 50% pea protein isolates extended less and were tougher than the extrudates containing lesser PPI values ($p<0.05$) [94]. Unexpected outcome was noticed in a research using several whole beans flours such lima, faba, pinto and red kidney beans. These flours had been extruded using process parameters of temperature ranging from 120 to 160°C, content of the moisture ranging from 17 to 25% and screw speed ranging between 150 and 250 rpm [95]. According to these findings, this impact might be attributed to interactions between protein and amylose or amylopectin (a component of cellulose) in faba beans. Researchers speculated that it might be related to protein interactions and the amylose and amylopectin ratio in faba beans. Nonetheless, more study is required to approve these research findings.

4.5.2 WATER ABSORPTION INDEX (WAI) AND WATER SOLUBILITY INDEX (WSI)

Numerous functional parameters such as water solubility index, water absorption capacity, water absorption index, water-oil absorption index, protein solubility index, and emulsion solubility index to name a few, are used qualitatively to characterize extruded goods [86]. To predict the material behavior of extruded products, the hydration properties such as WAI and WSI are crucial. The changes occurred in the hydration properties can be attributed to structural changes in the higher molecular weight entities like starch or in the low molecular weight entities like lipids/fatty acids and proteins [73]. The above-mentioned properties are strongly influenced by the structure of starch granules and particle size.

Water absorption index (WAI) measures the volume occupied by the starch polymer after swelling in the excess of water. At room temperature, the water absorption index value of extruded starch surpasses that of native starch granules because gelatinized starch absorbs more moisture than the natural starch granules [70]. Having said that, if dextrinization occurs or in simple words, starch thawing outweighs gelatinization, WAI will drop. Higher WAI is observed when the extrusion temperature and moisture content is more and the screw speed is low. Starch polymers are sensitive to higher screw speeds as it affects its gelatinization ability and water-binding potential, making them more susceptible to breaking at higher screw speeds [96].

The water solubility index (WSI) determines the amount of polysaccharides or soluble components released from the granule during extrusion process. The parameters like elevated temperature and higher screw speeds increase water solubility index as these conditions lead to larger degree of dextrinization [80]. Also, Higher feed moisture (for its plasticizing effect) and fat content decreases starch break-down [83]. The changes in WAI can be achieved by partial addition of legume fiber. The fiber-rich extrusions tend to absorb more water and retain it in well-established polysaccharide-protein network. Upon addition of 30% germinated chickpeas and tomato residue to corn extruder with extrusion parameters of 16% moisture content, 180°C temperature and screw speed of 250 rpm, the water absorption index increased and water solubility index reduced [89]. The other study was conducted using brown rice flour and pinto beans, where the contradictory results were reported. Stating that the WAI decreased with an increase in WAI upon increasing legume or bean composition [45]. The reduction in water absorption index was attributed to increased interactions between hydrophilic groups of starch and protein caused by the addition of legume flour, which inhibited the water binding potential of starch granules.

4.5.3 SENSORY ATTRIBUTES

The biggest challenge at an industrial scale is ensuring the quality of grain-based extrusion product that is partially or completely replaced with legumes during its development phase. It is essential to maintain similar characteristics such as color, texture, taste, flavor, and aroma in the modified product when compared to the original product. The product firmness, crispy and sticky nature are important market factors. The color variations can be attributed to legume flour at first, and then to the heat-induced Maillard reaction, which decreases the lysine concentration as previously stated [97]. A study conducted by Philipp et al. [94] reported that with the increase of Pea protein isolate (PPI) content in rice flour resulted in the higher pea flavor, firmer and crispier texture, lesser uniform shape and size, and darker color. The changes in the textural aspects can be assessed with a texture analyzer by microstructural analysis or sensory panel [94]. The usual hallmark of an extruded snack is a higher ER and stiffness [83].

The most studied parameters for indicating a sound relationship between human perception and instrumental analysis are hardness and bulk density (BD) that are related to brittle and crispy nature of the extruded product. In several studies, acceptability testing was performed for instance and the legume cracker snacks have recorded the higher level of acceptance than the traditional commercial corn-based snacks [64]. Nonetheless, the decline in overall acceptability of maize extrusions with additional legume flour (except at 30%) can be achieved provided all the process parameters are properly set and monitored. According to a study conducted by Shah et al. [90], supplementing cornmeal with soybean and chickpea flour resulted in a less puffed product that is less palatable to customers, especially at higher supplementation levels of 40/100 g [90]. Rathod and Annapure [64] produced extruded lentils with acceptable sensory acceptance at 18% moisture content and 200 rpm screw speed. The cowpea extrusion with a moisture content of 16% and a die temperature of 180°C at 200 rpm screw speed was found to be satisfactory, as validated by sensory assessment and statistical analysis [64]. The extended snack food that are made from natural or pre-treated protein isolates and legume concentrates like lentils, faba beans as a single component or in the combination with protein rich pseudo-cereals showed satisfying texture and sensory characteristics [98]. Consumers enjoy eating snacks produced from legumes using chickpeas and green peas in varied amounts of flour and bran. The fortified puff snacks made from beans are said to be crunchy, crispy, and nutritious [99]. Researchers were able to alter extrusion parameters in these trials to achieve the necessary quality attributes of extrusion with promising results. Though more work is required for developing new products on the pilot scale. Research activities like these will make way for the creation of a new line of extruded snack food products that themselves function as functional foods. Important points to be considered in the production of legume based extrudates are mentioned in the Figure 4.2.

4.6 PACKAGING AND FOOD SAFETY STANDARDS FOR LEGUME BASED RTE SNACKS

RTE stands for Ready-to-eat food which is intended for direct consumption without the need for cooking or other processing. As soon as the packaging is opened, it should be ready to be consumed in delicious and appetizing way. For an extruded snack food, basic packaging requirement include; oxygen barrier properties to prevent rancidity of fats and oils, shielding from loss of crispiness, packaging strength and integrity of seal [100]. The extruded snacks quickly absorb moisture due to their hygroscopic nature even in relatively less humid conditions causing the loss of crispiness and eventually making it less desirable to customers. In order to select a suitable packaging material, some of the most important factors to be considered are; water

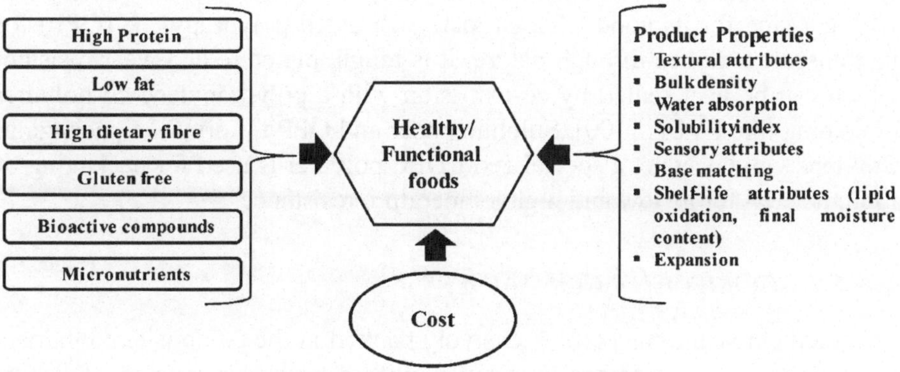

FIGURE 4.2 Nutritional highlights and important elements for legumes in extruded snacks.

vapor transmission rate (WVTR), oxygen transmission rate (OTR), food product characteristics such as; the type of food and its composition, form and shape of the product and the nature of the food. Factors that contribute to the deterioration of foods include; biological and microbial contamination, chemical degradation by oxidation, hydrolysis or enzymatic reactions. The food processing parameters such as enzymatic reactions, processing temperature and the environmental factors like sunlight, relative humidity, and temperature, must be taken into consideration. The usage or consumption time and shelf-life of a particular type of food product also plays a significant role in the selection of a packaging material [100].

4.6.1 PACKAGING MATERIALS FOR RTE SNACK FOODS

4.6.1.1 FLEXIBLE PLASTICS

If we look at the products in the market today, almost all kinds of snacks are packed in flexible bags. In the Indian market, a wide variety of flexible materials are used for snacks and waffles packaging that are branded and sometimes non-branded. Most of the short shelf-life and non-branded snacks are packed in low-density polyethylene (LDPE) and polypropylene pouches. The branded snacks and nuts are packed in films that have animated and printed material. Some of the extensively used materials include; Biaxially oriented polypropylene (BOPP) which has an ideal printing surface, Metallized polyester (PET met) for its great vapor barrier property, Low density polyethylene (LDPE) for its water and acid resistant property, Aluminum foil for its heat bearing capacity and Low-density polyethylene or casted polypropylene for their durability and strength are often used as a sealant layer. The packaging material used for crisps, chips and frites include; Polyvinylidene chloride (PVDC) coated with glassine for its good OTR property, Oriented polypropylene (OPP) for its transparent or see-through nature, it is tough, puncture and wear resistant which can be heat sealed by co-extrusion with a polypropylene co-polymer or coating like PVDC [100]. Sulphite paper and OPP for dry food packaging. Ethylene vinyl acetate, a specialized LDPE polymer is used for packaging of food material for its low and high temperature resistance.

4.6.1.2 COMPOSITE CONTAINERS

Extruded chips and nuts are extensively packed in the composite containers or canisters. The container is circular, with a body constructed of paper-polyethylene or plastic-aluminum, and the top pull tab is made up of aluminum or re-sealable plastic. The canister base is solidly molded, but the lid has a membrane which is pierced to consume the food product [100]. A novel technique is designed to prepare composite containers; the outer body is made up of light-weight high impact resistant polystyrene of the thickness ranging from 0.6 to 1.2 mm coated with a plastic film and the inner part of the body is made of aluminum or the combination of plastic and aluminum foil to have optimal barrier characteristics against oxygen and moisture. A wide range of options such as PVDC or Ethylene vinyl alcohol (EVOH), are available in film forms. In order to provide good barrier qualities, the base, the lid, and the re-closable caps are usually made up of high impact polystyrene (HIPS) covered with polyester [100, 101].

4.6.1.3 TINPLATE CONTAINERS

Tinplate containers are made up of steel with a thin coating of tin for corrosion protection. Often, a layer of white enamel is coated to some food containers to protect the contents and a re-closable plastic or tin ring is provided for easy access. The famous roasted and salted nuts are packed in laminated, rigid, and circular tin-plate containers filled with an inert gas such as nitrogen for increased shelf life [102].

4.6.2 FOOD SAFETY REGULATIONS AS PER BUREAU OF INDIAN STANDARDS (BIS) AND FOOD SAFETY AND STANDARDS AUTHORITY OF INDIA (FSSAI)

As discussed above, the ready-to-eat snacks are prepared by a process called extrusion cooking where the raw food material is subjected to high temperature and a short duration cooking procedure, making the product of plastic consistency. As the food product is extruded and passed through a die, it passes from a high-pressure zone to a lower one giving it a puffed texture. Hence, proper hygiene must be practiced to avoid contamination of the food to keep the diseases at bay. BIS states that the shape and size of the extruded product should be relatively uniform, must be completely cooked, and free from all kinds of contamination.

4.6.2.1 QUALITY REQUIREMENTS OF THE INGREDIENTS USED

The ingredients such as cereals and pulses should be de-husked; wheat flour, semolina, edible oils, salt, ghee, spices, garnishes, and condiments to be used, should comply with the MS numbers specified in the document. The RTE extruded snacks should give pleasant aroma and taste to the consumers and should not have rancid, bitter, and foul smell instead; they are expected to have an aroma of the spices used in their formulation. The characteristics of RTE snacks should not surpass the microbiological limits prescribed by the BIS.

The moisture content should not exceed the limit of 6% m/m, fat on dry basis to be below 25% m/m, peroxide value of not more than 10, total-aflatoxin content to be below 10 ppb and the total-fumonisin content which is restricted to 2 ppm. These specified limits are checked using Annex B and C, AOAC and ISI 16050 test methods as to not compromise on the quality of the food products. Similarly, the type of microbe and the count of the same is also mentioned in the document; which states that the total viable count to not exceed 10 CFU/g of the product, *Salmonella, E. coli, Shigella, Coliform* and *Staphylococcus aureus* should be absent, whereas the permeable limit for molds and yeast per gram to be less than 50. The concentration of microbial growth is checked using different ISO prescribed tests such as ISO 4833, ISO 21567 and ISO 6888, respectively [103].

4.6.2.2 REGULATIONS ON PACKAGING AND LABELING

RTE extruded snacks should be packed in food grade containers or packages that shield their hygiene and organoleptic qualities. The packaging material should not leech toxins and unwanted substances altering the actual flavor and aroma of the food products. Also, it should be sealed properly to keep the product intact and fresh. The label should compulsory be printed in either English or Hindi with a state spoken dialect for the convenience of consumers, being prominent, crystal clear and readable. It should not contain any information that is untrue, deceptive, or misleading in anyway. The vital nutritional information per 100 g or serving of the product such as, energy value in Kcal; protein, carbohydrates and fats in grams or milliliter should be mentioned clearly. If the product is vegetarian or a non-vegetarian, a proper diagrammatic representation or a logo is required to be printed. Food additives must be mentioned with their respective class or categories and if food colors are added, then a statement quoting, "Contains permitted colors," is expected to be mentioned below the ingredients' list. Similarly, a statement saying, "Contains added flavors," is printed if such external flavoring agents are used. The label should and must have the name and the complete address of the manufacturer, with net quantity of the product; exclusive of the weight of the packaging material, batch code, place of origin, date of manufacture or packing with an expiry date or expected shelf life and detailed instructions of storage and usage [91, 101, 102].

4.7 FUTURE OF LEGUME-BASED EXTRUDED SNACKS

Agriculture began more than 1,000 years ago and since then legumes have been an irreplaceable part of the human diet. Legumes are recognized globally as the most nutritious yet inexpensive food source making them one of the staple foods consumed by the humans. They are a great source of micro and macro nutrients which are no less than miracle foods. Because of the excellent nutritional value, these protein rich foods are replacing almost all kinds of snacks that are binge eaten carrying a baggage of guilt. The fast-paced environment demands ready-to-eat foods with no compromise on taste and quality. This is exactly when the extrusion technology to produce RTE comes into picture.

Already existing RTE snacks are too high in sugars, salt and fat content being regarded as calorie dense foods with low to nil nutritional value. However, this issue is being addressed by replacing conventional ingredients with protein and fiber rich substitutes called legumes, which will change the game of RTE snacks forever by giving a relief to consumers from digestive issues like hyperacidity, bloating and being light on their pockets. According to a survey report, the global market for extruded snacks in the year 2019, was estimated to be around US $ 48.3 billion, accounting for the compound annual growth rate (CAGR) of 4.4%. It is said that the production of legume-flour will reach a whopping 2.5 million tons globally by the end of 2025. Among the legumes, chickpea flour has contributed a remarkable sales share of 25.1% in the European market in the year 2018 and a growth rate of 5.2% is expected due to a shooting demand in health alternatives to extruded RTE snacks [102, 103]. As discussed above, incorporation of legume-flour is a wise choice but a wide range of research activities are necessary to make these healthy alternatives equally appealing when compared to already existing RTE snacks, in terms of esthetics, aroma and shelf-life.

4.8 CONCLUSION

This review paper was able to discuss the benefits of using legumes to prepare extruded snacks keeping in view the nutritional value these super foods possess, basics of extrusion technology, optimization of the process and simultaneously throwing a light on the properties of legume based RTE snacks, the kind of packaging along with the food safety standards to be followed as laid down by FSSAI and BIS. As discussed so far, the legumes are undoubtedly one of the notable alternatives to produce healthy extruded RTE snacks because of their impeccable nutritional value. Legume-based extruded RTE snacks will minimize the dependence of humans on animal-source foods for their nutritional fulfillment that will eventually help bring harmony into the ecosystem. Vast laboratory studies indicate that the addition of legume flour will enhance the sensory properties and the quality of the extrudates. Having said that, further research work is necessary to build consumer desirable food products that will keep the humankind healthy and fit for the fast-paced world of developments and prosperity. This review will assist the food and agro researchers to study and understand functional properties of different grains and pulses to create food products that will be sustainable, appetizing, and harmless to our mother nature.

ACKNOWLEDGEMENT

This endeavors would have not been possible without the constant support, access to facilities and guidance from resourceful people from the Annasaheb Dange college of engineering & technology, Astha-India, the department of biotechnology at Gowdara Mallikarjunappa institute of technology, Davangere-India, the department of polymer science and technology, Sri jayachamarajendra college of engineering (SJCE), JSSSTU, Mysore-India and the Centre for incubation, innovation, research and consultancy (CIIRC), Jyothy institute of technology, Bengaluru-India.

KEYWORDS

- **extrusion cooking**
- **food safety**
- **lentils**
- **nutrition**
- **packaging**

REFERENCES

1. Pasqualone, A., Costantini, M., Coldea, T. E., & Summo, C. (2020). Use of legumes in extrusion cooking: A review. *Foods, 9*(7), 958. https://doi.org/10.3390/foods9070958.
2. Moscicki, L. (2011). *Extrusion-Cooking Techniques: Applications, Theory and Sustainability*. John Wiley & Sons.
3. Anton, A. A., Fulcher, R. G., & Arntfield, S. D. (2009). Physical and nutritional impact of fortification of corn starch-based extruded snacks with common bean (*Phaseolus vulgaris* L.) flour: Effects of bean addition and extrusion cooking. *Food Chemistry, 113*(4), 989–996. https://doi.org/10.1016/j.foodchem.2008.08.049.
4. Félix-Medina, J. V., Montes-Ávila, J., Reyes-Moreno, C., Perales-Sánchez, J. X., Gómez-Favela, M. A., Aguilar-Palazuelos, E., & Gutiérrez-Dorado, R. (2020). Second-generation snacks with high nutritional and antioxidant value produced by an optimized extrusion process from corn/common bean flours mixtures. *LWT Food Science & Technology, 124*, 109172. https://doi.org/10.1016/j.lwt.2020.109172.
5. Shah, F. U., Sharif, M. K., Bashir, S., & Ahsan, F. (2019). Role of healthy extruded snacks to mitigate malnutrition. *Food Reviews International, 35*(4), 299–323. https://doi.org/10.1080/87559129.2019.1628375.
6. Tharanathan, R. N., & Mahadevamma, S. (2003). Grain legumes—A boon to human nutrition. *Trends in Food Science & Technology, 14*(12), 507–518. https://doi.org/10.1016/j.tifs.2003.08.005.
7. Ray, H., Bett, K., Tar'an, B., Vandenberg, A., Thavarajah, D., & Warkentin, T. (2014). Mineral micronutrient content of cultivars of field pea, chickpea, common bean, and lentil grown in Saskatchewan. *Canadian Journal of Plant Science, 94*(4), 1698–1708. https://doi.org/10.4141/cjps2013-311.
8. Saadat, S., Akhtar, S., Ismail, T., Sharif, M. K., Shabbir, U., Ahmad, N., & Ali, A. (2020). Multilegume bar prepared from extruded legumes flour to address protein energy malnutrition. *Italian Journal of Food Science, 32*(1). https://doi.org/10.15586/ijfs.v32i1.1783.
9. García-Segovia, P., Igual, M., Noguerol, A. T., & Martínez-Monzó, J. (2020). Use of insects and pea powder as alternative protein and mineral sources in extruded snacks. *European Food Research and Technology, 246*(4), 703–712. https://doi.org/10.1007/s11483-020-09678-7.
10. Sharma, R., Sharma, S., Makroo, H. A., & Dar, B. N. (2022). Role of pulses to modulate the nutritive, bioactive and technological functionality of cereal-based extruded snacks: A review. *International Journal of Food Science & Technology, 57*(7), 3882–3891. https://doi.org/10.1111/ijfs.16147.
11. Wójtowicz, A., Oniszczuk, A., Kasprzak, K., Olech, M., Mitrus, M., & Oniszczuk, T. (2020). Chemical composition and selected quality characteristics of new types of precooked wheat and spelt pasta products. *Food Chemistry, 309*, 125673. https://doi.org/10.1016/j.foodchem.2019.125673.
12. Patil, S. S., Brennan, M. A., Mason, S. L., & Brennan, C. S. (2016). The effects of fortification of legumes and extrusion on the protein digestibility of wheat-based snack. *Foods, 5*(2), 26. https://doi.org/10.3390/foods5020026.
13. Riaz, M. N. (2000). *Extruders in Food Applications*. CRC Press.

14. Koksel, F., & Masatcioglu, M. T. (2018). Physical properties of puffed yellow pea snacks produced by nitrogen gas assisted extrusion cooking. *LWT Food Science & Technology, 93*, 592–598. https://doi.org/10.1016/j.lwt.2018.03.074.
15. Chan, E., Masatcioglu, T. M., & Koksel, F. (2019). Effects of different blowing agents on physical properties of extruded puffed snacks made from yellow pea and red lentil flours. *Journal of Food Processing Engineering, 42*(3), e12989. https://doi.org/10.1111/jfpe.12989.
16. Frohlich, P., Boux, G., & Malcolmson, L. (2014). Pulse ingredients as healthier options in extruded products. *Cereal Foods World, 59*(3), 120–125. https://doi.org/10.1094/CFW-59-3-0120.
17. Abu-Ghannam, N., & Gowen, A. (2021). Pulse-based food products. In *Pulse Foods* (pp. 369–391). Academic Press.
18. Pelembe, L. A., Erasmus, C., & Taylor, J. R. (2002). Development of a protein-rich composite sorghum–Cowpea instant porridge by extrusion cooking process. *LWT Food Science & Technology, 35*(2), 120–127. https://doi.org/10.1016/S0023-6438(01)00129-5.
19. Ghumman, A., Kaur, A., Singh, N., & Singh, B. (2016). Effect of feed moisture and extrusion temperature on protein digestibility and extrusion behavior of lentil and horse gram. *LWT Food Science & Technology, 70*, 349–357. https://doi.org/10.1016/j.lwt.2016.03.036.
20. Tas, A. A., & Shah, A. U. (2021). The replacement of cereals by legumes in extruded snack foods: Science, technology and challenges. *Trends in Food Science & Technology, 116*, 701–711. https://doi.org/10.1016/j.tifs.2021.08.006.
21. Beacom, E., Bogue, J., & Repar, L. (2021). Market-oriented development of plant-based food and beverage products: A usage segmentation approach. *Journal of Food Products Marketing, 27*(4), 204–222. https://doi.org/10.1080/10454446.2021.1939782.
22. Kumar, R. A. (2020). *Strategy for Using Sensory Analysis for Category Appraisal to Develop New and Improved Products*. Kansas State University.
23. Escobedo, A., & Mojica, L. (2021). Pulse-based snacks as functional foods: Processing challenges and biological potential. *Comprehensive Reviews in Food Science and Food Safety, 20*(5), 4678–4702. https://doi.org/10.1111/crf2.12892.
24. Boukid, F., Zannini, E., Carini, E., & Vittadini, E. (2019). Pulses for bread fortification: A necessity or a choice? *Trends in Food Science & Technology, 88*, 416–428. https://doi.org/10.1016/j.tifs.2019.04.010.
25. Sá, A. G., Moreno, Y. M., & Carciofi, B. A. (2020). Plant proteins as high-quality nutritional sources for the human diet. *Trends in Food Science & Technology, 97*, 170–184. https://doi.org/10.1016/j.tifs.2020.01.005.
26. Marinangeli, C. P., Curran, J., Barr, S. I., Slavin, J., Puri, S., Swaminathan, S., Tapsell, L., & Patterson, C. A. (2017). Enhancing nutrition with pulses: Defining a recommended serving size for adults. *Nutrition Reviews, 75*(12), 990–1006. https://doi.org/10.1093/nutrit/nux045.
27. Marimuthu, M., & Gurumoorthi, P. (2013). Phytochemical screening and FT-IR studies on wild and common South Indian legumes. *Asian Journal of Pharmaceutical and Clinical Research, 6*(2), 141–144. https://doi.org/10.22159/ajpcr.2013.v6i2.6002.
28. Kumar, S., & Pandey, G. (2020). Biofortification of pulses and legumes to enhance nutrition. *Heliyon, 6*(3), e03682. https://doi.org/10.1016/j.heliyon.2020.e03682.
29. Kocira, A., Kocira, S., Bronowicka-Mielniczuk, U., Kornas, R., & Kozłowicz, K. (2017). Foliar application of bio stimulants and the antioxidant properties of soybean

seeds. In *Farm Machinery and Processes Management in Sustainable Agriculture* (pp. 164–169). International Scientific Symposium.

30. Daubech, B., Remigi, P., Doin de Moura, G., Marchetti, M., Pouzet, C., Auriac, M. C., Gokhale, C. S., Masson-Boivin, C., & Capela, D. (2017). Spatio-temporal control of mutualism in legumes helps spread symbiotic nitrogen fixation. *eLife, 6*, e28683. https://doi.org/10.7554/eLife.28683.
31. Ma, X., & Jiao, S. (2020). A study on the comprehensive utilization of water resources from the perspective of water footprint and water ecological footprint: A case study of Anyang City, China.
32. Singh, S., Gamlath, S., & Wakeling, L. (2007). Nutritional aspects of food extrusion: A review. *International Journal of Food Science & Technology, 42*(8), 916–929. https://doi.org/10.1111/j.1365-2621.2006.01370.x.
33. Naveena, B., & Singh, M. (2022). Effects of extrusion parameters on physical & textural properties of sorghum, barley, and chickpea-based composite extruded products. *Journal of Food Processing & Preservation, 46*(9), e16564. https://doi.org/10.1111/jfpp.16564.
34. Gulati, P., Brahma, S., & Rose, D. J. (2020). Impacts of extrusion processing on nutritional components in cereals and legumes: Carbohydrates, proteins, lipids, vitamins, and minerals. In *Extrusion Cooking* (pp. 415–443). Woodhead Publishing.
35. Karun, G., Sukumar, A., Nagamaniammai, G., & Preetha, R. (2022). Development of multigrain ready-to-eat extruded snack and process parameter optimization using response surface methodology. *Journal of Food Science and Technology, 5*, 1–1.
36. Shelar, G. A., & Gaikwad, S. T. (2019). Extrusion in food processing: An overview. *Journal of Pharmaceutical Innovation, 8*(2), 562–568. https://doi.org/10.1007/s12247-019-09451-5.
37. Jozinović, A., Šubarić, D., Ačkar, Đ., Babić, J., Orkić, V., Guberac, S., & Miličević, B. (2021). Food industry by-products as raw materials in the production of value-added corn snack products. *Foods, 10*(5), 946. https://doi.org/10.3390/foods10050946.
38. Nosworthy, M. G., Medina, G., Franczyk, A. J., Neufeld, J., Appah, P., Utioh, A., Frohlich, P., & House, J. D. (2018). Effect of processing on the *in vitro* and *in vivo* protein quality of beans (*Phaseolus vulgaris* and *Vicia faba*). *Nutrients, 10*(6), 671. https://doi.org/10.3390/nu10060671.
39. Huang, R., Pan, X., Lv, J., Zhong, W., Yan, F., Duan, F., & Jia, L. (2018). Effects of explosion puffing on the nutritional composition and digestibility of grains. *International Journal of Food Properties, 21*(1), 2193–2204. https://doi.org/10.1080/10942912.2018.1477874.
40. Kaur, S., Sharma, S., Singh, B., & Dar, B. N. (2015). Effect of extrusion variables (temperature, moisture) on the antinutrient components of cereal brans. *Journal of Food Science and Technology, 52*, 1670–1676. https://doi.org/10.1007/s11494-013-1123-4.
41. Diouf, A., Sarr, F., Sene, B., Ndiaye, C., Fall, S. M., & Ayessou, N. C. (2019). Pathways for reducing anti-nutritional factors: Prospects for *Vigna unguiculata*. *Journal of Nutritional Health & Food Science*, *7*(2), 1–10.
42. Fapojuwo, O. O., Maga, J. A., & Jansen, G. R. (1987). Effect of extrusion cooking on *in vitro* protein digestibility of sorghum. *Journal of Food Science*, *52*(1), 218–219.
43. Senthil, A., Ravi, R., Bhat, K. K., & Seethalakshmi, M. K. (2002). Studies on the quality of fried snacks based on blends of wheat flour and soya flour. *Journal of Food Quality*, *13*(5), 267–273.

44. Okoye, J. I., Nkwocha, A. C., & Ogbonnaya, A. E. (2008). Production, proximate composition and consumer acceptability of biscuits from wheat/soybean flour blends. *Continental Journal of Food Science and Technology*, *2*, 6–13.
45. Sumargo, F., Gulati, P., Weier, S. A., Clarke, J., & Rose, D. J. (2016). Effects of processing moisture on the physical properties and *in vitro* digestibility of starch and protein in extruded brown rice and pinto bean composite flours. *Food Chemistry*, *211*, 726–733.
46. Poutanen, K. S., Kårlund, A. O., Gómez-Gallego, C., Johansson, D. P., Scheers, N. M., Marklinder, I. M., Eriksen, A. K., Silventoinen, P. C., Nordlund, E., Sozer, N., & Hanhineva, K. J. (2022). Grains–A major source of sustainable protein for health. *Nutrition Reviews*, *80*(6), 1648–1663.
47. Maskan, M., & Altan, A. (2012). *Advances in Food Extrusion Technology*. Taylor and Francis Group.
48. Toews, R., & Wang, N. (2013). Physicochemical and functional properties of protein concentrate from pulses. *Food Research International*, *52*(2), 445–451.
49. Bajaj, R., Singh, N., Kaur, A., & Inouchi, N. (2018). Structural, morphological, functional and digestibility properties of starches from cereals, tubers and legumes: A comparative study. *International Journal of Food Science and Technology*, *55*, 3799–3808.
50. Toutounji, M. R., Farahnaky, A., Santhakumar, A. B., Oli, P., Butardo Jr, V. M., & Blanchard, C. L. (2019). Intrinsic and extrinsic factors affecting rice starch digestibility. *Trends in Food Science & Technology*, *88*, 10–22.
51. Ma, M., Wang, Y., Wang, M., Jane, J. L., & Du, S. K. (2017). Physicochemical properties and *in vitro* digestibility of legume starches. *Food Hydrocolloids*, *63*, 249–255.
52. McClure, J., Ahn-Jarvis, J., Wilde, P., Saibene, D., Linter, B., & Warren, F. (2020). The effects of extrusion processing on the digestibility and functionality of pulse starches. *Journal of the Academy of Nutrition and Dietetics*, *120*(9), A55.
53. Sharma, S., Singh, N., & Singh, B. (2015). Effect of extrusion on morphology, structural, functional properties and *in vitro* digestibility of corn, field pea and kidney bean starches. *Starch-Stärke*, *67*(9–10), 721–728.
54. Jeong, D., Han-Ja, A., Liu, Q., & Chung, H. J. (2019). Effect of processing, storage, and modification on *in vitro* starch digestion characteristics of food legumes: A review. *Food Hydrocolloids*, *90*, 367–376.
55. Delamare, G. Y., Butterworth, P. J., Ellis, P. R., Hill, S., Warren, F. J., & Edwards, C. H. (2020). Incorporation of a novel leguminous ingredient into savory biscuits reduces their starch digestibility: Implications for lowering the glycaemic index of cereal products. *Food Chemistry: X*, *5*, 100078.
56. Edwards, C. H., Ryden, P., Pinto, A. M., Van der Schoot, A., Stocchi, C., Perez-Moral, N., Butterworth, P. J., Bajka, B., Berry, S. E., Hill, S. E., & Ellis, P. R. (2020). Chemical, physical and glycaemic characterization of PulseON®: A novel legume cell-powder ingredient for use in the design of functional foods. *Journal of Functional Foods*, *68*, 103918.
57. De Pilli, T., Legrand, J., Derossi, A., & Severini, C. (2015). Effect of proteins on the formation of starch-lipid complexes during extrusion cooking of wheat flour with the addition of oleic acid. *International Journal of Food Science and Technology*, *50*(2), 515–521.
58. Thachil, M. T., Chouksey, M. K., & Gudipati, V. (2014). Amylose-lipid complex formation during extrusion cooking: Effect of added lipid type and amylose level on

corn-based puffed snacks. *International Journal of Food Science and Technology*, *49*(2), 309–316.

59. Hall, C., Hillen, C., & Garden Robinson, J. (2017). Composition, nutritional value, and health benefits of pulses. *Cereal Chemistry*, *94*(1), 11–31.
60. Conti, M. V., Guzzetti, L., Panzeri, D., De Giuseppe, R., Coccetti, P., Labra, M., & Cena, H. (2021). Bioactive compounds in legumes: Implications for sustainable nutrition and health in the elderly population. *Trends in Food Science & Technology*, *117*, 139–147.
61. Yadav, N., Kaur, D., Malaviya, R., Saini, P., & Anjum, S. (2019). Enhancement in mineral bioavailability of extruded pulses with reduced antinutrients. *British Food Journal*, *121*(11), 2967–2978.
62. Gulati, P., & Rose, D. J. (2018). Effect of extrusion on folic acid concentration and mineral element dialyzability in Great Northern beans (*Phaseolus vulgaris* L.). *Food Chemistry*, *269*, 118–124.
63. Moreno-Valdespino, C. A., Luna-Vital, D., Camacho-Ruiz, R. M., & Mojica, L. (2020). Bioactive proteins and phytochemicals from legumes: Mechanisms of action preventing obesity and type-2 diabetes. *Food Research International*, *130*, 108905.
64. Rathod, R. P., & Annapure, U. S. (2016). Effect of extrusion process on antinutritional factors and protein and starch digestibility of lentil splits. *LWT–Food Science and Technology*, *66*, 114–123.
65. Arribas, C., Cabellos, B., Cuadrado, C., Guillamón, E., & Pedrosa, M. M. (2019). The effect of extrusion on the bioactive compounds and antioxidant capacity of novel gluten-free expanded products based on carob fruit, pea and rice blends. *Innovative Food Science & Emerging Technologies*, *52*, 100–107.
66. Ciudad-Mulero, M., Fernández-Ruiz, V., Cuadrado, C., Arribas, C., Pedrosa, M. M., Berrios, J. D., Pan, J., & Morales, P. (2020). Novel gluten-free formulations from lentil flours and nutritional yeast: Evaluation of extrusion effect on phytochemicals and non-nutritional factors. *Food Chemistry*, *315*, 126175.
67. Singh, N. (2017). Pulses: An overview. *Journal of Food Science and Technology*, *54*, 853–857.
68. Kamalasundari, S., Babu, R., & Umamaheswari, T. (2019). Effect of domestic processing methods on anti-nutritional factors and its impact on the bioavailability of proteins and starch in commonly consumed whole legumes. *Journal of Dairy Foods & Home Science*, *38*(1), 67–72.
69. Costantini, M., Sabovics, M., Galoburda, R., Kince, T., Straumite, E., Summo, C., & Pasqualone, A. (2021). Effect of die configuration on the physico-chemical properties, anti-nutritional compounds, and sensory features of legume-based extruded snacks. *Foods*, *10*(12), 3015.
70. Tiwari, A., & Jha, S. K. (2017). Extrusion cooking technology: Principal mechanism and effect on direct expanded snacks—An overview. *International Journal of Food Studies*, *6*(1).
71. Bouvier, J. M., & Campanella, O. H. (2014). *Extrusion Processing Technology: Food and Non-Food Biomaterials*. John Wiley & Sons.
72. Kristiawan, M., Della Valle, G., & Berzin, F. (2022). Extrusion simulation for the design of cereal and legume foods. *Foods*, *11*(12), 1780.
73. Patil, S. S., & Kaur, C. (2018). Current trends in extrusion: Development of functional foods and novel ingredients. *Food Science and Technology Research*, *24*(1), 23–34.

74. Baskar, G., & Aiswarya, R. (2016). Role of extrusion technology in food processing and its effect on nutritional values. *International Journal of Modern Trends in Science and Technology*, *1*(1), 1–4.
75. Offiah, V., Kontogiorgos, V., & Falade, K. O. (2019). Extrusion processing of raw food materials and by-products: A review. *Critical Reviews in Food Science and Nutrition*, *59*(18), 2979–2998.
76. Smith, J., & Hardacre, A. (2011). Development of an extruded snack product from the legume *Vicia faba* minor. *Procedia Food Science*, *1*, 1573–1580.
77. Aprodu, I., Vasilean, I., Muntenită, C., & Patrascu, L. (2019). Impact of broad beans addition on rheological and thermal properties of wheat flour-based sourdoughs. *Food Chemistry*, *293*, 520–528.
78. Yousouf, M., Hussain, S. Z., Kanojia, V., Qadri, T., Naseer, B., Shafi, F., & Jabeen, A. (2023). Development of third generation protein-rich snacks from lentil and egg powder through microwave-assisted extrusion cooking. *Nutrition and Food Science*, *53*(2), 285–300.
79. Koksel, F., & Masatcioglu, M. T. (2018). Physical properties of puffed yellow pea snacks produced by nitrogen gas-assisted extrusion cooking. *LWT–Food Science and Technology*, *93*, 592–598.
80. Wani, S. A., & Kumar, P. (2016). Development and parameter optimization of health-promising extrudate based on fenugreek, oat, and pea. *Food Bioscience*, *14*, 34–40.
81. Narayana, P. K., Bueno, E., Baur, A., Ahmed, S., & Von Wettberg, E. J. (2022). Fenugreek, A legume spice and multiuse crop adapted to a changing climate. In *Developing Climate Resilient Grain and Forage Legumes* (pp. 105–123). Singapore: Springer Nature Singapore.
82. Sidhu, J. S., Zafar, T., Benyathiar, P., & Nasir, M. (2022). Production, processing, and nutritional profile of chickpeas and lentils. In *Dry Beans and Pulses: Production, Processing, and Nutrition* (pp. 383–407).
83. Sharma, C., Singh, B., Hussain, S. Z., & Sharma, S. (2017). Investigation of process and product parameters for physicochemical properties of rice and mung bean (*Vigna radiata*) flour-based extruded snacks. *International Journal of Food Science and Technology*, *54*, 1711–1720.
84. Rico, D., Cano, A. B., & Martín-Diana, A. B. (2021). Pulse-cereal blend extrusion for improving the antioxidant properties of a gluten-free flour. *Molecules*, *26*(18), 5578.
85. Gandhi, N., Kaur, K., Kaur, A., Singh, B., & Singh, P. (2016). Technological innovations in third generation snack foods. *Progressive Research: An International Journal*, *11*, 4948–4959.
86. Ye, J., Hu, X., Luo, S., Liu, W., Chen, J., Zeng, Z., & Liu, C. (2018). Properties of starch after extrusion: A review. *Starch-Stärke*, *70*(11–12), 1700110.
87. Brennan, M. A., Derbyshire, E., Tiwari, B. K., & Brennan, C. S. (2013). Ready-to-eat snack products: The role of extrusion technology in developing consumer acceptable and nutritious snacks. *International Journal of Food Science and Technology*, *48*(5), 893–902.
88. Jozinović, A., Šubarić, D., Ačkar, Đ., Babić, J., & Miličević, B. (2016). Influence of spelt flour addition on properties of extruded products based on corn grits. *Journal of Food Engineering*, *172*, 31–37.
89. Hegazy, H. S., El-Bedawey, A. E., Rahma, E. H., & Gaafar, A. M. (2017). Effect of extrusion process on nutritional, functional properties and antioxidant activity of

germinated chickpea incorporated corn extrudates. *American Journal of Food Science and Nutrition Research*, *4*(1), 59–66.
90. Shah, F. U., Sharif, M. K., Butt, M. S., & Shahid, M. (2017). Development of protein, dietary fiber, and micronutrient enriched extruded corn snacks. *Journal of Texture Studies*, *48*(3), 221–230.
91. Rathore, B., Krishnappa, P., & Mehta, S. U. (2016). Legal analysis of information displayed on dental material packages: An exploratory research. *Journal of the Indian Association of Public Health Dentistry*, *14*(2), 224.
92. Rangira, I., Gu, B. J., Ek, P., & Ganjyal, G. M. (2020). Pea starch exhibits good expansion characteristics under relatively lower temperatures during extrusion cooking. *Journal of Food Science*, *85*(10), 3333–3344.
93. Guldiken, B., Yovchev, A., Nosworthy, M. G., Stone, A. K., House, J. D., Hood-Niefer, S., & Nickerson, M. T. (2020). Effect of extrusion conditions on the physical properties of desi chickpea-barley extrudates and quality attributes of their resulting flours. *Journal of Texture Studies*, *51*(2), 300–307.
94. Philipp, C., Oey, I., Silcock, P., Beck, S. M., & Buckow, R. (2017). Impact of protein content on physical and microstructural properties of extruded rice starch-pea protein snacks. *Journal of Food Engineering*, *212*, 165–173.
95. Gu, B. J., Masli, M. D., & Ganjyal, G. M. (2020). Whole faba bean flour exhibits unique expansion characteristics relative to the whole flours of lima, pinto, and red kidney beans during extrusion. *Journal of Food Science*, *85*(2), 404–413.
96. Yağcı, S., Altan, A., & Doğan, F. (2020). Effects of extrusion processing and gum content on physicochemical, microstructural and nutritional properties of fermented chickpea-based extrudates. *LWT–Food Science and Technology*, *124*, 109150.
97. Monnet, A. F., Laleg, K., Michon, C., & Micard, V. (2019). Legume enriched cereal products: A generic approach derived from material science to predict their structuring by the process and their final properties. *Trends in Food Science & Technology*, *86*, 131–143.
98. Martin, A., Schmidt, V., Osen, R., Bez, J., Ortner, E., & Mittermaier, S. (2022). Texture, sensory properties and functionality of extruded snacks from pulses and pseudocereal proteins. *Journal of the Science of Food and Agriculture*, *102*(12), 5011–5021.
99. Natabirwa, H., Nakimbugwe, D., Lung'aho, M., Tumwesigye, K. S., & Muyonga, J. H. (2020). Bean-based nutrient-enriched puffed snacks: Formulation design, functional evaluation, and optimization. *Food Science & Nutrition*, *8*(9), 4763–4772.
100. Rao, B. D., Kiranmai, E., Srenuja, D., & Tonapi, V. A. (2022). Holistic value chain approach in finger millet. In *The Finger Millet Genome* (pp. 277–296). Cham: Springer International Publishing.
101. Pande, R., Gavaravarapu, S. M., & Kulkarni, B. (2020). Front-of-pack nutrition labelling in India. *The Lancet Public Health*, *5*(4), e195.
102. Paul, A. A., Kumar, S., Kumar, V., & Sharma, R. (2020). Milk analog: Plant-based alternatives to conventional milk, production, potential and health concerns. *Critical Reviews in Food Science and Nutrition*, *60*(18), 3005–3023.
103. Rafiq, S. I., Garg, S., Balaji, G., Irshaan, S., Hikhmathunnisa, S. N., & Rafiq, S. M. (2020). Food quality and food safety assessment in the supply chain. In *Shelf Life and Food Safety* (pp. 77–92). CRC Press.

CHAPTER 5

RECENT ADVANCES IN FLAKING, PUFFING/PARCHING, POPPING, AND INSTANT TECHNOLOGIES FOR PRODUCTION OF READY-TO-EAT CEREALS

RUBEKA IDRISHI,[1] SIDDHARTHA SINGHA,[1] and LATHA RANGAN[2]

[1]*School of Agro and Rural Technology, Indian Institute of Technology, Guwahati, Assam, India*

[2]*Department of Biosciences and Bioengineering, Indian Institute of Technology, Guwahati, Assam, India*

ABSTRACT

Technologies like flaking, puffing/parching, and popping has been in use since centuries to process cereals such as rice, millets, corn, and other nutri-cereals to make them ready to eat/ quick snacking or breakfast options. There had been various technological interventions in the machinery and the process to make these technologies friendlier and less hectic and to reduce the time and labor. Also, there are many non-traditional cereals or grains and nutri-cereals which have not been tested with respect to these technologies, black rice being one of them. In this chapter various processing techniques had been discussed to understand the feasibility and applicability of these technologies to black rice processing. Various published literature had been visited with regard to the technologies used for the processing of different types of cereals with a focus on rice processing. The advancements in these technologies from time to time were found to be increasing but not drastic.

Ready-to-Eat Snacks: Emerging Technologies for Production and Safety.
Ravi Pandiselvam and Gülşah Çalışkan Koç

This chapter discusses the traditional dry heat application for RTE processing like flaking, puffing/parching and popping technologies and the recent advancements in the basic processes of producing flakes, puffs, popped cereals and instant technologies. The optimization parameters have also been discussed for each process with regard to develop ready to eat products from cereals, pseudo-cereals, and nutri-cereals like millets. Few parts of this chapter had been focused on the recent developments in these technologies utilizing the black rice, processing them into ready to eat products by the incorporation of the advanced technologies used for white rice. The findings of this study can be employed to develop low-cost machinery for existing traditional processes, black rice processing, weaning food formulations, RTE food/snacks, value added health products and convenient foods from other cereals by utilizing their properties and modernized methods.

5.1 INTRODUCTION

In order to produce ready to eat cereal based snacks, cereals have to go under common steps of primary processing which involves cleaning of grains, grading, sorting and milling [1, 2]. After this the secondary processing of the cereals decides their ultimate fate which primarily involves value addition in the already milled cereals. The primary production outcomes such as whole grains, grits, flour and fractions of flour is obtained to increase its usability and shelf life. These technologies make the cereals ready to eat (RTE) by expansion [3–5] and gelatinization processes. The secondary processes should be used as an opportunity towards improving the nutrient profile of the end product. A distinctive difference between puffing and popping is the use of the starting material, in puffing a pre-treated or alike material is pre-treated followed by roasting at high temperature whereas popping is employed on moistened whole kernel of a grain after some pre-treatment such as brining and drying [6].

5.2 TRADITIONAL METHODS OF FLAKING, PUFFING/PARCHING, POPPING OF CEREALS

5.2.1 FLAKING

Flaking process for cereals has been used since ages in order to make them more palatable and easier to cook. Flaking refers to crushing raw or partially cooked cereal pellets between rollers. As the pellets pass through the rollers, they become flat resulting into thin, wide and elongated shape that are said as flakes [7]. The flaking process differs for different kinds of raw material and thus the processing parameters like feed rate, fractions of the rollers surfaces, number of cycle or runs, pellet shape and size [7]. The traditional flaking process involves following steps depicted in Figure 5.1.

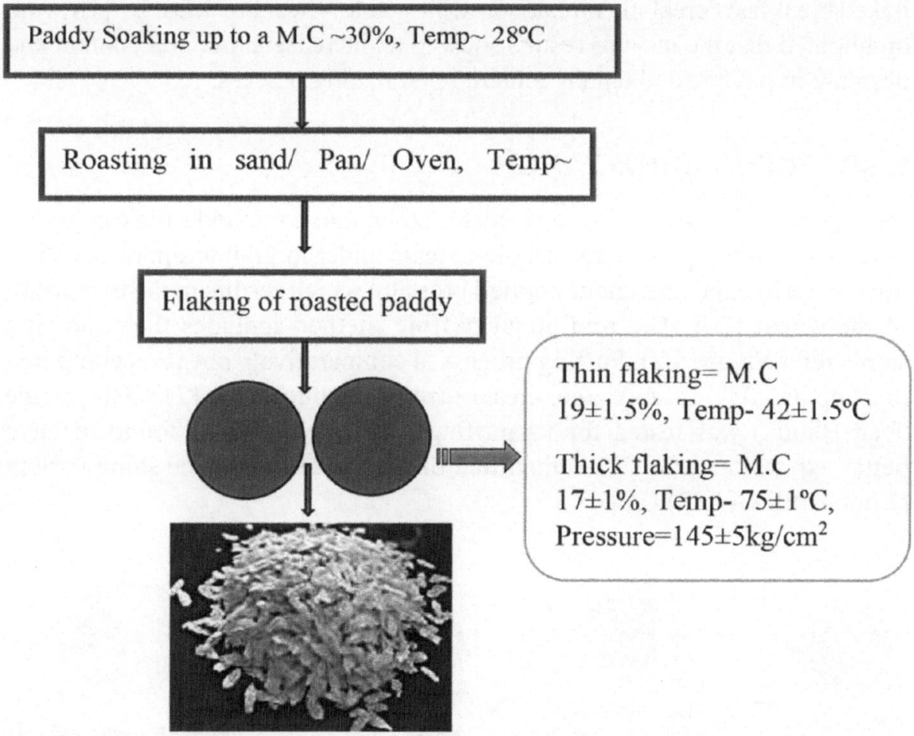

FIGURE 5.1 Traditional process of producing cereal flakes (thick and extra thin) the process suggested by [8–11].

Source: Adapted from Ref. [8-11]

Flaked products are widely popular for breakfast snacks. Flaked rice also known in India as Avalakki, Atakulu, Poha, and Chirwa [8], is a whole flattened rice grain product prepared from pre-roasted paddy [12] and is very popular among economically disadvantageous class of people. Beaten rice can be used to prepare various kind of crispy products combined with other ingredients like seasonings, fried herbs, oilseeds and pulses. Flakes made of black rice are not as popular as other cereals such as rice flakes, oat and rye flakes [13], corn flakes and millet flakes [14, 15], etc. Flaked rice is often called beaten rice as well. [9]used the traditional process suggested by [8, 16] with a little modification. Apart from these products, there has been various types of traditional rice flakes of different countries like Pinipig of Philippines, Simping of Malaysia, Hurum of Assam [17], India which is made from waxy rice varieties such as Kanhuli bora and Misri bora [10] which follows similar basic steps with little modifications. Three varieties of whole sorghum were used to make thin flaked breakfast cereal in comparison with whole wheat breakfast by [18] who highlighted that the in-vitro results showed an increase in mineral content and decrease in phytate and tannin content upon flaking process.

5.2.2 PUFFING/ PARCHING

Puffing of cereals is another low-cost method and a method of texturization [19] in which the pre-treated whole cereals undergo high temperature short time (HTST) exposure under applied pressure which results in the expansion of the cereal [20]. The traditional puffing method includes the following steps (refer Figure 5.2). Puffing process is comparatively cheap as compared to popping, flaking and other instantization technologies [21]. Black rice (Var. Bhutia) was tested for its puffing quality which was found to have better expansion ratio (7.17) than that of the white rice of the same variety (Bhutia-White-4.82) [22].

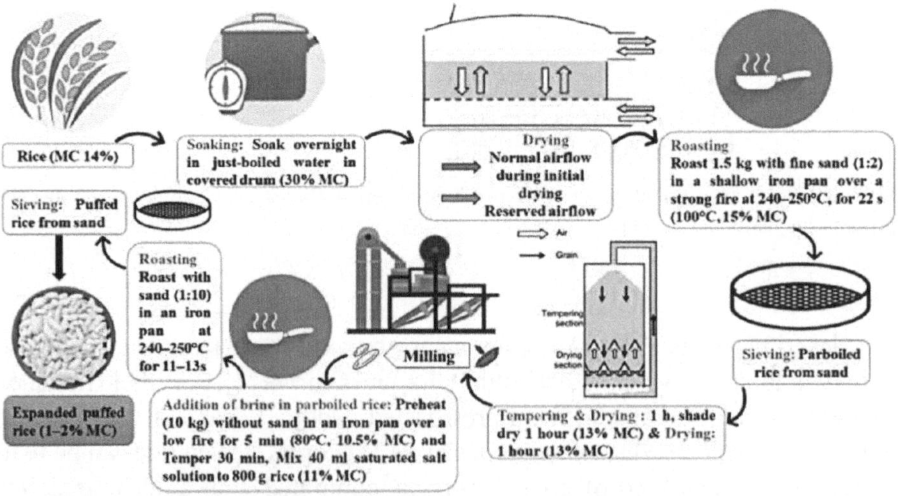

FIGURE 5.2 Traditional puffing technology for rice [9, 10, 22–24].

Source: Adapted from Ref. [9, 10, 22-24]

Black rice puffs were developed by [9] with the same traditional method depicted in Figure 5.2, The only difference was that the brine solution was of 10–12% sand roasted along with parboiled rice for about 15–20 seconds with continuous stirring and then the puffs were sieved to separate them from the sand. Four black rice varieties were taken namely Chakha (CH), Kalobhat (KB), Mamihunger (MA) and Manipuri Black (MN) along with a white rice variety called Swarna Sub–1(SW). It was found that the puffed and the beaten black rice products were highly liked by the age group of 21–35 and more by females. In India puffed products is a preferable snacking choice in low-income people and is a good traveling food option when tempered with some spices, groundnuts and curry leaves. Similarly, [25] developed puffs from three different non-colored/pigmented rice and highlighted that the quality of all three puffs were significantly similar because of the similar origin.

5.2.3 POPPING

Popping of rice is usually practiced in Asian subcontinents and largely specific to India. In India popped paddy is a popular snack and is associated with religious food practices as well. The traditional process includes direct heating the pretreated paddy of the cereal which results in the popping of the rice [10]. Figure 5.3 depicts the flowchart of traditional process of popping the rice. The pre-drying of paddy to 9% M.C (moisture content) is required to promote the lemma-palea interlocking mechanism whereas 14% M.C promotes popping. The pre-soaking step helps to loosen the adherence between husk and kernel.

The roasting step is common for almost all traditional process discussed above for flaking, parching/puffing and popping. The most common traditional ways of roasting are pan, oven and sand roasting [10]. Amongst these pan and sand method is widely popular in villages and semi urban areas. The sand is replaced partially or fully with different types of salt also to impart a little salty taste to the cereal. Sand roasting is widely practiced globally for making RTE puffed rice, popped rice, flaked rice, roasted maize, barley, oats, wheat, millets, and legumes [23].

5.3 PHYSICOCHEMICAL CHARACTERISTICS OF BLACK RICE WITH RESPECT TO VALUE ADDITION

Physicochemical properties of the grains affect the final popping and puffing qualities. The husk tightness and interlocking found to have a positive

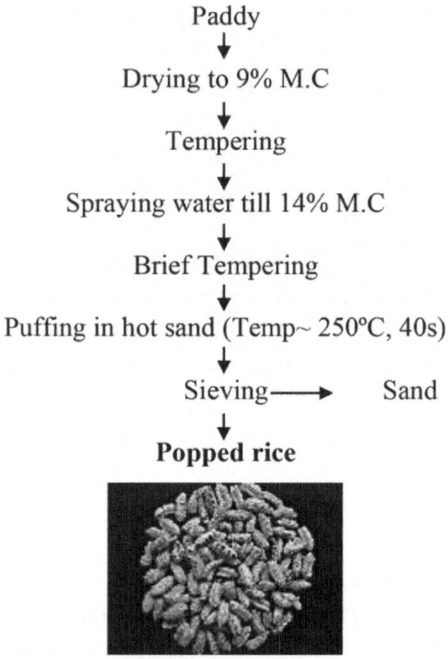

FIGURE 5.3 Traditional popping technology for rice.
Source: Adapted from Ref. [10, 23]

correlation in hot sand bed puffing, softness of the grain positively influence expansion ratio of the puffed product [6]. There are significant differences in the physical and chemical properties of black rice as compared to brown, red and white rice. Since black rice varieties lies more on the waxier side than that of chalky and mealy [26], they can serve as an ideal grain for RTE processing characteristics with a slight difference from other rice varieties. To preserve the outer layer of all colored rice polishing step is avoided and it is done till dehusking and separation [27], bran and germ are left intact [28]. Ponnappan, Thangavel, and Sahu, [1] compared milling and physical characteristics of white, black and red rice and found that varieties of red rice have better hulling strength, milling percentage, kernel recovery rate than that of black rice (Chettinadu origin) followed by white rice, higher the milling strength of black rice higher is the processing potential. The physical characteristics such as 1,000 kernel weight and volume were recorded the least for black rice as compared to red and white rice varieties, whereas black rice from Burma (Myanmar) and Manipur (India) was found to have a better physical parameter [29], physical characteristics such as 1,000 kernel weight, length/breadth ratio, etc., provides a predictable parameter of the final shape of the flaked/ puffed/ popped rice. Table 5.1 lists various physical parameters of black rice.

TABLE 5.1 Physical Characteristics of Black Rice

Varieties	1,000 Kernel Weight (g)	Grain Volume (ml)	Bulk Density (g/ml)	True Density (g/ml)	References
Chettinad black rice	17.50±0.04	21.20±0.03	0.96±0.01	1.92±0.03	[1]
Manipur Poreitonchakhao (deep purple pigment)	16.56±0.02	–	0.54±0.03	1.09±0.01	[30]
Manipur black rice Chakhaoamubi (deep red pigment)	22.32±0.00	–	0.53±0.01	1.03±0.01	[30]
Burma black rice	20.36±0.06	60.83±0.28	0.82 ± 0.003	–	[29]
Athikaraya (Karnataka pigmented rice)	26.00±0.10	–	0.50±0.01	0.90±0.01	[31]

Thermal degradation of phytochemicals of black rice was studied by [32] which highlighted that the heat liability of bioactive compounds such as anthocyanins, antioxidants, polyphenols and flavonoids greatly depends on the degree of sieving and grinding and out of all the components anthocyanins were more prone to thermal degradation which directly affects the antioxidant activity [33]. Processing black rice into different value-added products can affect the nutritional profile especially the anthocyanins, therefore choosing the processing method and customizing it to the type of black rice to be processed is very important in order to treat them in the gentlest way possible. [9] processed and studied four different types of black rice along with one white rice (CH, KB, MA, MN, SW) along with a white rice variety Swarna Sub–1 (SW) white rice into boiled rice, beaten rice/ flaked rice, popped rice, puffed rice, and raw rice. The results highlighted that CH, MN and KB retained anthocyanin best in flaking process after raw rice, followed by popping process, puffing and boiling.

5.4 RECENT ADVANCEMENTS

5.4.1 ADVANCEMENTS IN FLAKING TECHNOLOGY

Flaked cereals are quite versatile in nature and are enjoyed in almost every continent. There have been various advancements in the flaking process and the types of flaked ready to eat products and formulations. The roasting process is common to many methods. The advanced methods of roasting have been explained in a review by [34] which enlists microwave, superheated steam, infrared hot air, revtech roaster, and forced convection continuous tumble as advanced methods of roasting of flakes, puffs and popped cereals. These HTST methods of roasting provides a product with better shelf stability, functional property, less energy consumption, easy process control and results in increased consumer acceptance of the final product. In a study conducted by [10] few improvements in the traditional process of rice flaking (refer Figure 5.1) was suggested which included following steps (refer Figure 5.4).

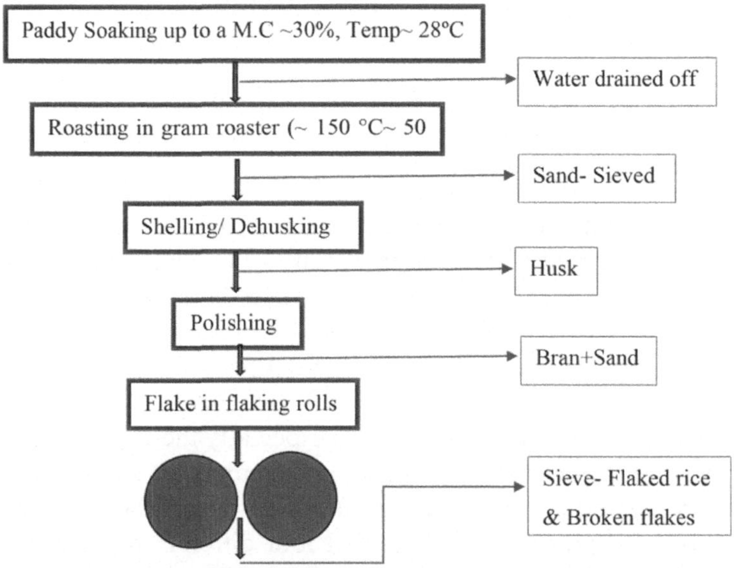

FIGURE 5.4 Advancement in traditional flaking technology for rice.

From section 5.4.1.1 to 5.4.1.4, the advancements in the type of flaking have been discussed.

5.4.1.1 CUT AND BAKED COMPOSITE FLAKE PRODUCTION

There had been very little advancement in the flaking process and specifically black rice flakes. The limitation of the traditional flaking process is that it does not allows a composite matrix processing and in the recent times flaking a mixture or a composite mix is becoming widely popular because of its nutrient profile and versatile nature in terms of incorporating ingredients of choice. A formulation of flakes made from millet flour and snakehead fish koya was developed by [15], In a similar study by [35] flakes were developed by mocaf [36], black rice, tapioca, soy, jack bean flour as a breakfast alternative high in protein and dietary fiber. The processes followed in both was on the basis of the trials based on randomized factorial design which followed the sequence (refer Figure 5.5) which is used in case of many components and is an advanced method in flake processing. Sorghum -pigeon pea flakes were developed by [37] by the same method that highlighted that combinations like these are helpful in children and adult feeding patterns. Combinations of two naturally different ingredients such as cereal and vegetable are easy to flake by this process because it requires customized flaking tool, an example of this was recently followed up by [38] where they developed flakes from a mixture of orange sweet potato and Indonesian red rice (Cempoabang). One great advantage of this method is that it also utilizes the broken of grains where grits blend of two or more grain broken can be made for the production of flakes, one similar study was done by [39] where broken rice, green gram and barnyard millet blend for the development of nutritious snacks with modified texture.

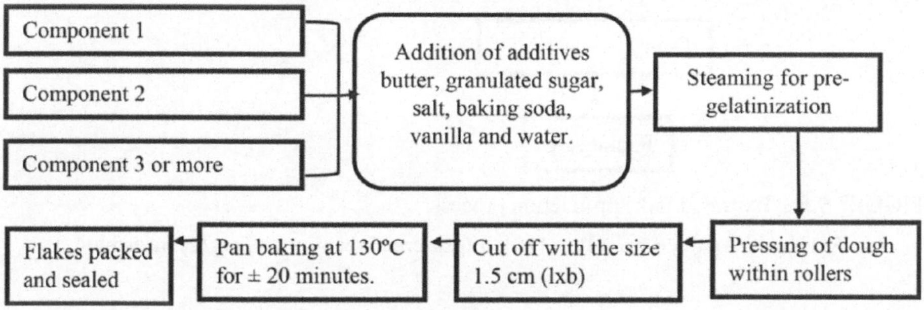

FIGURE 5.5 Flake processing with many components.

5.4.1.2 EXTRUDED FLAKE PRODUCTION

Extrusion technology is becoming popular for texturizing flakes. Extruded [40–42] flake processing omits steps such bas cooking and delumping of the traditional processing. This technology has taken over the breakfast cereals market and other instant products such as noodles, pasta, macaroni, etc. [43]. Thielecke and colleagues in their study of value addition of cereals quoted the modification made by Ministry of Health of Brazil[44]which followed the following steps (Refer Figure 5.6) omitting the cooking and delumping step [45].

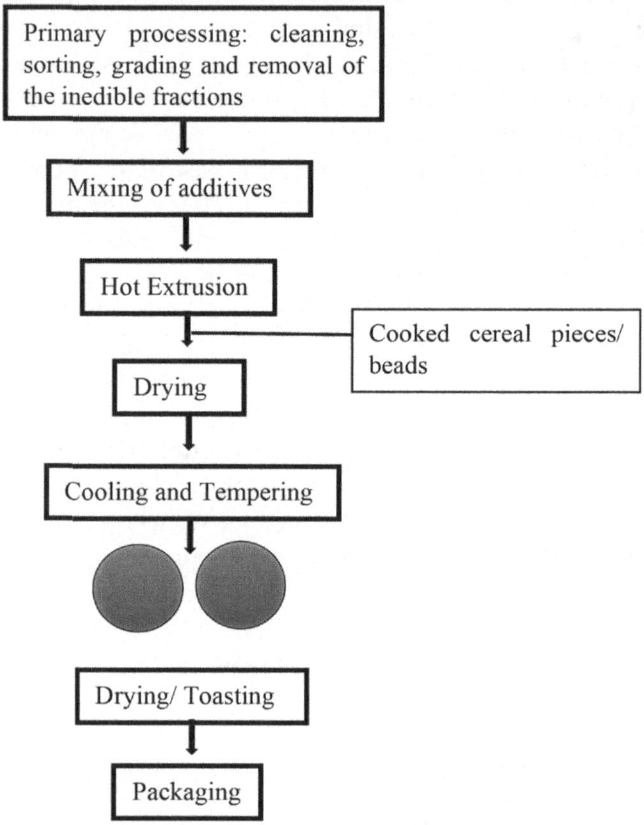

FIGURE 5.6 Extruded flake production process.

Source: Reprinted/Adapted from Ref. [45]. https://creativecommons.org/licenses/by/4.0/

The toasting step after flaking may be done on an open pan or can also be done in microwave as done by [14] who developed multigrain flakes from black gram flour, rice flour and flaxseed flour. The process followed was extrusion, flaking and then microwave roasting.

5.4.1.3 COATED RTE FLAKES

A coating or a composite mix of many ingredients can be coated on to the usual cereal flakes. Kumar and group [46] developed and optimized RTE roasted rice flakes coated with chocolate as an instant breakfast option.

The coating mix consisted of jaggery, dark chocolate, guar gum, skim milk powder, and ferrous sulfate (as iron fortificant). The process was as followed:

Traditional flaking method

↓

Roasting of flakes

↓

Coating of roasted rice flakes with the coating mix in a continuous tablet coating machine

(Temperature of hot air blower 70±2°C, Speed- 60 rpm/min, Time- 10 min)

↓

Drying in hot air oven (50±2°C for 2 h)

5.4.1.4 YOUNG RICE FLAKES

Methods discussed in the above sections requires the developed rice flakes to go a pre-roasting step, sun drying, baking or oven drying of the ready flakes. Piyarach et al. [47] in their study highlighted that flake when developed from young and/or immature harvested rice found to have more nutritional value. Since the rice flakes produced from immature harvested paddy will have shorter shelf life, poor color retention and are prone to have quick microbial contamination they optimized the drum drying process for Young Thai Jasmine Rice Flakes. The optimum treatments they came up with was with young rice grain to water ratio at 1:5 w/w with 0.5 rpm drum rotation speed, which was found to have the potential to retain the nutrition.

Apart from the technologies discussed in section 5.4.1 to make flakes from cereals, the degree of the easiness of the process can be manipulated/modified according to the desired end product. For example, omitting steps like developing flakes from paddy, the milled and partially cooked rice can be directly flattened under rollers. Similar technique was adapted by [48] who developed non-traditional and gluten free flakes from black and red rice (Czech Republic origin), white rice, teff and quinoa. Following steps were followed to develop flakes from the same.

1. Cooking of milled cereals (separately) in plain water (95–96°C for 15 to 20 min for rice grains for 8 to 9 min for quinoa grain and 17 minutes for teff).
2. Post cooking (indication-complete gelatinization) water was drained off and grains were conditioned in a covered vessel (rice for 45 min and teff and quinoa for 25 min).

3. Air drying at room temperature for 100–120 minutes
4. Flaking of rice and quinoa in flake roller of Combi-star grinder and manual flaking of teff by wooden roller.

Similarly, [49] developed flakes from Indonesian red, black and white glutinous rice with slight modifications in the traditional processes in which they soaked the grains for 24 hours at room temperature (1:2 ratio), filtering the water, cooking the grains by roasting in clay pot and then flaking by flattening the grains in mortar and pestle. The results highlighted that flakes made out of black and glutinous rice were nutritionally better than that of non-pigmented glutinous rice. Similar results were reported for another value-added product from Indonesian black rice [50].

5.4.2 ADVANCEMENTS IN PUFFING/ PARCHING TECHNOLOGY

The traditional process mentioned in the section 5.2.2 is the widely used method for puffing but there has been an increase in the technologies in order to automate this traditional process and to scale up the production capacity. Following are the technologies where innovation has been seen with regard to puffing technology in the recent years.

5.4.2.1 OIL PUFFING

Oil Puffing is becoming recently popular for puffing various cereals and seeds like lotus seed because when puffing is done in the presence of oil or some lubricant the volume increases but when it comes to commercializing the oil-based puffs it gains less consumer acceptance because of its higher calorific value as compared to other methods discussed further in this section. [24] attempted to develop rice puffs by using coconut oil. Puffing was carried out for 4–8 sec at 210–220°C, it was found that removal of excess oil was easy for small batch by straining and dabbing the puffs onto paper towels but is a tedious process if scaled up. Hence a better and sustainable approach can be looked upon for this method.

5.4.2.2 OVEN PUFFING

Oven puffing method is almost similar to that of traditional but with an exception that instead of treated paddy properly milled rice is pressure cooked with other ingredients such as malt/sugar/salt and/or seasonings which is then lightly flattened in rollers, dried, mixed with flavors and then toasted in oven at high temperatures. In oven puffing method the advancements have been seen in the type of products which can be manufactured by this technology like crispies and RTE snacks where various flavorings, seasonings and additives can be added to make them more acceptable.

5.4.2.3 GUN PUFFING

Gun puffing method is found to be the most cost-time efficient and economical method of puffing the cereals where the milled rice is directly puffed. It works on the principal of pressure building and release causes the grain to expand. One advancement being in the same is use of superheated steam into the gun puffing chamber.

5.4.2.4 MICROWAVE PUFFING

Microwave puffing is another method which is convenient, cheap and can provide a good alternative to traditional puffing in urban households because of availability of microwave assisted equipment. [52] studied the process parameters and optimized the rice puffing method in domestic microwave. The optimized parameters were highlighted as power level 780–942 W, heating time of 26.2–33.65 sec and 3.5% of salt concentration in the pretreated rice, respectively. In this method further improvement is possible to make it less self-limiting.

5.4.2.5 FLUIDIZED BED PUFFING

Fluidized bed puffing is another advanced method of puffing where the grains are uniformly exposed to hot air in fluidized state. This method prevents the kernels from charring/burning due to fluidization and heat uniformity and is found to be less exposed to contamination unlike sand puffing. This method is costlier as compared to other methods discussed above.

A comparison of the traditional and advanced methods of puffing was done by [24] who highlighted that out of six methods of puffing gun puffing achieved maximum expansion ratio of 17.4 followed up by fluidized air puffing (10), hot air puffing (9.71), hot sand bed puffing (7.6), microwave puffing (7.5) and lastly oil puffing (5.9).

5.4.3 ADVANCEMENTS IN POPPING TECHNOLOGY

Advancements of puffing technology can be employed to popping as well but with a clear understanding of the optimization parameters. High temperature-short time (HTST) fluidized bed is one of the methods used for popping, the fluidization phenomenon stands the same as discussed in section 4.2 where the grains behave as in a fluid motion but under hot air and suspended into the same. A fluidized bed popping machine employs approximately 2,450 MHz (Megahertz) of microwave frequency [53] with a temperature range of 240°C–270°C with an exposure time of 7–9.7 sec.

5.4.3.1 MICROWAVE POPPING

Microwave popping is majorly employed for microwaved popcorns [20, 23, 53, 54]. In this technology microwave expansion takes place because of the moisture present in the kernels (refer Figure 5.7). Coatings like oil, salt, butter and sodium bicarbonate helps to increase the yield of popped kernels in microwave popping [53, 54]. Solanki, Indore, and Nanda, [53] compared the conventional popping method with microwave popping of popcorn and concluded that 20% more popping yield was achieved by microwave popping with the first pop starting at 60 sec and the process completed at 1.5 min.

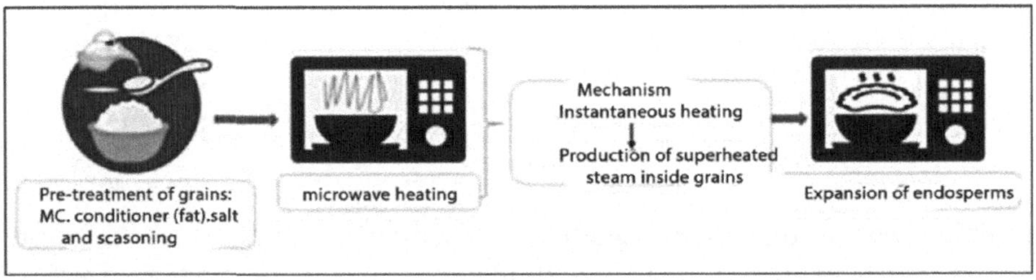

FIGURE 5.7 Mechanism of microwave popping.

5.4.3.2 EXTRUSION TECHNOLOGY

Extrusion technology is another method to develop porous popped snacks. These popped snacks are light weight, porous texture and expanded to a greater degree as compared to conventional popped products. This technology comparably produces less waste, more palatable and digestible, easy to season with spices and is very much liked by almost every age group because of its taste and texture [55]. This process induces better dispersion of fat and tocopherols, phenolic profile with significant reduction in readily digestible carbohydrates which improves the overall *in-vitro* glycemic profile of the extruded product. Apart from producing RTE products [56], extrusion also allows production of pre-gelatinized flour, instant flour with providing possibility of alteration in techno-functionality of the flour. In this process the extrudates are milled after extrusion cooking and the flour obtained shows better functionality suitable to be used as thickening agents, instant food and physically modified flour [57]. The extrudates characteristic depends upon the variation of feed composition and the process variables and affects qualities of the final product such as bulk density, expansion ratio, water solubility index, water absorption index compression force and cutting strength [58–60] studied effect of different extrusion temperature on bioactive compounds and extrusion behavior of black rice extrudates and their study concluded that anthocyanin decreased with increasing temperature, the total phenolic acid contents gradually increased with increase in temperature in decrease in die pressure.

5.4.3.3 ATMOSPHERIC RADIO FREQUENCY PLASMA (ARFP) JETS

Atmospheric radio frequency plasma (ARFP) jets are often used for surface activation, decontamination, inhibit microbial growth on food surfaces and enzyme activation such as for polyphenol oxidase and peroxidase to reduce/avoid enzymatic browning [61, 62, 63, 64] studied the effect of ARFP treatment on rice popping characteristics. Unhusked rice was exposed to ARFP in a glass container at 20 W and 40 W for about 90 s. This plasma treated rice was then popped in an electric pan at 200°C for 90 seconds. The results of this study highlighted that atmospheric ARFP improved the popping characteristics and had better nutritional quality, amino acid profile and trace elements as compared to an electric pan popping and microwave popping. Therefore, cold plasma can be studied more with regard to its applicability in RTE cereal technologies.

5.4.3.4 INFRARED ASSISTED HOT AIR POPPING

Infrared assisted hot air popping method was employed by [65] and nutritional changes were studied. Cleaned raw sorghum with a M.C of 16±0.5% (w.b) was popped in popping machine with the air temperature of 200–300°C of a velocity 3.2–5.5 m/s for less than one minute. The machine was equipped with ceramic infrared heaters emitting 1,000 W of heat which was circulating inside the chamber in order to pop the sorghum. The study highlighted that infrared assisted popping technology is safe and reliable. The temperature monitoring was critical in this method.

5.4.3.5 RTE PRODUCT FORMULATIONS WITH POPPED CEREALS

Rice, maize, sorghum, wheat, ragi, amaranth seeds, barley, kamut, buckwheat, proso, kodo, barnyard and foxtail millet are the mostly popped cereals from these technologies. RTE breakfast cereal was formulated by [66] by mixing pearl millet pops, with wheat puff, sunflower seeds, popped amaranth, raisins and flax seeds with a syrup mixture (water, sugar, honey, and sunflower oil). This mix was baked, toasted, cooled and packaged in metallized pouches. This breakfast cereal was found out to have a shelf life of three months with negligible effect of sensory qualities. Similarly, gluten free cereal bars were developed by [67] by using popped sorghum made by microwave method, the sorghum pop were mixed with dates, banana, buckwheat flakes, coconut flakes, flax seeds, almonds and vanilla sugar. The bars were found to have good nutritional and textural profile. The optimum level of sorghum pop was 6% by weight of the raw material [9].

5.5 OPTIMIZATION PARAMETERS OF EACH TECHNOLOGY

5.5.1 FLAKING

The optimization parameters of flakes are based upon the process parameters such as parboiling[68], [69], moisture content, temperature and time during processing, the pressure applied, speed and type of rollers, additives, cutting, roasting/toasting and/or baking parameters of flakes (in case of flaking with two or more components). These parameters will ultimately affect the physical, textural and nutritional quality of the flakes which leads to the final acceptance of the flaked cereal such as crunchiness in milk or during rehydration [15]and its sensory properties. As highlighted by [14] that an inlet feed with more M.C (moisture content) produced denser grits/ beads with lesser diameter a high extrusion temperature of the barrel increased the bead diameter, hence producing larger flakes.

5.5.2 PUFFING AND POPPING/ PARCHING

Factors which affects the popping and puffing quality has been stated by [6] which are moisture content, physicochemical properties of grains, storage time, degree of milling, salt content and parboiling. The degree of parboiling greatly affects the puffing and popping volume [21]. The moisture content in the milled parboiled rice also responsible for the quality of puffs. It should be around 10.5% (w.b) for dry roasting. The degree of milling is also responsible for cracked and broken rice kernels these cracks and breaks in kernels reduces the expansion ratio while tempering and roasting. The amount of salt added in the brine or in dried form is also a factor responsible for the final product quality because the appropriate quality and quantity of salt increases the expansion ratio whereas the soaking in plain water lowers the expansion ratio at all M.C as compared to soaking in brine. The age of the paddy has little effect on the puffing and popping ratio where the mid aged paddy is preferable for puffing because puffing ratio increases with the age but decreases later. Pre-drying of rice paddy positively affects the quality of the final product where pre heating the paddy to 9% M.C increases overall expansion ratio which does not gets affected by the drying method except that the temperature of the drying air should not be more than 70°C. This gelatinization within rice is essential for the crispy texture of puffed rice which is assisted by parboiling the paddy [24]. Puffing and popping temperature and time are critical factors as it facilitates the phase transition and expansion. The compositional characteristics of the cereal also affect the popping and puffing process. The protein, lipids and the starch characteristics as stated by [70] that relationship between the starch and expansion of grain is very complicated, the amount of starch can either increase or decrease the expansion ratio during puffing. In fluidized bed popping method MC of the grain and of the heating medium affects the degree of popping [54].

Combining these methods can have totally unexpected outcomes like adding brine to pre-dried paddy leads to the precipitation on the grain and results in decreased nutritional value and expansion ratio [22]. The type of cereal also affects the final quality of popped and puffed product, usually waxy and low amylose varieties are preferred over the cereals which are otherwise non-waxy and have higher amylose content. The protein content is inversely related to the expansion ratio of the puffed and popped product.

5.6 A COMPARISON BETWEEN CHARACTERISTICS OF BLACK RICE FLAKES, PUFFS, AND POPPED BLACK RICE

Indian rice possesses good physicochemical, rheological and thermal properties to be processed into instant products, reformed rice and RTE breakfast snacks [71]. Various different types of rice products from different types of Indian black rice varieties namely CH[72–74], KB, MA, MN and SW was developed and evaluated for their properties by [9] who highlighted that the SW (white rice) flakes was found to be the hardest followed by, MN black rice flakes and MA. Lower hardness of flaked rice than raw rice is due to breakdown and liquefication protein-starch matrix during flaking process [75]. In other textural parameters like cohesiveness and springiness both popped and beaten rice had negligible difference. Black rice puffs have a considerable number of antioxidants and anthocyanins in them although the study done by [9] revealed that the antioxidant capacity in puffed and beaten/flaked black rice decreased as compared to that of popped and boiled black rice. In terms of physical characteristics of black rice products, the popped rice was found to have the highest volume elongation ratio (VER) and puffed recording the minimum. Whereas the length (L), breadth (B) and elongation ratio (ER) of puffed rice was recorded more. KB black rice flakes was found to have the lowest thickness than that of CH, MN, MA and SW. The adhesiveness in black rice puffed products was found to be the lowest as compared to beaten and boiled rice. Out of all five variety of puffs MN black rice puffs had the highest hardness followed by white rice SW, MA, CH and KB. Waxy black rice provides better puffing characteristics than that of the non-waxy which includes mealy and chalky black rice [10].

There are various parameters to judge the quality of the expanded grains which are puffing/ popping percentage and yield [52], percentage of unexpanded kernels, expansion ratio, expansion volume [76, 77], bulk density, size of flakes/ puffs and popped grain, change in microstructure, sensory quality and effect of processing technique on bioactive compounds [60].

5.7 CHALLENGES AND SCOPE ASSOCIATED WITH THESE TECHNOLOGIES

1. Traditional processes lack processing controls like temperature, time, pressure and contamination of external materials like dirt, chaffs, sand, etc., if not done properly, hence should be preferable for small scale production only.

2. Many advanced methods working with composite matrix or formulations used manual handling, scaling up such processes would require specialized equipment for those specific purposes.
3. For many of the processes like roasting, there is a gap of commodity specific equipment which makes the process energy consuming, laborious with worker being directly exposed to heat and the smoke or pollution causing due to burning of wood or coal, etc. [23].
4. There lies a huge scope in development of cost-efficient small-scale and industrial large-scale equipment for RTE processing of cereals and legumes as well which is cost and energy efficient, provides better output with less labor and have processing parameter controls. Technologies like microwave dryers, fluidized bed dryers/roasters, automatic parboilers, puffing machines, flake cutters and extruders and instantization equipment can be possible interventions to make these processes better.
5. Modernizing the traditional machinery can increase the production efficiency drastically by also improving the working conditions, provides quality like in traditional processes and the runs can be longer as compared to traditional. Similar attempt was made by [78] who modernized the traditional rice puffing machine and found out that the production efficiency increased by 1,100% approximately as compared to traditional puffing method.
6. Advances in rice science and rice mechanization [79, 80] also affects the quality of the RTE products developed from cereals with development of high yielding and short duration varieties [79], there is a need to study the botany, production and crop improvements done in the field of all kinds of rice and the product developed from the same [81, 82].
7. Further research is needed to assess the proximate quality, micronutrients availability, digestibility and dietary fiber content in order to develop better and healthier snacking options and weaning foods for infants [54, 83].
8. Optimization of black rice processing, effect of processing on different other varieties of black rice.
9. Germination, fortification and heat and mass transfer studies of pigmented rice can be explored further because they provide a good option for preservation of vital components such as minerals, dietary fiber and vitamins through the discussed methods of flaking, popping, puffing and RTE processing[6, 84–86].

10. Advancements in RTE processing can be coupled with its health impact, microbial safety of the final product until its shelf stability [51].

5.8 CONCLUSION

Since past few years, there has been a rise in the new product development made from non-staple cereals, Nutri-cereals such as major and minor millets, pseudo cereals such as amaranth, buckwheat and other nontraditional cereals like pigmented rice (red rice, black rice, purple rice, etc.), quinoa, teff and many other. As compared to the health concerned products from cereals such as gluten free and less carb claiming products, there are many advancements done in the area of RTE cereals as well but the products are more centered towards the breakfast cereal category rather than instant healthy meal or snacks. Cereal processing techniques like flaking, parching, frying, roasting and popping is known to mankind since ages and yet it is very much utilized to process the cereals in quick consumption preparations. Various advancements in these technologies had been reported to present healthier choices to the consumers around the globe. Majority of the advancements owing to incorporate more cereals in one matrix was found to be the highest which has given rise to technologies like extruded, coated, baked flakes/puffs/popped products and their RTE mixes, which also valorizes the bi products of food sectors. Apart from that the technological advancements have been reported in the areas of heating medium such as microwave, plasma and infrared assisted technologies, superheated steam, HTST fluidization, etc. In order to develop products with better nutritional quality than that of the existing one's changes in the choice of raw material has also been addressed. The process optimization of these technologies in the recent years has played a crucial role in increasing the yield of the product, ease of processing, increasing nutritional and textural quality, reducing the chances of contamination, manual labor and direct exposure to heat. In this chapter products developed from black rice has been given a special importance because of the underutilization of the pigmented rice's in these technologies. Various researchers attempted to develop product with black rice with or without combining other cereals which highlighted the need of RTE processing of such underutilized and nutritionally rich cereals and it should be encouraged like the staple cereals as they have all the necessary processing features as that of the staple ones.

KEYWORDS

- **black rice**
- **flaking**
- **food processing**
- **instant technology**
- **parching**
- **popping**
- **puffing**
- **ready-to-eat (RTE)**

REFERENCE

1. Ponnappan, S., Thangavel, A., & Sahu, O. (2017). Milling and physical characteristics of pigmented rice varieties. *International Journal of Food Chemistry*, *1*(1). https://doi.org/10.11648/j.ijfc.20170101.15.
2. le Brun, O., Berra, M., & Dubat, A. (2019). Impact of rice varieties, milling processes of rice flour and final bakery products quality. *ResearchGate*. https://doi.org/10.13140/RG.2.2.15822.51525.
3. Ubwa, S. T., Abah, J., Asemave, K., & Shambe, T. (2012). Studies on the gelatinization temperature of some cereal starches. *International Journal of Chemistry*, *4*(6), 4–11. https://doi.org/10.5539/ijc.v4n6p22.
4. Faruq, G., Hadjim, M. O., & Meisner, C. A. (1560). Inheritance of gelatinization temperature in rice. *International Journal of Agriculture and Biology*, *8530*(August), 6–5.
5. Biliaderis, C. G., Maurice, T. J., & Vose, J. R. (1980). Starch gelatinization phenomena studied by differential scanning calorimetry. *Journal of Food Science*, *45*(6), 1669–1674. https://doi.org/10.1111/j.1365-2621.1980.tb07586.x.
6. Swarnakar, A. K., Mohapatra, M., & Das, S. K. (2022). A review on processes, mechanisms, and quality influencing parameters for puffing and popping of grains. *Journal of Food Processing and Preservation*. https://doi.org/10.1111/jfpp.16891.
7. Levine, L., Campanella, O. H., Corvalan, C. M., Okos, M. R., & Symes, S. T. (2002). A model for the formation of multiple flakes during cereal flaking. *Cereal Foods World*, *47*(6), 210–223.
8. Kumar, S., Shiv Kumar, C., & Prasad, K. (2018). Changes in the characteristics of indica rice on the process of flaking. *International Journal of Chemical Studies*, *6*(2).
9. Pal, S., et al. (2019). Evaluation of sensory, physicochemical properties and consumer preference of black rice and their products. *Journal of Food Science and Technology*, *56*(3), 1484–1494. https://doi.org/10.1007/s13197-019-03634-8.

10. Bhattacharya, K. R. (2011). *Rice Quality: A Guide to Rice Properties and Analysis.* Woodhead Publishing. https://doi.org/10.1533/9780857092793.
11. Sruthi, N. U., Premjit, Y., Pandiselvam, R., Kothakota, A., & Ramesh, V. (2021). An overview of conventional and emerging techniques of roasting: Effect on food bioactive signatures. *Food Chemistry*, *348*, 129088. https://doi.org/10.1016/j.foodchem.2021.129088.
12. Thakur, A. K., & Gupta, A. K. (2006). Water absorption characteristics of paddy, brown rice and husk during soaking. *Journal of Food Engineering*, *74*(2), 185–190. https://doi.org/10.1016/j.jfoodeng.2005.04.014.
13. Contini, G. K., et al. (2022). High nutritional value muffins produced with whole meal rye (*Secale cereale* L.) and whole meal bean (*Phaseolus vulgaris* L.) flour mix. *Journal of Food Science, 2061.*
14. Rani, P., Kumar, A., Purohit, S. R., & Rao, P. S. (2020). Development of multigrain extruded flakes and their sensory analysis using fuzzy logic. *Journal of Food Measurement and Characterization*, *14*(1), 411–424. https://doi.org/10.1007/s11694-019-00303-4.
15. Anandito, R. B. K., Siswanti, & Purnamayati, L. (2021). Breakfast cereal in flakes form based on millet flour and snakehead fish koya. *IOP Conference Series: Earth and Environmental Science*, *750*(1), 012050. https://doi.org/10.1088/1755-1315/750/1/012050.
16. Kumar, S., ul Haq, R., & Prasad, K. (2018). Studies on physico-chemical, functional, pasting and morphological characteristics of developed extra thin flaked rice. *Journal of the Saudi Society of Agricultural Sciences*, *17*(3), 259–267. https://doi.org/10.1016/j.jssas.2016.05.004.
17. Wong, J. H., et al. (2009). Digestibility of protein and starch from sorghum (*Sorghum bicolor*) is linked to biochemical and structural features of grain endosperm. *Journal of Cereal Science*, *49*(1), 73–82. https://doi.org/10.1016/j.jcs.2008.07.013.
18. Wu, G., Ashton, J., Simic, A., Fang, Z., & Johnson, S. K. (2018). Mineral availability is modified by tannin and phytate content in sorghum flaked breakfast cereals. *Food Research International*, *103*, 509–514. https://doi.org/10.1016/j.foodres.2017.09.050.
19. Singh, N., Kaur, L., Singh Sodhi, N., & Singh Sekhon, K. (2005). Physicochemical, cooking and textural properties of milled rice from different Indian rice cultivars. *Food Chemistry*, *89*(2), 253–259. https://doi.org/10.1016/j.foodchem.2004.02.032.
20. Subramani, D., Tamilselvan, S., Murugesan, M., & Shivaswamy, M. S. (2020). Optimization of sand puffing characteristics of quinoa using response surface methodology. *Current Research in Nutrition and Food Science*, *8*(2), 504–515. https://doi.org/10.12944/CRNFSJ.8.2.16.
21. Luh, B. S. (1991). *Rice Production* (Vol. 2, p. 1127). https://doi.org/10.2307/920283.
22. Hoke, K., Housova, J., & Houška, M. (2004). Optimum conditions of rice puffing. *Czech Journal of Food Science, 23*(1), 1–12. Retrieved from: ResearchGate.
23. Kora, A. J. (2019). Applications of sand roasting and baking in the preparation of traditional Indian snacks: Nutritional and antioxidant status. *Bulletin of the National Research Centre, 43*(1). https://doi.org/10.1186/S42269-019-0199-2.
24. Saha, S., & Roy, A. (2020). Puffed rice: A materialistic understanding of rice puffing and its associated changes in physicochemical and nutritional characteristics. *Journal of Food Process Engineering, 43*(9). https://doi.org/10.1111/jfpe.13479.

25. Khan, H., Devi, S., Aparna, K., & Kameswari, S. L. (2017). Nutritional properties of three selected varieties of puffed rice: WGL44, WGL283, and RNR 2458. *The Pharma Innovation Journal, 6*(9), 458–460. Retrieved from: Buzzle.
26. Yang, D. S., Lee, K., & Kays, S. (2010). Characterization and discrimination of premium-quality, waxy, and black-pigmented rice based on odor-active compounds. *Journal of the Science of Food and Agriculture, 90*(14), 2595–2601. https://doi.org/10.1002/jsfa.4126.
27. Corredor Sanchez, N., et al. (2018). Optimization of the paddy rice husking process, increasing the useful life of the rollers in Florhuila Plant Campoalegre Mills. Retrieved from: RI Publication.
28. S, R. P. T., Raeboline, A., Eliazer, L., Ravichandran, K., & Antony, U. (2019). Nutritional and functional properties of coloured rice varieties of South India: A review. *Journal of Food Science, 3*, 1–11.
29. Chanu, C. S., Yenagi, N. B., & Math, K. K. (2016). Nutritional and functional evaluation of black rice genotypes. *Journal of Food Science and Technology, 29*(1).
30. Saikia, S., Dutta, H., Saikia, D., & Mahanta, C. L. (2012). Quality characterization and estimation of phytochemicals content and antioxidant capacity of aromatic pigmented and non-pigmented rice varieties. *Food Research International, 46*(1), 334–340. https://doi.org/10.1016/j.foodres.2011.12.021.
31. Jayaraman, R., Uluvar, H., Khanum, F., & Singh, V. (2019). Influence of parboiling of red paddy varieties by simple hot soaking on physical, nutrient, phytochemical, antioxidant properties of their dehusked rice and their mineral, starch, and antioxidants' bio accessibility studies. *Journal of Food Biochemistry, 43*(7). https://doi.org/10.1111/jfbc.12839.
32. Bolea, C. A., Grigore-gurgu, L., Aprodu, I., & Vizireanu, C. (2019). Main bioactive of black rice flour sieving fractions.
33. Shao, Y., Xu, F., Sun, X., Bao, J., & Beta, T. (2014). Identification and quantification of phenolic acids and anthocyanins as antioxidants in bran, embryo, and endosperm of white, red, and black rice kernels (*Oryza sativa* L.). *Journal of Cereal Science, 59*(2), 211–218. https://doi.org/10.1016/j.jcs.2014.01.004.
34. Sruthi, N. U., Premjit, Y., Pandiselvam, R., Kothakota, A., & Ramesh, S. V. (2021). Food bioactive signatures. *Food Chemistry, 348*, 1–4. https://doi.org/10.1016/j.foodchem.2021.129088.
35. Agustia, F. C., Rosyidah, S., Subardjo, Y. P., Ramadhan, G. R., & Betaditya, D. (2019). Formulation of flakes made from mocaf-black rice-tapioca high in protein and dietary fiber by soy and jack bean flour addition. *IOP Conference Series: Earth and Environmental Science, 255*(1), 1–10. https://doi.org/10.1088/1755-1315/255/1/012019.
36. Hamidi, M. A., & Banowati, E. (2019). Utilization of mocaf flour (modified cassava flour) for revitalization of the use of tapioca flour in communities for empowering hulu-hilir human resources in Wonogiri Regency. In *IOP Conference Series: Earth and Environmental Science* (Vol. 243, No. 1). https://doi.org/10.1088/1755-1315/243/1/012081.
37. Mbaeyi, I., & Onweluzo, J. (2011). Effect of sprouting and pregelatinizing on the composition and sensory properties of flaked breakfast cereal produced from sorghum-pigeon pea blends. *Agro-Science, 9*(3). https://doi.org/10.4314/as.v9i3.65758.
38. Jati, I. R. A. P., Darmoatmodjo, L. M. Y. D., Suseno, T. I. P., Ristiarini, S., & Wibowo, C. (2022). Effect of processing on bioactive compounds, antioxidant activity, physicochemical, and sensory properties of orange sweet potato, red rice, and their application for flake products. *Plants, 11*(3). https://doi.org/10.3390/plants11030440.

39. Nisha, R., Nickhil, C., Pandiarajan, T., Pandiselvam, R., Jithender, B., & Kothakota, A. (2023). Chemical, functional, rheological and structural properties of broken rice–barnyard millet–green gram grits blend for the production of extrudates. *Journal of Food Processing and Preservation, 46(2)*. https://doi.org/10.1111/jfpe.14324.
40. Bhattacharya, S., States, N., & Transitions, G. (2017). Learn more about extruded food extrusion technology and glass transition. *Extrusion Technology*.
41. Dalbhagat, C. G., Mahato, D. K., & Mishra, H. N. (2019). Effect of extrusion processing on physicochemical, functional and nutritional characteristics of rice and rice-based products: A review. *Trends in Food Science & Technology*. https://doi.org/10.1016/j.tifs.2019.01.001.
42. Idrishi, R. (2020). Development of multimillet vermicelli by cold extrusion technology: A novel value-added millet product. In *UGC Sponsored National Conference on Food Safety, Nutritional Security and Sustainability* (pp. 193–197).
43. Thomas, B., Pulissery, K., Saranya, S., Kothakota, A., Pandiselvam, R., & Joseph, M. (2022). Development of protein enriched cold extruded pasta products using hybrid dried processed mushroom powder and defatted flours: A study on nutraceutical, textural, color and sensory attributes. *LWT, 170*, 113991. https://doi.org/10.1016/j.lwt.2022.113991.
44. Report: Brazil. Ministry of Health of Brazil. Secretariat of Health Care. Primary Health Care Department. Dietary Guidelines for the Brazilian population/Ministry of Health of Brazil, Secretariat of Health Care, Primary Health Care Department; translated by Carlos Augusto Monteiro. Brasília : Ministry of Health of Brazil, 2015. 1–150 p. Access mode: World Wide Web: http://189.28.128.100/dab/docs/portaldab/publicacoes/guia_alimentar_populacao_ingles.pdf. ISBN 978-85-334-2242-1
45. Thielecke, F., Lecerf, J. M., & Nugent, A. P. (2021). Processing in the food chain: Do cereals have to be processed to add value to the human diet? *Nutrition Research Reviews, 34*(2), 159–173. https://doi.org/10.1017/S0954422420000207.
46. Kumar, S., et al. (2021). Optimization and development of ready to eat chocolate coated roasted flaked rice as instant breakfast food. *Foods, 10*(7). https://doi.org/10.3390/FOODS10071658.
47. Piyarach, K., Sutham, P., Pongsiri, K., Randah, A., & Teerawan, S. (2021). Production and quality evaluation of young Thai jasmine rice flake using drum dryer. *E3S Web of Conferences, 302*, 02008. https://doi.org/10.1051/e3sconf/202130202008.
48. Šťastná, M., Mrázková, M., Sumczynski, D., Cındık, B., & Yalçın, E. (2019). The nutritional value of non-traditional gluten-free flakes and their antioxidant activity. *Antioxidants, 8*(11). https://doi.org/10.3390/antiox8110565.
49. Amrinola, W., Sitanggang, A. B., Kusnandar, F., & Budijanto, S. (2022). Characterization of pigmented and non-pigmented flakes glutinous rice (ampiang) on chemical compositions, free fatty acids compositions, amino acids compositions, dietary fiber content, and antioxidant properties. *Food Science and Technology (Brazil), 42*. https://doi.org/10.1590/fst.86621.
50. dela Rosa, J. G. L., & Medina, P. M. B. (2022). Philippine rice wine (Tapuy) made from Ballatinao black rice and traditional starter culture (Bubod) showed high alcohol content, total phenolic content, and antioxidant activity. *Food Science and Technology (Brazil), 42*. https://doi.org/10.1590/fst.45120.
51. Temgire, S., Borah, A., Kumthekar, S., & Idate, A. (2021). *Recent Trends in Ready to Eat / Cook Food Products: A Review*, *10*(5), 211–217.

52. Mohapatra, M. (2011). Effect of process parameters and optimization on microwave puffing performance of rice. *Research Journal of Chemistry and Environment*. Retrieved from: ResearchGate.
53. Solanki, C., Indore, N., & Nanda, K. (2018). Microwave vs conventional popping system: A comparative evaluation for maize popping. *International Journal of Chemical Studies, 6*(6), 176–181.
54. Joshi, D. C., Mishra, G., Joshi, D. C., & Kumar Panda, B. (2014). Popping and Puffing of Cereal Grains: A Review. *Journal of Grain Processing and Storage*. Retrieved from https://www.researchgate.net/publication/283355236.
55. Murtaza, G., Huma, N., Sharif, M. K., & Zia, M. A. (2022). Probing a best suited brown rice cultivar for the development of extrudates with special reference to physico-chemical, microstructure and sensory evaluation. *Food Science and Technology (Brazil), 42*, 1–9. https://doi.org/10.1590/fst.103521.
56. Indian Institute of Food Processing Technology/NIFTEM, Thanjavur (n.d.). *Handbook of Ready To Eat Extruded Snacks*.
57. Bender, D., & Sch, R. (2021). Recent developments and knowledge in pseudocereals including technological aspects. *Acta Alimentaria, 50*(4), 583–609. https://doi.org/10.1556/066.2021.00136.
58. Pandiselvam, R., Manikantan, M. R., Sunoj, S., Sreejith, S., & Beegum, S. (2019). Modeling of coconut milk residue incorporated rice-corn extrudates properties using multiple linear regression and artificial neural network. *Journal of Food Process Engineering, 42*(e12981), 1–10.
59. Preethi, P., Mangalassery, K., Shradha, K., Pandiselvam, R., & Manikantan, M. R. (2021). Cashew apple pomace powder enriched the proximate, mineral, functional and structural properties of cereal-based extrudates. *LWT, 139*, 110539. https://doi.org/10.1016/j.lwt.2020.110539.
60. Hu, Z., et al. (2018). Effects of different extrusion temperatures on extrusion behavior, phenolic acids, antioxidant activity, anthocyanins and phytosterols of black rice. *RSC Advances, 8*(13), 7123–7132. https://doi.org/10.1039/c7ra13329d.
61. Doona, C. J., Kustin, K., & Feeherry, F. E. (2010). *Case Studies in Novel Food Processing Technologies: Innovations in Processing, Packaging, and Predictive Modelling*. https://doi.org/10.1533/9780857090713.
62. Urbano, O., et al. (2020). Cost-effective implementation of a temperature traceability system based on smart RFID tags and IoT services. *Sensors (Switzerland), 20*(4), 1163. https://doi.org/10.3390/s20041163.
63. Alfian, G., et al. (2017). Integration of RFID, wireless sensor networks, and data mining in an e-pedigree food traceability system. *Journal of Food Engineering, 212*, 65–75. https://doi.org/10.1016/j.jfoodeng.2017.05.008.
64. Puangjinda, K., Matan, N., & Nisoa, M. (2016). Effects of atmospheric radio-frequency plasma treatment on popping characteristics of popped rice and its nutritional evaluation. *Innovative Food Science and Emerging Technologies, 35*, 119–124. https://doi.org/10.1016/j.ifset.2016.04.010.
65. Anjitha, P. K., Baskaran, N., Venkatachalapathy, N., & Anand, M. T. (2021). Nutritional changes of sorghum after popping by a developed infrared assisted hot air popping machine. *International Journal of Current Microbiology and Applied Sciences, 10*(1), 3620–3627. https://doi.org/10.20546/ijcmas.2021.1001.427.

66. Singh, R., Singh, K., Singh Nain, M., & Singh, R. (2020). Storage stability of popped pearl millet based ready to eat breakfast cereal. *Indian Journal of Agricultural Sciences, 90*(10), 1915–1920.
67. Aleksandrova, A., Mykolenko, S., Tymchak, D., & Aliieva, O. (2021). Effect of pop sorghum on the quality of gluten-free cereal bars. *Science Rise*, *6*, 3–10. https://doi.org/10.21303/2313-8416.2021.002199.
68. Dutta, H., & Mahanta, C. L. (2012). Effect of hydrothermal treatment varying in time and pressure on the properties of parboiled rices with different amylose content. *Food Research International, 49*(2), 655–663. https://doi.org/10.1016/j.foodres.2012.09.014.
69. Bora, P., Ragaee, S., & Marcone, M. (2019). Effect of parboiling on decortication yield of millet grains and phenolic acids and *in vitro* digestibility of selected millet products. *Food Chemistry, 274*(July 2018), 718–725. https://doi.org/10.1016/j.foodchem.2018.09.010.
70. Mahanta, C. L., & Bhattacharya, K. R. (1989). Thermal degradation of starch in parboiled rice. *Starch–Stärke, 41*(3), 91–94. https://doi.org/10.1002/star.19890410305.
71. Bhati, D., Singh, B., Singh, A., Sharma, S., & Pandiselvam, R. (2022). Assessment of physicochemical, rheological, and thermal properties of Indian rice cultivars: Implications on the extrusion characteristics. *Journal of Texture Studies, 53*(6), 854–869. https://doi.org/10.1111/jtxs.12678.
72. Asem, I. D., et al. (2019). Phenotypic characterization, genetic variability and correlation studies among ten Chakhao (scented) rice of Manipur. *International Journal of Current Microbiology and Applied Sciences, 8*(2), 612–618. https://doi.org/10.20546/ijcmas.2019.802.070.
73. Borah, N., Athokpam, F. D., Semwal, R. L., & Garkoti, S. C. (2018). Chakhao (Black Rice; *Oryza sativa* L.): A culturally important and stress tolerant traditional rice variety of Manipur. *Indian Journal of Traditional Knowledge, 17*(4), 789–794.
74. Roy, S., et al. (2014). Chakhao (delicious) rice landraces (Oryza sativa L.) of North-east India: Collection, conservation and characterization of genetic diversity. *Plant Genetic Resources: Characterization and Utilization, 12*(3), 264–272. https://doi.org/10.1017/S1479262113000580.
75. Corrêa, P. C., da Silva, F. S., Jaren, C., Afonso, P. C., & Arana, I. (2007). Physical and mechanical properties in rice processing. *Journal of Food Engineering, 79*(1), 137–142. https://doi.org/10.1016/j.jfoodeng.2006.01.037.
76. Chakraborty, P., & Banerjee, S. (2009). Optimization of extrusion process for production of expanded product from green gram and rice by response surface methodology. *Journal of Scientific and Industrial Research (India), 68*(2), 140–148.
77. Nascimento, T. A., Calado, V., & Carvalho, C. W. P. (2017). Effect of brewer's spent grain and temperature on physical properties of expanded extrudates from rice. *LWT–Food Science and Technology, 79*, 145–151. https://doi.org/10.1016/j.lwt.2017.01.035.
78. Bhattacharyya, M. K. R., & Bhattacharyya, M. (2016). Development of an innovative eco-friendly rice puffing machine for modernizing a traditional cottage industry. *African Journal of Science, Technology, Innovation and Development, 8*(5, 6), 371–376. https://doi.org/10.1080/20421338.2016.1219465.
79. Kumar Guru, P., Saha, S., Tiwari, P., Shrivastava, A. K., & Khandai, S. (2022). Advances in Rice Mechanization in India Introduction. *SATSA Mukhapatra-Annual Technical Issue* (Vol. 26).

80. Tadesse, T., Atnaf, M., Alemu, D., Tadesse, T., & Shiratori, K. (2019). *Advances in Rice Research in Ethiopia.*
81. Ponnusamy, K. A., & Karthikeyen, C. (2017). *Advances in Rice Cultivation: A Complete Guide on Rice*. [Online]. Available: https://www.researchgate.net/publication/357173061 (accessed on 9 August 2024).
82. Maiti, R., Sarkar, N. C., Rodriguez, H. G., Kumari, A., Begum, S., & Rajkumar, D. (2021). *Advances in Rice Science*. Apple Academic Press, T&F.
83. Guha, M., & Ali, S. Z. (2011). Changes in rheological properties of rice flour during extrusion cooking. *Journal of Texture Studies, 42*(6), 451–458. https://doi.org/10.1111/j.1745-4603.2011.00306.x.
84. Wattananapakasem, I., Penjumras, P., Malaithong, W., Nawong, S., Poomanee, W., & Kinoshita, H. (2021). Effect of heat–moisture treatment of germinated black rice on the physicochemical properties and its utilization by lactic acid bacteria. *Journal of Food Science and Technology, 58*(12), 4636–4645. https://doi.org/10.1007/s13197-020-04951-z.
85. Alavi, S., et al. (2008). *Rice Fortification in Developing Countries: A Critical Review of the Technical and Economic Feasibility*. 1–74.
86. Konwatchara, T., & Ahromrit, A. (2014). Effect of cooking on functional properties of germinated black glutinous rice (KKU-ULR012). *Songklanakarin Journal of Science and Technology, 36*(3), 283–290.

CHAPTER 6

HOT-EXTRUSION TECHNOLOGY FOR PRODUCTION OF CEREAL-BASED READY-TO-EAT SNACKS

HILAL ISLEROGLU

Tokat Gaziosmanpasa University, Faculty of Engineering and Architecture, Department of Food Engineering, Tokat, Turkey

ABSTRACT

Hot-extrusion is the process of thermo-mechanical conversion of material under high temperature and pressure. In the food industry, hot-extrusion is known as extrusion-cooking, in which the heating of food is usually carried out at a temperature higher than 100°C. Hot-extrusion (extrusion-cooking) is the shaping of a material containing starch and protein which is brought to the appropriate moisture. This process is carried out by passing the material through a die under mechanical shear stress, high pressure, and temperature which are created with the help of screw/screws rotating in a barrel. Extrusion-cooking is a simultaneous transfer of heat, mass, and momentum. Extrusion-cooking is often preferred in the production of ready-to-eat snacks because it has many advantages compared to other traditional food production methods. In this chapter, firstly, the principles of the hot-extrusion (extrusion-cooking) process and the process parameters that affect the product quality are briefly explained. Then, studies on cereal-based ready-to-eat snacks produced by extrusion-cooking process are reviewed, and how extrusion parameters affect the quality of a typical extruded product is discussed.

Ready-to-Eat Snacks: Emerging Technologies for Production and Safety.
Ravi Pandiselvam and Gülşah Çalışkan Koç

6.1 INTRODUCTION

Extrusion is defined as a process in which many processes such as mixing, heating, and shearing are combined, and the dough is shaped by pushing it from a die. Extrusion is a transport process consisting of the flow of materials in the system, thermal energy transfers between these materials and mass transfers in the process. Extrusion, a well-known technology in the plastics industry, is also widely used in the agro-food processing industry today. Compared to other food processing technologies, extrusion technology has become a highly preferred technology in recent years thanks to its versatility, high efficiency, low cost, energy efficiency and small amount of waste water generation [1, 2].

The starting point of the extruders is based on the Archimedes continuous screw, which is used to convey the feeding material from one end of the screw to the other. Today, extruders have been modified to allow the raw material to be cooked at the same time as it is being conveyed. The use of piston and pump was based on the construction of the first food extruders. Simple screw type extruders used for cutting by forcing soft food to pass through the die are the first applications. In Italy, in the mid-1930s, single screw extruders began to be used for pasta products. In the same period, similar extruders started to be used in the production of breakfast cereals in the USA. Extruders for curled corn snacks were developed commercially by Adam Corporation in 1946. Then, in the 1950s, extrusion was used in the production of pet foods, and from the 1960s, textured vegetable protein and ready-to-eat cereals were produced. Until the 1970s, single-screw extruders were used for all these applications, and after this date, twin-screw extruders began to be used in the food industry [3].

6.2 EXTRUSION TECHNOLOGY

6.2.1 PRINCIPLE OF EXTRUSION PROCESS AND EXTRUDER TYPES

The extrusion mechanism can be considered as a simple piston consisting of a cylindrical body with a die at the end of which the shaping is done. The material is loaded into the cylinder, the piston is pushed forward to create a pressure in the die, thus the material is removed from the die. As a result, a product of the desired shape is obtained. To make the extrusion process continuous, the piston in the cylinder has been replaced with a screw. In this screw extrusion, the material is pushed into the inlet groove and conveyed forward by the screw. As the material, namely the extrudate, approaches the head, significant increases in pressure and temperature occur in the cylinder. Thanks to the die placed at the end of the cylinder (barrel), a desired product is obtained [4].

The barrel, screw(s) and die are the three main components of an extruder. In addition, an extrusion production line includes other equipment such as motor, pump, cutting and feeding systems, collector, etc. The extruder screw is located in a barrel. Typically, the barrel is divided into sections of equal size. Thus, during the extrusion process, different parts of the extruder can be easily heated or cooled. Heating can be done with the help of steam or hot water in the jackets around the barrel, or with heating plates. Similarly, the cooling process can be applied with cold water or air circulating in the jacket. Screws used in extrusion systems have different configurations and profiles. The profile of the screws varies depending on the pitch, stagger angle, length, and locations. The screw profile affects the size and shape of the shear force acting on the material. In this case, it is a very important parameter for the extrusion process as it affects the quality of the final product. An extruder has 3 sections: conveying, compression/mixing and metering/melting. In the conveying section, the material is continuously pushed forward and delivered to the compression section. In the compression/mixing section, the next part of the extruder, the floury material is compressed to remove air while mixing the liquid and flour to develop a dough-like consistency. To perform this operation, the screw has shorter pitches in the part corresponding to this section. If this part is designed well by adjusting the screw configuration, the dough consistency will develop properly and thus the final product consistency will be uniform. At this stage, heat is released due to the shear that develops between the dough and screw and/or barrel. Metering/melting section is the last section where the dough is transferred to the die. In this section, due to shear and pressure (6.9×10^3–17.2×10^3 kPa), the material turns into a molten state with the heat released. Here, residence time is quite short (10–30 s). At the end of this section the die is settled. The die is a very critical part of the extruder, as it provides restriction and preferred shape of the product. After the die, there is a cutter system that ensures the size reduction after the product comes out [3].

Food extruders, which are increasingly used in food applications, are divided into two main groups: (i) Forming extruders: They are extruders with low shear force and pressure, and heat is not used for cooking. They are mainly used to form pellets with minimum energy input to the product. Application areas are generally pasta products, pretzels, and similar types.

These products come out of the die in the form of dough and are then cooked to give a harder structure. (ii) Cooking extruders: Cooking extruders are extruders in which the feed material is heated in the barrel and the feed exits the die under high pressure. They are used in cooking as well as creating puffed products [1, 4, 5].

Extruders, which basically consist of a continuous screw rotating in a barrel, can normally perform raw material feeding, cooking, and shaping operations with a single screw. In addition to single-screw extruders, twin-screw extruders are also frequently used due to their advantages.

In the food industry, the most preferred type of extruders is single-screw extruder. The variety of products obtained with single-screw extruders varies over a wide range such as fully cooked high-porosity corn cookies and shaped pastas. The design of these extruders is quite simple, and the main component is a screw. The design of the screw affects the working efficiency of the extruder. Single-screw extruders designed with increased screw diameter are the most commonly used types. The task of the screw is to mix the material, melt it, plasticize it and convey it to the die under pressure. In order for the screw to move the raw material, the raw material should not stick to the surface. Screw speed is an important parameter of the extrusion technology. The slow speed of the screw causes the dough to be exposed to more heat in the barrel. If it is too fast, it causes the material to not heat up enough. The mixing ability of single-screw extruder might be insufficient. To overcome this problem, the feed is usually premixed and preconditioned with water before extrusion.

Twin-screw extruder is more complex in design. There are two screws inside the barrel. Twin-screw extruders are classified according to the direction of rotation of the screws as co-current twin-screw extruders and counter-current twin-screw extruders. These extruders (co-rotating and counter-rotating) are also grouped in themselves as fully intermeshing and non-intermeshing types. Twin-screw extruders are versatile extruders with the ability to process numerous materials containing viscous and rigid ones [5]. Although the twin-screw extruders were first used in the plastics industry, they have now become very popular in the food industry with the increasing consumer demand for innovative products [3]. Twin-screw extruders are drag flow machines, similar to single-screw extruders. However, a positive displacement is a more dominant mechanism in these extruders. The type of transport is the main differences between these two extruders. In single-screw extruders, frictional and viscous drag flows are the main mechanisms of the transport in the conveying zone and the melt-conveying zone, respectively.

In twin-screw extruders, transportation is less related to the frictional properties of the material. Because in the intermeshing part, a positive displacement transport is provided by the movement of the other screw [4]. However, non-intermeshing types have limited positive displacement capacity as they do not have closed/semi-closed chambers. The intermeshing design of the screws offers many advantages. In twin-screw extruders with a fully intermeshing screw design, maximum positive displacement is achieved as material backflow is largely prevented. Since the length of the dough in the extruder is constantly disrupted by the flight of the screws, the formation of a continuous bed of material over the entire length of the extruder is prevented. This ensures high mixing efficiency. Also, fully intermeshing types have self-wiping/cleaning ability [6]. Counter-rotating twin-screw extruders generate high pressure as the screws rotate in opposite directions. Therefore, they are operated at a lower speed than co-rotating ones. Counter-rotating types also have better positive displacement ability due to the closed C-shaped chambers formed by intermeshing and reverse helices. However, the mixing ability is limited as the flow profile is deteriorated in these C-shaped closed chambers. Compression of the material in the intermeshing areas causes the screws to be pushed towards the barrel and abrasions occur. Shear stress distribution around the screw and temperature distribution are not uniform. Product deterioration may occur due to high pressure. Counter-rotating types are more often used in the plastics industries than in the food industry, where high pressure is required to process materials. Co-rotating types can operate at higher screw speeds than counter-rotating ones because they create lower pressure. Therefore, the product output speed is high. They convert mechanical energy into thermal energy better, and the temperature distribution control is better [7].

6.2.2 HOT-EXTRUSION (EXTRUSION-COOKING)

The use of extrusion technology in food applications began in the mid-1940s. These applications first started in order to shape pastas with cold extrusion. Subsequently, single screw extruders were used to produce cornmeal-based crunchy snack products. Extruders, which initially entered the food industry with limited use, are now seen as bioreactors operating on the high-temperature short-time (HT-ST) principle, used in the production of cereal-based foods, protein supplements, baby foods, crackers and snack foods, and converting waste raw materials into different products [8, 9].

The extrusion-cooking process is basically the process of both cooking and shaping mixtures of starch and protein with appropriate moisture content and particle size under mechanical shear stress, high pressure and temperature created by the screw rotating in a cylindrical channel. Because of the high pressure, temperature, and shear applied during extrusion, many changes occur in the food composition that lead to protein denaturation, starch gelatinization and complexes formation among the carbohydrate, lipid, protein and water molecules [10]. It is known that the nutritional and functional properties of food components change during the extrusion-cooking. As a result of the heat treatment, the starch granules swell by taking water into their structure and become gelatinized when the appropriate temperature is reached. On the other hand, as proteins are denatured, their digestibility increases and thus the structural properties of the extruded products are improved. The dough, which contains gelatinized starch in the barrel, expands when it reaches atmospheric conditions (rapid pressure and temperature drops) and gains a hard structure with the effect of sudden cooling [8]. The extrudate obtained by extrusion-cooking has a honeycomb structure shaped with molten protein fibers. As soon as the melt leaves the extruder, it takes the shape of the die. Extrudates suitable for the purpose can be produced by using die with different properties. Extrudates with very different shapes such as star, bowtie, ring, cone, ball, alphabet letter, etc., can be obtained. The length of the extrudate is adjusted as desired by adjusting the speed of the cutting system placed outside the die. The product, which is shaped as a result of the material coming out of the die, is then dried below 10% moisture content, and then the cooling process is carried out. Drum dryers or belt dryers with hot air circulation are generally used for the drying process. The cooling process is usually carried out with the air passed over the perforated belt at ambient temperature. If it is desired to add flavor to the final product or if any coating is to be made, this step can be applied at the drum dryer stage [5].

The extrusion-cooking process has many advantages compared to traditional cooking methods. Extruders are versatile tools that enable the cooking and processing of formulations prepared by combining raw materials with different properties in a single equipment. They have high productivity as they have higher production capacities than other cooking systems. With their modern process controls, they have less manpower requirements for production operations and lower costs per production unit. Products with very different shapes are obtained that cannot be easily produced with other cooking methods. With the application of high temperature and short-time

(HT/ST) heat, most microorganisms and pests are destroyed, while the nutritional value is preserved, thus high-quality products are obtained. It changes the digestibility and flavor properties of food by changing the structure of proteins and starch. This enables the production of new products. In addition, the extrusion-cooking is a hazardous material-free process with little or no waste [11].

The most important feature that distinguishes the extrusion-cooking process from convectional heat treatments is that it is a continuous process. This process operates in a dynamic and stable state where the system inputs (raw materials, additional ingredients, flavor and aroma substances, water, etc.,) and outputs (final product, water vapor, etc.,) are in balance. In the extrusion-cooking, heat, mass, and momentum transfer occur simultaneously. In addition, the retention time of the raw material significantly changes the physico-chemical properties of the final product. During the process, the cooking temperature may rise to 180–190°C, but the material's residence time in the barrel is quite short (~20–40 s). Accordingly, extrusion-cooking is defined as a HT/ST process [3, 4].

6.2.3 HOT-EXTRUSION PROCESS PARAMETERS

Extrusion, as a process that includes all mass, energy, and momentum transfers, is a very difficult process to evaluate together due to the interactions during the transfers and the complex physico-chemical transformations that affect the final product properties.

The most important parameters in the extrusion process are mass flow rate, moisture content, screw speed and screw configuration, die geometry, barrel temperature, and feed formulation.

6.2.3.1 MASS FLOW RATE

Mass flow rate in the extruder is the factor that directly determines the extrudate capacity, and its amount should be adjusted according to the openings of the die. Otherwise, it may cause undesired blockages in the die. When the mass flow rate is increased, the material's residence time in the barrel is shortened. In this case, sufficient temperature increase cannot be achieved, and therefore, the desired level of melt does not occur. As a result, the melt viscosity remains high and the dough cannot expand enough when it comes out of the die, resulting in a high density extrudate. Also, the specific mechanical energy decreases with increased mass flow rate. Therefore, it is one of the parameters that directly affects the quality of the final product and should be adjusted according to the extruder type, die configuration and thermo-physical properties of the material [4].

6.2.3.2 SCREW SPEED AND SCREW CONFIGURATION

In the extrusion process, shear increases at high screw speeds and mechanical energy also increases. The high mechanical energy and low motor torque cause the temperature of the melt and the pressure gradient to increase, which increases the expansion of the product. In addition to the screw speed, the configuration of the screw inside the barrel is also important in the extrusion process. Extrusion pressure can be increased by using fully filled screws with a large length/diameter ratio. This allows the material to plasticize more easily, and backflow is prevented. Generally, it is desirable to have a small gap between the barrel surface and the screw flights. If this gap decreases (screw diameter increases), motion of the material is restricted, friction force and pressure decrease. This causes the extruder to not work efficiently. In fact, the product output is reduced or completely stopped because the flow is blocked [4].

6.2.3.3 MOISTURE CONTENT

It is known that moisture of the conveying dough in the barrel is an important process parameter during the extrusion process. In extrusion-cooking, the moisture content behaves as a plasticizer in the biomaterial. Moisture content of the conveying dough lowers the glass transition temperature (T_g) and melting temperature. Less energy is required for processing as high moisture content lowers viscosity. Thus, a decrease in specific mechanical energy and motor torque is observed. Decreased specific mechanical energy and torque cause a decrease in temperature and die pressure. In this case, products with high bulk density are obtained. Since high moisture will also reduce friction, and the friction coefficient gradient between the barrel inner surface and the screw channel decreases. In this case the conveying principle becomes more important with single-screw extruders that depend on this gradient [5].

6.2.3.4 BARREL TEMPERATURE

Theoretically, the melt viscosity decreases with increasing barrel temperature, and accordingly the specific mechanical energy and motor torque also decrease. These changes, which trigger each other, cause a decrease in die pressure and the product density is expected to decrease. However, higher barrel surface temperature lowers the friction factor, resulting in more slippage and greater degree of fill. As a result, the torque actually increases with the increasing temperature and there is a higher input of mechanical energy. According to the product needs, the barrel can be cooled when necessary. The cooling process increases the friction in the material and increases the viscosity. Thus, it can contribute to the extruder performance. Thanks to the thermocouples placed in the barrel, the temperature can be monitored. However, the temperature of the material inside the extruder during the extrusion process is very difficult to determine exactly [4, 5].

6.2.3.5 DIE GEOMETRY

Die is the apparatus settled at the end of the barrel and restricting the open end of the extruder. It can have different geometries that give shape to the product to be obtained in the extrusion process. The die pressure, which determines many quality characteristics of the final product, depends on the geometry of the die. Small holes in the die cause higher resistance and increase pressure more. During the extrusion process, there may sometimes be blockages in the die holes. This causes a sudden increase in pressure and backflow. In such cases, it is useful to quickly drill the holes with a thin tool. If this intervention is not sufficient, the extruder must be stopped, and the die removed and cleaned [12].

6.2.3.6 FORMULATION

A food product must contain different components to meet the need for a healthy and balanced diet. Snack foods are usually made from grains. Cereals are ideal to produce expanded snacks by extrusion-cooking because they contain starch in their structure. Starch is the essential ingredient to produce a low-density, porous, and crunchy snack. However, cereals may be insufficient in terms of some essential compounds such as vitamins B_{12}, A, D and, zinc, calcium, and iron. Therefore, to produce healthier snacks in terms of nutrition and to meet consumer demands, different components can be added to the formulation as well as grains as a feed material in the extrusion-cooking process [13, 14]. However, the ingredients added to the formulation to increase the nutritional value may inhibit the starch gelatinization and ultimately impair some physicochemical and sensory properties of the snacks [15]. Refined flours are often used to obtain quality snacks in terms of textural properties. This can reduce the amount of dietary fiber taken with the diet. When extrusion is made with whole grain flour, the microstructure and texture of the product are adversely affected. Adding bran to the feed limits the starch gelatinization required to expand snacks and often results in higher density, lower expansion rates, firmer texture, less brittleness, and therefore less preferred products. This situation can be improved by adjusting the extrusion parameters or by adjusting the raw material formulation. There are studies to enhance the textural and rheological properties of the product, as well as the nutritional content, by adding components such as legumes, fruits and vegetables, different herbal extracts, gums, casein, egg white powder, emulsifiers, etc. [12, 16].

In addition to these independent process parameters, there are also system parameters that cannot be adjusted directly during the extrusion process, but change depending on the process parameters. They (residence time distribution and specific mechanical energy) change according to the rheological and physico-chemical properties of the material during processing [5].

6.2.3.7 SPECIFIC MECHANICAL ENERGY

The extent of required energy per unit mass during extrusion process is known as specific mechanical energy (SME, kJ/kg) and can be calculated by Eqn. (1)

$$SME = 2\pi \times T \times N / m \tag{1}$$

Where; T is the motor torque (Nm); m is the mass flow rate (g/min); and N is the screw speed (rpm) [14].

The magnitude of the specific mechanical energy has a direct effect on the reactions about starch and proteins namely gelatinization and denaturation during the extrusion process. These reactions that occur in these macromolecules depend on the extent of energy transferred to material. The higher the specific mechanical energy, the higher the gelatinization degree of starch. It is also associated with the melt viscosity in the barrel. Low specific mechanical energy is released due to the decrease in viscosity at high temperature and high feed moisture [17, 18].

6.2.3.8 RESIDENCE TIME DISTRIBUTION

The time that the material stays in the extruder is describe the residence time distribution. Knowing the distribution of time, the material spends in the extruder, provides information on mixing efficiency, understanding the effects of process parameters, used in scale-up and equipment design, and also allows process control. Particulates, liquid dyes, or molecular tracers are used as tracers to determine the residence time distribution. The detectability, rheological behavior, and miscibility of the tracer to be used with the material in the extruder are very important. While residence time distribution is significantly affected by the process variables namely mass flow rate, screw speed and screw configuration, it is not much affected by moisture content, die geometry and barrel temperature [5].

6.3 CEREAL-BASED READY-TO-EAT SNACKS PRODUCED BY EXTRUSION-COOKING

Snacks are foods and beverages that are usually consumed in small amounts between main meals, and they are often preferred because they are products that can be consumed quickly in daily life. Snacks can be classified according to the time they are consumed during the day. Snacks with low fat, sugar and salt content while containing the essential nutritional elements are considered healthy. However, snack products with high energy and poor nutritional components are also available in the market [19]. Snacks can be considered as ready-to-eat (RTE) foods which are a popular food group for many reasons such as ease of preparation and consumption, ease of storage, and attractive appearance in different shapes, colors and textures. RTE foods are defined as foods that can be consumed cold or hot, without the need for further processing. Extrusion-cooking has taken its place in the industry as an important technology in the production of RTE snacks such as breakfast cereals, cereal bars, extruded cereals, and biscuits [20].

Extrusion-cooking technology has been popular for a long time, especially in the production of cereal-based RTE snacks, as it has many advantages compared to other traditional food production methods. A typical feed to be used in the extrusion process consists of starch, protein, lipid, water, and additives. The effect of all these ingredients on the final product is very important. Along with starch, other components can also cause physicochemical interactions and rheological changes in the extruded product. Cereals contain different amounts of starch, protein, fat, and fiber as well as small amounts of sugar, vitamins, and minerals depending on the type. Starch is the main ingredient in providing the desired expanded structure and texture in the final product. For this reason, cereals rich in starch are preferred as the main raw material in the production of extruded RTE snacks [21].

Cereals are considered as valuable food and energy sources in human nutrition due to their rich composition. Cereals have been extremely important to people from past to present. They are used as the basic raw material of many foods around the world. The main component of cereals is starch, and they also contain vitamins, minerals and antioxidant components that are beneficial for human health. Therefore, products obtained from cereals, which are ideal for the extrusion process, have a rich composition. The extrusion-cooking process is a changeable and versatile operation that ensures the development of new snacks by adjusting different process parameters according to the product. During the extrusion-cooking process, a series of structural and chemical transformations occur in the feed material and these transformations determine the properties of the final product. Therefore, the use of different grain types, with different properties compatible with extrusion technology in terms of composition, in the versatile extrusion process offers unlimited alternatives for the development of new products and bringing them to the market. Today, the demand for RTE snacks is increasing day by day due to the changing dietary habits, the formation of different mealtimes with extended working hours, the increase in the number of single-person households, and the fact that individual family members offer different food options. Although wheat, oat, rice, and corn are generally used in the production of snack foods, there are products that use other cereals and their mixtures with different components. Cereal-based snacks are generally consumed by children. Therefore, it is very important to increase the nutritional value of such snacks. In this section, studies on cereal-based extruded products produced using different grains in recent years have been reviewed.

6.3.1 MAIZE (CORN)

Corn (*Zea mays*), commonly known as maize, is a grass variety belonging to the *Poaceae* family and is one of the widely grown plants. It is the staple food of many countries, especially in Africa and Latin America. Global corn production is increasing day by day as corn and corn flour are used in the production of many different foods. Apart from the food industry, corn is also used as a raw material in the feed and ethanol industry [22]. Depending on the concentration of beta-carotene and beta-cryptoxanthin it contains, there are different types of corn that vary in color from white to yellow, from red to black [23]. The nutritious part of the plant is the kernel, and it contains 72% carbohydrates, 8.84% protein and 4.5% fat. Carbohydrates found in corn consist of 25% amylose and 75% amylopectin. The main protein found in corn is zein and it does not contain gluten. In addition, corn is a source of vitamins and minerals and is rich in carotenoids [24]. Corn is a low-cost cereal used especially to produce expanded snack products and is widely used in the production of gluten-free snack extruded products [25].

In the production of RTE snacks, corn starch, corn flour or corn grits are often preferred with their superior expansion properties. However, since maize products are not rich enough in bioactive components, research has mostly focused on creating raw material formulations that will improve corn extrudates in terms of nutritional properties and optimizing extrusion conditions. Although there are many studies involving corn in extrusion-cooking technology, some of the studies related to corn and corn-based extruded products are given in Table 6.1.

6.3.2 WHEAT

Wheat (*Triticum* spp.) is one of the most used cereals in the world and many wheat classes exist. Different wheat classes are used in the processing of different products such as bread, pasta, confectionery products, and noodles. The large part of the global wheat production consists of bread and durum wheats. Wheat is generally considered primarily a source of calories. However, it has an important place in human nutrition having healthy nutritional components in its structure. The composition of wheat differs depending on genetic and environmental factors and their interactions [40]. Due to its high protein content, wheat can be harmful for those with celiac disease. For this reason, studies are carried out to decrease or remove the allergenic properties of wheat flour. It has been reported that extrusion process can be used for this purpose. With the effects of heat, shear force and pressure applied during the extrusion process, the proteins in wheat undergo denaturation and thus the fragmentation of the proteins can be achieved [41].

Studies in which wheat is integrated into extrusion technology are generally aimed at improving the quality of wheat flour. However, there are also studies targeting the production of wheat and wheat-based extruded snacks. Some of these studies, which are generally aimed at improving the nutritive properties of the extrudate, are given in Table 6.2.

TABLE 6.1 Studies on Corn and Corn-Based Extruded Snack Production

Product Type and Feed Formulation	Aim of the Study	Extrusion Type and Extrusion-Cooking Conditions				
		Type	T (°C)	SS (rpm)	Feed Rate	References
Puffed corn-fish snack Corn grits: 84.5% Minced fish meat: 15% Magnesium silicate: 0.5%	Optimize the extrusion conditions	Twin-screw	105–125	146–208	810–1,206 g/min	[26]
Functional corn snacks Corn grits: 90% Jerusalem artichoke tubers:10% Amaranth seeds: 10% Pumpkin flesh flour: 10%	To develop a functional corn-based snack rich in polyphenols with antioxidant properties by adjusting the formulation and process parameters	Twin-screw	120, 140 and 160	200 and 300	–	[15]
Expanded nixtamalized corn snacks Blue corn Moisture content: 14, 16.5, and 20.5%	To evaluate the effect of extrusion conditions	Single-screw	(first, second, third and fourth zone): 60/80/110/140	120	–	[27]
Corn snacks enriched with soybean sprout Corn grits Soybean sprout: 1.75%, 2.63% and 3.5% Moisture content: 12%, 15% and 18%	To produce corn-based snacks enriched with ferritin and to determine the effects of different moisture contents and temperatures on the product in the extrusion process	Single-screw	110, 140 and 170	90	–	[28]
Expanded corn snacks Mixture of corn and bean Corn: 60–70–84–100% Bean: 0–16–30–40%	To develop corn-based snacks using common bean flours with high nutritional/anti-oxidant value and to optimize the extrusion process parameters	Single-screw	120–170	50–240	–	[29]

TABLE 6.1 *(Continued)*

Product Type and Feed Formulation	Aim of the Study	Extrusion Type and Extrusion-Cooking Conditions				
		Type	T (°C)	SS (rpm)	Feed Rate	References
Corn snack Corn grits: 90–100% Defatted hemp cake: 0, 5, and 10% Moisture contents: 15, 20, and 25%	To produce value-added corn snacks using pressed hemp cake, a food industry by-product	Single-screw	150, 165, and 180	100	–	[30]
Expanded corn snack Blue corn: 66% Black bean: 33% Sweet chard: 1% Moisture contents: 20, 25, 30, 35, and 40%	To produce an expanded snack with high phytochemical content, increased functionality, and textural properties, and to optimize process conditions	Twin-screw	102, 110, 122, 134, and 142	96, 111, 133, 157, and 171	–	[31]
Corn snacks Corn grits: 86.7% Inulin-type fructans (ITFs): 13.3% (w/w) Favoring: bacon flavor and salt	To develop RTE corn snacks rich in fiber and to investigate ITF addition on the physico-chemical and sensory attributes of the product	Single-screw	90	398	–	[32]
Corn-based snacks Corn grits: 60% Garbanzo flour: 20–40% Dried distiller's grains with solubles: 0–20% Moisture contents: 14%–20% (wet basis)	To determine the effect of extrusion parameters on the dependent system variables such as apparent viscosity and SME, and to create a different usage area for dry distiller's grains	Single-screw	100–140	100–200	–	[33]
Expanded corn-based snacks with cheese flavor Corn grits Butyric acid: 0.4 g/100 g Cysteine: 0.2 g/100 g Moisture content: 15% (dry basis)	To develop an expanded snack with improved nutritional value, healthier, adequate physical and sensory quality characteristics	Single-screw	90	192	265 g/min	[34]

TABLE 6.1 *(Continued)*

Product Type and Feed Formulation	Aim of the Study	Extrusion Type and Extrusion-Cooking Conditions				
		Type	T (°C)	SS (rpm)	Feed Rate	References
Corn snack Corn grit Beetroot powder: 0%, 5%, 10% and 15% Moisture content: 16%	To enhance the nutritional content of the traditional extrudates	Twin-screw (CR)	125, 150 and 175	300	18 kg/h	[35]
Corn-based snacks Corn grits Date kernel flour: 0%–20% Moisture content: 14%	To improve the physicochemical properties of snacks and to optimize extrusion parameters	Twin-screw	120–160	120–220	40 kg/h	[36]
Corn-based extrudates 100% corn flour 75% corn + 25% pearl millet flour 75% corn + 25% red sorghum flour 75% corn + 25% quinoa flour 75% corn + 25% canary seed flour Moisture content: 28% (wet basis)	To determine the effects of the mixing different flours on the physical, mechanical, and microstructural properties of corn-based extrudates	Single-screw	135	160	–	[37]
Corn-based extrudates Corn: tomato pomace 95:5%–75:25% Moisture content: 12–20%	To develop corn-based extruded healthy, fiber and lycopene-rich ready-to-eat snacks and to optimize the process parameters	Twin-screw (CR)	125–185	200–600	–	[38]
Whole-grain corn-based breakfast cereals Whole corn flour Honey: 5–15% Moisture content: 14 to 18%	To optimize the process variables by replacing refined sugar with honey, which contains many bioactive compounds and is beneficial for health	Twin-screw (CR)	120–180	–	10 kg/h	[39]

T: Temperature, SS: Screw speed, CR: Co-rotating.

TABLE 6.2 Studies on Wheat and Wheat-Based Extruded Snack Production

Product Type and Feed Formulation	Aim of the Study	Extrusion Type and Extrusion-Cooking Conditions				
		Type	T (°C)	SS (rpm)	Feed Rate	References
Wheat-based Snack Wheat grain Different legumes (lentil, yellow pea, green pea chickpea pulses): 0, 5, 10, and 15% Moisture content: 12%	To produce wheat-based snacks using different legumes and to investigate whether the extrusion process will affect protein digestibility in vitro	Single-screw	180	210 ± 5	–	[42]
Whole wheat snacks Whole wheat flour Crude lycopene: 50–100 mg/100 g Tomato powder: 2–4 g/100 g Saffron extracts: 50–100 mg/100 g	To determine the effect of feed formulation and extrusion on the antioxidant capacity of the RTE snacks	Twin- screw (CR)	100	300	–	[13]
Wheat flour-based expanded snacks Whole wheat flour Honey: 5–15% Moisture Content: 14–20%	To enhance the functional and nutritional properties of extruded snacks using whole wheat flour and honey instead of table sugar, and to produce sweetened RTE extruded snacks	Twin-screw (CR)	120–180	500	10 kg/h	[43]
Expanded wheat flour extrudates Wheat flour Moisture content: 20%	To achieve the nitrogen assisted extrusion at different pressures (100 and 200 kPa) adapted to conventional extrusion process, to estimate the effectiveness of nitrogen as a blowing agent by determining both the physical and functional properties of the extrudates	Twin-screw	110	150	–	[44]

TABLE 6.2 *(Continued)*

Product Type and Feed Formulation	Aim of the Study	Extrusion Type and Extrusion-Cooking Conditions				
		Type	T (°C)	SS (rpm)	Feed Rate	References
Waxy-wheat snack Waxy-Pen wheat flour Bran (white, red, and purple bran) addition: 12.5, 25, and 37.5% Moisture content: 15%	To evaluate the efficacy of different colored wheat brans addition, and to formulate wheat and bran mixtures for producing wheat-based snacks having antioxidant properties	Twin-screw	140	250	2.96 kg/h	[45]
Colored-grain wheat extrudate Ten colored-grain wheat varieties Moisture content: 20%	To reveal the effect of the extrusion process on carotenoid degradation and to produce wheat-based extrudates	Twin-screw	130	200	–	[46]
Expanded extruded wheat Sprouted soft wheat variety Sprouting temperature and time: 20°C/1, 3, 5, 7, or 9 days Moisture content: 27%	To examine the effect of sprouting degree on extrudate properties	Twin-screw	140	200	6 kg/h	[47]
Wheat-corn-based extrudates Wheat flour: 40% Corn polenta: 48% Salt: 1% Calcium carbonate: 1% Citrus fiber and sugar: 0–10% Moisture content: 20%	To produce healthier snacks enriched with citrus fiber, and to investigate the ideal formulation for a high-quality product	Twin-screw (CR)	140	275	5.7 kg/h	[48]
Expanded extrudates Wheat starch or wheat flour: 96% Oleic acid: 4% Moisture Content: 18–24%	To produce expanded wheat-based snacks containing oleic acid, and to investigate the effects of occurred starch-lipid and lipid-protein complexes on the textural properties of the products	Twin-screw (CR)	72–128	110	Wheat flour: 29.90 kg/h, Wheat starch: 2.70 kg/h	[49]

TABLE 6.2 *(Continued)*

Product Type and Feed Formulation	Aim of the Study	Extrusion Type and Extrusion-Cooking Conditions				
		Type	T (°C)	SS (rpm)	Feed Rate	References
Expanded extruded products Soft durum wheat flour Moisture content: 16, 20, and 24% (wet basis)	To investigate the expansion properties of soft durum wheat, and to create a new use for soft durum wheat	Twin-screw (CR)	120, 140, and 160	150, 200, and 250	3.0 kg/h	[50]
Wheat-based extruded products Wheat flour: 38–100% Tef: 0–35% Tomato powder: 0–5%. Sesame protein concentrate: 0–25% Moisture content: 17%	To develop nutritionally rich wheat-based RTE snacks, and to optimize feed mixture ratios for an improved snack formulation with high added value	Twin- screw (CR)	160	185	~46 g/min	[51]

T: Temperature, SS: Screw speed, CR: Co-rotating

6.3.3 BARLEY

Barley (*Hordeumvulgare* L.) is a durable crop that can adapt to harsher environmental conditions with its high drought tolerance compared to other cereal grains [52]. Although it has a wide distribution around the world, the usage area of barley in bakery products has remained lower than other grains due to its poor cooking performance and low sensory quality of the products obtained. However, nowadays, it has become more and more preferred due to its high dietary fiber [53]. The starch, protein, and oil that barley has are mainly found in the endosperm. The starch content is about 54–75%, of which 27% is amylose and 73% is amylopectin. The protein ratio of barley is 10–20% and about 75% of the protein is gluten. Therefore, barley is not a suitable grain for celiac patients. While the fat content is 2–3%, the dietary fiber content is quite high, 11–34% [54]. It has been reported that due to its β-glucan content, which is about 10 times more than wheat, it has immune-enhancing, anti-hypertensive effects, as well as protection against cardio-vascular disease and type 2 diabetes [52, 55]. With its beneficial nutritional content for health, barley is a very advantageous grain for developing high fiber ready-to-eat snacks. Studies on barley and barley-based extruded snack production are given in Table 6.3. These studies are presented on the basis of product formulation, study aim, and extrusion-cooking process parameters.

6.3.4 RICE

Rice (*Oryza sativa* L.) is an important food for people in many countries, especially in Asia, and is the main cereal crop grown worldwide after wheat and corn. There are white, brown, and black forms. It has been reported that 100 g of brown rice contains 71.1 g of carbohydrates, 7.3 g of protein and 2.2 g of fat [65]. Especially brown rice is known to be rich in dietary fiber, ferulic acid, vitamins, and gamma-aminobutyric acid (GABA). Black rice is also beneficial in the prevention of chronic diseases with its phytochemicals and high anthocyanin content [66, 67]. Because of its very good textural, nutritional, and sensory properties, rice is incorporated into the formulation of a wide variety of foods including cereal-based snacks. In addition, it is an indispensable cereal in the production of gluten-free extrudates because it does not contain gluten [12, 25].

Due to its good expansion ability, rice is often preferred in the extrusion of expanded snacks in different shapes. It is also used as a complement to

TABLE 6.3 Studies on Barley and Barley-Based Extruded Snack Production

Product Type and Feed Formulation	Aim of the Study	Extrusion Type and Extrusion-Cooking Conditions				
		Type	T (°C)	SS (rpm)	Feed Rate	References
Barley-based extruded snacks Barley flour with spice mix (powders of the ginger, onion, garlic, chili and citric acid) Fortification with ferric pyrophosphate: 16 mg/100 g of flour Moisture 10%–22%	To develop barley-based snacks rich in iron and having low-glycemic index	Twin-screw (CR)	95–160	215–385	–	[56]
Barley-based unexpanded extrudates Barley flour (naked): 45% Black rice flour: 20% Oat flour: 10% Colored kidney bean flour: 5% Black sesame flour: 5% Gum Arabic: 5% Moisture content: 25% to 35%	To develop barley-based and mixed-grain extruded products, and to produce diabetic-appropriate low-glycemic index snacks	Twin-screw	65–85	70–80	–	[57]
Barley extrudates Pearled barley flour (9.10% moisture, 0.77% ash, 8.90% protein, 0.95% oil, 66.7% starch, 0.08% resistant starch, wet basis) Moisture content: 20–24% (wet basis)	To determine the quality characteristics of hull-less barley snacks affected by extrusion-cooking process parameters effects	Twin-screw (CR)	120–150	260–340	14 kg/h	[58]
Barley RTE snacks Hull-less waxy barley: 40% fixed Brown rice: 0–60% Corn: 0–60% Moisture content: 15% (wet basis)	To optimize the extrusion-cooking conditions	Single-screw	110, 120, and 130	200 and 300	–	[59]

TABLE 6.3 *(Continued)*

Product Type and Feed Formulation	Aim of the Study	Extrusion Type and Extrusion-Cooking Conditions				
		Type	T (°C)	SS (rpm)	Feed Rate	References
Expanded barley extrudates Five barley varieties (three hull-less and two hulled) Moisture content: 15%, 20%, and 25% (wet basis)	To investigate the extrusion properties of whole flours obtained from hull-less and hulled barley, and to produce snacks containing nutritious whole barley flours	Twin-screw (CR)	120, 140, and 160	150, 200, and 250	100 g/min	[60]
Expanded barley-based snack Barley flour: Carrot pomace (90:10, 82.5:17.5 and 75:25)	To produce RTE expanded barley-based snacks rich in fiber, and to optimize process conditions	Twin- screw (CR)	120, 145, and 170	160	40 kg/h	[16]
Barley-corn snack Barley: 0–100% Corn: 0–100% Moisture content: 38%- 45%	To develop barley-corn snacks, and to identify the quality characteristics to assess whether it is a healthy product	Single-screw	130–180	20	–	[61]
Barley- chickpea extrudates Pearled barley flour: chickpea flour (40:60) Moisture content: 16–20%	To investigate the effect of extrusion-cooking on the physical properties of barley-chickpea snacks	Twin-screw	150–170	150–170	14 kg/h	[62]
Hulless barley extrudates The hulless barley flour Moisture content: 12, 15, and 18%	To optimize extrusion-cooking process parameters to produce hulless barley extrudates	Single-screw	100, 110, and 120	300, 350, and 400	–	[63]
Barley extrudates Six different cultivars of barley Moisture content: 30%	To investigate how extrusion processing factors affect system properties as well as physical properties of the product	Twin-screw (CR)	150 and 180°C	400 rpm	20 kg/h	[64]

T: Temperature, SS: Screw speed, CR: Co-rotating

many rice-based product formulations, with its low protein content limiting non-enzymatic browning reactions and its bland taste. However, studies have been carried out to develop different mixture formulations for ready-to-eat snacks due to their high calorie content, low fiber, protein, and bioactive components and to optimize extrusion process parameters. Studies on extruded rice and rice-based products are summarized in Table 6.4.

6.3.5 MILLET

Millet is another cereal belonging to the *Poaceae* family. Millet, the oldest known cereal grain to be used by humans, is mostly grown in places such as Asia and Africa where the annual precipitation is low. Millet is also a pest and disease resistant grain. There are seven varieties of millet (proso, pearl, foxtail, finger, barnyard, kodo, and little millet) in different colors, shapes, and sizes. It contains approximately 65–75% carbohydrates, 7–12% proteins, 2–5% fats and 15–20% dietary fibers in its structure. It also contains vitamins, minerals and numerous phenolic compounds [76, 77]. Due to the polyphenols, flavonoids and antioxidants in its structure, it has nutraceutical properties that reduce cancer risk, regulate blood pressure, lower cholesterol, prevent stomach problems, and protect against cardiovascular diseases. It is a useful grain for diabetic and obese patients with its low glycemic index. Also, millet does not contain gluten. Hence, it can be consumed by celiac patients and gluten-sensitive people. With these properties, millet is added to different flour mixtures and used in the production of puffed/popped products, flakes, and different extruded products [14]. Studies conducted for different purposes, examining the effects of feed formulation and process parameters in millet and millet-based extruded snack production since 2016, are presented in Table 6.5.

6.3.6 OAT

Oat (*Avena sativa* L.) differs in composition from other cereal products with its high protein, lipid and soluble β-glucan content. It is known that oats contain ~ 60% carbohydrates, 14% proteins, 7% fats and 4% fibers (β-glucan) [83]. In addition to this unique composition, thanks to the antioxidant compounds known as avenanthramides (AVNs) in its structure, it has cholesterol-lowering, protective effects in cardiovascular diseases and regulating blood pressure [84]. Oat, which is generally preferred in breakfast

TABLE 6.4 Studies on Rice and Rice-Based Extruded Snack Production

Product Type and Feed Formulation	Aim of the Study	Extrusion Type and Extrusion-Cooking Conditions				
		Type	T (°C)	SS (rpm)	Feed Rate	References
Thiamine-rich snack Brown rice flour with spice mix (sugar, powders of the ginger, onion, cheese, garlic, chili and citric acid) Moisture content: 10–22%	To produce a ready-to-eat snack using brown rice with an increased vitamin B1 content compared to white rice and high sensory quality and to optimize processing conditions	Twin- screw (CR)	95–160	215–385	-	[68]
Rice-based shrimp snack Rice grits: polished rice grains (1:1) Shrimp (whole shrimp-with shell-converted into flour):16–184 g/kg NaCl: 3% Moisture content: 106.4–173.6 g/kg	To produce snacks with increased protein and fiber content and improved nutritional value and to determine the effects of extrusion-cooking process on functional and sensory properties of the extruded rice-shrimp snacks	Single-screw	63.2–96.8	175	290 g/min	[69]
Expanded snack Fermented rice: black gram flour (3:1) Solid-state fermentation: flour mix+ yeast suspension (3% yeast, 4% sugar in 85% water) Moisture content: 10–15%.	To develop a RTE puffed snack based on fermented rice-black gram flour, to determine the extrusion-cooking conditions on the physicochemical and bioactive properties of the product, and to optimize the extrusion process	Twin-screw (CR)	100, 110, 120	250, 300, 350	11 kg/h	[70]
Rice-based snack Rice flour: 65–85% Defatted soy flour: 7.5–17.5% Carrot pomace powder: 7.5–17.5% Cauliflower trimmings powder: 7.5–17.5% Salt	To develop a RTE snack rich in fiber using vegetable by-products, and to optimize the extrusion process	Twin-screw (CR)	125–175	300–500	–	[71]

TABLE 6.4 *(Continued)*

Product Type and Feed Formulation	Aim of the Study	Extrusion Type and Extrusion-Cooking Conditions				
		Type	T (°C)	SS (rpm)	Feed Rate	References
Rice-mushroom snacks Broken jasmine rice: 60.5% Dried oyster mushroom: 15% Soybean flour: 18% Vegetable oil: 1% $CaCO_3$: 0.5% Modified starch: 5% Moisture content: 12–18% (wet basis)	To formulate an enriched snack with health-promoting components of oyster mushroom and high protein content of soy flour, and to propound the quality parameters of the rice-based snacks after extrusion-cooking	Twin-screw (CR)	130, 140, 150	275, 350, and 425	17 kg/h	[72]
Gluten-free-rice-based extruded snacks Commercial white rice: 50–100% Cowpea: 0–30% Whey protein concentrate: 9–20% Moisture content: 9.10- 11.42%	To produce a new, RTE rice-based snack rich in fiber and protein, and to evaluate the effects of extrusion on antinutrients and digestibility of proteins	Twin-screw (CR)	110	251–253	–	[73]
Expanded rice pellets Rice flour Moisture content: 27, 29, and 31% (wet basis)	To evaluate the effect of temperature and feed moisture content on the microstructural properties of the product (third generation (3G) snack)	Twin-screw (CR)	110, 135, and 150	100	–	[74]
Rice-based extruded snack Modified rice flour (resistant starch: 6.20%, glycemic index: 75.10, and glycemic load: 50) Moisture content: 10%–35%	To demonstrate that the use of modified rice flour in snack production is a new way to produce healthier snacks that do not raise blood sugar levels, and to investigate the effects of extrusion process parameters on resistant starch, physical and sensory properties of snacks	Twin-screw (CR)	130–180	150–200	–	[75]

T: Temperature, SS: Screw speed, CR: Co-rotating

TABLE 6.5 Studies on Millet and Millet-Based Extruded Snack Production

Product Type and Feed Formulation	Aim of the Study	Extrusion Type and Extrusion-Cooking Conditions				
		Type	T (°C)	SS (rpm)	Feed Rate	References
Finger millet- based extruded snack Finger millet (ragi) flour: 40% Rice flour: 30% Corn flour: 30% Banana powder: 1–4% Cheese: 3% (constant for all formulations)	To produce finger millet-based RTE snacks with banana powder and determine the effects of extrusion-cooking process	Twin-Screw	118–122	345–355	–	[78]
Millet-maize based snacks Whole grain pearl millet: 50% Maize grits: 50% Moisture content: 11–20% After extrusion: Flavoring: 2% Salt: 2% Soybean oil: 10%	To develop snacks with low-cost ingredients namely millet and maize, to propound the good quality and health benefits of the extruded snacks	Single-screw	69–120	250	300 g/min	[79]
Multigrain ready-to-eat snack Pearl millet: 30% Corn: 30% Rice: 25% Green gram: 8% Cowpea bean:7% After extrusion Salt: 2% Moisture content: 8–12%	To improve multigrain RTE snack food by extrusion-cooking, and to optimize the process parameters	Twin screw (CouR)	117–121	300–350	–	[80]

TABLE 6.5 *(Continued)*

Product Type and Feed Formulation	Aim of the Study	Extrusion Type and Extrusion-Cooking Conditions				
		Type	T (°C)	SS (rpm)	Feed Rate	References
Iron-rich extruded snacks Banyard millet: Defatted soy flour blend ratio (2:1 to 6:1) Amla (Indian gooseberry): 10–14% Rice flour: 20% fixed Moisture content: 14% (dry basis)	To develop iron and protein rich extruded ready-to-eat snacks for anemic population and to optimize the process variables for better consumer acceptability	Twin-screw (CR)	100–115	350	10 kg/h	[81]
Extruded RTE products with millet Composite flour: Finger millet: 20–30% Maize: 40–55% Defatted soy: 10–25% Elephant foot yam: 10% fixed Moisture content: 15% (wet basis)	To determine the optimum blend level of millet, maize, and soy to develop ready-to-eat extruded product	Twin-screw	100	300	5 kg/h	[21]
Millet extrudates Dehulledproso millet flour Moisture content: 17–25%	To determine the effects of extrusion-cooking parameters on quality and antioxidant properties of millet extrudates	Twin-screw (CR)	90–150	170–250	76 g/min	[82]

T: Temperature, SS: Screw speed, CR: Co-rotating, CouR: Counter-rotating cereals and snack bars, is recommended to be added to many different products due to these positive effects on health [85]. However, the high fiber content, which is considered beneficial for health, and the high amount of oil in its structure may adversely affect the rheology and physicochemical stability of extruded products. For this reason, it is generally preferred to be used in extrusion processes as a mixture with products such as rice, corn, and wheat flour [86]. With its excellent lipid profile and high soluble fiber content, oat has started to be increasingly preferred in extruded snack compositions, and related studies are listed in Table 6.6.

6.3.7 SORGHUM

Sorghum (*Sorghum bicolor* L. *Moench*) is a grain belonging to the *Poaceae* family and is the fifth widely grown cereal in the world. It is a very drought tolerant grain that is easily grown in arid and semi-arid regions. Containing a large amount of starch (55.6–70.0%) in its structure, sorghum has 7–15% protein. The oil content is higher than rice and wheat and is at the level of 3% [91]. It is also rich in phenolic acids, flavonoids and anthocyanins. Due to its bioactive components, it has cholesterol-lowering, anti-diabetic, anti-inflammatory, antithrombotic, antioxidation and anti-cancer properties [92, 93]. In addition to its beneficial components for health, it is a grain suitable for gluten intolerant and celiac patients as it does not contain gluten. It is frequently included in formulations in extruded snack products and is available in ready-to-eat forms especially for celiac patients [25]. Studies in which sorghum was included in the formulation to produce extruded snacks are given in Table 6.7.

6.3.8 RYE

Rye (*Secalecereale* L.) belongs to the grass family of *Poaceae* and has been cultivated since ancient times. The starch content, which is the main component of rye, is slightly lower than that of wheat and is between 55 and 65%. Protein content varies between 8 and 15%. Rye is frequently used in the production of products such as bread, biscuits, and flakes in the food industry, as it is very rich in gluten. The fat content of rye is 2–3%, and it ranks first among cereals in terms of fiber content (15–25%) [98]. While arabinoxylan constitutes most of the dietary fibers in its structure, there are also fructans, β-glucans and lignins. In addition, rye contains bioactive components with

TABLE 6.6 Studies on the Production of Extruded Snacks Containing Oat

Product Type and Feed Formulation	Aim of the Study	Extrusion Type and Extrusion-Cooking Conditions				References
		Type	T (°C)	SS (rpm)	Feed Rate	
Oat fiber-corn-based snack Corn grit: 68.7% Oat fiber: 30.0% NaCl: 0.3% $CaCO_3$: 1% Powdered turmeric extract: 1.5% (w/w) Moisture content: 21, 28, and 35%	To produce oat fiber-corn-based extruded snacks enriched with curcuminoids, to produce a functional food with health benefits of curcuminoids and dietary fiber by extrusion technology, to investigate extrusion conditions to enhance the sensory properties of extruded snacks rich in curcumin	Twin-screw (CR)	80/90/100/110 (the barrel has four zones)	200 and 300	~4 kg/h	[87]
Nutritionally enriched snacks Oats: 21.72% Green pea flour: 8.62% Fenugreek seed: 1.78% Fenugreek leave powder: 0.66% Rice +corn flour (80:20): 67.22% Moisture content: 12%	To produce snacks rich in nutritional compounds, and to examine the extrusion-cooking process effects on the developed snacks	Twin-screw	110	200	–	[88]
Porridges-like extrudates White rice flour + whole oat flour (60/40) Moisture content: 11.4%	To obtain ready-made extrudates from a mixture of whole oat flour and rice flour using the extrusion- cooking technique	Twin-screw (CR)	90–187	109–187	250 kg/h Water feed rate: 40 L/h	[86]
Tef-oat-based healthy snack Oat flour (Avena sativa): 0, 15, and 30% Tef flour (E. tef): 70, 85, 100% Moisture content: 12, 14, and 16% (wet basis)	To produce a value-added low glycemic index healthy snack, to investigate the effect of extrusion-cooking parameters for tef and oat flour formulation on extrudate physical and protein quality, and to optimize the process	Single-screw	80, 110, and 140	80, 140, and 200	–	[89]

TABLE 6.6 *(Continued)*

Product Type and Feed Formulation	Aim of the Study	Extrusion Type and Extrusion-Cooking Conditions				References
		Type	T (°C)	SS (rpm)	Feed Rate	
Expanded snacks Oat fiber: 40% Pea starch: 50% Pea protein: 10% Moisture content: 11.2–16.8% (dry basis)	To produce fiber and protein rich snacks, to investigate how the extrusion-cooking affects the expansion of a fiber-protein rich mixture	Twin-screw	146–175	200 rpm	0.68 kg/h	[90]

T: Temperature, SS: Screw speed, CR: Co-rotating

TABLE 6.7 Studies on the Production of Extruded Snacks Containing Sorghum

Product Type and Feed Formulation	Aim of the Study	Extrusion Type and Extrusion-Cooking Conditions				References
		Type	T (°C)	SS (rpm)	Feed Rate	
Puffed snacks Whole grains of maize: rice: sorghum flours (70:15:15, 15:70: 15, 15:15:70, 42.5:42.5:15, 42.5:15:42.5, 15:42.5:42.5, 33.3–33.3–33.3) Moisture content: 14% (wet basis)	To produce crispy puffed nutritious RTE snacks with high dietary fiber and protein content using sorghum, corn, and rice.	Single-screw	130	200	4.5 kg/h	[94]
Expanded composite extruded products Sorghum: barley: chickpea grains (70:15:15, 65:20:15, 60:25:15, 55:30:15, 50:35:15) Moisture content: 8, 10, 12, 14, 16%	To develop multi-ingredient sorghum-based extruded snacks, and to evaluate the quality characteristics of the final product	Twin-screw	120–200 at five level	120–200 at five level	–	[95]
Sorghum-barley extrudate Red sorghum: brown barley (60:40) Moisture content: 25, 30 and 40%	To investigate the effects of extrusion-cooking on digestibility of starch and quality properties of sorghum-barley extrudate	Twin-screw (CR)	140	150, 220 and 300	2 kg/h	[96]
Sorghum-based extrudates Corn flour: Sorghum flour: apple pomace (10:90:10, 10:80:20, 10:70:30) Moisture content: 25, 30 and 35% (wet basis)	To incorporate sorghum and industrial waste apple pomace into human diet to produce nutritious, healthy, and gluten-free snacks for celiac patients	Single-screw	80, 110 and 140	100, 150 and 200	–	[97]

T: Temperature, SS: Screw speed, CR: Co-rotating

antioxidant properties, and it is rich in vitamins and minerals [99, 100]. Some studies on extruded products produced using rye are summarized in Table 6.8.

6.3.9 *TRITICALE*

Triticale (x*Triticosecale* Wittmack) was developed by crossing rye and wheat with the aim of combining the positive aspects of these two grains. The resistance of rye to harsh conditions and the versatility of wheat with high yield and good grain quality in food applications are combined. Therefore, triticale has the characteristics of wheat and rye. However, its chemical composition is more similar to wheat [101]. Carbohydrates make up most of the dry weight of triticale, which is a source of starch, protein and dietary fiber. Its dietary fiber content is similar to wheat but contains higher amounts of soluble arabinoxylans. It also has phenolic acids, flavonoids, alkylresorcinols, anthocyanins and vitamins with antioxidant activity [102]. Although triticale has several applications, its use in the food industry has been limited compared to other grains. The main reasons for this situation are the low stability of its technological properties, high amylolytic activity and low rheological properties of the dough, as it contains less gluten than wheat and has a lower gluten quality [103]. Hence, only one study has been found in which triticale extrudate production has been performed since 2016 (Table 6.8).

6.3.10 *PSEUDO-CEREALS*

Pseudo-cereals are defined as fruit and seed-producing plants that can be used as flour for bread and other foods but do not belong to the grass family. Pseudo-cereals are similar to real cereals in their functional properties and composition. The best-known fake cereals are buckwheat, amaranth and quinoa. Pseudo-cereals do not contain gluten, but are high in protein. Pseudo-cereals have positive effects on human health thanks to the flavonoids, phenolic acids, trace elements, fatty acids, saponins and vitamins they contain. In addition to being rich in health-friendly components, the fact that they are an alternative for people with gluten sensitivity increases the work on the commercialization of these products [110]. Studies on buckwheat, amaranth and quinoa used in the production of different extruded snacks are given in Table 6.9.

TABLE 6.8 Studies on the Production of Extruded Snacks Containing Rye and Triticale

Product Type and Feed Formulation	Aim of the Study	Extrusion Type and Extrusion-Cooking Conditions				References
		Type	T (°C)	SS (rpm)	Feed Rate	
Triticale extrudates Triticale flour Moisture content: 16 and 20% (wet basis)	To produce snacks from whole grain triticale, and to investigate how the extrusion-cooking changes the macroscopic and molecular properties of the extrudate	Single-screw	135 and 175	80	25 kg/h	[104]
Rye extrudates Whole grain rye flour Bilberry press cake powders: 10% and 25% Moisture content: 11–12.5%	To obtain a value-added extruded product with high nutritional value and flavor with the phenolic compounds found in Bilberry press cake.	Twin-screw (CR)	135–128–89–69	360	72 g/min	[105]
Rye extrudates Whole grain rye flour NaCl: 0.8%	To investigate the effects of different processes including extrusion-cooking affects the microstructure and composition of rye extrudates, their digestive properties	Twin-screw (CR)	124	400	50 g/min	[106]
High dietary fiber rye extrudates Rye flour Fermented rye bran obtained from rye kernels: 20% and 40%	To demonstrate how fermentation affects the behavior of rye bran in extrusion process	Twin-screw (CR)	120	345	60 g/min	[107]
Puffed rye extrudates Mixture of rye endosperm flour (70%) and corn starch (30%) Rye bran: 15 or 30%	To produce rye bran-enriched extrudates and to investigate the structural and digestive properties	Twin- screw (CR)	110	500	67 g/min	[108]

TABLE 6.8 *(Continued)*

Product Type and Feed Formulation	Aim of the Study	Extrusion Type and Extrusion-Cooking Conditions				References
		Type	T (°C)	SS (rpm)	Feed Rate	
High fiber rye extrudates Endosperm rye flour Rye bran (fermented with a dextran-producing culture): 32 g/100 g	To produce extrudates high in dietary fiber and to investigate the effects of bran modification, acidification, and exopolysaccharides in extrusion	Twin-screw	120	350	60 g/min Water feed rate: 3.5 and 4.5 g/min	[109]

T: Temperature, SS: Screw speed, CR: Co-rotating.

TABLE 6.9 Studies on the Production of Extruded Snacks Containing Pseudo-Cereals

Product Type and Feed Formulation	Aim of the Study	Extrusion Type and Extrusion-Cooking Conditions				References
		Type	T (°C)	SS (rpm)	Feed Rate	
Extruded snacks Mixture of peanut flour and buckwheat flour (1:2) (PBF) PBF: 20–40% Rice flour: 60–80% Moisture content: 16–24%	To develop extruded snacks with different flours, and to determine the appropriate extrusion-cooking parameters for a quality snack	Twin-screw	110–150	*10–30 Hz	–	[111]
Extruded snacks Buckwheat-corn flour mixtures containing 0, 10, 20 and 30% buckwheat flour Moisture content: 16%	To improve the corn-based extrudate with buckwheat in terms of nutritional value, and to establish the addition of buckwheat, and extrusion temperature on the quality of the extrudates	Twin-screw (CR)	130, 150 and 170	500	28 kg/h	[112]
Expanded snacks Quinoa, amaranth, buckwheat as pseudo-cereals-Lentil, lupin, faba bean as protein sources-Wheat starch, maize starch, pea starch as starch sources Moisture content: 33% (dry basis)	To produce expanded snacks with protein content of 30–70% using protein isolates and concentrates from pulses and pseudo-cereals flours	Twin-screw (CR)	140 and 160	300	–	[113]
Amaranth-based extruded snack Amaranth: shallot: soy cake 55:25:20%, 75:14.81:10.19%, 60.43:25:14.57%, 70:25:5% (Four different mixture ratios) Moisture content: 20%	To develop amaranth-based extruded snack rich in shallot, and to examine the diabetic properties	Single-screw	100	145	–	[114]
Gluten-free extruded snacks Amaranth flour: 20, 35 and 50% or Quinoa flour: 20, 35 and 50% Corn flour: 50, 65 and 80% Moisture content: 14–18%	To obtain expanded extrudates despite 50% amaranth and quinoa content, and to produce gluten-free snacks with increased nutritional content	Twin-screw	140–160	200–500	86 g/min	[115]

TABLE 6.9 *(Continued)*

Product Type and Feed Formulation	Aim of the Study	Extrusion Type and Extrusion-Cooking Conditions				References
		Type	T (°C)	SS (rpm)	Feed Rate	
Instant-extruded products Amaranth: 18.7–33.1% Flaxseed: 6.6–9.3% Maize grits: 55.6–67.3% Minor ingredients (sucralose, cocoa, and cinnamon): 4.7% Moisture content: 22%	To produce extrudate rich in fibers, and to determine its physicochemical and functional properties	Twin- screw (CR)	150	272	7.5 kg/h	[116]
Extruded snack Biofortified maize flour with provitamin A Amaranth leaf powder: 0, 1 and 3% Moisture content: 20%	To produce an extruded snack with high nutritional content and consumer acceptability by adding Amaranth leaf powder to maize flour fortified biologically with provitamin A	Twin-screw (CR)	140.7	400	–	[117]
Expanded snacks Amaranth: 20, 35, 50% Quinoa: 20, 35, 50% Kañiwa: 20, 35, 50% The remainder is corn flour Moisture content: 14%	To examine the effect of ingredients on sensory attributes of the expanded snacks	Twin-screw (CR)	140	500	86 g/min	[118]
Puffed quinoa Quinoa seeds Moisture content: 14%	To characterize the physical, chemical and nutritional properties of expanded quinoa	Single-screw	-	500	43 kg/h	[119]
Quinoa snack Quinoa flour Moisture content: 5%	To develop the use of quinoa in ready-to-eat traditional Peruvian products and to examine the effect of extrusion on protein properties	Single-screw	95	149	90 g water/ 1 kg quinoa flour	[120]

T: Temperature, SS: Screw speed, CR: Co-rotating, *screw speed was given in different unit

6.4 EFFECT OF EXTRUSION PROCESSING ON QUALITY PARAMETERS OF CEREAL-BASED READY-TO-EAT SNACKS

As can be seen from the studies in the literature, the effects of optimization process parameters on the quality properties of the final product were generally investigated in the production of cereal-based ready-to-eat snacks with extrusion-cooking. In these studies, quality criteria were generally chosen as dependent variables and their maximization was aimed. Except for the products produced for very specific purposes, the expansion ratio, bulk density, water adsorption index and water solubility index were examined as quality criteria for extruded snacks, and also the sensory properties and nutritional properties of the final product were generally evaluated.

6.4.1 EXPANSION RATIO

Expansion is one of the important quality criteria as it affects the crispness of extruded snack products. The expansion of cereal-based extruded snacks is mainly related to the starch content of the grain or grain mixture used. The addition of components with high starch content to the feed mixture increases the expansion ratio. The expansion ratio provided in the final product and the different textural properties that change accordingly vary depending on the difference in amylose content in the raw material, the amylose/amylopectin ratio, the ratio of long/short chains in amylopectins, particle size, degree of gelatinization, gelatinization temperature and process parameters [121]. The expansion ratio in extruded products is also related to the interactions of starch and protein molecules in the material. It can change and improve the expansion ratio by affecting the number and distribution of internal air spaces in the product by starch-protein interaction [9]. Guldiken et al. [62] determined that the protein content significantly affects the expansion ratio in extruded products and the addition of protein reduces the expansion ratio and results in a harder structured product. In addition to proteins, adding fiber, sucrose or salt to the feed mixture can also alter the expansion ratio. It is known that the fibers bind with free water and reduce the expansion capacity of the dough [14, 29]. On the other hand, the addition of some high protein and crude fiber flours such as quinoa or canary seed has been reported to increase expansion ratios in cereal-based extrudates. Studies show that the type of protein and fiber added affects the expansion ratio [37, 122].

In addition to the raw material content, process parameters such as barrel temperature, feed moisture, screw speed also affect the expansion ratio. High barrel temperature, high screw speed and low feed moisture cause high shear. High shear also produces more changes in starch molecules and promotes gelatinization of starch. Thus, higher expansion ratio is obtained. In addition, low feed moisture combined with high barrel temperature causes a pressure increase in the barrel. This creates an increasing pressure gradient in front of and behind the die, thus increasing the expansion ratio. As the material leaves the die, expansion occurs in the hot viscoelastic material with the sudden evaporation of water vapor. Bubbles form, grow larger and eventually break off from each other, creating a large number of voids in the product. Thus, extruded products with high expansion ratio and high porosity are obtained [123]. In a study with finger millet-based extruded snacks, an increase in the expansion ratio was observed when the barrel temperature was increased at high feed moisture content. This increase in expansion ratio was explained by the increase in gelatinization degree of starch with increasing barrel temperature at high moisture content. In addition, increased die pressure with overheating provided high expansion ratio and high porosity in the final product [78]. Similar results were observed for extruded products prepared with pearl millet, kodo millet, barnyard millet, proso millet, fox-tail millet and pearl millet [14, 18, 82, 124].

Oliveira et al. [79] reported that the moisture content of the extruded snacks obtained with the maize and pearl millet mixture affects the expansion ratio of the extruded final product. The highest expansion values were obtained in the raw materials under less moisture conditions. Especially at high humidity values, when the temperature is not high enough to evaporate more water from the raw material, the expansion decreases and a high-density product with collapsed cells is obtained. The researchers also emphasized that the sizes of different particles should be equal, especially in feeds prepared with a mixture of different materials, and reported that the larger the particle, the lower the expansion. Similarly, Zambrano et al. [74], in their study where they produced rice-flour pellets by extrusion process, they determined a significant increase in the expansion ratio when the moisture content of the feed decreased at constant temperature. This was explained by the fact that high feed moisture content reduces the viscoelasticity of the dough in the extruder, resulting in lower expansion. Rani et al. [70] produced extruded snacks using fermented rice-black gram flour at different die diameters, screw speeds and barrel temperatures. They determined that the expansion ratios of the snacks varied between 2.7 and 4.1, and they observed higher puffing at low die diameter (3 mm) compared to high die diameter (3.5 mm). This is due to the greater pressure generated inside the extruder at a die opening of 3 mm. In addition, it was observed that the increase in screw speed affected the expansion ratio positively. The increased pressure gradient due to the increase in screw speed caused more water to evaporate during the exit of the molten raw material from the die, and as a result, high expansion was achieved. Gulzar et al. [75] reported that the expansion ratios of extruded snacks produced with modified rice flour ranged from 2.21 to 3.75. The majority of the studies in the literature on cereal-based extrudate production are aimed at optimizing different extrusion parameters. In most of these studies, the expansion ratio was considered as the process response, and it was seen that different values were obtained for the same study under different processing conditions. For example, expansion ratios were determined as 2.66–3.16 for mushroom-rice snacks [72], 1.32–2.10 for millet-corn based RTE snacks composed of multi-grains [80], 0.56–1.31 for soybean-rice snacks [71], 1.17–1.40 for teff-oat snacks [89], 1.48–2.46 for oat-rice snacks [86], 1.22–2.42 for whole wheat flour-based snacks[43], 1.46–2.37 for soft durum wheat snacks[50], and 1.4–2.5 for waxy wheat flour-bran snacks [45].

6.4.2 BULK DENSITY

Bulk density is an indicator of the porosity of the material and determines the packaging conditions [12]. Bulk density and expansion ratio are inversely related, and these are important quality criteria for the extruded snack products. If a snack has a high expansion ratio, its bulk density is low. In the extrusion process, pressure increase is observed in the barrel when low moisture content, high screw speed and high temperature are applied. As soon as the material comes out of the die, the water evaporates, and expansion occurs. Expansion increases with increasing pressure gradient, while bulk density decreases [67, 125, 126]. There are studies in which it has been determined that the bulk density decreases with increased screw speeds [18, 124]. High barrel temperature and screw speed create high shear on the dough, the starch network is disrupted, and the bulk density decreases with the increase in expansion [123]. The elastic properties of the starch-containing material depend on its moisture content. The high feed moisture content creates a plasticizing effect on the amylopectin structure, resulting in lower viscosity of melt and dough elasticity. With the decreasing viscosity, the friction inside the barrel decreases. Thus, the expansion decreases, and the bulk density increases due to the low pressure formed in the interior and the decreasing pressure gradient [127].

Gulati et al. [82] determined the bulk densities between 0.20 and 0.48 g/cm^3 in their study where they produced proso millet extrudates with the extrusion-cooking. The highest bulk density was obtained at the highest feed moisture (25%) and the lowest bulk density was achieved at the lowest feed moisture content (17%). They found that while moisture content has a strong effect on bulk density, temperature has little effect. While an increase in bulk density was observed at screw speeds increasing from 170 rpm to 210 rpm, a decrease was observed at screw speeds higher than 210 rpm. Bhat et al. [68] stated that the bulk density values (0.065–0.188 g/cm^3) of the snacks they obtained in their extrusion study with brown rice increased with increasing feed moisture content and decreased with increasing temperature.

Menis-Henrique et al. [34] determined that the bulk density values of the product they obtained in the study they developed corn-based snacks having cheese flavor changed in the range of 0.09–0.13 g/cm^3. It was observed that the bulk density of snacks decreased with increasing temperature and decreasing sunflower oil. The researchers attributed the decrease in bulk density with increasing temperature to air bubbles formed in the melt, which promote moisture loss as it exits the die at high temperature. Félix-Medina et al. [29] made a very similar explanation, and in their study with corn and common bean flours, they determined that the bulk density values of the samples varied between 0.168 g/cm^3 and 0.225 g/cm^3. In addition, the bulk density decreased with increasing bean flour amount.

Bulk density values determined in different studies are 0.062–0.162 g/cm^3 for mushroom-rice snacks [72], 0.087–0.144 g/cm^3 for millet-corn based RTE snacks composed of multi-grains [80], 0.082–0.423 g/cm^3 for iron-rich snacks composed of millet, soy flour and amla [81], 0.112–0.365 g/cm^3 for soybean-rice snacks [71], 0.614–0.734 g/cm^3 for teff-oat snacks [89], 0.128–0.303 g/cm^3 for whole wheat flour-based snacks [43], 0.32–0.70 g/cm^3 for soft durum wheat snacks [50] and 0.4–0.8 g/cm^3 for waxy wheat flour-bran snacks [45].

6.4.3 WATER ADSORPTION INDEX AND WATER SOLUBILITY INDEX

Water absorption index (WAI) and water solubility index (WSI) are substantial quality characteristics of extruded snacks. In addition to determining the quality, especially the water absorption index has an economic effect. From a practical point of view, the more water the product absorbs, the less dry matter will be needed. Especially breakfast cereals are usually consumed by adding water or milk. Therefore, hydration properties are very important in such products. The extrusion process is known to improve the hydration properties of products. More hydrophilic groups, which arise as a result of the structural opening of starch and proteins during the extrusion process, increase the hydration properties of the product [128].

The amount of water retained by extruded products is expressed as the water absorption index (WAI) and is directly related to the starch content of the product. The water solubility index (WSI) is an indicator of the soluble components of the product after the extrusion process and is a measure of the amount of molecules released in water. WSI depends on the degree of dextrinization of starch in starchy raw materials and is an indicator of the formation of soluble polysaccharides such as amylose with the deterioration of the starch network during the extrusion process, while it is related to protein degradation in protein-containing raw materials. WSI can increase under increasing temperature and decreasing moisture conditions. It is reported that the best WSI values are obtained in cases with high thermal and mechanical energy input [78]. There is a strong correlation between WAI and viscosity. Because the damaged starch absorbs water and swells and increases the viscosity. WAI correlates with the starch gelatinization degree and indicates the stability of retrograde starch during the extrusion process. The presence of hydrophilic components has a positive effect on WAI. It has been determined that extrudates with high amylose content have higher WAI values than those with low amylose content [12].

Román-Gutiérrez et al. [61] determined the WAI and WSI values of the snacks obtained from the barley and corn mixture with the extrusion-cooking process (130–180°C). It was observed that the WAI values of the snacks varied between 2.06 and 7.25 g/g sample, and the WSI values between 1.81% and 7.77%. It was observed that WAI values increased with decreased moisture content and increased barley ratios. The increase in WAI values was associated with the exposure of more polar groups due to protein denaturation during extrusion. The decrease in WAI when the corn content increases is explained by the decrease in molecular size of starch found at a higher ratio in corn during extrusion and the decrease in the water holding capacity of the molecules. WSI values were determined to be high at high moisture content (43–45%) and relatively low temperatures (130–140°C), and low at low moisture content (38%) and low temperature (130°C). This is explained by the fact that the protein is more soluble as a result of less denaturation at low temperature. WSI increased due to an increase in the distribution of amylose and amylopectin because of the gelatinization at moderate extrusion conditions (130–140°C). In another study, it was determined that the WAI values of the extrudates developed using fermented rice and black gram flours varied in the range of 6.1–10.5 g/g. It was observed that WAI values decreased due to the thermal breakdown of starch into shorter chains having low water holding capacity at high temperature with a smaller die diameter. It was determined that the WSI values of these samples increased with the increasing temperature and screw speed at the same die diameter. Under these conditions, the degradation of starch increased and caused the formation of more easily soluble molecules [70]. Bhat et al. [68], in their study aiming to produce thiamine-rich snacks using brown rice, determined that the WAI values of the snacks varied between 4.72 and 7.81 g/g, and the WSI values varied between 5.43 and 14.32%. It was determined that WAI changed directly proportional to moisture content and inversely proportional to barrel temperature and screw speed. Also, while WSI is directly proportional to screw speed and barrel temperature, it is inversely proportional to moisture content.

In different studies, there are similar explanations about how process parameters affect WAI and WSI. Optimization studies have shown that these values vary even for the same raw material when the process conditions are changed. In some studies of cereal-based extruded products WAI and WSI values were 4.33–6.87 g/g and 1.0–12.69% for rice-based shrimp snacks [69], 3.8–4.5 g/g and 3.7–8.1% for proso millet-based snacks [82], 5.07–5.87 g/g and 13.06–21.08% for soybean-rice snacks [71], 2.7–3.6 g/g and 0.8–5.0% for teff-oat snacks [89], and 3.76–4.82 g/g and 5.32–12.06% for soft durum wheat snacks [50], respectively.

6.4.4 SENSORY PROPERTIES

The main criterion in the preference of extruded snack products by consumers is their sensory properties. In the extrusion-cooking process, with the shear, high temperature and pressure, the starch is broken down, the proteins are denatured, and a lot of changes occur in the physico-chemical structure due to starch-protein interactions. All these changes affect the appearance, color, aroma, and texture of the extrudate.

The degree of Maillard reactions occurring during extrusion-cooking is determined by the composition of the raw material, extrusion temperature and residence time, and these reactions are especially effective on the flavor and color characteristics of the product. However, since the extrusion process is the HT/ST process, typical cooked flavors that are exposed in traditional cooking processes may not be produced in the extrusion process. In addition, as soon as the dough comes out of the die, volatile aromas can be lost with water vapor due to sudden evaporation. For this reason, extruded products are generally coated with materials having different aromas according to the product need, to bring the lost or non-developing aromas to the product. In the food industry, post-extrusion flavoring is the traditional way. This process is usually carried out by sprinkling salt and other additives on the surface with a lipid carrier such as vegetable oil. The lipid, salt and caloric value of extruded snacks that are flavored in this way increase. In recent years, studies have been carried out to realize a healthier flavoring process with the pre-extrusion flavoring process. In the pre-extrusion flavoring method, commercial aroma, or volatile compounds, encapsulated volatile compounds and flavor precursors are added to the raw material before the extrusion process. This process ensures a uniform distribution of flavor compounds compared to conventional post-extrusion flavoring and increases the stability of these compounds against oxidation as they are encapsulated in the matrix of the product. However, volatile compounds may still be lost and costs may increase [34].

One of the most important features affecting consumer preference is color in extruded products as it is in all foods. The product color changes depending on the feed composition and the process variables. In addition to the Maillard reaction, caramelization, and changes in the color pigments of the raw material determine the color of the extruded final product. It is known that low moisture and high extrusion temperature increase the Maillard reactions responsible for the formation of melanoidins [129]. Karun et al. [80] reported that the lightness values of ready-to-eat snacks varied with barrel temperature, moisture content and particle size. They determined that the lightness decreased when the particle size is low and temperature is high and increased with increasing feed moisture content.

The most important and preferred feature of the expanded extruded products is that they have a fragile and brittle texture. These textural properties also depend on the parameters affecting the expansion ratio and porosity during the process, and notably depend on the raw material content and the type of additives used. Starch is the most important compound imparting the desired textural properties in extruded snacks, and starch gelatinization produces expanded snacks with typical porous structure, low density, and crunchy texture. On the other hand, replacing starch with fiber, protein, sugar, or fat significantly affects the expansion and texture (crispiness) of snacks [90].

In a study with a mixture of barley and maize, the extrudates with the highest hardness values were obtained with high moisture content and medium barley ratio. The decrease in viscosity and plasticization with high moisture caused a decrease in mechanical energy, so the expansion decreased, and the hardness increased. The lowest hardness values were obtained when the moisture and barley content decreased. With the decrease in the amount of barley, the amount of fiber in the mixture decreased. Fiber in the environment can form protein-fiber and starch-fiber complexes by combining depolymerized fragments of protein and starch. These denser complexes also cause the hardness of the product to increase [61]. In the study in which rice-flour pellets were obtained by extrusion-cooking, it was determined that the hardness values of the products increased with increased moisture content of the feed (from 27% to 31%) [74]. Rani et al. [70], in their study with fermented rice-black gram flour, found that the crispness of extrudates decreased with increasing temperature and screw speeds.

6.4.5 NUTRITIONAL PROPERTIES

The extrusion-cooking process can positively or negatively change the structure of different food components such as carbohydrates, proteins, dietary fiber, amino acid profile, mineral and vitamin content and some bioactive compounds at varying rates associated with the processing conditions and raw material composition. In addition to improving the digestibility of protein and starch, antinutritional degradation, increasing soluble dietary fiber, inactivation of enzymes that cause deterioration and reduction of lipid oxidation are positive effects [88, 130]. At low moisture content and high temperature, loss of lysine is one of the most important consequences of the Maillard reaction, which is very important in cereals with limited levels of lysine, an essential amino acid. However, protein quality can be improved by optimizing the process conditions in extrusion-cooking [89]. In cereal-based extruded products, starch gelatinization and molecular breakdown of polymer starch take place during extrusion. It is known that the extrusion-cooking process increases the resistant starch content and decreases the glycemic index value of the products. In a study, it was found that the resistant starch content of extruded barley extrudates increased from 0.62% to 1.70%, and the glycemic index values decreased from 45.21 to 40.37 [56]. In the studies carried out to date, it has been determined that the starch digestibility of most cereal grains is increased by the extrusion cooking process [75]. Degradation of anti-nutrient compounds and an increase in the soluble dietary fiber are observed. High temperature causes breakdown of glycosidic bonds of polysaccharides and oligosaccharides are released [130]. Gong et al. [131] showed that the extrusion process with germinated and ungerminated whole grain corn increased the nutritional properties of the product. After the process, an increase was observed in the amount of soluble dietary fiber, while the amount of insoluble dietary fiber decreased. In addition to the positive effects of the extrusion-cooking on the nutritional properties of the product, some nutritional elements may be lost depending on the high temperature used in the process. Phongthai et al. [132] emphasized that high extrusion temperature (135–150°C) and high moisture content formed protein aggregates by cross-linking of disulfide bonds, and this situation markedly affected heat-stable proteins. They stated that the structure of alkali-soluble proteins was disrupted.

Knowing how the extrusion-cooking parameters affect the antioxidant activity of the products is important for developing snacks for consumers who are conscious of healthy nutrition. It is known that phenolic compounds with antioxidant properties generally decrease in high temperature applications. However, in many extrusion-cooking studies, an increase in the total amount of phenolic substances was observed. Rani et al. [70] performed the optimization of the extrusion-cooking process with fermented rice-black gram flour and determined that the total amount of phenolic substance, which was 63.47 mg GAE/100 g in fermented dough, increased to 210.3 mg GAE/100 g at 300 rpm and 100°C.It was stated that although the amount of phenolic substance decreased at high extrusion temperatures, the amount of phenolic substance could be increased by optimizing the processing conditions. This increase in phenolic compounds is explained by the breakdown of the cell wall during processing and the release of bound phenolic compounds. Bangar et al. [64] determined that the extrusion process increased the antioxidant capacity, metal chelating activity and ABTS+ scavenging activity in their study where they produced barley extrudates. However, at 180°C barrel temperature, the total phenolic content decreased by 34% and the total flavonoid content decreased by 38%. Bobade et al. [43] produced whole wheat flour-based snacks with the addition of honey. Researchers stated that the addition of honey and the temperature increased the antioxidant activity. It was determined that the snacks extruded at 14% feed moisture, 180°C extrusion temperature and 15% honey level showed the highest antioxidant activity. A 19% increase was observed in the antioxidant activity when the temperature changed from 120 to 180°C.This increase was explained by the release of some Maillard reaction products with antioxidant activity at high temperatures.

Bhat et al. [68], in their study using brown rice to produce thiamine-rich snacks, they determined that the amount of vitamin B_1 that was 0.54 mg/100 g in flour decreased to 0.45 mg/100 g in the extruded product. This decrease was attributed to the low moisture and high temperature conditions used in the study. In addition, a small increase in the mineral content of the extrudates was determined in this study. This increase was explained by the degradation of anti-nutritional components because of the high temperature in extrusion-cooking. Suri et al. [81] used mixture of the barnyard millet, soy flour and amla as a feed in the extrusion of RTE millet-based snacks rich in iron for the anemic people and examined how the snack quality properties influenced by the extrusion parameters. The researchers reported that the ascorbic acid amount of the snacks varied between 28 and 34.25 mg/100 g, the raw material mixture ratios influenced ascorbic acid, and the amount of the final products' ascorbic acid increased significantly with increasing amla ratio. In addition, the researchers determined a linear trend in the ascorbic acid amount at increasing barrel temperature.

6.5 CONCLUSION

Cereals are considered as valuable food and energy sources in human nutrition due to their rich composition. The main component of cereals is starch, but they also contain considerable amounts of protein, fiber, fat, and other nutrients (i.e., vitamins, minerals, and bioactive compounds). In recent years, the demand for RTE snacks is increasing due to the changing dietary habits. Due to the increasing trend towards healthier nutrition and the higher consumption of cereal-based snacks especially by children, efforts are being made to produce healthier forms of these products. In this context, extrusion-cooking process is a flexible and versatile process that ensures the development of new cereal-based RTE snacks by adjusting different process parameters according to the product. Refined flours obtained from cereals, whole grain flours or their fractions such as starch, protein and bran are used in the extrusion-cooking process. However, scientific studies are not limited to these. In cereal-based snack production, studies are carried out to develop new products by preparing different formulations with a wide variety of ingredients such as legumes, different food industry by-products, herbal extracts, etc. Extrusion-cooking parameters can be optimized depending on the end product requirements and the characteristics of the feed material used. Thus, with the extrusion-cooking technology, it is possible to develop a wide variety of cereal-based RTE snacks with the desired quality characteristics.

KEYWORDS

- **cereal-based**
- **expansion ratio**
- **hot-extrusion**
- **nutritional properties**
- **physico-chemical structure**
- **ready-to-eat snacks**

REFERENCES

1. Qamar, S. Z. (2018). Extrusion of metals, polymers and food products. In Ajita, T. (ed.), *Extrusion Cooking Technology: An Advance Skill for Manufacturing of Extrudate Food Products* (pp. 197–210). London, United Kingdom, Intechopen.
2. Perdon, A. A., Schonauer, S. L., & Poutanen, K. S. (2020). Breakfast cereals and how they are made. In Kristiawan, M., Chaunier, L., Sandoval, A. J., & Valle, G. D. (eds.), *Extrusion-Cooking and Expansion* (pp. 141–168). Woodhead Publishing and Cereals & Grains Association, Cambridge, United Kingdom.
3. Ganjyal, G. M. (2020). Extrusion cooking cereal grains processing. In Ek, P., & Ganjyal, G. M. (ed.), *Basics of Extrusion Processing* (pp. 1–28). Woodhead Publishing, Cambridge, United Kingdom.
4. Ganjyal, G. M. (2020). Extrusion cooking cereal grains processing. In Yacu, W. (ed.) *Extruder Screw, Barrel, and Die Assembly: General Design Principles and Operation* (pp. 73–117). Woodhead Publishing, Cambridge, United Kingdom.
5. Ganjyal, G. M. (2020). Extrusion cooking cereal grains processing. In Alvarez, J. C. M. (ed.), *Engineering Aspects of Extrusion: Extrusion Processing as a Multiple-Input and Multiple Output System* (pp. 29–71). Woodhead Publishing, Cambridge, United Kingdom.
6. Martin, C. (2016). Twin screw extruders as continuous mixers for thermal processing: A technical and historical perspective. *AAPS PharmSciTech, 17*(1), 3–19. https://doi.org/10.1208/s12249-016-0485-3.
7. Dhaval, M., Sharma, S., Dudhat, K., & Chavda, J. (2022). Twin-screw extruder in pharmaceutical industry: History, working principle, applications, and marketed products: An in-depth review. *Journal of Pharmaceutical Innovation, 17*(3), 294–318. https://doi.org/10.1007/s12247-020-09520-7.
8. Alam, M. S., Kaur, J., Khaira, H., & Gupta, K. (2016). Extrusion and extruded products: Changes in quality attributes as affected by extrusion process parameters: A review. *Critical Reviews in Food Science and Nutrition, 56*(3), 445–473. https://doi.org/10.1080/10408398.2013.779568.
9. Mosibo, O. K., Ferrentino, G., Alam, M. R., Morozova, K., & Scampicchio, M. (2022). Extrusion cooking of protein-based products: Potentials and challenges. *Critical*

Reviews in Food Science and Nutrition, 62(5), 2526–2547. https://doi.org/10.1080/10408398.2020.1854674.

10. Offiah, V., Kontogiorgos, V., & Falade, K. O. (2019). Extrusion processing of raw food materials and by-products: A review. *Critical Reviews in Food Science and Nutrition, 59*(18), 2979–2998. https://doi.org/10.1080/10408398.2018.1480007.
11. Pilli, T. D., & Alessandrino, O. (2018). Effects of different cooking technologies on biopolymers modifications of cereal-based foods: Impact on nutritional and quality characteristics review. *Critical Reviews in Food Science and Nutrition, 60*(4), 556–565. https://doi.org/10.1080/10408398.2018.1544884.
12. Dalbhagat, C. G., Mahato, D. K., & Mishra, H. N. (2019). Effect of extrusion processing on physicochemical, functional and nutritional characteristics of rice and rice-based products: A review. *Trends in Food Science & Technology, 85*, 226–240. https://doi.org/10.1016/j.tifs.2019.01.001.
13. Bhat, N. A., Wani, I. A., Hamdani, A. M., & Gani, A. (2019). Effect of extrusion on the physicochemical and antioxidant properties of value-added snacks from whole wheat (*Triticum aestivum* L.) Flour. *Food Chemistry, 276*, 22–32. https://doi.org/10.1016/j.foodchem.2018.09.170.
14. Yadav, G. P., Dalbhagat, C. G., & Mishra, H. N. (2022). Effects of extrusion process parameters on cooking characteristics and physicochemical, textural, thermal, pasting, microstructure, and nutritional properties of millet-based extruded products: A review. *Journal of Food Processing and Engineering, 45*(6), e14106. https://doi.org/10.1111/jfpe.14106.
15. Peksa, A., Kita, A., Carbonell-Barrachina, A. A., Miedzianka, J., Kolniak-Ostek, J., Tajner-Czopek, A., Rytel, E., Siwek, A., Miarka, D., & Drozdz, W. (2016). Sensory attributes and physicochemical features of corn snacks as affected by different flour types and extrusion conditions. *LWT–Food Science and Technology, 72*, 26–36. https://doi.org/10.1016/j.lwt.2016.04.034.
16. Shirazi, S. L., Koocheki, A., Milani, E., & Mohebbi, M. (2020). Production of high fiber ready-to-eat expanded snack from barley flour and carrot pomace using extrusion cooking technology. *Journal of Food Science and Technology, 57*(5), 2169–2181. https://doi.org/10.1007/s13197-020-04252-5.
17. Jafari, M., Koocheki, A., & Milani, E. (2017). Effect of extrusion cooking on chemical structure, morphology, crystallinity and thermal properties of sorghum flour extrudates. *Journal of Cereal Science, 75*, 324–331. https://doi.org/10.1016/j.jcs.2017.05.005.
18. Kharat, S., Medina-Meza, I. G., Kowalski, R. J., Hosamani, A., Ramachandra, C. T., Hiregoudar, S., & Ganjyal, G. M. (2019). Extrusion processing characteristics of whole grain flours of select major millets (Foxtail, Finger, and Pearl). *Food Bioprocess Technology, 114*(1), 60–71. https://doi.org/10.1016/j.fbp.2018.07.002.
19. Almoraie, N. M., Saqaan, R., Alharthi, R., Alamoudi, A., Badh, L., & Shatwan, I. M. (2021). Snacking patterns throughout the life span: Potential implications on health. *Nutrients, 91*(3), 81–94. https://doi.org/10.1016/j.nutres.2021.05.001.
20. Shah, F. H., Sharif, M. K., Bashir, S., & Ahsan, F. (2019). Role of healthy extruded snacks to mitigate malnutrition. *Food Reviews International, 35*(3), 299–323. https://doi.org/10.1080/87559129.2018.1542534.
21. Sahu, C., & Patel, S. (2021). Optimization of maize–millet-based soy fortified composite flour for preparation of rte. extruded products using d-optimal mixture design.

Journal of Food Science and Technology, 58(7), 2651–2660. https://doi.org/10.1007/s13197-020-04771-1.

22. Ekpa, O., Palacios, R. N., Kruseman, G., Fogliano, V., & Linnemann, A. R. (2019). Sub-Saharan African maize-based foods-processing practices, challenges and opportunities. *Food Reviews International, 35*(6), 609–639. https://doi.org/10.1080/87559129.2019.1588290.
23. Priya, T. S. R., & Manickavasagan, A. (2021). Characterizing corn grain using infrared imaging and spectroscopic techniques: A review. *Journal of Food Measurement and Characterization, 15*(2), 3234–3249. https://doi.org/10.1007/s11694-021-00898-7.
24. Garzón, R., Rosell, C. M., Malvar, R. A., & Revilla, P. (2017). Diversity among maize populations from Spain and the United States for dough rheology and gluten-free breadmaking performance. *International Journal of Food Science and Technology, 52*(4), 1000–1008. https://doi.org/10.1111/ijfs.13364.
25. Mushtaq, B. S., Zhang, W., Al-Ansi, W., Haq, F. U., Rehman, A., Omer, R., Khan, I. M., Niazi, S., Ahmad, A., Mahdi, A. A., Al-Maqtari, Q. A., Walayat, N., & Wang, L. (2021). A critical review on the development, physicochemical variations and technical concerns of gluten free extrudates in food systems. *Food Reviews International, 37*(6), 1–29. https://doi.org/10.1080/87559129.2021.1976793.
26. Shahmohammadi, H. R., Jamilah, B., Russly, A. R., & Noranizan, M. A. (2016). Optimization of puffed corn-fish snack extrusion conditions using response surface methodology. *International Food Research Journal, 23*(4), 1685–1693.
27. Escalante-Aburto, A., Ponce-Garcia, N., Ramirez-Wong, B., Santiago-Ramos, D., Veles-Medina, J. J., & Cardenas, J. D. F. (2016). Effect of extrusion factors and particle size on starch properties of nixtamalized whole blue corn snacks. *Starch–Stärke, 68*(11, 12), 1111–1120. https://doi.org/10.1002/star.201500316.
28. Makowska, A., Zielinska-Dawidziak, M., Niedzielski, P., & Michalak, M. (2018). Effect of extrusion conditions on iron stability and physical and textural properties of corn snacks enriched with soybean ferritin. *International Journal of Food Science and Technology, 53*(1), 296–303. https://doi.org/10.1111/ijfs.13585.
29. Félix-Medina, J. V., Montes-Ávila, J., Reyes-Moreno, C., Perales-Sánchez, J. X. K., Gómez-Favela, M. A., Aguilar-Palazuelos, E., & Gutiérrez-Dorado, R. (2020). Second-generation snacks with high nutritional and antioxidant value produced by an optimized extrusion process from corn/common bean flours mixtures. *LWT–Food Science and Technology, 124*, 109172. https://doi.org/10.1016/j.lwt.2020.109172.
30. Jozinović, A., Ačkar, Đ., Jokić, S., Babić, J., Balentić, J. P., Banožić, M., & Šubarić, D. (2017). Optimization of extrusion variables for the production of corn snack products enriched with defatted hemp cake. *Czech Journal of Food Sciences, 35*(6), 507–516. https://doi.org/10.17221/83/2017-CJFS.
31. Neder-Suarez, D., Quintero-Ramos, A., Melendez-Pizarro, C. O., Zazueta-Morales, J. J., Paraguay-Delgado, F., & Ruiz-Gutierrez, M. G. (2021). Evaluation of the physicochemical properties of third-generation snacks made from blue corn, black beans, and sweet chard produced by extrusion. *LWT–Food Science and Technology, 146*, 111414. https://doi.org/10.1016/j.lwt.2021.111414.
32. Capriles, V. D., Conti-Silva, A. C., & Areas, J. A. G. (2021). Effects of ligofructose-enriched inulin addition before and after the extrusion process on the quality and postprandial glycemic response of corn-snacks. *Food Bioscience, 43*, 101263. https://doi.org/10.1016/j.fbio.2021.101263.

33. Singha, P., Muthukumarappan, K., & Krishnan, P. (2018). Influence of processing conditions on apparent viscosity and system parameters during extrusion of distiller's dried grains-based snacks. *Food Science and Nutrition, 6*(1), 101–110. https://doi.org/10.1002/fsn3.534.
34. Menis-Henrique, M. E. C., Janzantti, N. S., Monteiro, M., & Conti-Silva, A. C. (2020). Physical and sensory characteristics of cheese-flavored expanded snacks obtained using butyric acid and cysteine as aroma precursors: Effects of extrusion temperature and sunflower oil content. *LWT–Food Science and Technology, 122*, 109001. https://doi.org/10.1016/j.lwt.2019.109001.
35. Singh, J. P., Kaur, A., Shevkani, K., Singh, N., & Singh, B. (2016). Physicochemical Characterization of corn extrudates prepared with varying levels of beetroot (*Beta vulgaris*) at different extrusion temperatures. *International Journal of Food Science and Technology, 51*(5), 911–919. https://doi.org/10.1111/ijfs.13051.
36. Asgharipour, S., Noshad, M., Nasehi, B., Jooyandeh, H., & Beiraghi-Toosi, S. (2018). Optimization of physicochemical and functional properties of corn-based snacks containing date kernel flour. *Journal of Food Processing and Preservation, 42*(6), e13821. https://doi.org/10.1111/jfpp.13821.
37. Rolandelli, G., García-Navarro, Y. T., García-Pinilla, S., Farroni, A. E., Gutiérrez-López, G. F., & Buera, M. P. (2020). Microstructural characteristics and physical properties of corn-based extrudates affected by the addition of millet, sorghum, quinoa and canary seed flour. *Food Structure, 25*, 100140. https://doi.org/10.1016/j.foostr.2020.100140.
38. Jabeen, A., Naik, H., Jan, N., Hussain, S. Z., Amin, T., & Rafiq, A. (2021). Studying the effect of tomato pomace incorporation on physicochemical, nutritional and storage characteristics of corn-based extrudates using response surface approach. *British Food Journal, 124*(11), 3705–3723. https://doi.org/10.1108/BFJ-05-2021-0483.
39. Bobade, H., Singh, A., Sharma, S., Gupta, A., & Singh, B. (2022). Effect of extrusion processing on techno-functional, textural and bioactive properties of whole-grain corn flour-based breakfast cereals sweetened with honey. *Journal of Texture Studies, 53*(5), 672–683. https://doi.org/10.1111/jtxs.12708.
40. Shewry, P. R. (2018). Do ancient types of wheat have health benefits compared with modern bread wheat? *Journal of Cereal Science, 79*, 469–476. https://doi.org/10.1016/j.jcs.2017.11.010.
41. Wu, Y., Xiao, Z., Jiang, X., Lv, C., Gao, J., Yuan, J., Shan, L., & Chen, H. (2022). Effect of extrusion on the modification of wheat flour proteins related to celiac disease. *Journal of Food Science and Technology, 59*(6), 2655–2665. https://doi.org/10.1007/s13197-021-05285-0.
42. Patil, S. S., Brennan, M. A., Mason, S. L., & Brennan, C. S. (2016). The effects of fortification of legumes and extrusion on the protein digestibility of wheat-based snack. *Foods, 5*(2), 26. https://doi.org/10.3390/foods5020026.
43. Bobade, H., Singh, A., Sharma, S., Gupta, A., & Singh, B. (2022). Effect of extrusion conditions and honey on functionality and bioactive composition of whole wheat flour-based expanded snacks. *Journal of Food Processing and Preservation, 46*(5), e16132. https://doi.org/10.1111/jfpp.16132.
44. Li, X., Masatcioglu, M. T., & Koksel, F. (2019). Physical and functional properties of wheat flour extrudates produced by nitrogen injection assisted extrusion cooking. *Journal of Cereal Science, 89*, 102811. https://doi.org/10.1016/j.jcs.2019.102811.

45. Fleischman, E. F., Kowalski, R. J., Morris, C. F., Nguyen, T., Li, C., Ganjyal, G., & Ross, C. F. (2016). Physical, textural, and antioxidant properties of extruded waxy wheat flour snack supplemented with several varieties of bran. *Journal of Food Science, 81*(11), E2726–E2733. https://doi.org/10.1111/1750-3841.13511.
46. Paznocht, L., Burešová, B., Kotíková, Z., & Martinek, P. (2021). Carotenoid content of extruded and puffed products made of colored-grain wheats. *Food Chemistry, 340*, 127951. https://doi.org/10.1016/j.foodchem.2020.127951.
47. Krapf, J., Arysanto, A., Walther, G., & Flöter, E. (2019). Effect of sprouting conditions on the properties of direct expanded extruded wheat. *Journal of Food Process Engineering, 42*(2), e13123. https://doi.org/10.1111/jfpe.13123.
48. Pitts, K. F., McCann, T. H., Mayo, S., Favaro, J., & Day, L. (2016). Effect of the sugar replacement by citrus fiber on the physical and structural properties of wheat-corn based extrudates. *Food and Bioprocess Technology, 9*(12), 1803–1811. https://doi.org/10.1007/s11947-016-1764-4.
49. Pilli, T. D., Giuliani, R., Buléon, A., Pontoire, B., & Legrand, J. (2016). Effects of protein–Lipid and starch–Lipid complexes on textural characteristics of extrudates based on wheat flour with the addition of oleic acid. *International Journal of Food Science and Technology, 51*(5), 1063–1074. https://doi.org/10.1111/ijfs.13070.
50. Gu, B. J., Kerr, C. J., Morris, C. F., & Ganjyal, G. M. (2021). Soft durum wheat as a potential ingredient for direct expanded extruded products. *Journal of Cereal Science, 98*, 103184. https://doi.org/10.1016/j.jcs.2021.103184.
51. Sisay, M. T., Emire, S. A., Ramaswamy, H. S., & Workneh, T. S. (2018). Effect of feed components on quality parameters of wheat–tef–sesame–tomato based extruded products. *Journal of Food Science and Technology, 55*(7), 2649–2660. https://doi.org/10.1007/s13197-018-3187-x.
52. Geng, L., Li, M., Zhang, G., & Ye, L. (2022). Barley: A potential cereal for producing healthy and functional foods. *Food Quality and Safety, 6*(1), 1–13. https://doi.org/10.1093/fqsafe/fyac012.
53. Pejcz, E., Czaja, A., Wojciechowicz-Budzisz, A., Gil, Z., & Spychaj, R. (2017). The potential of naked barley sourdough to improve the quality and dietary fiber content of barley enriched wheat bread. *Journal of Cereal Science, 77*, 97–101. https://doi.org/10.1016/j.jcs.2017.08.007.
54. Farag, M. A., Xiao, J., & Abdallah, H. M. (2022). Nutritional value of barley cereal and better opportunities for its processing as a value-added food: A comprehensive review. *Critical Reviews in Food Science and Nutrition, 62*(7), 1092–1104. https://doi.org/10.1080/10408398.2020.1835817.
55. Geng, L., Li, M. D., Xie, S. G., Wu, D., Ye, L., & Zhang, G. (2021). Identification of genetic loci and candidate genes related to β-glucan content in barley grain by genome-wide association study in international barley core selected collection. *Molecular Breeding, 41*(1), 6. https://doi.org/10.1007/s11032-020-01199-5.
56. Reshi, M., Shafiq, F., Hussain, S. Z., Naseer, B., & Amin, T. (2020). Physicochemical properties of iron-fortified, low glycemic index (gi) barley-based extruded ready-to-eat snacks developed using twin-screw extruder. *Journal of Food Processing and Preservation, 44*(4), e14606. https://doi.org/10.1111/jfpp.14606.
57. Yang, Y., Jiao, A., Liu, Q., Xu, E., Chen, Y., & Jin, Z. (2020). Functional and physical properties of naked barley-based unexpanded extrudates: Effects of low temperature.

International Journal of Food Properties, 23(8), 1886–1898. https://doi.org/10.1080/10942912.2020.1826511.

58. Yovchev, A., Stone, A., Hood-Niefer, S., & Nickerson, M. (2017). Influence of the extrusion parameters on the physical properties of chickpea and barley extrudates. *Food Science and Biotechnology, 26*(2), 393–399. https://doi.org/10.1007/s10068-017-0054-x.
59. Chou, C. F., & Hsu, S. C. (2021). Effects of extrusion parameters on the physicochemical characteristics of extruded barley ready-to-eat snacks. *Journal of Food Process Preservation, 45*(5), e15788. https://doi.org/10.1111/jfpp.15788.
60. Baidoo, E. A., Murphy, K., & Ganjyal, G. M. (2019). Hulled varieties of barley showed better expansion characteristics compared to hull-less varieties during twin-screw extrusion. *Cereal Chemistry, 96*(3), 391–404. https://doi.org/10.1002/cche.10138.
61. Román-Gutiérrez, A. D., Barrera-Barrera, D. I., Morales-Sánchez, E., Castro-Rosas, J., Gómez-Aldapa, C. A., Falfán-Cortés, R. N., Rodríguez- Marín, M. L., & Guzmán-Ortiz, F. A. (2022). Modelling and optimization of the extrusion process in a snack of barley/corn and nutritional evaluation of the optimized product. *Cereal Chemistry, 99*(5), 556–567. https://doi.org/10.1002/cche.10518.
62. Guldiken, B., Yovchev, A., Nosworthy, M. G., Stone, A. K., House, J. D., Hood-Niefer, S., & Nickerson, M. T. (2020). Effect of extrusion conditions on the physical properties of desi chickpea-barley extrudates and quality attributes of their resulting flours. *Journal of Texture Studies, 51*(2), 300–307. https://doi.org/10.1111/jtxs.12470.
63. Chang, C., & Lin, J. (2017). Comparison between collet and cooking extrusions on physicochemical properties of whole grain barley. *Journal of Food Process Engineering, 40*(2), E12480. https://doi.org/10.1111/jfpe.12480.
64. Bangar, S. P., Sandhu, K. S., Trif, M., Rusu, A., Pop, I. D., & Kumar, M. (2022). Enrichment in different health components of barley flour using twin-screw extrusion technology to support nutritionally balanced diets. *Frontiers in Nutrition, 8*, 823148. https://doi.org/10.3389/fnut.2021.823148.
65. Sen, S., Chakraborty, R., & Kalita, P. (2020). Rice–Not just a staple food: A comprehensive review on its phytochemicals and therapeutic potential. *Trends in Food Science & Technology, 97*, 265–285. https://doi.org/10.1016/j.tifs.2020.01.022.
66. Laishram, B., & Das, A. B. (2017). Effect of thermal pretreatments on physical, phytochemical, and antioxidant properties of black rice pasta. *Journal of Food Process Engineering, 40*(4), E12553. https://doi.org/10.1111/jfpe.12553.
67. Hu, Z., Tang, X., Zhang, M., Hu, X., Yu, C., Zhu, Z., & Shao, Y. (2018). Effects of different extrusion temperatures on extrusion behavior, phenolic acids, antioxidant activity, anthocyanins, and phytosterols of black rice. *RSC Advances, 8*(12), 7123–7132. https://doi.org/10.1039/C7RA13329D.
68. Bhat, T., Hussain, S. Z., Naseer, B., Rather, A. H., & Mir, S. A. (2021). Development of thiamine-rich snacks from brown rice using extrusion technology. *British Food Journal, 123*(7), 1732–1757. https://doi.org/10.1108/BFJ-08-2020-0732.
69. Lourenço, L. F. H., Tavares, T. S., Araujo, E. A. F., Pena, R. S., Joele, M. R. S. P., & Carvalho, A. V. (2016). Optimization of extrusion process to obtain shrimp snacks with rice grits and polished rice grains. *CYTA–Journal of Food, 14*(3), 340–348. https://doi.org/10.1080/19476337.2015.1114025.
70. Rani, P., Kumar, A., Purohit, S. R., & Rao, P. S. (2021). Extrusion of fermented rice-black gram flour for development of functional snacks: Characterization, optimization,

and sensory analysis. *Journal of Food Science and Technology, 58*(2), 494–509. https://doi.org/10.1007/s13197-020-04558-4.

71. Alam, M. S., Pathania, S., & Sharma, A. (2016). Optimization of the extrusion process for development of high fiber soybean-rice ready-to-eat snacks using carrot pomace and cauliflower trimmings. *LWT–Food Science and Technology, 74*, 135–144. https://doi.org/10.1016/j.lwt.2016.07.031.
72. Kantrong, H., Charunuch, C., Limsangouan, N., & Pengpinit, W. (2018). Influence of process parameters on physical properties and specific mechanical energy of healthy mushroom-rice snacks and optimization of extrusion process parameters using response surface methodology. *Journal of Food Science and Technology, 55*(9), 3462–3472. https://doi.org/10.1007/s13197-018-3271-2.
73. Dilrukshi, H. N. N., Torrico, D. D., Brennan, M. A., & Brennan, C. S. (2022). Effects of extrusion processing on the bioactive constituents, *in vitro* digestibility, amino acid composition, and antioxidant potential of novel gluten-free extruded snacks fortified with cowpea and whey protein concentrate. *Food Chemistry, 389*, 133107. https://doi.org/10.1016/j.foodchem.2022.133107.
74. Zambrano, Y., Contardo, I., Moreno, M. C., & Bouchon, P. (2022). Effect of extrusion temperature and feed moisture content on the microstructural properties of rice-flour pellets and their impact on the expanded product. *Foods, 11*(2), 198. https://doi.org/10.3390/foods11020198.
75. Gulzar, B., Hussain, S. Z., Naseer, B., Shikari, A. B., Nazir, N., & Gani, G. (2021). Investigation of process and product parameters on physical attributes, resistant starch, and *in vitro* starch digestibility of modified rice flour-based extruded snacks. *Journal of Food Process Preservation, 45*(4), E15953. https://doi.org/10.1111/jfpp.15953.
76. Srilekha, K., Kamalaja, T., Maheswari, K. U., & Rani, R. N. (2019). Nutritional composition of little millet flour. *International Research Journal of Pure and Applied Chemistry, 20*(4), 1–4. https://doi.org/10.9734/irjpac/2019/v20i430140.
77. Yousaf, L., Hou, D., Liaqat, H., & Shen, Q. (2021). Millet: A review of its nutritional and functional changes during processing. *Food Research International, 142*, 110197. https://doi.org/10.1016/j.foodres.2021.110197.
78. Sukumar, A., & Athmaselvi, K. A. (2019). Optimization of process parameters for the development of finger millet-based multigrain extruded snack food fortified with banana powder using RSM. *Journal of Food Science and Technology, 56*(2), 705–712. https://doi.org/10.1007/s13197-018-3527-x.
79. Oliveira, D. P. L., Júnior, M. S. S., Bento, J. A. C., Dos Santos, I. G., & Ferreira, T. A. P. C. (2021). Influence of extrusion conditions on the physical and nutritional properties of snacks from maize and pearl millet. *Journal of Food Process Preservation, 45*(5), E15215. https://doi.org/10.1111/jfpp.15215.
80. Karun, G., Sukumar, A., Nagamaniammai, G., & Preetha, R. (2023). Development of multigrain ready-to-eat extruded snack and process parameter optimization using response surface methodology. *Journal of Food Science and Technology, 60*(4), 947–957. https://doi.org/10.1007/s13197-022-05390-8.
81. Suri, S., Dutt, A., Shahi, N. C., Raghuvanshi, R. S., Singh, A., & Chopra, C. S. (2020). Numerical optimization of process parameters of ready-to-eat (rte) iron rich extruded snacks for anemic population. *LWT–Food Science and Technology, 134*, 110164. https://doi.org/10.1016/j.lwt.2020.110164.

82. Gulati, P., Weier, S. A., Santra, D., Subbiah, J., & Rose, D. J. (2016). Effects of feed moisture and extruder screw speed and temperature on physical characteristics and antioxidant activity of extruded proso millet (*Panicum Miliaceum*) Flour. *International Journal of Food Science and Technology, 51*(1), 114–122. https://doi.org/10.1111/ijfs.12974.
83. Zhang, K., Dong, R., Hu, X., Ren, C., & Li, Y. (2021). Oat-based foods: Chemical constituents, glycemic index, and the effect of processing. *Foods, 10*(6), 1304. https://doi.org/10.3390/foods10061304.
84. Tang, Y., Li, S., Yan, J., Peng, Y., Weng, W., Yao, X., Gao, A., Cheng, J., Ruan, J., & Xu, B. (2022). Bioactive components and health functions of oat. *Food Reviews International, 1*(20). https://doi.org/10.1080/87559129.2022.2029477.
85. Paudel, D., Dhungana, B., Caffe, M., & Krishnan, P. (2021). A review of health-beneficial properties of oats. *Foods, 10*(11), 2591. https://doi.org/10.3390/foods10112591.
86. Sandrin, R., Caon, T., Zibetti, A. W., De Francisco, A. (2018). Effect of extrusion temperature and screw speed on properties of oat and rice flour extrudates. *Journal of the Science of Food and Agriculture, 98*(9), 3427–3436. https://doi.org/10.1002/jsfa.8855.
87. Sayanjali, S., Sanguansri, L., Ying, D., Buckow, R., Gras, S., & Augustin, M. A. (2019). Extrusion of a curcuminoid-enriched oat fiber-corn-based snack product. *Journal of Food Science, 84*(2), 284–291. https://doi.org/10.1111/1750-3841.14432.
88. Wani, S. A., & Kumar, P. (2016). Effect of extrusion on the nutritional, antioxidant, and microstructural characteristics of nutritionally enriched snacks. *Journal of Food Process Preservation, 40*(1), 166–173. https://doi.org/10.1111/jfpp.12593.
89. Gopirajah, R., & Muthukumarappan, K. (2018). Effect of extrusion process conditions on the physical properties of tef-oat healthy snack extrudates. *Journal of Food Process Preservation, 42*(6), E1355. https://doi.org/10.1111/jfpp.13559.
90. Carmo, C. S., Varela, P., Poudroux, C., Dessev, T., Myhrer, K., Rieder, A., Zobel, H., Sahlstrom, S., & Knutsen, S. H. (2019). The impact of extrusion parameters on physicochemical, nutritional, and sensorial properties of expanded snacks from pea and oat fractions. *LWT–Food Science and Technology, 112*, 108252. https://doi.org/10.1016/j.lwt.2019.108252.
91. Khoddami, A., Messina, V., Venkata, K. V., Farahnaky, A., Blanchard, C. L., & Roberts, T. H. (2023). Sorghum in foods: Functionality and potential in innovative products. *Critical Reviews in Food Science and Nutrition, 63*(7), 1170–1186. https://doi.org/10.1080/10408398.2021.1960793.
92. Lopes, O., De Lima, S., Da Silva, P., Toledo, L., De Castro, M., Anunciação, P. C., Walter, E. H. M., Carvalho, C. W. P., Queiroz, V. A. V., & Ribeiro, A. Q. (2018). Evaluation of the health benefits of consumption of extruded tannin sorghum with unfermented probiotic milk in individuals with chronic kidney disease. *Food Research International, 107*, 629–638. https://doi.org/10.1016/j.foodres.2018.03.004.
93. Li, Z., Zhao, X., Zhang, X., & Liu, H. (2021). Bioactive compounds and biological activities of sorghum grains. *Foods, 10*(11), 2868. https://doi.org/10.3390/foods10112868.
94. Toledo, V. C. S., Carvalho, C. W. P., Vargas-Solórzano, J. W., Ascheri, J. L. R., & Comettant-Rabanal, R. (2020). Extrusion cooking of gluten-free whole grain flour blends. *Journal of Food Processing Engineering, 43*(2), E13303. https://doi.org/10.1111/jfpe.13303.

95. Naveena, B., & Singh, M. (2022). Effects of extrusion parameters on physical & textural properties of sorghum, barley, and chickpea-based composite extruded products. *Food Process Preservation, 46*(5), E16564. https://doi.org/10.1111/jfpp.16564.
96. Koa, S. S., Jin, X., Zhang, J., & Sopade, P. A. (2017). Extrusion of a model sorghum-barley blend: Starch digestibility and associated properties. *Journal of Cereal Science, 75*, 314–323. https://doi.org/10.1016/j.jcs.2017.04.007.
97. Lohani, U. C., & Muthukumarappan, K. (2017). Effect of extrusion processing parameters on antioxidant, textural, and functional properties of hydrodynamic cavitated corn flour, sorghum flour, and apple pomace-based extrudates. *Journal of Food Processing Engineering, 40*(4), E12424. https://doi.org/10.1111/jfpe.12424.
98. Nemeth, R., & Tömösközi, S. (2021). Rye: Current state and future trends in research and applications. *Acta Alimentaria, 50*(4), 620–640. https://doi.org/10.1556/066.2021.00162.
99. Jonsson, K., Andersson, R., Bach Knudsen, K. E., Hallmans, G., Hanhineva, K., Katina, K., & Landberg, R. (2018). Rye and Health–Where do we stand and where do we go? *Trends in Food Science & Technology, 79*, 78–87. https://doi.org/10.1016/j.tifs.2018.06.018.
100. Kaur, K., Sandhu, K. S., Purewal, S. S., Kaur, M., & Singh, K. S. (2021). Rye: A wonder crop with industrially important macromolecules and health benefits. *Food Research International, 150*, 110769. https://doi.org/10.1016/j.foodres.2021.110769.
101. Zhu, F. (2018). Triticale: Nutritional composition and food use. *Food Chemistry, 241*, 468–479. https://doi.org/10.1016/j.foodchem.2017.09.009.
102. Makowska, A., Waskiewicz, A., & Chudy, S. (2020). Lignans in triticale grain and triticale products. *Journal of Cereal Science, 93*, 102939. https://doi.org/10.1016/j.jcs.2020.102939.
103. Fras, A., Gołebiewska, K., Gołebiewski, D., Mankowski, D. R., Boros, D., & Szecowka, P. (2016). Variability in the chemical composition of triticale grain, flour and bread. *Journal of Cereal Science, 71*, 66–72. https://doi.org/10.1016/j.jcs.2016.06.016.
104. Makowska, A., Baranowska, H. M., Michniewicz, J., Chudy, S., & Kowalczewski, P. L. (2017). Triticale extrudates–Changes of macrostructure, mechanical properties and molecular water dynamics during hydration. *Journal of Cereal Science, 74*, 250–255. https://doi.org/10.1016/j.jcs.2017.03.002.
105. Höglund, E., Eliasson, L., Oliveira, G., Almli, V. L., Sozer, N., & Alminger, M. (2018). Effect of drying and extrusion processing on physical and nutritional characteristics of bilberry press cake extrudates. *LWT–Food Science and Technology, 92*, 422–428. https://doi.org/10.1016/j.lwt.2018.02.042.
106. Johansson, D. P., Gutiérrez, J. L. V., Landberg, R., Alminger, M., & Langton, M. (2018). Impact of food processing on rye product properties and their *in vitro* digestion. *European Journal of Nutrition, 57*, 1651–1666. https://doi.org/10.1007/s00394-017-1450-y.
107. Nikinmaa, M., Alam, S. A., Raulio, M., Katina, K., Kajala, I., Nordlund, E., & Sozer, N. (2017). Bioprocessing of bran with exopolysaccharide producing microorganisms as a tool to improve expansion and textural properties of extruded cereal foams with high dietary fiber content. *LWT–Food Science and Technology, 77*, 170–177. https://doi.org/10.1016/j.lwt.2016.11.041.
108. Alam, S. A., Jarvinen, J., Kokkonen, H., Jurvelin, J., Poutanen, K., & Sozer, N. (2016). Factors affecting structural properties and *in vitro* starch digestibility of extruded starchy foams containing bran. *Journal of Cereal Science, 71*, 190–197. https://doi.org/10.1016/j.jcs.2016.08.018.

109. Nikinmaa, M., Kajala, I., Liub, X., Nordlund, E., & Sozer, N. (2020). The role of rye bran acidification and *in situ* dextran formation on structure and texture of high fiber extrudates. *Food Research International, 137*, 109438. https://doi.org/10.1016/j.foodres.2020.109438.
110. Mir, N. A., Riar, C. S., & Singh, S. (2018). Nutritional constituents of pseudo cereals and their potential use in food systems: A review. *Trends in Food Science & Technology, 75*, 170–180. https://doi.org/10.1016/j.tifs.2018.03.016.
111. Li, H., Xiong, Z., & Li, X. (2019). Optimization of the extrusion process for the development of extruded snacks using peanut, buckwheat, and rice blend. *Journal of Food Process Preservation, 43*(3), E14264. https://doi.org/10.1111/jfpp.14264.
112. Singh, J. P., Kaur, A., Singh, B., Singh, N., & Singh, B. (2019). Physicochemical evaluation of corn extrudates containing varying buckwheat flour levels prepared at various extrusion temperatures. *Journal of Food Science and Technology, 56*(5), 2205–2212. https://doi.org/10.1007/s13197-019-03703-y.
113. Martin, A., Schmidt, V., Osen, R., Bez, J., Ortner, E., & Mittermaier, S. (2022). Texture, sensory properties and functionality of extruded snacks from pulses and pseudocereal proteins. *Journal of the Science of Food and Agriculture, 102*(12), 5011–5021. https://doi.org/10.1002/jsfa.11041.
114. Omoba, O. S., Olagunju, A. I., Akinrinlola, F. O., & Oluwajuyitan, T. D. (2022). Shallot-enriched amaranth-based extruded snack influences blood glucose levels, hematological parameters, and carbohydrate degrading enzymes in streptozotocin-induced diabetic rats. *Journal of Food Biochemistry, 46*(3), E14098. https://doi.org/10.1111/jfbc.14098.
115. Diaz, J. M. R., Sundarrajan, L., Kariluoto, S., Lampi, A. M., Tenitz, S., & Jouppila, K. (2017). Effect of extrusion cooking on physical properties and chemical composition of corn-based snacks containing amaranth and Quinoa: Application of partial least squares regression. *Journal of Food Processing Engineering, 40*(5), E12320. https://doi.org/10.1111/jfpe.12320.
116. Tobias-Espinoza, J. L., Amaya-Guerra, C. A., Quintero-Ramos, A., Pérez-Carrillo, E., Núñez-González, M. A., Martínez-Bustos, F., Meléndez-Pizarro, C. O., Báez-González, J. G., & Ortega-Gutiérrez, J. A. (2019). Effects of the addition of flaxseed and amaranth on the physicochemical and functional properties of instant-extruded products. *Foods, 8*(6), 183. https://doi.org/10.3390/foods8060183.
117. Beswa, D., Dlamini, N. R., Siwela, M., Amdnsdu, E. D., & Kdlanisi, U. (2016). Effect of amaranth addition on the nutritional composition and consumer acceptability of extruded provitamin a-biofortified maize snacks. *Food Science and Technology Campinas, 36*(1), 30–39. https://doi.org/10.1590/1678-457X.6813.
118. Ramos-Diaz, J. M., Rinnan, A., & Jouppila, K. (2019). Application of NIR imaging to the study of expanded snacks containing amaranth, quinoa, and Kanawa. *LWT–Food Science and Technology, 102*, 8–14. https://doi.org/10.1016/j.lwt.2018.12.029.
119. Zapana, F., De Bruijn, J., Vidal, L., Melin, P., Gonzalez, M. E., Cabrera, G., Williams, P., & Borquez, R. (2020). Physical, chemical and nutritional characteristics of puffed Quinoa. *International Journal of Food Science and Technology, 55*(1), 313–322. https://doi.org/10.1111/ijfs.14290.
120. Kuktaite, R., Repo-Carrasco-Valencia, R., De Mendoza, C. C. H., Plivelic, T. S., Hall, S., & Johansson, E. (2022). Innovatively processed Quinoa (*Chenopodium quinoa* Willd.) Food: Chemistry, structure and end-use characteristics. *Journal of the Science of Food and Agriculture, 102*(13), 5065–5076. https://doi.org/10.1002/jsfa.11214.

121. Chusak, C., Ying, J. A. Y., Zhien, J. L., Pasukamonset, P., Henry, C. J., Ngamukote, S., & Adisakwattana, S. (2019). Impact of *Clitoria ternatea* (Butterfly Pea) flower on *in vitro* starch digestibility, texture and sensory attributes of cooked rice using domestic cooking methods. *Food Chemistry, 295*, 646–652. https://doi.org/10.1016/j.foodchem.2019.05.157.
122. Philipp, C., Oey, I., Silcock, P., Beck, S. M., & Buckow, R. (2017). Impact of protein content on physical and microstructural properties of extruded rice starch–pea protein snacks. *Journal of Food Engineering, 212*, 165–173. https://doi.org/10.1016/j.jfoodeng.2017.05.024.
123. Yadav, G. P., Dalbhagat, C. G., & Mishra, H. N. (2021). Development of instant low glycemic rice using extrusion technology and its characterization. *Journal of Food Process Engineering, 45*(5), e16077. https://doi.org/10.1111/jfpp.16077.
124. Narang, N., Kaur, N., Singh, P., Saran, R., Singh, B., & Sharma, S. (2018). Development of extruded flakes from pearl millet. *Pharma Innovation Journal, 7*(8), 502–506.
125. Sharma, C., Singh, B., Hussain, S. Z., & Sharma, S. (2017). Investigation of process and product parameters for physicochemical properties of rice and mung bean (*Vigna radiata*) flour-based extruded snacks. *Journal of Food Science and Technology, 54*(6), 1711–1720. https://doi.org/10.1007/s13197-017-2606-8.
126. Singha, P., Singh, S. K., Muthukumarappan, K., & Krishnan, P. (2018). Physicochemical and nutritional properties of extrudates from food-grade distiller's dried grains, garbanzo flour, and corn grits. *Food Science and Nutrition, 6*(8), 1914–1926. https://doi.org/10.1002/fsn3.769.
127. Pardhi, S., Singh, B., Nayik, G. A., & Dar, B. (2019). Evaluation of functional properties of extruded snacks developed from brown rice grits by using response surface methodology. *Journal of the Saudi Society of Agricultural Sciences, 18*(1), 7–16. https://doi.org/10.1016/j.jssas.2016.11.006.
128. Patil, S. S., & Kaur, C. (2018). Current trends in extrusion: Development of functional foods and novel ingredients. *Food Science and Technology Research, 24*(1), 23–34. https://doi.org/10.3136/fstr.24.23.
129. Akande, O. A., Nakimbugwe, D., & Mukisa, M. (2017). Optimization of extrusion conditions for the production of instant grain amaranth-based porridge flour. *Food Science and Nutrition, 5*(5), 1205–1214. https://doi.org/10.1002/fsn3.513.
130. Wang, P., Fu, Y., Wang, L., Saleh, S., Cao, H., & Xiao, Z. (2017). Effect of enrichment with stabilized rice bran and extrusion process on gelatinization and retrogradation properties of rice starch. *Starch–Stärke, 69*(3–4), 1600201. https://doi.org/10.1002/star.201600201.
131. Gong, K., Chen, L., Li, X., Sun, L., & Liu, K. (2018). Effects of germination combined with extrusion on the nutritional composition, functional properties, and polyphenol profile and related *in vitro* hypoglycemic effect of whole grain corn. *Journal of Cereal Science, 83*, 1–8. https://doi.org/10.1016/j.jcs.2018.07.002.
132. Phongthai, S., D'Amico, S., Schoenlechner, R., Homthawornchoo, W., & Rawdkuen, S. (2017). Effects of protein enrichment on the properties of rice flour-based gluten-free pasta. *LWT – Food Science and Technology, 80*, 378–385. https://doi.org/10.1016/j.lwt.2017.02.044.

CHAPTER 7

HOT-EXTRUSION TECHNOLOGY FOR THE PRODUCTION OF MILLET-BASED READY-TO-EAT SNACKS

RAHUL RATHEE, SUSHIL KUMAR SINGH, and POONAM SINGHA

Department of Food Process Engineering, National Institute of Technology, Rourkela, Odisha, India

ABSTRACT

This chapter focuses on the process development to produce millet based ready to eat (RTE) snacks by hot-extrusion process. Millets are also classified as "Nutri-cereals" because of their predominant source of proteins with balanced amino acids, fiber, fatty acids, carbohydrates, minerals, phytochemicals, and antioxidants. Millet-based foods are typically prepared using several conventional processing techniques. Millets are used for making RTE snacks due to their high nutritional profile. Now-a-days RTE products are popular among most consumers, especially the young population, and the demand for quick snacks with appealing sensory and textural features are expected to expand substantially over the next 5 years. Extrusion technology has been widely utilized for the development of RTE snacks because of how simple it is to use and its capacity to create different textures and shapes that appeal to customers. There are lot of millet based RTE snacks in the market like ragi flakes, millet and flax cinnamon chips and millet pasta. The present chapter concentrates on the technology of producing RTE snacks from millets as well as the elements that influence the snack properties during processing. The nutritional quality of millet-based RTE products is also discussed in this chapter.

Ready-to-Eat Snacks: Emerging Technologies for Production and Safety.
Ravi Pandiselvam and Gülşah Çalışkan Koç

7.1 INTRODUCTION

Over the decades, consumption of extruded snacks has been increasing, particularly among the younger generations, due to their attractive appearance, taste, texture, availability in various shapes and sizes, as well as their lower cost [1–3]. However, most of these snacks are developed from starch-based sources like corn, rice, potatoes, wheat, and sweet potatoes. Most of the snacks available in the market are fried. However, excess consumption of these fried snacks could cause various diseases like obesity, high blood pressure, heart-related problems, and cancer in human health [4]. Therefore, the demand for functional snacks that provide nutritional value along with improved functionality has been increasing.

Extrusion processing is flexible and economical, high-temperature short-time (HTST) method, which includes several unit operations like cooking and plasticizing starch, protein based raw materials in an enclosed barrel with single or twin screws [5, 6]. In addition, it has become a multiple input/output process because of the interplay between flour characteristics and process parameters that as speed of screw, moisture of feed, barrel temperature, and screw arrangement [7, 8]. The extrusion technique has facilitated the development of several cereal-based foods, protein supplements, as well as sausages. It is currently used to develop pasta, bread, biscuits, breakfast cereals, baby meals, crackers, snack foods, croutons, chewing gum, confectionery goods, modified starch, texturized vegetable protein, and pet foods, etc. [9]. During the extrusion process, melting of the ingredients inside the barrel casing and expansion of the products coming out of the die are two important processes. These modifications are linked to physicochemical changes in the starch, protein, and other biopolymers. The brief idea about the rheological and extrusion features of millet can help to alter the textural and sensory characteristics of the expanded extrudates. Generally, refined flour is used for making the breakfast cereals. However, nowadays researcher/industries are trying to use whole grain flour but there are many difficulties. Several difficulties need to be addressed when whole grains are substituted for refined flours in extruded products. Poor expansion, enzymatic browning, lipid oxidation, and darker color due to bran, after extrusion are some of the factors that have an impact on the physicochemical characteristics s of whole-grain flour extrudates [10].

Major millet's physicochemical and nutritional features can be helpful for global use and produce an extensive range of value-added food stuffs, dietary foods, and RTE snacks. The use of millet in the form of both grain and flour in extruded snacks has been increasing rapidly in the food industry. Only millets or mixture of different millets and mixture of millets with other cereal grains can be used to make extruded snacks. Few researchers have concluded that fortification of cereal-based snacks with millets can help in improving the nutritional value of existing snacks. For example, Kharat, Medina-Meza [7] assessed the protein content of millet-cowpea blends. In another study, Onyango, Noetzold [11] investigated the digestibility of fermented, extruded maize-finger millet and cassava-finger millet blends. Furthermore, it was found that millets can be used as stabilizing agents for food texture [12]. Whole grains can increase the nutritional value of various foods, but it is important to consider the processing effects. Understanding how the factors of extrusion influenced the nutritional content, as well as extruded qualities of millet is important for making nutritious RTE snacks.

This chapter discusses various millet-based extrudates and various processing factors that affect the physical, sensory, structural, and functional characteristics of the extrudates.

7.2 MILLETS: PRODUCTION, NUTRITION, AND PROCESSING

The global production of cereals in 2021 was 2,800.88 million tons [13]. However, the world is currently dealing with several concerns, including a growing population, climatic changes, rising food prices, lack of fresh water, environmental damage, and other socioeconomic effects. Millet can be an alternative nutritious food, which can fulfill the nutritional needs of a growing population [14]. Before the Green Revolution, millets and pseudo-cereals were major crops but later, they were replaced with other major cereals. Now-a-days, millets are getting more attention due to their high nutritive value and climatic resilience. India contributed 41% of the total millet grown in the world in the year 2021–22 [15].

7.2.1 MILLET PRODUCTION

Asia and Africa dominate the global millet production. In 2020, Asia accounted for approximately 51.66% of total millet production, estimated at 15,737,396 tons, and Africa (45.31%) with a total production of 13,806,107 tons, followed by the European Union (2.19%) with a total production of about 669,103 tons, America (0.70%) with 214,600 tons whereas the Australian subcontinent contributed only 36,436 tons or 0.119% of total global production [16]. According to USDA, pearl millet is the largest producing millet, which is about half of the global millet production.

Millet is a necessary crop to ensure food security for several drought-prone regions in the world. Small-scale farmers can grow nearly all millet on marginal soils for commercial and household consumption. For millet species, only a few national and international bodies adopt a separate documentation pattern. Most millet species have poor and fragmented statistical records. Millets are grouped alongside sorghum and cereals by some authorities, and they are classified as 'coarse grains' by others. According to USDA, in 2021, Asia dominated the total global output of millet with area under cultivation of 11,630,005 ha with a yield of 13,532 hg/ha, and Africa with 13,806,107 ha with a yield of 7,000 hg/ha, followed by the European union with 528,988 ha with a yield of 12,649 hg/ha and Australia with 35,782 ha with a yield of 10,183 hg/ha. The country-by-country analysis of millet statistics revealed India's supremacy in global production, with 41% of total millet output estimated at 11,500,000 tons, followed by China (10.00%) with 2,70,000 tons and Niger (8.00%) with 2,142,000 tons. According to USDA, 39.30% of total millet is produced in African countries (~ 11,185,551 tons). India has been the world's leading millet grower for several decades. According to USDA data on area under cultivation for millet, India leads with 9,714,019 hectares, followed by Niger (6,743,482 hectares) and Nigeria (2,000,000 hectares).

Millet's prominence as a staple food has diminished since the Green Revolution, particularly in India, due to a variety of causes such as people's eating patterns, rising income, increased reliance on staple cereals, migration from rural to urban areas, and government initiatives. More than 90% of India's population relies on primary basic crops (i.e., rice, wheat, and maize) to meet their nutritional needs. India is a densely populated country, making it difficult to feed nutrient-dense food to such a large population. As a result, after decades of neglect, these nutrient-dense grains are making a significant comeback in Indian cuisine. The government has revised its strategy and is focusing on these crops as an alternative to major grains to address future hunger and hidden hunger problems while also addressing sustainability concerns. These crops were once known as 'coarse cereals,' but nowadays are popular as "super cereals." In India millet are grown in around 21 states including Madhya Pradesh, Kerala, Andhra Pradesh, Punjab, Rajasthan, Uttarakhand, Jharkhand, Maharashtra, Uttar Pradesh, and Haryana. Millets are grown in states like Meghalaya, Manipur, and Nagaland since the tribal population in these areas relies heavily on millets as the main food. Sorghum, finger, foxtail, pearl, kodo millets are the major millets grown in India.

7.2.2 MILLET NUTRITION

Millet is a generic name for cereals with small grains because of their grain texture, which is like rice and wheat but difficult to handle and cook in a suitable manner. They are currently known as 'Nutri cereals,' and even 'Super foods' due to their nutritional value and health-promoting properties. Because of their ability to survive in difficult agricultural settings, they are important for food security in many countries especially in the African and Asian continents. The millet seeds are composed of the endosperm, seed coat, and germ. Millet is not only a good source of carbohydrate, protein, fat, it also is a good source of vitamins and minerals (Figure 7.1). Millet is also a predominant source of antioxidants including glycosylated flavonoids and phenolic acids. Additionally, millet meals are considered as potential prebiotics that may improve the survivability of probiotics and have positive health effects [17]. A summary of nutritional profile of the varieties of millets are shown in Table 7.1. Some varieties of millet, like finger millet is low in protein but high in dietary fiber and phenolic content.

Millets have several health advantages. According to epidemiological studies, eating millets regularly lowers the chance of developing type II diabetes, improves digestion, reduce the risk of cancer, boosts immunity, and detoxifies the body, reduces the chances of Parkinson's disease and respiratory health issues [18–21]. Additionally, millets also possess antimicrobial properties. By preventing the growth of *Fusarium oxysporum*, and *Macrophominaphaseolina*, millet protein extract showed antifungal action [22, 23]. For habitual millet eaters, the proverb "Let food be thy medicine and not medicine thy meal" is more appropriate.

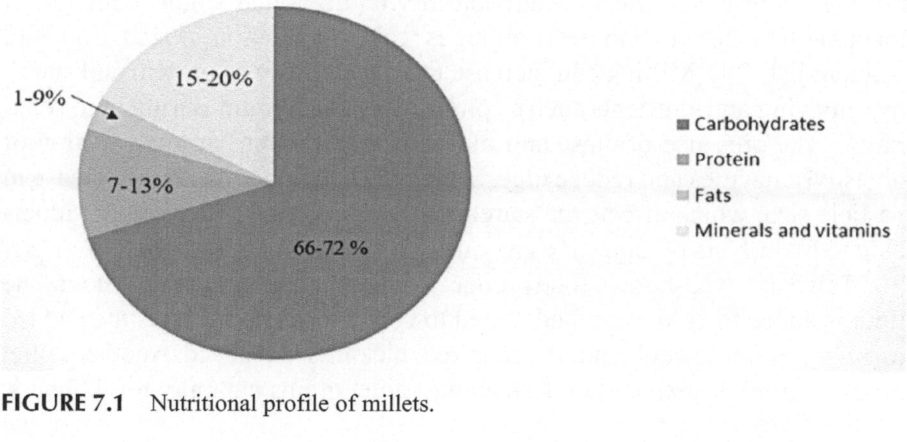

FIGURE 7.1 Nutritional profile of millets.

TABLE 7.1 Nutritional Profile of Millets

Name	Protein %	Carbohydrate%	Fat %	Ash %	References
Sorghum	11.36	71.95	4.70	3.17	[24]
Pearl millet	10.50	68.08	8.56	2.00	[25]
Finger millet	7.3	68	2.73	2.2	[26]
Foxtail millet	12.3	60.9	4.3	–	[22]
Porso millet	12.5	70.4	1.1	–	[22]

7.2.3 MILLET PROCESSING

Processing is essential for any material as it reduces the anti-nutritional factors and enhances the nutritional bioavailability and organoleptic properties. In food processing, the raw material is converted into a consumable product with extended shelf life and improved taste and texture. Food processing also brings diversity to diet and also enhances the nutritional quality but it can't be achieved by just one processing method for that we have to do multiple processing like we do in traditional practice like dehulling, milling, soaking, drying, cooking, germination, roasting, and fermentation [27]. The improved quality of processed millet makes it suitable for the development of new foods, consequently boosting regional food security. There can be many changes occur in the food's functional, physical, sensory, and nutritional properties due to the different processing methods [21, 28]. Milling reduces the size of particles because of a mechanical force for separating the endosperm, germ, and bran and facilitating refined flour production. Milling changes the chemical composition of millets, there is a loss of nutrients, minerals, and vitamins that are found in bran [21]. It is found that during milling, there is significant reduction in vitamin B and slight reduction of tocopherol (vitamin E) in pearl millet as well as reduction in zinc, iron, and calcium [21, 29]. Milling can increase the digestibility of protein and starch by removing anti-nutrients such as phytic acid and tannins during bran separation. The presence of these anti-nutrients may prevent proper function of digestive enzymes and reduce digestibility [21]. Another factor is decrease in particle size, which affects the starch granules and make them more vulnerable to hydrolysis by digestive enzymes [21, 30].

'Tuwo' is a well-known food product in Nigeria made from millet flour. The flour is added to cold water and mixed to create a slurry, which is then added to boiling water to cook until the required thickness is reached. Another millet product, gruel, is prepared by fermenting millet grains naturally for 72 hours.

Wet milled fermented millet grains were cleaned, and they were filtered through muslin fabric. This slurry is allowed to dry out before being prepared into pap, ogi, etc. [21, 31].

7.3 EXTRUSION PROCESSING

In extrusion processing, the food materials are forced to pass through die opening at a predetermined screw speed and die pressure to produce various products of different shape and size. There is an evolution in extrusion technology from over 30 years. The food extruders are considered as high-temperature, short-time (HTST) bioreactors capable of transforming raw material into final product. Previously, food extruders were only used to mixing and make ready-to-eat cereal pellets and macaroni. Mixing, kneading, shearing, shaping, and molding are just a few unit operation in extrusion [32]. Extrusion is a multifaceted technology in which wet expansible starch and protein material are modified by a variety of process factors, allowing a one-of-a-kind opportunity to create a novel product. It alters the product's microstructure, chemical characteristics, and macroscopic shape by gelatinizing the starch, breaking down the protein, and creating complexes between proteins, lipids, and starch [33]. Breakfast cereals, crispy flatbread, savory snacks, textured proteins, and cereal-based products like baby food are the results of this technique. Extrusion can be classified based on method of operation as well as based on type of construction. On the basis of method of operation, extrusion can be classified as hot and cold extrusion and based on the type of construction it can be classified as single screw extruder (SSE) and twin screw extruder (TSE) [34]. With recent technological advancements, it is now feasible to regulate the extruder and achieve the desired extruded product properties. Extrusion technology is widely used since it produces fewer effluents and waste. It is favorable compared to other processing in terms of high capacity, continuous processing, high production, energy efficient, automatic control, and low cost. One of the easiest ways to understand the extrusion process is the MIMO (multiple input and multiple output) system. It helps in the breakdown of complex processes into simpler forms, and it can assist to take the small to large scale machine and new product development. The high temperature generated under high pressure and shear force inside the barrel converts the feed material into melt form, the product expands due to the pressure drop occurring after the product comes out from the die outlet and this process is known as direct expansion.

7.3.1 SCREW CONFIGURATION

The screw is one of the critical elements of the extruder. Conveying, metering, and kneading the raw material is accomplished by screw within the barrel. The screw configuration, also known as screw profile, is a critical feature of the extruder because it affects the amount of shear that is applied to the material being extruded. The shear extent is closely related to the degree of material change and, eventually, to the quality of extrudates [35–37]. One of the independent parameters in the MIMO (multiple input multiple output) description of extrusion process is screw arrangement. Single-screw extruders often have an unmodulated screw; therefore, the manufacturer predetermines screw designs. If a different screw configuration is required, then the operator must alter the entire screw. The designs may differ depending on the machine's manufacturer.

Most of the TSE contain modified screws, allowing for an infinite number of screw profile designs. Different parts are put on a series of screw shafts to make the screws. Screw profile can be adjusted while operating the extruder. From the beginning to the end of the shaft, these screw parts have diverse roles. Based on design, different screw element can be used for conveying, kneading, and mixing. The raw materials are taken in by the feeding element and can be transported using conveying components. Materials are mixed and kneaded using kneading blocks demonstrates several screw components. A screw profile is defined by the pitch, length, stagger angle, and position of screw parts, which are key features that determine extrudate characteristics during extrusion [38]. Apart from having single or double screws, the primary distinction between single and twin-screw system is that the uses of the variety of raw materials as well as operating parameters of extruder and TSE have a broad range of operational capabilities.

7.3.1.1 SINGLE SCREW EXTRUDER (SSE)

In SSE (with only one screw), the material is transported to the transition region by a single screw rotating in a melt barrel with either an axial or helical groove on the inner surface when the screw channel grows shallower, and the material becomes compact. The use of SSE to process millet flour is straightforward but controlling the properties of the resulting product is more difficult. The barrel and screw both contribute significantly to shear produced in an SSE. Shear is produced when raw material particles rub against one another as well as the surfaces of the barrel and the screw.

Figure 7.2 is the schematic diagram of an SSE having three zones [39]. The first one is the feeding zone followed by the melting zone and then cooking zone. Figure 7.3 is the schematic diagram of extruder die confirmation and shows the position of pressure and temperature transducers [40]. Based on the type of screw, SSE can be classified into two types of screws, namely, parallel screws, in which the screw's diameter is the same throughout its entire length; and the other is conical screws, in which the screw's diameter increases along its length, with the largest diameter at the end of the screw, i.e., towards the die.

The materials for kneading and mixing are compressed due to the changing diameter design. The barrel shape is more important in the case of uniform diameter screws to achieve compression and mixing. Compression ratio is used to describe various SSE. The compression ratio is the proportion of the flow area in the feeding segment to the metering section. The range of this ratio is 1:1 to 5:1 [41]. In transition region, compression increases quickly, dissipating majority of the mechanical energy needed to turn screw and quickly raising the temperature of the food ingredients. Feed rate, melt properties, screw speed, and viscosity all influences screw fill. Flexible use and

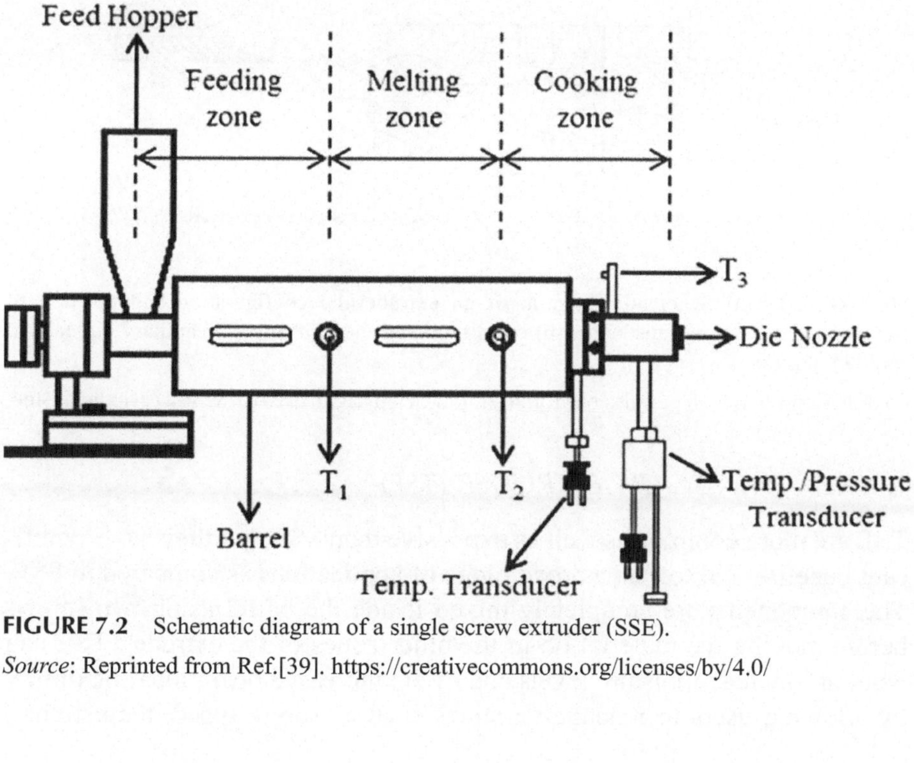

FIGURE 7.2 Schematic diagram of a single screw extruder (SSE).
Source: Reprinted from Ref.[39]. https://creativecommons.org/licenses/by/4.0/

working of SSE depends on some factors like adjustment of screw, the ratio of barrel length (L) to diameter (D) and drive motor capacity. The restricted mixing of the material is a drawback of single-screw food extruders, even though they are generally simple to use and maintain. If the combination has many components, they must be thoroughly combined before feeding. Due to their generally poor mixing ability, SSE are typically supplied with premixed materials that are frequently preconditioned with additional moisture [41]. Products including pasta, carbohydrates rich food, corn curls, pet food, and breakfast cereals are frequently produced using single-screw extruders.

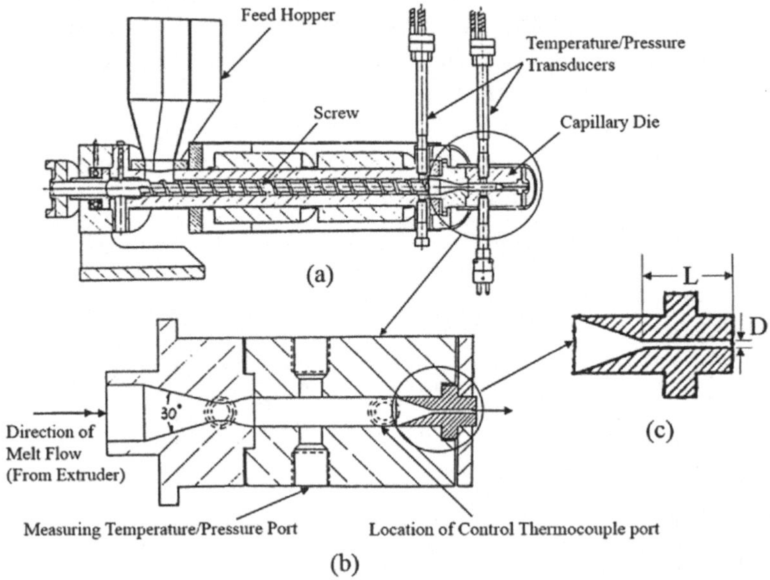

FIGURE 7.3 (a) Schematic diagram of an extruder/die configuration and position of temperature / pressure transducers (b) cross sectional view of a round capillary die and (c) capillary nozzle.

Source: Reprinted with permission from Ref. [42]. Copyright © 2016. Wiley Periodicals, Inc.

7.3.1.2 TWIN SCREW EXTRUDER (TSE)

TSE are more complex as well as expensive than SSE, but they have significant benefits. TSE offers a broad range of applications as compared to SSE. The components are completely mixed inside the barrel using two screws before moving the material on to the other zones of the extruder. TSE can work at low feed moisture levels and offers increased operational flexibility by allowing users to manage variables such as screw speed, temperature, and time [43]. Directly expanded, co-extruded and texturized products are formed using TSE. Various millet-based extrudates are made utilizing single and twin-screw extruders under various extrusion process conditions. Breakfast cereals, puffed snacks, and texturized proteins are all produced using twin-screw extruders, but their greatest utility is in the production of specialty food items. TSE can process raw materials with finer particle sizes, higher levels of proteins and fibers, and products that call for high levels of quality [41, 44]. While a TSE's barrel lining is typically smooth, most of the shear is produced because of particles rubbing up against one another when the material is transported along the two screws. In a dual screw extruder, the potential for shear creation is enormous. TSE produces more variety of extrudates than SSE because they are more efficient at producing shear [45]. Additionally, TSE are simpler to regulate and can maintain a consistent level of product quality. The food business makes extensive use of both types of extruders.

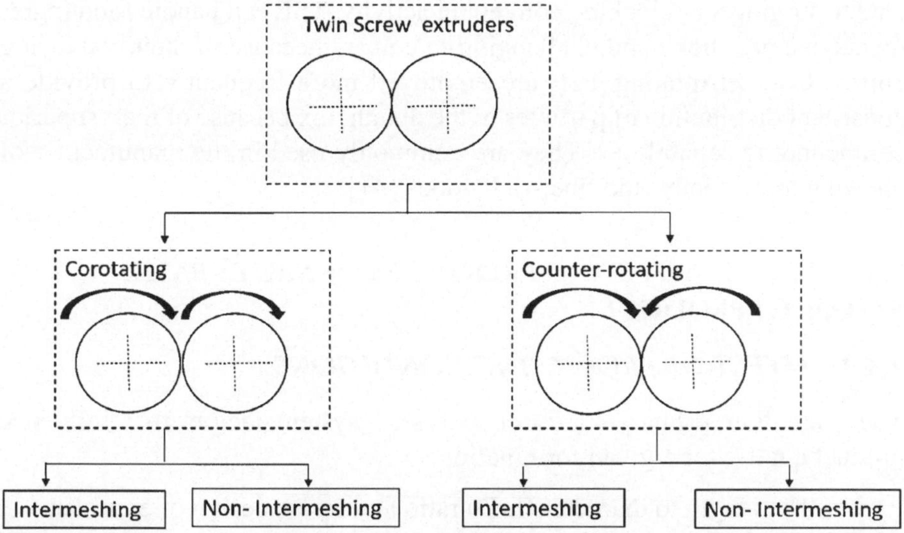

FIGURE 7.4 Classification of twin screw extruders.

Despite their relatively complex design, twin-screw food extruders have a wide range of processing capabilities. Due to their versatility and effectiveness in mixing and kneading, TSE can be used for a variety of materials. TSE is classified in two types based on rotating direction of screws. First, one is corotating TSE and second is counter rotating TSE. In corotating TSE, screws rotate in same direction while in counter rotating TSE screws rotate in opposite direction (as shown in Figure 7.4). Further, they are divided into intermeshing and non-intermeshing screw. In the intermeshing corotating design, each screw wipes the other to move the product ahead while the screws work together to squeeze the product forward in the intermeshing counter-rotating extruder. In the majority of TSE, the screws are modularly constructed and installed on the screw shafts. The operator can manage how material behaves inside the extruder and affect the range of chemical as well as physical processes in the extrusion cooking by properly setting up the screw parts. Most extruder manufacturers offer a wide variety of screw elements. In food processing industry, corotating TSE also known as self-wiping extruders have gained popularity. A corotating twin-screw extruder has specialized parts on its screws that carry out the various tasks required for the extrusion process. Screw sections come in fully intermeshing and self-wiping varieties, with either square or circular channels. Self-wiping screws stop materials from accumulating on channel surfaces, which can lead to surging and blocking conveying activity. TSE can handle food ingredients that are sticky and challenging to convey because of their conveying ability. Counter-rotating TSE are employed more frequently to provide a consistent distribution of particles in the dough mix because of their superior compounding capabilities. They are commonly used in the manufacture of chewing gum, candy, and fiber-rich goods [41].

7.4 FACTORS AFFECTING PROPERTIES OF MILLET BASED EXTRUDED PRODUCT

7.4.1 EFFECT OF PROCESSING CONDITIONS

There are four primary kinds of extrusion system factors that influence product quality for a given formulation.

i. The length to diameter (L/D) ratio and screw design of extruder.
ii. The design of revolving knife.
iii. The process factors that can be changed during the extrusion run.
iv. The die diameter, land length, shape, and number of dies.

The amount of mixing, the amount of shear forces injected, the heat produced by friction, and residence time distribution (RTD) are all impacted by the extruder screw arrangement. Degree of cooking, and product quality, melt rheology and homogeneity are all impacted by these variables. The L/D ratio influences the time of residence and subsequent degree of cooking.

Although majority of work is done in the barrel, shape, and size of extrudate, die pressure, rheology of melt, and pressure drop across die is influenced by die design. These elements affect the uniformity and quality of the result. The final size and shape of the extrudates result are influenced by rotating knife's design and speed. Even though the extruder's L/D, rotating knife design, die design, and screw arrangement all have an influence on the final product's quality, none of these elements can be altered while the extruder is in process. Factors that can be altered while the extruder is running are water injection rate, rotating knife speed, extruder barrel temperature, speed of screw, and feed rate, RTD, amount of frictional energy produced, amount of shear and the barrel fill are affected by the SS. The injection rate of water influences the moisture content of the barrel, which affects the frictional energy as well as the viscosity of the melt. Barrel fill, time of residence, and energy production are all affected by the feed rate. The product temperature is influenced by the barrel's temperature profile, which also influences the product's degree of cooking and melts viscosity. External heating systems regulate the barrel's temperature (depending on the kind of extruder, electrical induction, hot oil, or steam for heating can be used). Direct steam injection is used in some extruders to provide the product with thermal energy. If necessary, cooling water could be circulated to remove heat from the extruder barrel. Figure 7.5 illustrates the independent and process parameters as well as the product properties. The different process parameters affect the physical, functional, and sensory characteristics are following:

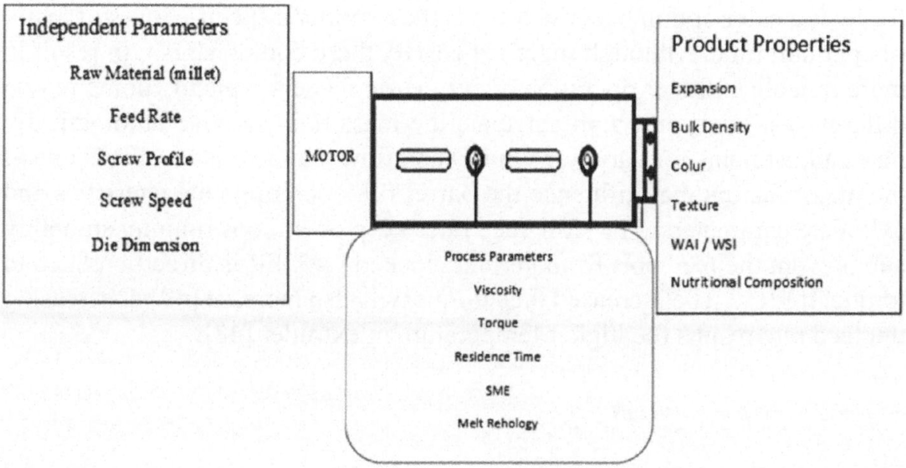

FIGURE 7.5 Extrusion processing: independent parameters, product properties, process parameters.

7.4.1.1 TEMPERATURE

Product temperature is also known as mass temperature. The melt temperature inside the barrel increases due to friction between screw and the feed, while passing through the screw flight. For measuring the temperatures of barrels, dies, and melts, J or K type thermocouples inserted with the extruder to measure the barrel temperature and die block. Measuring the melt temperature in barrel is not easy because thermocouples can't be inserted into the melt. Moisture also effects the temperature, due to the shearing process there is increase in temperature by 7–9°C while by lowering the moisture it get increased 18–20°C [46]. Specific mechanical energy will also increase by increasing the temperature. Expansion ratio (ER) will increase and water absorption index will decrease by increasing in temperature [47].

7.4.1.2 FEED RATE (FR)

A set volume of material is continuously fed into the extruder. A volumetric feeder is often used to measure the powder feed rate. However, minor changes in the feed rate due to variations in bulk density (BD) can significantly damage the extruder. There are several additional volumetric feeder kinds, that includes single screw feeder (screw speed frequently varies in direct proportion to volumetric rate), disk feeder (rotational speed affects the amount of material delivered, free-flowing materials hardly ever flood) and volumetric belt feeder (the belt's speed regulates the feed rate). A gravimetric feeder is a safer option because it feeds the extruder a specific mass of material per unit time. Although more expensive, these components will result in more reliable extruder operation. Gravimetric feeders typically come in two different varieties, and both calculate the mass rate and also automatically make adjustments for variations in BD and flow characteristics. FR is one of the major factors that influence the barrel fill, operating characteristics and extrudate parameters. The slow feed rates keep operations running smoothly and prevent the feed port from getting blocked [48]. FR is directly related to torque. If FR will be increased then torque will also increase [49]. Decreasing the feed rate results the higher temperature in extruder [48].

7.4.1.3 SCREW SPEED (SS)

The current frequency and screw speed are inversely correlated. Screw speed majorly affect bulk density, crispiness and hardness, while feeder speed affected expansion ratio, crispiness and hardness [50]. If there is increase in SS then mass flow rate and pellet durability will also increase but water absorption index (WAI) of the extrudate will decrease [47]. Screw speed is inversely related to the torque. If screw speed will increase then there will be reduction in torque [48]. SS is the major factor that influence the operating characteristics, barrel fill, and extrudate parameters [48]. Decrease in the FR and increase in the SS will result in the increase of SME and the gelatinization degree, higher the temperature [48]. It was also found that ER has a good linear correlation with screw speed and Additionally, SS and WAI had a negative correlation [51].

7.4.1.4 DIE PRESSURE AND TORQUE

Die pressure serves as a gauge for material flow resistance and its geometry has a big impact on the extrudate characteristics. It rises if more feed material gathers at the opening of die. The pressure at the die can be measured using pressure transducers inserted into the die plate. Pressure transducers need to be maintained, calibrated, and temperature compensated because they are prone to abrasion. The most used transducer measures angular deflection, which is directly interrelated to torque, using strain gauges. The extruder's torque serves as a gauge for the machine's safety as well as the amount of effort needed to thrust the extrudate out of the die hole during operation. Torque helps to (a) move the product, and (b) rotate the screw under full and no-load conditions.

The materials being extruded undergo physicochemical changes because of the heat and shear under the high pressure in the extruder [52]. Increase in SS impacts the shear rate and the barrel's heat exchange, which reduce material viscosity, this causes the extruder's torque to decrease [46]. As the high feed moisture enhances the melt plasticity and the high temperature will decrease the viscosity, the die pressure [35]. Torque is inversely proportional to the temperature and feed moisture, if temperature and feed moisture will increase then torque will decrease [53, 54].

7.4.1.5 SPECIFIC MECHANICAL ENERGY (SME)

SME is the ratio of mass flow rate of feed to the extruder to the total mechanical energy obtained from material conveyance owing to the screw rotation [55]. SME can also be defined as energy required per unit mass of the raw material. The mechanical energy used to extrude the feed material, which aids in rupturing intermolecular hydrogen bonds, determines the degree of starch conversion. Greater starch gelatinization and product expansion are caused by high SME levels [36, 46]. SME is a relationship between an expansion indication and a melt temperature. As starch will gelatinizes the viscosity of melt will decrease and solid mass flow changes to viscoelastic flow [56]. If there will be increase in temperature, then SME for all millets will decrease. At higher moisture, there is no significant change found in SME for foxtail millet but if screw speed and temperature will be increased then SME will also increase.

7.4.2 EFFECT OF MATERIAL PROPERTIES

The extrusion process and extrudates are greatly influenced by the material properties. The inherent characteristics of starch promote expansion during the extrusion process. This quality makes it ideal for producing extruded items with appealingly airy and crunchy textures for consumers. When mixed with starch, fibers and hydrocolloids alter the characteristics of extrudates [52]. Gelatinization occurs at substantially lower moisture levels (12–20% in extrusion cooking) than are required in other food operations, making this method of cooking rather distinctive. The rate of gelatinization tends to rise under processing settings that raise the temperature, shear, and pressure. Gelatinization is also effected by other food nutrients like dietary fiber, lipids, and salt [57]. The digestion will be increased even if complete gelatinization does not happen. Theoretically, extrusion might predigest starch. Amylopectin molecules have branches that are simple to cut in the barrel. Generally during extrusion amylose and amylopectin losses some molecular weight. Starch breakdown can be minimized or increased by adjusting the screw configuration. The moisture concentrations are typically relatively low throughout the extrusion process when producing direct-expanded goods, ranging from 14% to 20% on a wet basis [52]. The simultaneous occurrence of starch gelatinization, melting, and fragmentation is possible under these low moisture circumstances [52, 58]. Due to the shear and heat of the extrusion process, starch undergoes a melt phase with viscoelastic behavior at low moisture levels [52]. Heat and moisture as well as mechanical shear are two possible strategies to start physicochemical changes. Starch partially gelatinizes when exposed to heat and a lack of moisture. Multiple strategies can be used to alter the extrusion conditions to develop digestion-resistant starch (RS). As amylopectin molecules' branches are cut off, they may interact with other carbohydrates to form novel connections that our enzymes are unable to break down.

Unlike starch, fibers have a different behavior, specially insoluble fibers don't go through the minimal change in phase transfer during extrusion [52]. It might be because of some supramolecular linkages can break with heat and shear, the surface of insoluble fibers can change and become fractured [59, 60]. Based on their structure and affinity for water, soluble fibers and hydrocolloids may undergo phase changes in the presence of heat and shear. Each one of them affects how a raw material behaves during a phase change under heat and shear. Fibers added during extrusion typically result in less expansion, tougher textures, and less crispiness. As compare to soluble fibers, insoluble fibers have more significant effects [10]. It is found that the more the crude fiber results in higher bulk density [61].

Extrudates' water solubility index (WSI) and ER significantly reduced as feed moisture content increased while their WAI, BD, and hardness increased. With increasing barrel temperature, extrudates' WSI, BD, water holding capacity (WHC), and hardness were greatly reduced, while their WAI and ER were significantly increased [62]. By exposing enzyme-access sites through denaturation, extrusion increases the digestion of proteins. Denaturation causes most proteins to lose their ability to function as enzymes and enzyme inhibitors. Protein denaturation level is often determined by its solubility in water or other aqueous solutions. Nutritional analyzes of extruded foods have been prioritized because most extruded foods are low in protein. Maillard reactions are facilitated during extrusion by high temperatures of barrel and low moisture levels.

Although lipids are an intense source of energy, excessive dietary lipid consumption has been linked to chronic diseases like obesity, cancer, and heart disease. Typically, foods with fewer than 10% lipids are extruded since higher amounts of lipids diminish barrel slip, which makes extrusion challenging, especially for enlarged products. For removing the moisture and to improve the flavor and texture frying is done after the extrusion. Since a commercial product only contains 21% oil, the oil content ranged from 20–35%, indicating that more advancement are required. Oil is released throughout the heating and shearing processes; therefore, extrusion can be employed to help with oil extraction. Lipid oxidation is one of the main factors contributing to the loss of sensory as well as nutritional quality in foods. Lipid oxidation can happen during storage even though it is believed that it does not happen during extrusion because of the short residence duration. Higher pro-oxidant mineral concentrations are a result of screw wear. The development of air cells in enlarged products, which results in an increase in surface area, is another element that favors oxidation. However, during extrusion, lipolytic enzymes and other oxidation-promoting enzymes might be inactivated, and starch-lipid complexes produced in the barrel may be more oxidation-resistant. Extruded food packing in opaque containers with nitrogen or vacuum may offer further security. Feed moisture, temperature of barrel, and SS all have major influence on the extrudates moisture, carbohydrate, and protein contents, but not on the fat or crude fiber content [63].

Melt rheology is also an essential material property. There is not yet a reliable way to measure the rheology of an extrusion melt online. Rheological behaviors can be measured using techniques like capillary viscometers [37]; Carri-med rheometers, rapid visco analyzers (RVA), and micro-Fourier rheometer are all options. Expansion, texture, and overall quality of extrudate are influenced by the rheological characteristics of the melt, which also aid in understanding transport processes. Because it is not possible to measure, the melt viscosity within an extruder and very difficult to characterizing melt rheology directly in the extruder. Capillary, in-line slit and slit, or rotating rheometers are the offline rheometers. Off-line rheometers positioned at the ends of the dies are some of the methods used to determine the rheological characteristics of the melt off-line rheometers may be utilized under controlled circumstances; however, they frequently are unable to replicate the conditions found inside an extruder. The viscosity of the melt can only be measured in-line at the die end, and associated rheometer's measurements could be inaccurate since the melt's flow is restricted. To mimic the extrusion circumstances, laboratory rheometers have undergone several recent advancements [52].

7.5 PHYSICAL CHARACTERISTICS OF MILLET-BASED EXTRUDATES

7.5.1 EXPANSION RATIO (ER)

ER can be measured, by dividing mean diameter of 10 selected extruded samples, by the die diameter [7]. Extrudate's radial ER often affected by screw speed (SS) and feed moisture [64]. Increase in barrel temperature might reduce the expansion ratio. This might be due to starch breakdown at higher barrel temperature as a result in degradation of starch molecules and provide less expansion. Therefore, temperature of barrel, SS and moisture level of feed affected the expansion during extrusion cooking. In addition, feed rate affected the ER [50]. Furthermore, die pressure and vapor pressure have the major impact on product expansion. The quick release of water causes the material to expand as it comes out of the die opening because there is a severe pressure difference between the die and atmosphere [46, 64].

ER is the key quality parameter for snack products and puffed cereal as it is closely related to the eating preferences. It was found that the highest ER value of foxtail millet, was 4.41 when temperature was 130°C, feed moisture was 17.5% and SS was 400 rpm, whereas it was 3.24 when temperature was 130°C, feed moisture was 17.5%, and SS was 300 rpm for finger millet and 3.78 when temperature was 130°C, feed moisture was 20%, and SS was 350 rpm for pearl millet. The lowest ER value measured for foxtail millet was 3.58 when temperature was 110°C, 17.5%, and SS was 300 rpm, whereas the lowest ER value measured for finger millet was 2.81 at 110°C, feed moisture was 15% and SS was 350 rpm, and the lowest ER measured for pearl millet was 3.00 when temperature was 110°C, feed moisture was 17.5% and SS was 300 rpm [7]. At 150°C and 180 rpm of SS, recorded the greatest ER, i.e., 4.7 in an 80% blend of pearl millet and other flours [65]. Temperature, SS, feed moisture, increases result in an increase in ER. The best result was observed for foxtail millet as the maximum ER achieved at 130°C and it did not depend on the feed moisture range that was examined, allowing for greater control of process factors. The highest ER was found at both higher temperature and moisture content, respectively. For example, ER of pearl millet was moisture and temperature dependent. Finger millet demonstrated maximum ER with lower moisture content compared to the pearl and foxtail millet expansion. Foxtail millet also shown the best ER response when SS was considered. This outcome may be the result of the prolonged residence durations associated with low screw speed hence the material needs more shear for transformation, leading to better expansion. Additionally, water in the extruder serves as a heat sink, lubricant, and lowers shear strength, which results in improved expansion [66]. According to Adebowale, Afolabi [67] high temperature causes a greater extent of starch gelatinization and/or degradation. This finding also suggests that the increased starch gelatinization causes an increase in the extruded product's expansion. The best-expanded product was made by using foxtail millet flour, followed by finger millet and pearl millet [7]. These results demonstrate how radically different the millets' growth properties are, whereby, foxtail millet demonstrated the best expansion across the spectrum of extrusion conditions examined [7].

The expansion ratio of millet extrudate**:** extrudate was made up of little millet (10%), rice (45%) and maize flour (45%). Extrusion was run at temperature of 90 to 110°C and 320 to 350 rpm screw speed. The extruder used for making that product is TSE. The expansion ratio of the extrudate was 3.19 [68]. The ER of other millet-based extrudates is mentioned in Table 7.2.

7.5.2 BULK DENSITY (BD)

Bulk density reveals a product's porosity, which affects packaging design and material selection. Bulk density (BD) is calculated using the average diameter and length of around 10 extrudates pieces chosen randomly. The BD is obtained by dividing the mass of extrudates with their volume, and it is expressed in kg/m^3 [69]. When characterizing the amount of puffing in extrudates, BD is typically linked with the ER. In general, bulk density and ER have an inverse relationship. The extrudates with higher BD have more continuous as well as uniform protein matrix and poor hydration characteristics [7, 70]. Generally, the consumers did not prefer high BD extrudates. The BD values of pearl, foxtail, and finger millet ranged from 87 to 363 kg/m^3, 71.2 to 172.6 kg/m^3, and 117.7 to 226.7 kg/m^3, respectively. Bulk density often decreased as temperature and moisture increased [71]. During extrusion processing of pearl, foxtail and finer millet, the residence period and subsequent rheology would be affected by screw speed, the overall bulk density would not be significantly affected [72]. But during extrusion of kodo millet screw speed has major influence on bulk density [50]. High quantities of insoluble fiber likely encourage more nucleation near the die exit, which results in larger levels of BD [61]. The potential of starch to gelatinize and form complexes with lipids, as well as the relationship between BD and WSI [72]. When there was increased lipid content and higher enthalpy of gelatinization, finger millet displayed high BD [7].

The bulk density of millet extrudate: Extrudate was made up of little millet, rice and maize flour. The little millet was 10% and rice, maize was 45%, respectively. Extrusion was run at temperature of 90 to 110°C and 320 to 350 rpm screw speed. The extruder used for making that product is TSE. The BD of the extrudate was 0.16 g/cm^3 [68]. The BD of other millet-based extrudates is given in Table 7.2.

7.5.3 COLOR

The color of the extruded product has a major effect on the acceptability of customers. It was significantly influenced by the feed matrix composition and processing conditions. The extrudate's color is impacted by a variety of reactions that take place during extrusion. Color deterioration was one of the most common reactions due to non-enzymatic browning, which includes the caramelization and Maillard reaction. The color value is affected by moisture content in addition to temperature. The degree of thermal treatment has a major impact on the product's color as the high temperature increases the browning response.

Color analysis of sorghum based rich protein extruded snack was done by 75 group of consumers from different age group and it was found that snack is appealing to 100% adolescents, 92% adults, and 80% elder age group of consumers [73].

7.5.3 TEXTURE

The extruded product's structural characteristics determine its textural aspects, which are directly correlated to its sensory qualities. The extrusion cooking technique has a massive impact on the adhesiveness, hardness, chewiness, springiness, and crispness of the millet-based extrudates. The force need to break or collapse an extrudate is proportional to the extrudate's hardness. It is a crucial human sensory perception that affects the structure and growth of a product. The extrudate cell wall strength was weakened by protein material [73]. Porosity is a crucial factor influencing textural characteristics like crispness. Crushing force and crispness work changed differently according to different protein source [73]. Additionally, SS and feeder speed affect the hardness and crispness [50]. Extrudate made from little millet with maize and rice in twin screw extruder has the hardness value 16.83 and this product is found acceptable by consumers [68].

7.6 FUNCTIONAL CHARACTERISTICS OF MILLET-BASED EXTRUDATES

7.6.1 WATER ABSORPTION INDEX (WAI) AND WATER SOLUBILITY INDEX (WSI)

WAI is used to measure the starch's capacity to retain water after expansion because of excess of water [74]. It serves as a gauge to know the behavior of extrudate behave when it comes into contact with water during subsequent processing and starch modification [75]. Whereas, the water-soluble index (WSI) quantifies how much more soluble polysaccharide is produced when starch molecules are broken down during extrusion [76]. The WAI is influenced by the hydrophilic group's available in food material [77, 78]. Due to the breakdown of starch, the molecular size reduces, which lowers the molecules' ability to store water, lowering WAI. Furthermore, it is affected by the temperature, SS, and moisture content [47, 79]. WAI considerably reduces at high temperatures and low moisture, which can be described by the preference of dextrinization over gelatinization [80]. However, high temperature and SS increase the number of soluble solids by shortening the polymeric chains due to shear. High residence time enhances cooking at any temperature, boosting WAI. Higher WAI values were seen in extrudates with higher amylose levels than in those with lower amylose contents. This could be due to reduced levels of amylose, which lead to higher levels of starch breakdown during extrusion heating and higher WSI [81].

Generally, indicators of the deterioration of molecular components are WAI and WSI [82]. WAI values ranged from 3.51–4.18 (g/g), 1.85–2.69 (g/g), and 1.85–2.69 (g/g) for the foxtail, finger, and pearl millets, respectively. The highest value of WAI for foxtail millet was 4.18 (g/g) at temperature of 130°C, feed moisture and SS was 15% and 350 rpm, respectively. The maximum value of WAI for finger millet was 2.69 (g/g) at temperature of 130°C, feed moisture and SS was 17.5% and 400 rpm, respectively. The maximum value of water absorption index for pearl millet was 3.78 (g/g) at 130°C temperature, feed moisture and SS was 15% and 350 rpm, respectively. It was found that feed moisture does not show significance impact on WAI as compared to temperature and screw speed [7]. For finger, foxtail, and pearl millet, the extrudates' WSI ranged from 5.75 to 12.12%, 9.37 to 18.54%, and 12.72 to 19.39%, respectively. The WSI for foxtail millet varied from 9.37% at 110°C, 20% feed moisture, and 350 screw revolutions per minute to 18.50% at 130°C, 17.5% feed moisture, and SS is 400 rpm. The lowest WSI value for finger millet was 5.8 at 110°C, 20% moisture, and SS was 350 rpm, while the highest WSI value was 12.10%, obtained at 130°C, 17.5% moisture, and SS was 400 rpm [7]. The maximum value of WSI for pearl millet was 19.44% at 130°C temperature, feed moisture, SS was 17.5% and 400 rpm, respectively, and the minimum value of WSI for pearl millet value was 14.4% at 110°C, temperature, 15% moisture and SS was 350 rpm, respectively. Increased feed moisture results in higher WAI and lower WSI. High moisture content caused the shearing forces to have less influence on the material being processed, which in turn reduced the degree of dextrinization of the starch polymers and lowered the water solubility index [83]. In contrast to WSI, which seemed to increase as the temperature rose, the

WAI got decrease. This could be due to depolymerization of amylose and amylopectin caused by rising temperature, the product's ability to absorb water was decreased [84, 85]. Additionally, WAI is indirectly correlated with water-holding capacity, which showed impact on the stability of a product's storage [47]. WAI and WSI of pasta made up of sorghum and wheat with proportion of 50:50 is 21.27 and 11.28, respectively [86]. The WSI and WAI of other millet-based extrudates are given in Table 7.2.

7.6.2 PASTING PROPERTIES

The extruded product's pasting profile gives information on the product's macromolecule arrangement, the firmness of the starch granules, and amylose leaching [90, 91]. The pasting profile of the extrudate is used to express the pasting temperature (PT), hot paste viscosity (HPV), peak viscosity (PV), cold paste viscosity (CPV), setback viscosity, and breakdown viscosity. The temperature at which starch begins to gelatinize; molecules expand by absorbing water, is known as the pasting temperature [92]. By speeding up the screws, more gelatinization occurs. As the screw speed and barrel temperature rise, PV, HPV, and CPV values typically fall. The magnitude of PV might decrease because of the increased gelatinization and starch breakdown. The gelatinization of starch was reduced by increasing the high peak viscosity and gelatinization decreased with decreasing in HPV. The pasting characteristics of ingredients and how they behave in an extruder affect the formulation changes during extrusion. Raw material composition also affects the pasting properties of extrudate [46, 93–95]. A quick way to examine formulation variations is by comparing their pasting qualities.

Major millets are classified as having starch type B by Schoch [93] (starches that are moderately swollen but have a significantly lower pasting peak and substantially less thinning). Major millets starches appear to be comparable to hard wheat flour in terms of paste stability, according to pasting qualities. Although not always accurate, pasting qualities can aid in predicting how they would behave throughout the extrusion process. It is crucial to remember that the conditions within the extruder and the rapid Visco-analyzer are significantly different from one another. A viscoelastic melt expands during the expansion phase as result of moisture flashing at the outlet die, which is followed by bubble formulation and collapse. The viscosity values were lower at lower die pressure, SME and torque, which could be due to the reduced the resistance to the screw rotating and flowing

TABLE 7.2 Physical and Functional Properties of Millet-Based Extrudates

Raw Material	Expansion İndex	Bulk Density kg/cm3	WAI %	WSI %	Extruder Type	References
Foxtail millet (60%), amaranth (5%), rice (5%), Bengal gram (20%), cowpea (10%)	2.32±0.03	0.10±0.02	5.2±0.04	0.32±0.06	Twin-screw extruder	[69]
Finger millet (20%), maize (50%), rice (20%), and soybean (10%)	3.50±0.05	0.1618±0.03	5.08±0.04	8.89±0.10	Twin-screw extruder	[61]
Sorghum (90%), cocoa (5%), sugar (5%)	1.57	0.21	15.5	13.1	Twin-screw extruder	[88]
Maize (40%), defatted soy (20%), finger millet (30%), elephant foot yam (10%)	3.9	0.14	6.13	17.07	Twin-screw extruder	[89]
Pasta sorghum (50%), wheat (50%)	-	0.43 g	21.27±1.56	11.28±0.12%	Single-screw extruder	[86]

through the extruder die [96]. The amount of starch, the molecular makeup of the starch, the degree of crystallinity, and the length of amylose chain are few variables that could affected different flours behavior during pasting. Additionally, the characteristics and viscosity of the paste may fluctuate depending on the protein composition [97].

Pasting curves often offer details on the retrogradation and gelatinization of starch. Compared to the foxtail and finger millets, pearl millet has much lower viscosity during the period of high temperature holding, which suggests that pearl millet flour would be simpler to extrude. When temperature decreases the foxtail millet's final viscosity is much higher than the final viscosities of the finger and pearl millet [7].

7.7 NUTRITIONAL VALUE OF MILLET-BASED EXTRUDED SNACKS

The higher intake of starch-based snacks is responsible for various chronical diseases and health disorders. Therefore, development of millet-based and millet-fortified snacks can improve the nutritional status, which is even superior to the cereal-based extrudates. Nutritional profile of the various millet-based extrudates was summarized in Table 7.3.

- **Little Millet Based Extruded Product:** Extrudate was made up of little millet (10%), rice (45%) and maize flour (45%). Extruder was run at temperature 90 to 110°C and 320 to 350 rpm screw speed. Extruder used for making that product was twin screw extruder [68]. Nutritional composition of the extrudate is mentioned in Table 7.3. The expansion ratio of the final product was 3.19 and bulk density was 0.16 g/cm^3, hardness was 16.83 N and the overall sensory acceptability was 8.28 out of 9 (judged on nine point hedonic scale) [68].
- **Soghum Based Snacks:** Sorghum based extruded protein rick snack was made up of sorghum and corn flour and blend with the whey protein isolate and defatted soy flour with the proportion of 5:2:1.5:1.5, respectively. Twin-screw extruder was used for making snacks. Extruder was run at the SS and feed rate of 400 rpm and 80 kg/h, respectively. The screw diameter and L/D was 52 mm and 16:1, respectively with the circular opening die 3.2 mm [73]. Nutritional profile of the snack is given in Table 7.3.
- **Sorghum Based Pasta:** Raw material used for making the pasta was sorghum and wheat 50% each. Single screw extruder was used

TABLE 7.3 Nutritional Profile of Millet-Based Extrudates

Raw Material	Carbohydrate (%)	Fat (%)	Protein (%)	Ash (%)	Moisture (%)	Fiber (%)	Calorific Value (kcal)	References
Little millet (10%), rice (45%) and maize (45%)	87.35	1.05	3.27	1.27	5.03	1.93	372.33	[68]
Kodo millet-chickpea flour blend (70:30)	75.7	0.32	11.6	1.3	10.6	0.47	351.9	[50]
Sorghum (90%), cocoa (5%), sugar(5%)	78	1.0	5.6	1.9	2.0	1.2	–	[88]
Millet–cowpea based Fura	74.1–77.3	1.5–3.6	11.2–16.8	1.7–2.0	14	-	–	[66]
Maize-millet based extrudate fortified with soy	64.99–72.98	0.49–1.07	13.96–16.81	-	6.14–9.78	2.48–3.36	–	[63]
Pearl millet (85%), soy(25%)	61.3	4.4	16.0		18	1.7	388	[98]
Uji	88.87	0.95	8.19	2.04	8.19		402.72	[11]
Pearl millet (70%), Bengal gram (20%), green gram (10%)	65.2	4.70	14.70		18.00	1.20	362.00	[98]
Sorghum (50%), wheat(50%)	74.20	0.44	9.70	–	8.25	0.90	348.40	[86]
Sorghum and corn flour- based snack (5:2)	–	1.71	25.25	1.98	7.29	8.43	–	[73]

for making pasta. Extruder was run at the SS of 600 rpm and feed rate was 15 kg/h. Diameter of the die was 6.5 mm [86]. Nutritional profile of the sorghum pasta is given in Table 7.3.

- **Pearl Millet Based Extrudates:** Two different extrudates were made by using pearl millet. Formulation of the first one was done with soy and formulation of second one was done with Bengal gram and green gram. The first extrudate was made up of 85% of pearl millet and 15% of soy. The second extrudate was made up of 70% pearl millet, 20% of Bengal gram and 10% of green gram. Extruder was run at the 18% moisture and SS was 200 rpm at 150°C barrel temperature. The average BD and ER of the extrudates were 0.25 g/ml and 1.7, respectively [98]. Nutritional profile of both extrudates is given in Table 7.3.
- **Maize-millet Based Extrudate Fortified with Soy:** Snack was made up of finger millet, maize, defatted soy, and elephant foot yam. Corotating TSE was used for extrusion processing at barrel temperature 110°C, SS of 300 rpm with feed having 14% moisture content [63]. Nutritional profile of extrudate is given in Table 7.3.
- **Sorghum Based Extrudate:** Sorghum (90%) was blended with cocoa (5%) and sugar (5%). Corotating TSE was used for extrusion with a feed rate of 18 kg/h and L/D, diameter of the screw and screw speed was 20:1, 30 mm, 290 rpm, respectively [88]. Nutritional profile of extrudate is given in Table 7.3.

7.8 CHALLENGES OF MILLET BASED READY-TO-EAT SNACKS

7.8.1 SENSORY ATTRIBUTES

The scientific field of sensory analysis examines how the five senses of smell, taste, sight, touch, and hearing react to the characteristics of food material. The consumer's response to the physical and chemical components of extrudate when it is prepared and formulated form is generally referred to as the sensory quality of food. The acceptability of the product is known by sensory analysis, which is evaluated by 9-point hedonic scale, which shows the acceptability of the product by consumer. Extruded product's color, flavor, taste, hardness, and general acceptability were all taken into consideration during the sensory evaluation.

The extruded product's sensory analysis revealed that the treatments for the organoleptic attributes varied significantly. By combining millet flour with other high-acceptability flours and creating composite foods, millet-based goods' sensory acceptability can be increased [14]. Extrusion processing further enhances the flavor and aroma of pearl millet flours [99].

7.8.2 CONSUMER ACCEPTANCE

Sensory qualities like flavor, size, color, shape, texture, "bowl life" (for morning cereals), BD and microbiological safety make up the customer criteria.

The raw ingredients composition, extruder working parameters and post-extruder processes including drying, roasting, toasting and flavor addition all have an impact on the majority of consumer and regulatory requirements. The presence of vitamins is also influenced by the extrusion process, where regulating temperature and shear inside the extruders protects the vitamins present in the food ingredients in general. Controlling the loss of nutrients during extrusion is usually done in a variety of ways [100]. Product, nutritional, and microbiological quality are the three primary criteria that influence food use. It is necessary for specific nourishing foods, such as newborn kid and weaning nourishments. Because it has lowered calories during the extrusion process, it has high starch absorbability. In general, cold expulsion conditions will form nutritious nourishments because they produce products with a high digestibility. The microbiological nature of the food is one of the most important benchmarks in terms of customer demand. This approach was deemed effective because it involves extrusion at higher temperatures, causing the vegetative life forms to be pulverized [87].

Some millet extrudates that were accepted based their sensory evaluation are as follows:

1. Extrudate made from foxtail millet was acceptable by consumers and it was suggested that till a proportion of 70:30 we can add pluses in coarse millet grain for better sensory or acceptability of extrudate [69].
2. Sorghum based extruded protein rick snack was made up of sorghum and corn flour and blend with the whey protein isolate and defatted soy flour with proportion of 5:2:1.5:1.5, respectively has found

with acceptable sensory and almost every group of customers has accepted that snack. Sensory evaluation was done by 75 consumers with different age groups from 13 to 67 years, from 75 consumers, 25 are adolescent, 25 are adults, and 25 are from elder age group. Overall sensory analysis result was 4.33 (judged on the basis of 5 point hedonic scale) [73].

3. Sorghum pasta is made up of 50% sorghum and 50% wheat. Pasta was evaluated based on texture, color, bulkiness, taste, firmness, and overall acceptability by sensory evaluation and it was found that the pasta is acceptable [86].

Extruded snacks from millets are mentioned in Table 7.3 and sensory panels accepted all the products.

7.9 CONCLUSION

In recent years, extrusion technology has been used to develop numerous nutritional and healthy millet-based snacks. However, development of the good quality product mainly depends on the range of processing parameters and millet composition, which affects physicochemical, functional, and nutritional aspects of the millet-based extrudates. In this study, detailed information about the millets production, processing, type of the extruders and various process parameters and material properties on the various quality characteristics (physical, functional, and sensory characteristics) of the millet-based extrudates were evaluated. From this study, it was concluded that developed extrudates mainly depend on composition of the raw materials (type of millet and cereals as well as their proportion). In addition, process parameters play an important role to develop extrudates with better quality, sensory and nutritional profile. However, thermo-mechanical stress caused during extrusion process can degrade the micronutrients like vitamins, minerals, and bioactive compounds. Thus, the process parameters need to optimize to retain these compounds after the extrusion process. In addition, fortification of minerals and vitamins with millets can be studied to produce nutritional extrudates.

Globally, demand for RTE extruded snacks that are nourishing, appealing, and healthy with improved physicochemical and sensory properties is rising exponentially. Food formulations with an extrusion technique have been properly devised to satisfy consumer demand. There is a need to conduct numerous studies regarding millet and millet fortified extrudates.

KEYWORDS

- **extrusion**
- **functional properties**
- **nutritional quality**
- **physical properties**
- **process parameters**
- **RTE snacks**

REFERENCES

1. Bhati, D., Gupta, R., & Sharma, A. (2022). Assessment of physicochemical, rheological, and thermal properties of Indian rice cultivars: Implications on the extrusion characteristics. *Journal of Texture Studies, 53*(6), 854–869.
2. Pandiselvam, R., Kumar, S., & Sharma, S. (2018). Modeling of coconut milk residue incorporated rice–corn extrudates properties using multiple linear regression and artificial neural network. *Journal of Food Process Engineering, 42*(2), e12981. https://doi.org/10.1111/jfpe.12981.
3. Shameena Beegum, P. P., Rajendran, K., & Rathod, R. P. (2019). Optimization of processing variables for the development of virgin coconut oil cake based extruded snacks. *Journal of Food Process Engineering, 42*(4), e13058. https://doi.org/10.1111/jfpe.13058.
4. Gadiraju, T., & Babu, R. (2015). Fried food consumption and cardiovascular health: A review of current evidence. *Nutrients, 7*(10), 8424–8430. https://doi.org/10.3390/nu7105424.
5. Preethi, P., Chennaswamy, N., & Sriram, M. (2021). Cashew apple pomace powder enriched the proximate, mineral, functional and structural properties of cereal based extrudates. *LWT–Food Science and Technology, 139*, 110539. https://doi.org/10.1016/j.lwt.2021.110539.
6. Preethi, P., & Thakur, S. (2022). Synergistic effect of powdered cashew sprout cum cotyledon and cereals on improving the biochemical and physical properties of extrudates. *Journal of Food Processing and Preservation, 46*(11), e16938. https://doi.org/10.1111/jfpp.16938.
7. Kharat, S., & Suryawanshi, N. (2019). Extrusion processing characteristics of whole grain flours of select major millets (foxtail, finger, and pearl). *Food and Bioproducts Processing, 114*, 60–71. https://doi.org/10.1016/j.fbp.2019.03.002.
8. Ditudompo, S., & Shafie, M. (2013). The effect of temperature and moisture on the mechanical properties of extruded cornstarch. *Journal of Texture Studies, 44*(3), 225–237. https://doi.org/10.1111/j.2042-4568.2012.00201.x.

9. Chang, Y. H., & Ng, P. K. W. (2009). Effects of extrusion process variables on extractable ginsenosides in wheat–ginseng extrudates. *Journal of Agricultural and Food Chemistry, 57*(6), 2356–2362. https://doi.org/10.1021/jf803602x.
10. Robin, F., & Schilling, M. (2012). Process, structure and texture of extruded whole wheat. *Journal of Cereal Science, 56*(2), 358–366. https://doi.org/10.1016/j.jcs.2012.05.008.
11. Onyango, C., & Nyongesa, H. (2004). Proximate composition and digestibility of fermented and extruded uji from maize–finger millet blend. *LWT–Food Science and Technology, 37*(8), 827–832. https://doi.org/10.1016/j.lwt.2004.01.001.
12. Maitre, A., Rathod, R. P., & Annapure, U. S. (2017). Effect of hydrocolloids and process parameters on the extrusion behavior of pearl millet grits. *International Food Research Journal, 24*(4), 1538–1544.
13. Food and Agriculture Organization of the United Nations (FAO). (2022). *World Cereal Production*. FAO.
14. Kumar, A., Singh, R., & Sharma, M. (2018). Millets: A solution to agrarian and nutritional challenges. *Agriculture & Food Security, 7*(1), 1–15. https://doi.org/10.1186/s40066-018-0155-0.
15. U.S. Department of Agriculture (USDA). (2022). *USDA Millet Production Scenario*. USDA.
16. U.S. Department of Agriculture (USDA). (2022). *Millet 2020 Production Scenario*. USDA.
17. Sarita, S. E. A. (2016). Nutraceutical and food processing properties of millets: A review. *Journal of Nutrition and Food Science, 4*(1), 1077. https://doi.org/10.4172/2155-9600.1001077.
18. Pang, M., Zhang, Z., & Li, X. (2014). Physicochemical properties, antioxidant activities and protective effect against acute ethanol-induced hepatic injury in mice of foxtail millet (*Setaria italica*) bran oil. *Food Function, 5*(8), 1763–1770. https://doi.org/10.1039/c4fo00156h.
19. Ren, X., Liu, H., & Xu, L. (2016). *In vitro* starch digestibility and *in vivo* glycemic response of foxtail millet and its products. *Food & Function, 7*(1), 372–379. https://doi.org/10.1039/c5fo00772b.
20. Yang, R., Zhang, H., & Li, W. (2020). Inhibitory effects of bound polyphenol from foxtail millet bran on colitis-associated carcinogenesis by the restoration of gut microbiota in a mice model. *Journal of Agricultural and Food Chemistry, 68*(11), 3506–3517. https://doi.org/10.1021/acs.jafc.0c00264.
21. Yousaf, L., Khan, M. N., & Shahid, M. (2021). Millet: A review of its nutritional and functional changes during processing. *Food Research International, 142*, 110197. https://doi.org/10.1016/j.foodres.2021.110197.
22. Malleshi, N. G., & Desikachar, H. S. R. (2021). Nutritional quality and health benefits. In *Millets and Pseudo Cereals* (pp. 159–168).
23. Radhajeyalakshmi, R., & Rajagopal, R. (2003). Existence of Thaumatin-like proteins (TLPs) in seeds of cereals. *Acta Phytopathologica et Entomologica Hungarica, 38*(3–4), 251–257. https://doi.org/10.1556/APhyt.38.2003.3-4.7.
24. Mohapatra, D., & Misra, R. C. (2019). Effect of different processing conditions on proximate composition, anti-oxidants, anti-nutrients and amino acid profile of grain sorghum. *Food Chemistry, 271*, 129–135. https://doi.org/10.1016/j.foodchem.2018.07.040.

25. Kumari, R., & Pal, S. (2022). Effect of fermentation conditions on nutritional and phytochemical constituents of pearl millet flour (*Pennisetum glaucum*) using response surface methodology. *Applied Food Research, 2*(1), 100055. https://doi.org/10.1016/j.afres.2021.100055.
26. Mitharwal, S., Kumar, S., & Chauhan, K. (2021). Nutritional, polyphenolic composition and *in vitro* digestibility of finger millet (*Eleusine coracana* L.) with its potential food applications: A review. *Food Bioscience, 44*, 101382. https://doi.org/10.1016/j.fbio.2021.101382.
27. Aisoni, J. E., Yusha'u, M., & Orole, O. O. (2018). Processing effects on physicochemical and proximate composition of finger millet (*Eleusine coracana*). *Greener Journal of Biological Sciences, 8*(2), 014–020.
28. Urooj, A. (2017). Impact of household processing methods on the nutritional characteristics of pearl millet (*Pennisetum typhoideum*): A review. *MOJ Food Processing & Technology, 4*(1), 28–32. https://doi.org/10.15406/mojfpt.2017.04.00085.
29. Taylor, J. R. N., & Kruger, J. (2019). Sorghum and millets. In *Sorghum and Millets* (pp. 171–224).
30. Li, E., Dhital, S., & Hasjim, J. (2014). Effects of grain milling on starch structures and flour/starch properties. *Starch–Stärke, 66*(1–2), 15–27. https://doi.org/10.1002/star.201200199.
31. Oladejo, A. A., & Adejuyigbe, C. (2020). Nutritional composition, functional properties and food applications of millet grains. *Asian Food Science Journal, 20*(1), 9–19.
32. Singh Gujral, H., & Singh, N. (2002). Extrusion behaviour and product characteristics of brown and milled rice grits. *International Journal of Food Properties, 5*(2), 307–316. https://doi.org/10.1081/JFP-120005354.
33. Arribas, C., & Salvador, A. (2017). The impact of extrusion on the nutritional composition, dietary fiber and *in vitro* digestibility of gluten-free snacks based on rice, pea and carob flour blends. *Food & Function, 8*(10), 3654–3663. https://doi.org/10.1039/c7fo00848k.
34. Mishra, A., Mishra, H. N., & Rao, P. S. (2012). Preparation of rice analogues using extrusion technology. *International Journal of Food Science & Technology, 47*(9), 1789–1797. https://doi.org/10.1111/j.1365-2621.2012.03135.x.
35. Singha, P., & Muthukumarappan, K. (2017). Effects of processing conditions on the system parameters during single screw extrusion of blend containing apple pomace. *Journal of Food Process Engineering, 40*(4), e12513. https://doi.org/10.1111/jfpe.12513.
36. Singha, P., Muthukumarappan, K., & Krishnan, P. (2018). Influence of processing conditions on apparent viscosity and system parameters during extrusion of distiller's dried grains-based snacks. *Food Science & Nutrition, 6*(1), 101–110. https://doi.org/10.1002/fsn3.521.
37. Singh, S. K., Singha, P., & Muthukumarappan, K. (2019). Modeling and optimizing the effect of extrusion processing parameters on nutritional properties of soy white flakes-based extrudates using response surface methodology. *Animal Feed Science and Technology, 254*, 114197. https://doi.org/10.1016/j.anifeedsci.2019.114197.
38. Choudhury, G. S., & Gautam, A. (1999). Screw configuration effects on macroscopic characteristics of extrudates produced by twin-screw extrusion of rice flour. *Journal of Food Science, 64*(3), 479–487. https://doi.org/10.1111/j.1365-2621.1999.tb15636.x.
39. Singh, S. K., & Muthukumarappan, K. (2015). Single screw extrusion processing of soy white flakes based Catla feed. *Journal of Food Research, 4*(1), 1–9. https://doi.org/10.5539/jfr.v4n1p1.

40. Singh, S. K., & Muthukumarappan, K. (2017). Rheological characterization and CFD simulation of soy white flakes-based dough in a single screw extruder. *Journal of Food Process Engineering, 40*(2), e12368. https://doi.org/10.1111/jfpe.12368.
41. Ek, P., & Ganjyal, G. M. (2020). Basics of extrusion processing. In *Extrusion Cooking* (pp. 1–28).
42. Singh, S. K., & Muthukumarappan, K. (2017). A viscosity model for soy white flakes-based aquafeed dough in a single screw extruder. *Journal of Food Process Engineering, 40*(2), 1–7. https://doi.org/10.1111/jfpe.12322.
43. Singha, P. (2017). Understanding the impact of extrusion processing on rheological, textural and functional properties of high-protein, high-fiber extrudates. South Dakota State University.
44. Wang, S., Wang, Y., Zhang, M., & Xu, J. (2016). Impacts of the particle sizes and levels of inclusions of cherry pomace on the physical and structural properties of direct expanded corn starch. *Food and Bioprocess Technology, 10*(2), 394–406. https://doi.org/10.1007/s11947-016-1690-0.
45. Chang, K. L. B., & Halek, G. W. (1991). Analysis of shear and thermal history during co-rotating twin-screw extrusion. *Journal of Food Science, 56*(2), 518–531. https://doi.org/10.1111/j.1365-2621.1991.tb03926.x.
46. Dalbhagat, C. G., Mahato, D. K., & Mishra, H. N. (2019). Effect of extrusion processing on physicochemical, functional and nutritional characteristics of rice and rice-based products: A review. *Trends in Food Science & Technology, 85*, 226–240. https://doi.org/10.1016/j.tifs.2018.12.015.
47. Singh, S. K., & Muthukumarappan, K. (2016). Effect of feed moisture, extrusion temperature and screw speed on properties of soy white flakes based aquafeed: A response surface analysis. *Journal of the Science of Food and Agriculture, 96*(6), 2220–2229. https://doi.org/10.1002/jsfa.7447.
48. Yeh, A. I., & Jaw, Y. M. (1999). Effects of feed rate and screw speed on operating characteristics and extrudate properties during single-screw extrusion cooking of rice flour. *Cereal Chemistry Journal, 76*(2), 236–242. https://doi.org/10.1094/CCHEM.1999.76.2.236.
49. Fletcher, S. I., Richmond, P., & Smith, A. C. (1985). An experimental study of twin-screw extrusion-cooking of maize grits. *Journal of Food Engineering, 4*(4), 291–312. https://doi.org/10.1016/0260-8774(85)90037-1.
50. Geetha, R., Mishra, H. N., & Srivastav, P. P. (2012). Twin screw extrusion of kodo millet-chickpea blend: Process parameter optimization, physico-chemical and functional properties. *Journal of Food Science and Technology, 51*(11), 3144–3153. https://doi.org/10.1007/s11483-012-0514-2.
51. Li, C., Wu, Y., Zhang, Y., & Xu, X. (2017). Extrusion expansion characteristics of samples of select varieties of whole yellow and green dry pea flours. *Cereal Chemistry Journal, 94*(3), 385–391. https://doi.org/10.1094/CCHEM-01-17-0002-R.
52. Ek, P., Kowalski, R. J., & Ganjyal, G. M. (2020). Raw material behaviors in extrusion processing I (Carbohydrates). In *Extrusion Cooking* (pp. 119–152).
53. De Pilli, T., di Gioia, D., & Vignali, G. (2012). Starch–lipid complex formation during extrusion-cooking of model system (rice starch and oleic acid) and real food (rice starch and pistachio nut flour). *European Food Research and Technology, 234*(3), 517–525. https://doi.org/10.1007/s00217-012-1748-1.

54. Singh, B., Sekhon, K. S., & Singh, N. (2007). Effects of moisture, temperature and level of pea grits on extrusion behavior and product characteristics of rice. *Food Chemistry, 100*(1), 198–202. https://doi.org/10.1016/j.foodchem.2005.09.051.
55. Kantrong, H., Vongkhamheng, C., & Lee, S. K. (2018). Influence of process parameters on physical properties and specific mechanical energy of healthy mushroom-rice snacks and optimization of extrusion process parameters using response surface methodology. *Journal of Food Science and Technology, 55*(9), 3462–3472. https://doi.org/10.1007/s11483-018-2113-8.
56. Meng, X., Wu, W., & Liu, X. (2010). Effects of extrusion conditions on system parameters and physical properties of a chickpea flour-based snack. *Food Research International, 43*(2), 650–658. https://doi.org/10.1016/j.foodres.2009.10.021.
57. Jin, Z., Hsieh, F., & Huff, H. E. (1994). Extrusion cooking of corn meal with soy fiber, salt, and sugar. *Cereal Chemistry, 71*(3), 227–234. https://doi.org/10.1094/CCHEM.1994.71.3.227.
58. Kowalski, R. J., & Ganjyal, G. M. (2018). Waxy flour degradation–Impact of screw geometry and specific mechanical energy in a co-rotating twin screw extruder. *Food Chemistry, 239*, 688–696. https://doi.org/10.1016/j.foodchem.2017.07.029.
59. Rashid, S., Khalid, N., & Sheikh, M. A. (2015). Effects of extrusion cooking on the dietary fibre content and water solubility index of wheat bran extrudates. *International Journal of Food Science & Technology, 50*(7), 1533–1537. https://doi.org/10.1111/ijfs.12756.
60. Gu, B. J., Wolcott, M. P., & Ganjyal, G. M. (2018). Pretreatment with lower feed moisture and lower extrusion temperatures aids in the increase in the fermentable sugar yields from fine-milled Douglas-fir. *Bioresource Technology, 269*, 262–268. https://doi.org/10.1016/j.biortech.2018.08.054.
61. Sawant, A. A., Thakor, N. J., Swami, S. B., & Divate, A. D. (2013). Physical and sensory characteristics of ready-to-eat food prepared from finger millet-based composite mixture by extrusion cooking. *Agricultural Engineering International: CIGR Journal, 15*, 100–105.
62. Tadesse, S. A., Sintayehu, K. W., & Belayneh, M. (2019). Functional and physical properties of sorghum-based extruded product supplemented with soy meal flour. *Cogent Food & Agriculture, 5*(1), 1707608. https://doi.org/10.1080/23311932.2019.1707608.
63. Sahu, C., Kumar, A., & Sharma, S. (2022). Effect of feed and process variables on nutritional quality of maize-millet based soy fortified extruded product using response surface methodology. *Applied Food Research, 2*(2), 100139. https://doi.org/10.1016/j.afres.2022.100139.
64. Singha, P., Singh, S. K., Muthukumarappan, K., & Krishnan, P. (2018). Physicochemical and nutritional properties of extrudates from food grade distiller's dried grains, garbanzo flour, and corn grits. *Food Science & Nutrition, 6*(7), 1914–1926. https://doi.org/10.1002/fsn3.769.
65. Nkama, I., & Filli, K. B. (2006). Development and characterization of extruded fura from mixtures of pearl millet and grain legumes flours. *International Journal of Food Properties, 9*(2), 157–165. https://doi.org/10.1080/10942910600631737.
66. Filli, K. B., Nkama, I., Jideani, V. A., & Abubakar, M. (2012). The effect of extrusion conditions on the physicochemical properties and sensory characteristics

of millet-cowpea based fura. *European Journal of Nutrition & Food Safety, 66*(3), 283–289. https://doi.org/10.9734/EJNFS/2012/1012.
67. Adebowale, K. O., Afolabi, T. A., & Olu-Owolabi, B. I. (2005). Hydrothermal treatments of finger millet (*Eleusine coracana*) starch. *Food Hydrocolloids, 19*(6), 974–983. https://doi.org/10.1016/j.foodhyd.2005.01.003.
68. Yadav, R. S., & Kaur, K. C. (2018). Development and quality evaluation of little millet (*Panicum sumatrense*) based extruded product. *Journal of Pharmacognosy and Phytochemistry, 7*(3), 3457–3463.
69. Deshpande, H., & Poshadri, S. (2011). Physical and sensory characteristics of extruded snacks prepared from foxtail millet-based composite flours. *International Food Research Journal, 18*(2), 751–756.
70. Singh, S. K. (2016). Understanding the effect of extrusion processing parameters on physical, nutritional, and rheological properties of soy white flakes based aquafeed in a single screw extruder. Ann Arbor (252).
71. Subramanian, S., & Viswanathan, R. (2007). Bulk density and friction coefficients of selected minor millet grains and flours. *Journal of Food Engineering, 81*, 118–126. https://doi.org/10.1016/j.jfoodeng.2006.11.019.
72. Bhatnagar, S., & Hanna, M. A. (1994). Amylose-lipid complex formation during single-screw extrusion of various corn starches. *Cereal Chemistry, 71*(6), 582–587. https://doi.org/10.1094/CCHEM.1994.71.6.582.
73. Devi, N. L., Suneetha, V. R., & Singh, J. (2013). Development of protein-rich sorghum-based expanded snacks using extrusion technology. *International Journal of Food Properties, 16*(2), 263–276. https://doi.org/10.1080/10942912.2011.630630.
74. Rathod, R. P., & Annapure, U. S. (2017). Physicochemical properties, protein and starch digestibility of lentil-based noodle prepared by using extrusion processing. *LWT, 80*, 121–130. https://doi.org/10.1016/j.lwt.2016.12.060.
75. Alam, M. S., Pathania, S., & Sharma, A. (2016). Optimization of the extrusion process for development of high fiber soybean-rice ready-to-eat snacks using carrot pomace and cauliflower trimmings. *LWT, 74*, 135–144. https://doi.org/10.1016/j.lwt.2016.07.017.
76. Altan, A., McCarthy, K. L., & Maskan, M. (2008). Evaluation of snack foods from barley–Tomato pomace blends by extrusion processing. *Journal of Food Engineering, 84*(2), 231–242. https://doi.org/10.1016/j.jfoodeng.2007.06.016.
77. Pardhi, S. D., Rajput, P., & Yadav, A. (2019). Evaluation of functional properties of extruded snacks developed from brown rice grits by using response surface methodology. *Journal of the Saudi Society of Agricultural Sciences, 18*(1), 7–16. https://doi.org/10.1016/j.jssas.2018.05.004.
78. Arora, B., Singha, P., & Rizvi, S. S. H. (2021). Supercritical fluid extrusion: Die design and physicochemical properties of milk protein extrudates. *Innovative Food Science & Emerging Technologies, 68*, 102637. https://doi.org/10.1016/j.ifset.2021.102637.
79. Singha, P., & Muthukumarappan, K. (2018). Single screw extrusion of apple pomace-enriched blends: Extrudate characteristics and determination of optimum processing conditions. *Food Science and Technology International, 24*(5), 447–462. https://doi.org/10.1177/1082013218759494.
80. Gujral, H. S., Sandhu, K. S., & Kaur, A. (2012). Total phenolic content and antioxidant activity of extruded brown rice. *International Journal of Food Properties, 15*(2), 301–311. https://doi.org/10.1080/10942912.2010.485835.

81. Singh, S. K., & Muthukumarappan, K. (2016). Effect of feed moisture, extrusion temperature and screw speed on properties of soy white flakes based aquafeed: A response surface analysis. *Journal of the Science of Food and Agriculture, 96*(6), 2220–2229. https://doi.org/10.1002/jsfa.7494.
82. Kowalski, R. J., Ganjyal, G. M., & Lai, S. H. (2016). Extrusion processing characteristics of quinoa (Chenopodium quinoa Willd.) var. Cherry Vanilla. *Journal of Cereal Science, 70*, 91–98. https://doi.org/10.1016/j.jcs.2016.05.013.
83. Ganjyal, G. M., & Hanna, M. A. (2006). Role of blowing agents in expansion of high-amylose starch acetate during extrusion. *Cereal Chemistry, 83*(6), 577–583. https://doi.org/10.1094/CC-83-0577.
84. Sun, Y., & Muthukumarappan, K. (2002). Changes in functionality of soy-based extrudates during single-screw extrusion processing. *International Journal of Food Properties, 5*(2), 379–389. https://doi.org/10.1080/10942910209356947.
85. Aluwi, N. A., Ganjyal, G. M., & Hou, J. (2016). Impacts of scarification and degermination on the expansion characteristics of select quinoa varieties during extrusion processing. *Journal of Food Science, 81*(12), 2939–2949. https://doi.org/10.1111/1750-3841.13586.
86. Benhur, D. R., Kumar, V., & Sinha, N. K. (2015). Development and standardization of sorghum pasta using extrusion technology. *Journal of Food Science and Technology, 52*(10), 6828–6833. https://doi.org/10.1007/s11483-015-0924-1.
87. Hagenimana, A., Ding, X., & Fang, T. (2006). Evaluation of rice flour modified by extrusion cooking. *Journal of Cereal Science, 43*(1), 38–46. https://doi.org/10.1016/j.jcs.2005.06.001.
88. Dlamini, N. S., & Khan, S. W. K. (2016). Effect of ingredients ratio on physicochemical and sensory properties of sorghum, maize, and soya protein concentrate blend extrudates. *International Journal of Food Science and Nutrition Engineering, 6*(5), 112–120.
89. Sahu, C., & Patel, S. (2020). Optimization of maize-millet based soy fortified composite flour for preparation of RTE extruded products using D-optimal mixture design. *Journal of Food Science and Technology, 58*(7), 2651–2660. https://doi.org/10.1007/s11483-020-09772-8.
90. Kaushal, P., Kumar, V., & Sharma, H. K. (2012). Comparative study of physicochemical, functional, antinutritional and pasting properties of taro (*Colocasia esculenta*), rice (*Oryza sativa*) flour, pigeonpea (*Cajanus cajan*) flour and their blends. *LWT–Food Science and Technology, 48*(1), 59–68. https://doi.org/10.1016/j.lwt.2012.03.018.
91. Barbiroli, A., Dalla Rosa, M., & Bortolomeazzi, R. (2013). Process conditions affect starch structure and its interactions with proteins in rice pasta. *Carbohydrate Polymers, 92*(2), 1865–1872. https://doi.org/10.1016/j.carbpol.2012.11.066.
92. Marti, A., Ferrer, E., & de la Cueva, C. (2017). *In vitro* starch digestibility of commercial gluten-free pasta: The role of ingredients and origin. *Journal of Food Science, 82*(4), 1012–1019. https://doi.org/10.1111/1750-3841.13688.
93. Schoch, T. J., & Mears, M. (1968). Preparation and properties of various legume starches. *Cereal Chemistry, 45*, 564–573.
94. Kaur, L., Singh, N., & Singh, J. (2007). Physico-chemical, rheological and structural properties of fractionated potato starches. *Journal of Food Engineering, 82*(3), 383–394. https://doi.org/10.1016/j.jfoodeng.2007.02.004.
95. Aluwi, N. A., Murphy, K. M., & Ganjyal, G. M. (2017). Physicochemical characterization of different varieties of quinoa. *Cereal Chemistry, 94*(5), 847–856. https://doi.org/10.1094/CCHEM-11-16-0252-R.

96. Campanella, O. H., Li, P. X., Ross, K. A., & Okos, M. R. (2002). The role of rheology in extrusion. In J. Welti-Chanes, G. V. Barbosa-Canovas, & J. M. Aguilera (Eds.), *Engineering and Food for the 21st Century* (pp. 393–413). Technomic Publishing.
97. Ragaee, S., & Abdel-Aal, E. S. M. (2006). Pasting properties of starch and protein in selected cereals and quality of their food products. *Food Chemistry, 95*(1), 9–18. https://doi.org/10.1016/j.foodchem.2005.01.024.
98. Sumathi, A., Ushakumari, S. R., & Malleshi, N. G. (2009). Physico-chemical characteristics, nutritional quality and shelf-life of pearl millet-based extrusion cooked supplementary foods. *International Journal of Food Sciences and Nutrition, 58*(5), 350–362. https://doi.org/10.1080/09637480802109816.
99. Onyeoziri, I. O., Onyenekwe, P. C., & Okoro, R. (2021). Descriptive sensory analysis of instant porridge from stored wholegrain and decorticated pearl millet flour cooked, stabilized, and improved by using a low-cost extruder. *Journal of Food Science, 86*(9), 3824–3838. https://doi.org/10.1111/1750-3841.16321.
100. Camire, M. E. (1998). Chemical changes during extrusion cooking. In J. Welti-Chanes, G. V. Barbosa-Canovas, & J. M. Aguilera (Eds.), *Process-Induced Chemical Changes in Food* (pp. 109–121). Springer.

CHAPTER 8

HOT-EXTRUSION TECHNOLOGY FOR PRODUCTION OF FIBER ENRICHED READY-TO-EAT SNACKS

ANANDU CHANDRA KHANASHYAM,[1] NANCY ASWATI,[2] KARTIK SHAH,[3] VENKATA SARATH PAMU,[4] RAVI PANDISELVAM,[5] and MATHALA JULIET GUPTA[1]

[1]*Horticulture Division, ICAR–Central Coastal Agricultural Research Institute, Goa, India*

[2]*Lactalis USA, Buffalo, New York, USA*

[3]*Sargento Foods, Elkhart Lake, Wisconsin, USA*

[4]*Oklahoma State University, Stillwater, Oklahoma, USA*

[5]*Physiology, Biochemistry, and Post-Harvest Technology Division, ICAR–Central Plantation Crops Research Institute, Kasaragod, Kerala, India*

ABSTRACT

Extrusion technology has been widely employed in manufacturing ready-to-eat cereal snacks, owing to its simplicity and flexibility in creating diverse textures and shapes that attract consumers. However, numerous existing RTE products are considered to be low in nutritional value since they contain high levels of sugar and salt, resulting in calorie-dense but nutrient-poor foods. The demand for healthier snack options has led to the development of fiber-enriched snacks using hot-extrusion technology. Researchers are exploring the possibility of using fiber-rich raw materials such as whole grains, vegetables, fruits, and their by-products to develop high-fiber extruded

Ready-to-Eat Snacks: Emerging Technologies for Production and Safety.
Ravi Pandiselvam and Gülşah Çalışkan Koç

products with health benefits. Moreover, continuous investigations are being conducted to manipulate nutritionally rich components to create nutritionally balanced and functional formulations. The addition of dietary fiber alters extrusion parameters and modifies the nutritional and physiochemical properties of the finished product. This chapter discusses the use of hot-extrusion technology for the production of fiber-enriched RTE snacks. The chapter explores the effect of extrusion parameters such die shape, pressure, temperature, moisture content, screw speed, and composition of raw materials on retaining the nutrition, quality, functional and sensory attributes of the finished product. Overall, this chapter offers an insight into the potential of extrusion technology to develop novel fiber-rich food products.

8.1 INTRODUCTION

Foods that are ready to eat are becoming more and more popular today due to the consumers time saving needs of modern lifestyles. Several ready-to-eat items, such as crunchy expanded snacks (puffs, collets, etc.), breakfast cereals, meat substitutes, and sports nutrition products, are made using the extrusion cooking technique – (i) Customers are drawn to these foods due to their convenience, desirable appearance, and texture; (ii) Extrusion technology has revolutionized the food industry by providing many benefits over other processing methods. Extrusion cooking offers a significant advantage in the ability to produce a wide range of extruded products using inexpensive raw materials and a relatively short processing time; (iii). It provides a flexible platform for processing a wide range of products from different food groups by adjusting processing conditions and modifying minor or major ingredients; (iv) Extrusion cooking is a widely used food processing technique that involves the use of high pressure, temperature, and shear forces to transform raw materials into a variety of products. Extrusion cooking offers several key benefits in food production, including versatility; (v) efficiency; (vi) improved nutritional value (vii), extended shelf life and microbial safety (viii), and raw material utilization; ad (ix) This method can produce a diverse range of food products, with varying shapes, sizes, textures, and flavors. Additionally, it is a highly efficient process that enables continuous production of large quantities of food with minimal processing time, making it cost-effective for mass production. Extrusion cooking can also enhance the nutritional value of food products by increasing the bioavailability of nutrients and allowing for the addition of functional ingredients such as vitamins, minerals, and fiber. Furthermore, it can improve the shelf life of food products by reducing moisture content and minimizing bacterial growth, which can help to reduce food waste. Finally, extrusion cooking can utilize low-quality raw materials that would otherwise be discarded, contributing to a more sustainable food production system.

The worldwide market for extruded snacks had an estimated worth of 51.59 billion US dollars in 2019 and is projected to experience a growth rate of over 4% up until the year 2026 [9]. This increase can be attributed to changes in consumer dietary preferences and the escalating need for easily consumable foods, Cereals are a major component of the extruded foods market because they have good expansion qualities and are the most often utilized raw material for extrusion cooking. The cereal ready-to-eat (RTE) market is a critical industry that primarily comprises extruded snack products such as breakfast cereals, biscuits, and bars. These products originated from the Kellogg brothers' development of healthy vegetarian foods associated with human nutrition and the subsequent growth of the breakfast cereal market in the late 1800s and early 1900s in the United States [10]. Originally, most cereal ready-to-eat (RTE) products were prepared using whole-grain sources and were primarily flaked after steaming, making the grains pliable for the flaking process [2]. However, due to consumer expectations for white food products with improved taste and texture, and advancements in flour separation and refining techniques, modern extruded cereal snacks are typically made from a mixture of flour components rather than whole-grain sources. This shift has had a significant impact on the nutritional quality of these products, as the milling and refining process concentrates starch while removing the outer coats of the cereal grain [11]. The extrusion cooked items, like breakfast cereals and savory snacks, have considerable calorie content. To decrease the energy density of foods, the food industry is taking a strategic approach of increasing the dietary fiber content. However, boosting the dietary fiber content in extrusion cooked directly expanded cereals to meet these requirements is a challenging task. Usually, the addition of dietary fiber leads to lower expansion volumes, increased densities, firmer textures, and less crispiness, which makes them less desirable to consumers [12].

8.2 HISTORY AND PRINCIPLE OF EXTRACTION

Extrusion technology has a rich history, originating in the plastics industry during the 1930s [13]. However, it was not until the 1950s that this technology was adapted and modified for use in the food industry. The General Mills Company developed the first food extruder in the 1940s to produce dry breakfast cereals [14]. Since then, extrusion has become an essential tool for the food industry, allowing for efficient and versatile processing of various food products such as snacks, cereals, pet food, and pasta [15]. The progress made in extrusion technology has resulted in the creation of more advanced machinery, resulting in a wider range of food products that can be produced using the process [16]. The advancement of extrusion technology has given rise to the creation of single- and twin-screw extruders (Figure 8.1) with distinct configurations, operating at various speeds and temperatures. Co-extrusion is another innovation in extrusion technology that involves extruding multiple layers of different materials simultaneously [17]. In the 1960s, the twin-screw extruders were introduced, enabling enhanced flexibility and regulation during the extrusion process [18].

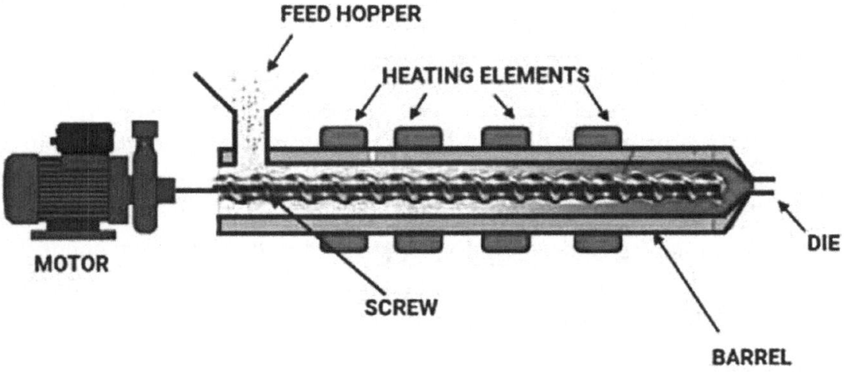

FIGURE 8.1 A schematic diagram of an extruder.

The extrusion process involves a series of steps, including feeding, preconditioning, extrusion, shaping, and cooling, and it requires the application of high temperatures and pressures to food ingredients to create a wide range of products with specific properties [13]. The feeder, conditioner, extruder barrel, die, and cutting system are the main components of an extruder and have specific functions critical to the success of the extrusion process [19]. Extruders can be classified into single-screw and twin-screw extruders, and their configurations can vary based on the specific product being produced [17]. Several variables can affect the quality of the final product, including feed composition, screw speed, temperature, residence time, die shape, moisture content, and pressure, among others [13]. Each of these parameters can have a significant impact on the final product quality, making it essential to carefully control and regulate them to achieve the desired characteristics of the product. The subsequent sections will provide detailed explanations of the extrusion parameters.

8.2.1 FEED COMPOSITION

The composition of feed in the extrusion process refers to the type and proportion of raw materials used. Different feed ingredients, such as grains, legumes, and vegetables, have varying characteristics that can impact the extrusion process, including their nutrient content, starch content, protein content, and moisture content [19]. High protein content in the feed can improve the textural properties of the extrudate and promote expansion, while low protein content can lead to poor expansion and poor textural quality [20]. The type of protein used can also influence the functional properties of the extrudate, such as water absorption and viscosity. Moreover, carbohydrates, particularly starch, are a crucial component of feed composition. The level of starch in the feed affects the degree of gelatinization, which can significantly impact the physical properties of the extrudate. Higher levels of starch in the feed can lead to better expansion and texture, while low levels of starch can result in dense and less palatable extrudates [13, 20].

8.2.2 MOISTURE

The moisture content in the extrusion process is a critical factor that can impact the final product's texture and expansion. High moisture content can cause gelatinization of starch and protein, resulting in better expansion and improved texture. However, excessive moisture content can lead to product sticking and blockages in the extruder. Conversely, low moisture content can result in extrudate brittleness and low expansion [21]. The ideal moisture content for extrusion varies depending on the specific product and desired characteristics. Generally, the moisture content for most extruded food products should be between 10% and 20% [22]. However, the optimal moisture content can also vary based on the specific ingredients used in the feed composition. For instance, if the feed composition has high amounts of starch, a higher moisture content may be necessary to ensure proper gelatinization of the starch during extrusion [23]. Furthermore, the moisture content of the raw materials used in the feed composition can also impact the extrusion process. If the raw materials are excessively wet, it can lead to a high level of steam formation during extrusion, resulting in poor expansion and lower product quality [24].

8.2.3 BARREL TEMPERATURE

Barrel temperature is a critical factor in the extrusion of food products as it affects the degree of gelatinization of starches, protein denaturation, and other physical properties of the final product. One of the primary effects of barrel temperature in extrusion is the degree of gelatinization of starches. Starches in food products must be gelatinized to achieve the desired texture and appearance of the final product. The degree of gelatinization increases with increasing barrel temperature, which leads to a more expanded and porous product with a crispy texture [25]. However, excessive barrel temperature can cause the product to become too brittle or hard. Another critical effect of barrel temperature in extrusion is the degree of protein denaturation. Proteins in food products can denature or unfold under high temperatures, affecting the texture, flavor, and nutritional quality of the final product. The extent of protein denaturation increases with increasing barrel temperature, which can lead to a tougher and less palatable product [26]. The ideal range of barrel temperatures for extrusion varies depending on the type of product being produced. For example, for snack products such as chips and puffs, the ideal barrel temperature range is typically between 150°C and 180°C [25]. This temperature range is sufficient to achieve the desired degree of gelatinization of starches while minimizing protein denaturation. However, for products such as breakfast cereals, which require a more significant degree of expansion, higher barrel temperatures between 200°C and 220°C may be necessary [25]. Moreover, the temperature profile along the barrel also affects the quality of the extruded products. A temperature profile that gradually increases along the barrel length can provide better control of the degree of gelatinization of starches and protein denaturation, resulting in more uniform and consistent products [26].

8.2.4 PARTICLE SIZE

The size of the particles is a crucial aspect of food extrusion as it can influence the sensory features and physical characteristics of the end product [27]. A smaller particle size can lead to a smoother texture and increased surface area, which can improve the digestibility of the product [28]. However, too small of a particle size can result in a reduced flow rate and increased die pressure, leading to processing difficulties [28]. The optimal particle size for extrusion varies depending on the type of food product being produced. For example, in the production of breakfast cereals, a particle size of 100–400 μm has been found to result in desirable textural properties [29]. In contrast, the production of textured vegetable protein requires a particle size of 3–5 mm to ensure proper hydration and extrusion [29]. The particle size can be controlled through the use of milling equipment, such as hammer mills or roller mills, which can break down larger particles into smaller sizes [30]. It is important to note that the milling process can also affect the nutritional and functional properties of the feed material, as it can result in the loss of nutrients and damage to the protein structure [31].

8.2.5 SCREW SPEED

The screw speed is a crucial process parameter in extrusion that determines the shear rate, residence time, and pressure in the extruder. These factors, in turn, affect the extent of cooking, gelatinization, and expansion of the food product. High screw speeds result in higher shear rates and shorter residence times, leading to less cooking and less gelatinization. On the other hand, low screw speeds result in longer residence times and more cooking, leading to higher gelatinization [32]. Nonetheless, the impact of screw speed on the extrusion process and the quality of the product is reliant on other process parameters such as moisture content, barrel temperature, and feed rate. Therefore, it is essential to optimize the screw speed in combination with these parameters to achieve the desired product characteristics. In general, higher screw speeds are suitable for extruding low-moisture products such as breakfast cereals, while lower screw speeds are suitable for high-moisture products such as meat analogs [23]. It is also important to note that high screw speeds result in higher motor power consumption and higher wear on the screw and barrel components, which can affect the efficiency and longevity of the extruder [33].

8.2.6 FEED RATE

Feed rate is a critical parameter in extrusion processing as it directly affects the residence time of the material in the extruder and thus the quality of the final product. A high feed rate can cause insufficient shear and heat transfer in the extruder, leading to incomplete cooking and inadequate gelatinization of starches, while a low feed rate can result in overcooking, causing discoloration and a bitter taste [34]. Therefore, it is crucial to optimize the feed rate to achieve the desired quality of the final product. Several factors affect the optimal feed rate, including the raw material characteristics, barrel temperature, moisture content, and screw speed [35]. For example, a high moisture content can increase the feed rate, as it reduces the viscosity of the material and makes it easier to extrude [36]. On the other hand, a low barrel temperature may require a slower feed rate to ensure sufficient cooking and gelatinization of the material [36]. Another consideration is the type of extruder used. Due to their extended residence time and elevated shear rates, single-screw extruders necessitate a reduced feed rate than twin-screw extruders [36]. Moreover, the design of the screw elements can affect the optimal feed rate. For instance, a screw with a deeper channel depth can accommodate a higher feed rate than a shallow channel depth screw [37].

8.2.7 PRESSURE

The extrusion process involves subjecting the feed material to high pressure within the barrel of an extruder, which results in a variety of changes to the physiochemical properties of the product. The extrusion process can be substantially impacted by pressure, which is one of the most critical process parameters that influences the quality and features of the extruded product [38]. One of the primary effects of pressure during extrusion is the formation of a dense and compact material structure, which is essential for achieving a desirable texture and mouthfeel in many food products. Higher pressures generally lead to greater density and hardness of the extruded product, which can be beneficial for products such as snacks, cereals, and pet food [39]. However, excessively high pressure can also have negative effects on the extruded product, such as causing the formation of large air pockets or voids, or inducing excessive shear forces that can lead to product degradation or undesirable changes in texture [38. Therefore, it is important to carefully control the pressure during extrusion to achieve the desired product quality. The ideal pressure for extrusion can fluctuate based on various factors, such as the feed material and the product being manufactured, in addition to other process parameters like screw speed and temperature. However, in general, pressures in the range of 20–60 MPa are commonly used for extrusion of food products [39].

8.2.8 TIME

The optimal extrusion time can vary depending on the type of feed and the desired characteristics of the product. In general, longer extrusion times can result in increased expansion and improved texture, but may also cause over-processing and a loss of nutrients [40]. Research has shown that increasing the residence time of the feed in the extruder barrel can improve the degree of gelatinization and increase the digestibility of starches [41]. However, prolonged exposure to high temperatures and pressures can also lead to the degradation of some nutrients, such as vitamins and amino acids [42].

8.2.9 DIE SHAPE

Die shape is a critical extrusion variable that influences the final product characteristics of extruded food products. The die shape determines the shape, size, and texture of the extrudate and influences the expansion ratio, bulk density, and water absorption capacity of the product [34]. The design and dimensions of the die orifice can have an impact on the material flow through the die, as well as the degree of shear and mixing during the extrusion process. A larger orifice size can result in increased expansion and a lower bulk density of the product [34]. Die shapes can vary from simple circular orifices to more complex shapes, such as rectangular, triangular, or star-shaped, depending on the desired product characteristics and processing conditions [34]. Different die shapes can be used to produce various products, such as snacks, breakfast cereals, and pet foods [34]. For example, a rectangular die can be used to produce expanded products with a crispy texture, while a circular die can be used to produce products with a more uniform texture. The optimal die shape and size can vary depending on the specific feed composition and processing conditions used. Therefore, careful selection of die shape and size is essential to achieve the desired product characteristics [26].

8.3 EFFECT OF EXTRUSION ON THE PHYSIO-CHEMICAL PARAMETERS OF EXTRUDED SNACKS

8.3.1 NUTRITIONAL PROPERTIES

8.3.1.1 PROTEIN AND AMINO ACID

Proteins, composed of a chain of amino acids, are known to be heat-sensitive. As a result, they are significantly impacted by the extrusion process [43].

The nutritional value of proteinaceous materials can be altered significantly by adjusting the screw speed during extrusion, which in turn alters the shear pressures. Extrusion causes the denaturation of proteins by subjecting them to elevated temperatures. Moreover, the extrusion process, characterized by high temperature and intense mechanical shear, accelerates structural changes and protein denaturation. These changes facilitate improved accessibility to proteolytic enzymes [44]. These Proteolytic enzymes stimulate protein breakdown and lead to protein loss. The higher temperature may also deactivate antinutritional factors like lectins and antitrypsin factors. For example, when soy protein is extruded, the undesirable volatile compounds and the flavor of bitterness are significantly reduced [45]. On the contrary, low protein digestibility and low lysine bioavailability have been linked to very high temperatures [46, 47]. The technology of extrusion allows for a greater degree of protein absorbability in the food product that has been extruded. During the extrusion process, the chemical cross-linking activity and molecular aggregation of soybean protein were examined under different moisture content levels. Protein extrudate's structure was found to be held together by a combination of hydrogen bonds, hydrophobic interactions, disulfide bonds, and their interactions, which were found to be effective independent of the extruder's position or the amount of moisture present [48].

8.3.1.2 CARBOHYDRATES

1. **Sugar:** An excellent source of instant energy is sugar, which can come in the form of sucrose, fructose, or lactose. Sugar is responsible for many extruded products' sweeter flavor profile. Some of the chemical processes that take place during the extrusion process are caused by sugars. In order to preserve the nutritional and sensory qualities of the extruded food product, it is crucial to effectively manage and regulate the sugar content throughout the extrusion process [49]. The Maillard reaction is a chemical process that occurs when protein and sugar interact, resulting in the conversion of sucrose to glucose and fructose. As a result of this, the process of extrusion may lose some sugar in the final product [50]. A study reported significant reduction in the amount of sucrose, ranging from 2% to 20%, was observed during process of making extruded biscuits [47]. Elimination of certain oligosaccharides results in an improvement in the quality of some processed food products that are based on legumes [46, 47].

2. **Starches:** Starch is primarily made up of amylose [20–25%] and amylopectin [75%–80%], both of which are essential for the starch's structural integrity. Amylose is a linear chain polysaccharide whereas, amylopectin is a branched polysaccharide made up of many glucose molecules connected by glycosidic bonds [43]. During the extrusion process, starches are subjected to high temperatures, high levels of mechanical stress as well as high pressures. When subjected to the extrusion process, starches undergo modifications in their structure, such as melting, gelatinization, and fragmentation. The extent of these alterations depends on several factors, including pressure, temperature, moisture content, and shearing force [51]. The gelatinization process involves the breaking of intermolecular hydrogen bonds, resulting in the loss of rigid structures. This process enhances the water-absorbing capacity of starch, leading to the expansion of extruded products. Moreover, when starch granules undergo the gelatinization process in excessive water, they enlarge and release amylose from within. These amylose molecules can then create a smoother surface on the exterior of the extruded granules. Following the extrusion process, both the amylose and the amylopectin molecules have a lower molecular weight. Because of its branched structure, amylopectin is more prone to being sheared [47]. The extrusion process also improves starch digestibility by making the amylase enzyme more easily accessible by shearing off the amylopectin's branching structures. The degradation of the structural integrity of starch molecules, their hydration, and partial solubilization are all factors that contribute to higher levels of starch digestibility [52]. The building blocks of starch are glucose molecules that are linked together via a glyosidic linkage. Amylopectin and amylose are the components of the prepared pre-extrudate that are responsible for its viscosity and its gelatinization.

8.3.1.3 VITAMINS AND MINERALS

Minerals are less susceptible to be destroyed by high temperatures and will not evaporate away in the steam condensate at the die. Increased mineral intake is possible through extrusion because it eliminates some of the barriers to mineral uptake [53]. The extrusion process significantly affects the stability of vitamins in extruded snack foods. For instance, lower feed moisture levels and higher barrel temperatures can result in the loss of ascorbic acid. A study concluded the impact of barrel length impacted the retention rate of B vitamin group where the authors concluded that improved retention rate with shorter barrel [54]. Similarly, the use of very high temperatures for extrusion procession can impact the stability of fat-soluble vitamins A and E [55].

8.3.1.4 FIBER

Fiber is typically defined as a type of polysaccharide that cannot be metabolized and is made up of cellulose, hemicelluloses, and lignin. Fruits and vegetables contain higher ratio of insoluble to soluble dietary fiber. Soluble dietary fiber is more attractive for health-related benefits compared to insoluble dietary fiber. The rise in soluble dietary fiber content observed during the extrusion process is probably due to the redistribution of insoluble dietary fiber into its soluble form, which occurs as a result of the degradation of insoluble dietary fiber during extrusion [56]. The product undergoes processing with a twin-screw extruder, which changes the amount of nutritional fiber that is retained after processing. For example, it was discovered that increasing the temperature will cause an increase in the amount of nutritional fiber that is found in wheat. Similarly, the process of extrusion heating results in increased levels of dietary fiber in the extruded barley product. Increasing the amount of total dietary fiber causes an increase in the soluble dietary fiber content of sticky barley [49]. Under mild extrusion conditions, the composition of dietary fiber remains largely unchanged. However, certain fiber components may experience increased solubility. In contrast, severe extrusion conditions can lead to an elevation in soluble dietary fiber and enzyme-resistant starch fractions, resulting in an overall increase in dietary fiber content [47].

8.3.1.5 LIPIDS

Lipids may be present in the ingredients used for extrusion or can be added externally as well. They modify extrudates' texture, stickiness, and storing stability as plasticizers or emulsifiers. The high temperatures during the extrusion process involve lipid release. Alternatively, some of the mechanical disruptions during the processing may also cause lipid release. As a result of the lipid's ability to lubricate the barrel, the torque is decreased. As a consequence, the product will expand poorly because there is insufficient pressure [49]. The effects of lubrication are most apparent in low moisture extrusion. The texture, shelf life, and digestibility of extrudates can undergo significant changes due to the creation of starch-lipid complexes during the extrusion process at high temperature [57]. Even though oats can contain as much as 10% oil, most cereals like wheat and maize have much lower oil content (2%). Up to half of the kernel weight of oilseeds like soybeans and cottonseed is oil. It is possible to use either full-fat or partially or completely defatted oilseed flours for extrusion. Extruding high-fat materials, especially expanded products, is discouraged because lipid concentrations exceeding 5–6% can diminish the efficiency of the extruder [53].

During extrusion, lipids undergo intricate physicochemical transformations that are influenced by the hydrophilic-lipophilic balance of the lipids, the specific type of lipids present, and the composition of the materials being extruded. Since lipase and lipoxygenase activity is inhibited at elevated temperatures, conditions that promote the formation of free fatty acids and their decomposition are reduced [58].

8.3.2 PHYSICOCHEMICAL/ FUNCTIONAL PROPERTIES

The starch used in the product, both its variety and its quantity, has a significant impact on the extrusion process. The amount of moisture present and the length of time the material is in the extruder both influence the physicochemical characteristics of the extruded material, which in turn influences the nutritive value and organoleptic parameters of the product. Extruded materials have several different physical characteristics, including bulk density, color, Water solubility index (WSI), expansion ratio, hardness, and Water Absorption Index (WAI).

8.3.2.1 EXPANSION RATIO

The moisture content of the pre-extrudate is a crucial factor that impacts various aspects of the extrusion process. As the moisture content of the feeding material decreases, the temperature of the barrel, the rotating speed of the screw, and the expansion ratio all tend to increase. Similarly, when the feed moisture content is increased, there is a significant reduction in the expansion ratio of the extruded food product [59]. The die pressure and the vapor pressure are the primary factors that determine the enlargement of the product. Water is released suddenly, causing the substance to expand as it leaves the die [60]. For example, as the differential pressure increases, the product expands to a greater extent. When feed material is extruded at high

Barrel temperature and high Screw speed with low Feed Moisture, structural starch degradation occurs due to high shearing, resulting in an increased Expansion Ratio and decreased Bulk Density [43]. The situation is inverted when extruding at low Barrel Temperature and low Screw Speed with high Feed Moisture.

8.3.2.2 BULK DENSITY

Bulk density is an important factor for extruded snacks since they indicate the porosity of the extruded snack. The final packaging of the product as well as the type of materials for the final product is influenced by bulk density [61]. Factors such as moisture level, temperature, and screw speed influence the density of an extruded food product in a manner that is inversely proportional to the expansion ratio. The reduction in bulk density of the extruded food product is caused by an increase in both the rotating speed of the screw and the temperature of the barrel. Additionally, the bulk density of the extruded food product was also impacted by the high moisture content of the extruded food product.

8.3.2.3 WATER ABSORPTION INDEX (WAI)

The water absorption index (WAI) is a measurement of the capability of starch to hold water when it has been exposed to an excessive amount of liquid [62, 63]. It provides information regarding the stability of the starch network in aqueous dispersion as well as the behavior of the extruded products during subsequent processing [62, 64]. The water absorption index (WAI) is associated with the presence of hydrophilic ingredients. It has been demonstrated that extrudates with a greater amylose content have a higher WAI in comparison to those that have a lower amylose content [64]. Because native starch is unlikely to absorb water at room temperature, the WAI can also be used to determine the degree of starch gelatinization [65].

8.3.2.4 WATER SOLUBILITY INDEX

The degree of dextrinization influences the water solubility index (WSI), which is a measurement of the amount of soluble polysaccharides. During the process of extrusion, the WSI gives an indication of the degree to which the starch network is degraded and converted into soluble polysaccharides, such as amylose [62, 63, 65]. The starch network degrades as a result of the extrusion process, which results in an increase in the WSI and a reduction in the WAI [64]. When moisture levels are low, starch is more likely to undergo structural breakdown, which leads to a reduction in molecular size and water-holding capacity, leading to an increase in soluble components. This phenomenon results into an increased water solubility index [63]. As temperature and screw rotation speed rise, moisture in the extrudate causes the water solubility index to decrease. Temperature plays a role in the gelatinization of starch, which in turn leads to an increase in the abundance of soluble starches and a subsequent increase in WSI.

8.3.2.5 OIL ABSORPTION INDEX

The ability of extrudates to retain oil, as indicated by the oil absorption index (OAI), is linked to the presence of hydrophobic groups within their structure, particularly proteins [56, 66]. Except for proteins, the chemical composition of some polysaccharides is associated with OAI. Several factors influence the binding process, such as charge density, surface properties, and the hydrophobic structure of fiber particles. In addition to the source of the materials used in processing the particle size, composition, and microstructure also have an impact on the oil absorption index (OAI). Porosity is an another important parameter to be considered when evaluating OAI of extruded products as high porous material tend to have a higher propensity to absorb oil [29, 67]. Due to the lubricating properties of oil in the extruder barrel, the specific mechanical energy input to the product is influenced. Thus, the use of Oil Absorption Index (OAI) is critical when working with high-protein materials, such as legumes, for texturization purposes. When dealing with texturization of high-oil content materials, it is necessary to alter the screw and die configuration [68].

8.3.2.6 TEXTURAL PROPERTIES

The highest peak force experienced by the material during the first compression is the measure of its hardness. It is connected to the growth and structure of the product and is considered an essential aspect of the human being's sensory perception [69, 70]. The level of feed moisture is discovered to have the greatest impact on the resulting extrudates' hardness [61]. At high feed moisture levels, the inability of starch granules to expand, attributed to the lubricating properties of water, results in a product that is less enlarged and has a dense intercellular structure, thus contributing to increased hardness [71]. A study reported increase in Hardness as the feed ratio increases. This observation was associated with a decrease in starch expansion and bubble formation, leading to a denser intercellular structure and an increase in hardness [72]. The study also reported reduction in hardness of kodo millet-based extrudates as the screw speed and barrel temperature increased. This phenomenon was attributed to superheated feed moisture turning into steam causing the extrudates to swell when screw speed and barrel temperature are high. Large numbers of bubbles are produced when the superheated vapor is suddenly flashed. As a consequence of the high shear, these bubbles expand and eventually collapse, leaving behind voids and pores [63, 73].

TABLE 8.1 Summary of Key Effects on Product Characteristics and Quality as a Result of Extrusion Processing

Product Characteristics and Quality	Key Effects of Extrusion Processing	References
Protein and amino acid	• Denaturation of proteins • Protein breakdown by proteolytic enzymes • Destruction of antinutritional factors	[44–46]
Carbohydrates	• Loss of sugar as a result of Maillard reaction and conversion of sucrose into glucose and fructose • Gelatinization of starch molecule resulting into expansion of molecule • Lower molecular weight of starch molecules (amylose and amylopectin) • Increased starch digestibility	[47, 50, 52]
Dietary fiber	• Increase in soluble fiber content • Increased retention of nutritional fiber	[49, 74]
Water solubility index	• Degradation of start network to soluble polysaccharides • Increase in soluble components dues to reduction in molecular size and water holding capacity	62, 63, 65]
Hardness	• Reduction in expansion and bubble development results into denser product with increased hardness • Water converting into superheated steam causes extrudates to swell	[63, 72, 73]
Oil absorption index	• Hydrophobic groups can keep oil within the structure • Increased porosity results in increased Oil absorption	[29, 66, 67, 74]

8.4 NOVEL PRODUCT DEVELOPMENT THROUGH EXTRUSION (FIBER RICH PRODUCTS)

Food extrusion technology is one of the major processing steps which have been widely used for making breakfast cereals, ready to eat snacks, value added snack and functional foods. As compared to traditional practices, extrusion cooking promotes a variety of advantages to the food and feed manufacturing industries. Numerous studies have directed on the effect of extrusion parameters such as fiber ratio, screw speed, extrusion temperature, die pressure, moisture content, and composition of raw material (particle size / proportion) on retaining the nutrition, quality, functional and sensory attributes with enhanced shelf- life of the finished product.

Initially to produce extruded products using this technology the milling and refining of cereal concentrate resulted in removal of outer coats of the grain [75] and thus this reduces the fiber and phytochemical content of the finished product. Refined flour, although abundant in starch, lacks sufficient bioactive ingredients [76], which adversely affects its nutritional profile. To address this issue, research is underway to investigate the feasibility of incorporating whole grains or a combination of whole grain components such as bran and high-fiber cereals into ready-to-eat (RTE) snacks. The aim is to enhance the nutritional value of these snacks and improve their overall health benefits [77–79]. Not only current studies are focusing on adding fiber rich cereals but also using vegetables, fruits and their by-products [80–83] that not only offer a rich source of fiber but also provide health benefits due to addition of functional bioactive components. This paper focuses on the role of extrusion in producing fiber rich food and various studies that have been conducted to produce novel food products using fiber rich food supplements (Table 8.2). The expanded extruded product is widely appealing and acceptable to the consumer due to its increased crunchiness or crispiness, however from an industrial viewpoint, expansion is relate in causing brittleness of the extruded product [84, 85]. Numerous authors have studied the relationship between various extrusion factors. Extensive investigation conducted by [84, 85] have demonstrated that protein and starch presence in the raw ingredients results into increased expansion of extruded product, although inclusion of dietary fibers (wheat barn) can lessen the density and expansion of the finished product. Certainly, several researchers have centered on the application of dietary fibers and polysaccharides to form the extruded ready to eat snacks and how inclusion of dietary fibers can modify the overall nutritional and rheological properties of the finished product [69, 86, 88]. In total, the integration of fiber in the raw ingredient blend usually result in alteration of extrusion parameters such as torque, energy consumption, pressure thrust, texture, structure and nutrient profile of the finished product [89] Martín-Cabrejas, Jaime [90] conducted a study to investigate the impact of extrusion on pretreated beans and observed that the total fiber content values were unaffected whereas insoluble fiber decreased by 25% with parallel increase in soluble fiber. Bashir, Sharif [91] investigated the extrusion of rice- soy crisps supplemented with spirulina and observed that increased fiber content with most acceptability in terms sensory when compared to snack extruded with supplementation of spirulina. Kharat, Medina-Meza [73] compared the extrusion processing attributes of three major whole grain flours (foxtail, finger, and pearl) individually and reported that in terms of expandability, foxtail extruded flour performed better than other two extruded products.

A recent study conducted by Lotfi Shirazi, Koocheki [78] investigated the impact of adding carrot pomace in barley flour to prepare high fiber ready to eat snack using extrusion technology and their results showed increase in soluble dietary fibers. Similarly, a study conducted by do Carmo, Varela [92] examined the extruded pea and oat bran fiber using die temperature of 146–175°C with screw speed of 200 rpm and demonstrated higher porosity with better sensory attributes. A study done by Stojceska, Ainsworth [93] demonstrated the direct effect of effect of increasing the water feed to 15% and have reported increase in total dietary fibers and bulk density of ready to eat snack made using wheat flour, red cabbage and brewer's spent grain extruded at feed rate of 25 kg/h and screw speed of 200 rpm%. Sumardiono, Budiyono [94] obtained analog rice snack using flour of mung beans, Sago and corn and demonstrated the change in crude fiber content of extruded product operated at two different temperatures (50 and 70°C) from 3.90% ± 0.04% to 4.39% ± 0.02% using hot extrusion technology. However, the application of Extrusion process is not only extended to the vegetable or cereals, but numerus studies have conducted extrusion technology to produce insect-based meat analogs. A study done by Smetana, Larki [95] supplemented protein concentrate from insect with fiber protein rich soy concentrates blended and extruded at 160–170°C and reported improved texture properties when soy fiber was added. However, there is still a lack of advance knowledge and understanding on how the fiber from various food sources and their interaction with different parameters and other ingredients affect the nutritional, sensorial and rheological properties of the extrudate finished product.

TABLE 8.2 High Fiber Rich RTE Snack Produced Using Extrusion Technology

Sl. No.	Food Matrix	Product Composition and Process Parameters	Major Findings	References
1	Rice-soy crisps and Spirulina	Rice flour, spirulina (2–8%) and soy flour, extruded at 100°C.	The extrusion process retained fiber and mineral content and in terms of preference it was scored highest by the sensory panel.	[91]
2	Pineapple by-products and corn starch based sweet whey protein	Sweet whey protein and pineapple by-product, extrusion temperature of 120–180°C	Optimized the parameters of extrusion technology reducing generation of waste and prevent environmental pollution	[82]
3	Fiber rich RTE produced using Apple pomace and Sorghum flour	Sorghum flour, Corn flour, and Apple pomace with barrel temperature of (80, 120, 95 and 160°C),	Extruded finished product with high dietary fiber, improved antioxidant, functional and antioxidant attributes when product extruded at 100°C extrusion temperature	[96]
4	Pea and oat fractions	Pea protein, pea starch, oat fiber and wheat bran extruded at Die temperature of 146–175°C	Scored high for sensory attributes with higher porosity and retained fiber rich attributes in finished product	[92]
5	Barley flour and carrot pomace high Fiber RTE snack	Carrot pomace and Barley Flour extruded at die temperature of 145 and 170°C	Increase in soluble dietary fiber with no change in overall dietary fiber content	[78]
6	Meat Substitute from insects	Soy protein concentrate, *Alphitobius diaperinus* protein concentrate soy fiber, Temperature (160–170°C)	Supplementation of soy fiber resulted into improved textural properties of the extruded meat analog.	[95]
7	RTE snacks from brewer's spent grain	wheat flour, Red cabbage and brewer's spent grain with feed entry and die temperature of 80 and 120°C	The level of total dietary content increased after increasing the water feed to 15% and it also effected other attributes such as color, total phenolic content and bulk density.	[93]

TABLE 8.2 *(Continued)*

Sl. No.	Food Matrix	Product Composition and Process Parameters	Major Findings	References
8	Analog rice snack using demonstrated the change in crude fiber content of extruded product operated at	Flour of mung beans, Sago and corn extruded to produce Rice based high fiber snack using two different temperatures (50 and 70°C)	Increased in crude fiber content of extruded product from 3.90% ± 0.04% to 4.39% ± 0.02% when product was extruded at 70°C than 50°C.	[94]
9	High-fiber snack bars	Oat bran (20 or 40%) and wheat or triticale flour, whey proteins, sugar, and dried milk, using twin-screw extrusion cooking of products containing at low (120°C) and high (140°C) temperatures	Temperature used showed no effect on textural properties, triticale products were harder than wheat products, and extruded products with 40% added fiber were harder than products with 20% added fiber.	[97]

8.4.1 HOT-MELT EXTRUSION

Initially employed in the plastic industry, hot-melt extrusion (HME) has been adapted for use in the food sector, particularly for the manufacturing of cereal products. HME is considered to be an effective technique for creating solid dispersions by modifying the mechanical, thermal, and physiochemical characteristics of drugs, foods, and food products. This process is aimed at enhancing the stability, firmness, and bioavailability of the final products. Due to the capability of enhancing solubility of the treated food, improving gelatinization of starches and denaturation of proteins, HME is now well recognized as an industrial process. The popularity of HME technology leans on the ability to improve various properties of food products such as digestibility and bioavailability and results into inactivation of microorganisms, enzymes and other anti-nutritional components [98].

A typical hot melt extruder consists of several essential components, including a platform for supporting the drive system, a barrel, a rotating screw mounted on a screw shaft, and an extrusion die for shaping the extruded food product. In hot-melt extrusion (HME), an extruder is utilized, which typically comprises one or two rotating screws that can rotate in the same or opposite directions within a stationary barrel. To minimize the residence time of the molten food material, the barrel used in HME is divided into sections [99].

For hot melt extrusion process, twin screw extruders are mostly used, and these extruders are mainly comprising of a metering, compression and a feed zone. All these three zones of extruders play different role such as uniform feeding and mixing, kneading, melting, homogenization, shearing of the extrudate and mix food under consistent temperature to produce food with similar composition. The main application of HME includes, microencapsulation of drugs, preparation of nanoparticles by enhancing the solubility of hydrophobic compounds [100].

In comparison to other technologies, such as co-crystallization, spray drying and freeze drying, HME technology is observed as a better option in preventing and preserving nutritional, sensory and quality attributes with improved taste masking properties. The product extruded using HME technology usually is usually more stable thermodynamically and have less inclination towards recrystallization [87]. Overall, extrusion have shown a positive impact on the protein and starch digestibility [102] with increased in the total fiber content [103].

8.4.2 REACTIVE EXTRUSION

Reactive extrusion includes using of reactor with continuous chemical reaction and is not a recent technique but rapidly evolving over past three decades. Now this technique is mostly used in food sector to produce functional foods and chemical modification of existing polymers. One of the advantages of using reactive extrusion is the absence of solvents as a reaction medium, therefore the ingredients with different viscosities can be processed using twin screw extruder [104]. In addition to this, reactive extrusion also provide opportunity for using more flexible extruding conditions such as high temperature, mixing conditions, modular geometry and so on. In comparison to traditional extruders, reactive extrusion enables effective heat removal due to larger reactor surface to volume ratio. The application of reactive extrusion has unveiled the avenues of managing reaction within an extruder [105].

However, the major challenge is controlling these reactions within the extruder therefore need advanced understanding research on impact of extruding parameters and reaction kinetics during the extrusion process. Additionally in comparison to traditional batch extrusion, reactive extrusion is a continuous process, which is another benefit towards environment. Suitably directed reactions are known to enhance the functionality of the extruded finished product such as improved emulsifying ability of extruded whey protein-pectin blends [106, 107]. The major drawback of using reactive extrusion is the residence time which is considerably short, ranging from few seconds to few minutes and therefore additional knowledge and understanding is required to use ingredients that can result in fast reaction. Another drawback of using this type of extruder is the limited cooling capacity therefore managing the exothermic reactions within the extruder can add up further challenges in producing product with desirable sensory, nutritional and functional characteristics.

8.5 CONSUMER DEMAND AND ACCEPTANCE OF EXTRUDED FIBER RICH RES

In recent decades, there has been a growing interest among consumers in ready-to-eat snacks, and this trend is expected to continue in the coming years. The popularity of these snacks can be attributed to their convenience and the consumer desire for enhanced sensory, nutritional, functional, and textural characteristics. Extrusion technology has gained significant popularity in the production of ready-to-eat (RTE) snacks due to its versatility in creating a wide range of shapes and textures, making it well-received by consumers. However, a concern with many existing RTE products is their high content of salt or sugar, which contributes to their high calorie content and poor nutritional profile. With increase in knowledge and advance understanding of consumer towards healthy and nutritious food, there is increase in demand for producing food that not only satisfy their hunger but also is healthy and nutritionally rich with low fat. Keeping the requirements of consumer aligned, this demand is steering manufacturers away from starch and sugar rich food to nutritionally balanced formulations with potentially valued added and functional ingredients. Thus, continuous investigations have been conducted to manipulate the nutritionally rich components by integrating bioactive ingredients such as dietary fibers.

8.6 CHALLENGES AND FUTURE TRENDS

Extrusion technology appeared in the food sectors about almost few decades back and the was first used in the plastic industry. This technology in food sector is now mainly used to produce ready to eat snacks, meat analog, jellies, baking dough, chocolates and many textured products. Extrusion technology is also one of the major components of the 3D printing technology. Although extrusion technique was initially launched in plastic industry but with advance in its ability to enhance the shelf life and preserving the nutritional and bioactive components of the extruded food product this technique is widely used in the industry. The use of extrusion technique processing application in the food manufacturing orbits around various objectives such as improved textural attributes, improved bioavailability, prevention of food wastage by increasing the shelf life of treated food, alleviating microbial spoilage, nutrition/health, transport suitability, and proficiency. Although, the modification and development of extruders for its application at industrial scale have changes drastically in the last few decades and thus have wide scope of using supercritical and cryogenic fluids for changes in structure of proteins, fibers, starches and other nutrients. As extrusion technology is linked with improve bioavailability with maximum retention of nutrient due to minimum generation heat during processing, therefore this technique can present as a promising emulsifying method that can substitute chemical emulsifiers like sodium caseinates.

Several efforts have been made by scientists and researchers worldwide to apply extrusion technology into the food sector, but because of some reasons, there still exist numerous hurdles for this technology to be broadly applied for all food types at much larger scale. The other major hurdles are the production efficiency, increase in production cost, continuous consistency to produce multi-flavor product with desired attributes, and requirement of trained personnel. Additional attention should be given during production of extruded food product with varying color and texture. For a wider purpose of this technique, the above-mentioned hurdles and challenges need to be carefully investigated and addressed. To increase the efficiency and to improve the rheological and sensory attributes of the treated product, the extrusion technology should be combined with other techniques such as cryogenic and supercritical fluid extrusion. The future trends of using this technology uncover the scope to develop extruded pellets by removing short chain fatty acids and thus snacks so produced can be expanded using microwave within fraction of seconds, therefore this would make it more attractive to consumers due to its easy preparation at home without have chemicals added to it. Though, a major challenge during super critical extrusion step is sustaining a temperature below 100°C and therefore this technology is still growing acceptance and popularity in the field of research and need addition modeling studies and experiments to advance technical expertise and better understanding before scaling up this technique for commercial application in the food industry.

8.7 CONCLUSION

The demand for ready-to-eat (RTE) snacks among consumers is on the rise, primarily due to the convenience of these products and their appealing texture and sensory properties. Extrusion processing, which involves passing a dough-like mixture through a high-pressure and high-temperature extruder, allows the food industry to produce RTE snacks from relatively inexpensive cereal-based ingredients that are acceptable to consumers. The global food industry has the potential to manipulate the nutritional content of extruded snacks, offering a variety of options to consumers. Starch plays a crucial role in determining the quality of these snacks, and the amount and type of starch in the ingredients significantly affect the formation of air cells during the extrusion process. The number and size of air cells directly impact the texture and crispness of the final product. However, the presence of fiber can hinder starch gelatinization, resulting in reduced air cell expansion and leading to low expansion, compactness, and an undesirably hard texture. The snack industry has faced challenges in creating extruded snacks that are both nutrient-rich and have appealing textures and flavors. Modifications of insoluble dietary fiber prior to extrusion may lead to significant improvement in its expansion properties and thus allow increasing applications in human nutrition.

In summary, extrusion processing offers the food industry an efficient means of producing RTE snacks from inexpensive cereal-based ingredients, and the nutritional status of these products can be manipulated to offer a range of nutritional profiles to consumers. Achieving the desired product quality of extruded snacks is heavily influenced by the quantity and type of starch present in the feed mixture. Fiber can hinder starch gelatinization and air cell expansion, but the presence of soluble fibers is desirable.

KEYWORDS

- **hot-melt extrusion**
- **oil absorption index**
- **ready-to-eat**
- **water absorption index**
- **water solubility index**

REFERENCE

1. Bhati, D., Gupta, R., & Sharma, A. (2022). Assessment of physicochemical, rheological, and thermal properties of Indian rice cultivars: Implications on the extrusion characteristics. *Journal of Texture Studies, 53*(6), 854–869.
2. Pandiselvam, R., Kumar, S., & Sharma, S. (2018). Modeling of coconut milk residue incorporated rice–corn extrudates properties using multiple linear regression and artificial neural network. *Journal of Food Process Engineering, 42*(2), e12981. https://doi.org/10.1111/jfpe.12981.
3. Shameena Beegum, P. P., Rajendran, K., & Rathod, R. P. (2019). Optimization of processing variables for the development of virgin coconut oil cake based extruded snacks. *Journal of Food Process Engineering, 42*(4), e13058. https://doi.org/10.1111/jfpe.13058.
4. Gadiraju, T., & Babu, R. (2015). Fried food consumption and cardiovascular health: A review of current evidence. *Nutrients, 7*(10), 8424–8430. https://doi.org/10.3390/nu7105424.
5. Preethi, P., Chennaswamy, N., & Sriram, M. (2021). Cashew apple pomace powder enriched the proximate, mineral, functional and structural properties of cereal based extrudates. *LWT–Food Science and Technology, 139*, 110539. https://doi.org/10.1016/j.lwt.2021.110539.
6. Preethi, P., & Thakur, S. (2022). Synergistic effect of powdered cashew sprout cum cotyledon and cereals on improving the biochemical and physical properties of extrudates. *Journal of Food Processing and Preservation, 46*(11), e16938. https://doi.org/10.1111/jfpp.16938.
7. Kharat, S., & Suryawanshi, N. (2019). Extrusion processing characteristics of whole grain flours of select major millets (foxtail, finger, and pearl). *Food and Bioproducts Processing, 114*, 60–71. https://doi.org/10.1016/j.fbp.2019.03.002.
8. Ditudompo, S., & Shafie, M. (2013). The effect of temperature and moisture on the mechanical properties of extruded cornstarch. *Journal of Texture Studies, 44*(3), 225–237. https://doi.org/10.1111/j.2042-4568.2012.00201.x.
9. Chang, Y. H., & Ng, P. K. W. (2009). Effects of extrusion process variables on extractable ginsenosides in wheat–ginseng extrudates. *Journal of Agricultural and Food Chemistry, 57*(6), 2356–2362. https://doi.org/10.1021/jf803602x.
10. Robin, F., & Schilling, M. (2012). Process, structure and texture of extruded whole wheat. *Journal of Cereal Science, 56*(2), 358–366. https://doi.org/10.1016/j.jcs.2012.05.008.
11. Onyango, C., & Nyongesa, H. (2004). Proximate composition and digestibility of fermented and extruded uji from maize–finger millet blend. *LWT–Food Science and Technology, 37*(8), 827–832. https://doi.org/10.1016/j.lwt.2004.01.001.
12. Maitre, A., Rathod, R. P., & Annapure, U. S. (2017). Effect of hydrocolloids and process parameters on the extrusion behavior of pearl millet grits. *International Food Research Journal, 24*(4), 1538–1544.
13. Food and Agriculture Organization of the United Nations (FAO). (2022). *World Cereal Production*. FAO.
14. Kumar, A., Singh, R., & Sharma, M. (2018). Millets: A solution to agrarian and nutritional challenges. *Agriculture & Food Security, 7*(1), 1–15. https://doi.org/10.1186/s40066-018-0155-0.

15. U.S. Department of Agriculture (USDA). (2022). *USDA Millet Production Scenario.* USDA.
16. U.S. Department of Agriculture (USDA). (2022). *Millet 2020 Production Scenario.* USDA.
17. Sarita, S. E. A. (2016). Nutraceutical and food processing properties of millets: A review. *Journal of Nutrition and Food Science, 4*(1), 1077. https://doi.org/10.4172/2155-9600.1001077.
18. Pang, M., Zhang, Z., & Li, X. (2014). Physicochemical properties, antioxidant activities and protective effect against acute ethanol-induced hepatic injury in mice of foxtail millet (*Setaria italica*) bran oil. *Food Function, 5*(8), 1763–1770. https://doi.org/10.1039/c4fo00156h.
19. Ren, X., Liu, H., & Xu, L. (2016). *In vitro* starch digestibility and *in vivo* glycemic response of foxtail millet and its products. *Food & Function, 7*(1), 372–379. https://doi.org/10.1039/c5fo00772b.
20. Yang, R., Zhang, H., & Li, W. (2020). Inhibitory effects of bound polyphenol from foxtail millet bran on colitis-associated carcinogenesis by the restoration of gut microbiota in a mice model. *Journal of Agricultural and Food Chemistry, 68*(11), 3506–3517. https://doi.org/10.1021/acs.jafc.0c00264.
21. Yousaf, L., Khan, M. N., & Shahid, M. (2021). Millet: A review of its nutritional and functional changes during processing. *Food Research International, 142*, 110197. https://doi.org/10.1016/j.foodres.2021.110197.
22. Malleshi, N. G., & Desikachar, H. S. R. (2021). Nutritional quality and health benefits. In *Millets and Pseudo Cereals* (pp. 159–168).
23. Radhajeyalakshmi, R., & Rajagopal, R. (2003). Existence of Thaumatin-like proteins (TLPs) in seeds of cereals. *Acta Phytopathologica et Entomologica Hungarica, 38*(3–4), 251–257. https://doi.org/10.1556/APhyt.38.2003.3-4.7.
24. Mohapatra, D., & Misra, R. C. (2019). Effect of different processing conditions on proximate composition, anti-oxidants, anti-nutrients and amino acid profile of grain sorghum. *Food Chemistry, 271*, 129–135. https://doi.org/10.1016/j.foodchem.2018.07.040.
25. Kumari, R., & Pal, S. (2022). Effect of fermentation conditions on nutritional and phytochemical constituents of pearl millet flour (*Pennisetum glaucum*) using response surface methodology. *Applied Food Research, 2*(1), 100055. https://doi.org/10.1016/j.afres.2021.100055.
26. Mitharwal, S., Kumar, S., & Chauhan, K. (2021). Nutritional, polyphenolic composition and *in vitro* digestibility of finger millet (*Eleusine coracana* L.) with its potential food applications: A review. *Food Bioscience, 44*, 101382. https://doi.org/10.1016/j.fbio.2021.101382.
27. Aisoni, J. E., Yusha'u, M., & Orole, O. O. (2018). Processing effects on physicochemical and proximate composition of finger millet (*Eleusine coracana*). *Greener Journal of Biological Sciences, 8*(2), 014–020.
28. Urooj, A. (2017). Impact of household processing methods on the nutritional characteristics of pearl millet (*Pennisetum typhoideum*): A review. *MOJ Food Processing & Technology, 4*(1), 28–32. https://doi.org/10.15406/mojfpt.2017.04.00085.
29. Taylor, J. R. N., & Kruger, J. (2019). Sorghum and millets. In *Sorghum and Millets* (pp. 171–224).

30. Li, E., Dhital, S., & Hasjim, J. (2014). Effects of grain milling on starch structures and flour/starch properties. *Starch–Stärke, 66*(1–2), 15–27. https://doi.org/10.1002/star.201200199.
31. Oladejo, A. A., & Adejuyigbe, C. (2020). Nutritional composition, functional properties and food applications of millet grains. *Asian Food Science Journal, 20*(1), 9–19.
32. Singh Gujral, H., & Singh, N. (2002). Extrusion behaviour and product characteristics of brown and milled rice grits. *International Journal of Food Properties, 5*(2), 307–316. https://doi.org/10.1081/JFP-120005354.
33. Arribas, C., & Salvador, A. (2017). The impact of extrusion on the nutritional composition, dietary fiber and *in vitro* digestibility of gluten-free snacks based on rice, pea and carob flour blends. *Food & Function, 8*(10), 3654–3663. https://doi.org/10.1039/c7fo00848k.
34. Mishra, A., Mishra, H. N., & Rao, P. S. (2012). Preparation of rice analogues using extrusion technology. *International Journal of Food Science & Technology, 47*(9), 1789–1797. https://doi.org/10.1111/j.1365-2621.2012.03135.x.
35. Singha, P., & Muthukumarappan, K. (2017). Effects of processing conditions on the system parameters during single screw extrusion of blend containing apple pomace. *Journal of Food Process Engineering, 40*(4), e12513. https://doi.org/10.1111/jfpe.12513.
36. Singha, P., Muthukumarappan, K., & Krishnan, P. (2018). Influence of processing conditions on apparent viscosity and system parameters during extrusion of distiller's dried grains-based snacks. *Food Science & Nutrition, 6*(1), 101–110. https://doi.org/10.1002/fsn3.521.
37. Singh, S. K., Singha, P., & Muthukumarappan, K. (2019). Modeling and optimizing the effect of extrusion processing parameters on nutritional properties of soy white flakes-based extrudates using response surface methodology. *Animal Feed Science and Technology, 254*, 114197. https://doi.org/10.1016/j.anifeedsci.2019.114197.
38. Choudhury, G. S., & Gautam, A. (1999). Screw configuration effects on macroscopic characteristics of extrudates produced by twin-screw extrusion of rice flour. *Journal of Food Science, 64*(3), 479–487. https://doi.org/10.1111/j.1365-2621.1999.tb15636.x.
39. Singh, S. K., & Muthukumarappan, K. (2015). Single screw extrusion processing of soy white flakes based Catla feed. *Journal of Food Research, 4*(1), 1–9. https://doi.org/10.5539/jfr.v4n1p1.
40. Singh, S. K., & Muthukumarappan, K. (2017). Rheological characterization and CFD simulation of soy white flakes-based dough in a single screw extruder. *Journal of Food Process Engineering, 40*(2), e12368. https://doi.org/10.1111/jfpe.12368.
41. Ek, P., & Ganjyal, G. M. (2020). Basics of extrusion processing. In *Extrusion Cooking* (pp. 1–28).
42. Singh, S. K., & Muthukumarappan, K. (2017). A viscosity model for soy white flakes-based aquafeed dough in a single screw extruder. *Journal of Food Process Engineering, 40*(2), 1–7. https://doi.org/10.1111/jfpe.12322.
43. Singha, P. (2017). Understanding the impact of extrusion processing on rheological, textural and functional properties of high-protein, high-fiber extrudates. South Dakota State University.
44. Wang, S., Wang, Y., Zhang, M., & Xu, J. (2016). Impacts of the particle sizes and levels of inclusions of cherry pomace on the physical and structural properties of direct expanded corn starch. *Food and Bioprocess Technology, 10*(2), 394–406. https://doi.org/10.1007/s11947-016-1690-0.

45. Chang, K. L. B., & Halek, G. W. (1991). Analysis of shear and thermal history during co-rotating twin-screw extrusion. *Journal of Food Science, 56*(2), 518–531. https://doi.org/10.1111/j.1365-2621.1991.tb03926.x.
46. Dalbhagat, C. G., Mahato, D. K., & Mishra, H. N. (2019). Effect of extrusion processing on physicochemical, functional and nutritional characteristics of rice and rice-based products: A review. *Trends in Food Science & Technology, 85*, 226–240. https://doi.org/10.1016/j.tifs.2018.12.015.
47. Singh, S. K., & Muthukumarappan, K. (2016). Effect of feed moisture, extrusion temperature and screw speed on properties of soy white flakes based aquafeed: A response surface analysis. *Journal of the Science of Food and Agriculture, 96*(6), 2220–2229. https://doi.org/10.1002/jsfa.7447.
48. Yeh, A. I., & Jaw, Y. M. (1999). Effects of feed rate and screw speed on operating characteristics and extrudate properties during single-screw extrusion cooking of rice flour. *Cereal Chemistry Journal, 76*(2), 236–242. https://doi.org/10.1094/CCHEM.1999.76.2.236.
49. Fletcher, S. I., Richmond, P., & Smith, A. C. (1985). An experimental study of twin-screw extrusion-cooking of maize grits. *Journal of Food Engineering, 4*(4), 291–312. https://doi.org/10.1016/0260-8774(85)90037-1.
50. Geetha, R., Mishra, H. N., & Srivastav, P. P. (2012). Twin screw extrusion of kodo millet-chickpea blend: Process parameter optimization, physico-chemical and functional properties. *Journal of Food Science and Technology, 51*(11), 3144–3153. https://doi.org/10.1007/s11483-012-0514-2.
51. Li, C., Wu, Y., Zhang, Y., & Xu, X. (2017). Extrusion expansion characteristics of samples of select varieties of whole yellow and green dry pea flours. *Cereal Chemistry Journal, 94*(3), 385–391. https://doi.org/10.1094/CCHEM-01-17-0002-R.
52. Ek, P., Kowalski, R. J., & Ganjyal, G. M. (2020). Raw material behaviors in extrusion processing I (Carbohydrates). In *Extrusion Cooking* (pp. 119–152).
53. De Pilli, T., di Gioia, D., & Vignali, G. (2012). Starch–lipid complex formation during extrusion-cooking of model system (rice starch and oleic acid) and real food (rice starch and pistachio nut flour). *European Food Research and Technology, 234*(3), 517–525. https://doi.org/10.1007/s00217-012-1748-1.
54. Singh, B., Sekhon, K. S., & Singh, N. (2007). Effects of moisture, temperature and level of pea grits on extrusion behavior and product characteristics of rice. *Food Chemistry, 100*(1), 198–202. https://doi.org/10.1016/j.foodchem.2005.09.051.
55. Kantrong, H., Vongkhamheng, C., & Lee, S. K. (2018). Influence of process parameters on physical properties and specific mechanical energy of healthy mushroom-rice snacks and optimization of extrusion process parameters using response surface methodology. *Journal of Food Science and Technology, 55*(9), 3462–3472. https://doi.org/10.1007/s11483-018-2113-8.
56. Meng, X., Wu, W., & Liu, X. (2010). Effects of extrusion conditions on system parameters and physical properties of a chickpea flour-based snack. *Food Research International, 43*(2), 650–658. https://doi.org/10.1016/j.foodres.2009.10.021.
57. Jin, Z., Hsieh, F., & Huff, H. E. (1994). Extrusion cooking of corn meal with soy fiber, salt, and sugar. *Cereal Chemistry, 71*(3), 227–234. https://doi.org/10.1094/CCHEM.1994.71.3.227.

58. Kowalski, R. J., & Ganjyal, G. M. (2018). Waxy flour degradation–Impact of screw geometry and specific mechanical energy in a co-rotating twin screw extruder. *Food Chemistry, 239*, 688–696. https://doi.org/10.1016/j.foodchem.2017.07.029.
59. Rashid, S., Khalid, N., & Sheikh, M. A. (2015). Effects of extrusion cooking on the dietary fibre content and water solubility index of wheat bran extrudates. *International Journal of Food Science & Technology, 50*(7), 1533–1537. https://doi.org/10.1111/ijfs.12756.
60. Gu, B. J., Wolcott, M. P., & Ganjyal, G. M. (2018). Pretreatment with lower feed moisture and lower extrusion temperatures aids in the increase in the fermentable sugar yields from fine-milled Douglas-fir. *Bioresource Technology, 269*, 262–268. https://doi.org/10.1016/j.biortech.2018.08.054.
61. Sawant, A. A., Thakor, N. J., Swami, S. B., & Divate, A. D. (2013). Physical and sensory characteristics of ready-to-eat food prepared from finger millet-based composite mixture by extrusion cooking. *Agricultural Engineering International: CIGR Journal, 15*, 100–105.
62. Tadesse, S. A., Sintayehu, K. W., & Belayneh, M. (2019). Functional and physical properties of sorghum-based extruded product supplemented with soy meal flour. *Cogent Food & Agriculture, 5*(1), 1707608. https://doi.org/10.1080/23311932.2019.1707608.
63. Sahu, C., Kumar, A., & Sharma, S. (2022). Effect of feed and process variables on nutritional quality of maize-millet based soy fortified extruded product using response surface methodology. *Applied Food Research, 2*(2), 100139. https://doi.org/10.1016/j.afres.2022.100139.
64. Singha, P., Sinha, N., & Muthukumarappan, K. (2018). Physicochemical and nutritional properties of extrudates from food grade distiller's dried grains, garbanzo flour, and corn grits. *Food Science & Nutrition, 6*(7), 1914–1926. https://doi.org/10.1002/fsn3.842.
65. Nkama, I., & Filli, K. B. (2006). Development and characterization of extruded fura from mixtures of pearl millet and grain legumes flours. *International Journal of Food Properties, 9*(2), 157–165. https://doi.org/10.1080/10942910600631737.
66. Filli, K. B., Nkama, I., Jideani, V. A., & Abubakar, M. (2012). The effect of extrusion conditions on the physicochemical properties and sensory characteristics of millet-cowpea based fura. *European Journal of Nutrition & Food Safety, 66*(3), 283–289. https://doi.org/10.9734/EJNFS/2012/1012.
67. Adebowale, K. O., Afolabi, T. A., & Olu-Owolabi, B. I. (2005). Hydrothermal treatments of finger millet (*Eleusine coracana*) starch. *Food Hydrocolloids, 19*(6), 974–983. https://doi.org/10.1016/j.foodhyd.2005.01.003.
68. Yadav, R. S., & Kaur, K. C. (2018). Development and quality evaluation of little millet (*Panicum sumatrense*) based extruded product. *Journal of Pharmacognosy and Phytochemistry, 7*(3), 3457–3463.
69. Deshpande, H., & Poshadri, S. (2011). Physical and sensory characteristics of extruded snacks prepared from foxtail millet-based composite flours. *International Food Research Journal, 18*(2), 751–756.
70. Singh, S. K. (2016). Understanding the effect of extrusion processing parameters on physical, nutritional, and rheological properties of soy white flakes based aquafeed in a single screw extruder. Ann Arbor (252).

71. Subramanian, S., & Viswanathan, R. (2007). Bulk density and friction coefficients of selected minor millet grains and flours. *Journal of Food Engineering, 81*, 118–126. https://doi.org/10.1016/j.jfoodeng.2006.11.019.
72. Bhatnagar, S., & Hanna, M. A. (1994). Amylose-lipid complex formation during single-screw extrusion of various corn starches. *Cereal Chemistry, 71*(6), 582–587. https://doi.org/10.1094/CCHEM.1994.71.6.582.
73. Devi, N. L., Suneetha, V. R., & Singh, J. (2013). Development of protein-rich sorghum-based expanded snacks using extrusion technology. *International Journal of Food Properties, 16*(2), 263–276. https://doi.org/10.1080/10942912.2011.630630.
74. Rathod, R. P., & Annapure, U. S. (2017). Physicochemical properties, protein and starch digestibility of lentil-based noodle prepared by using extrusion processing. *LWT, 80*, 121–130. https://doi.org/10.1016/j.lwt.2016.12.060.
75. Alam, M. S., Pathania, S., & Sharma, A. (2016). Optimization of the extrusion process for development of high fiber soybean-rice ready-to-eat snacks using carrot pomace and cauliflower trimmings. *LWT, 74*, 135–144. https://doi.org/10.1016/j.lwt.2016.07.017.
76. Altan, A., McCarthy, K. L., & Maskan, M. (2008). Evaluation of snack foods from barley–Tomato pomace blends by extrusion processing. *Journal of Food Engineering, 84*(2), 231–242. https://doi.org/10.1016/j.jfoodeng.2007.06.016.
77. Pardhi, S. D., Rajput, P., & Yadav, A. (2019). Evaluation of functional properties of extruded snacks developed from brown rice grits by using response surface methodology. *Journal of the Saudi Society of Agricultural Sciences, 18*(1), 7–16. https://doi.org/10.1016/j.jssas.2018.05.004.
78. Arora, B., Singha, P., & Rizvi, S. S. H. (2021). Supercritical fluid extrusion: Die design and physicochemical properties of milk protein extrudates. *Innovative Food Science & Emerging Technologies, 68*, 102637. https://doi.org/10.1016/j.ifset.2021.102637.
79. Singha, P., & Muthukumarappan, K. (2018). Single screw extrusion of apple pomace-enriched blends: Extrudate characteristics and determination of optimum processing conditions. *Food Science and Technology International, 24*(5), 447–462. https://doi.org/10.1177/1082013218759494.
80. Gujral, H. S., Sandhu, K. S., & Kaur, A. (2012). Total phenolic content and antioxidant activity of extruded brown rice. *International Journal of Food Properties, 15*(2), 301–311. https://doi.org/10.1080/10942912.2010.485835.
81. Singh, S. K., & Muthukumarappan, K. (2016). Effect of feed moisture, extrusion temperature and screw speed on properties of soy white flakes based aquafeed: A response surface analysis. *Journal of the Science of Food and Agriculture, 96*(6), 2220–2229. https://doi.org/10.1002/jsfa.7494.
82. Kowalski, R. J., Ganjyal, G. M., & Lai, S. H. (2016). Extrusion processing characteristics of quinoa (Chenopodium quinoa Willd.) var. Cherry Vanilla. *Journal of Cereal Science, 70*, 91–98. https://doi.org/10.1016/j.jcs.2016.05.013.
83. Ganjyal, G. M., & Hanna, M. A. (2006). Role of blowing agents in expansion of high-amylose starch acetate during extrusion. *Cereal Chemistry, 83*(6), 577–583. https://doi.org/10.1094/CC-83-0577.
84. Sun, Y., & Muthukumarappan, K. (2002). Changes in functionality of soy-based extrudates during single-screw extrusion processing. *International Journal of Food Properties, 5*(2), 379–389. https://doi.org/10.1080/10942910209356947.

85. Aluwi, N. A., Ganjyal, G. M., & Hou, J. (2016). Impacts of scarification and degermination on the expansion characteristics of select quinoa varieties during extrusion processing. *Journal of Food Science, 81*(12), 2939–2949. https://doi.org/10.1111/1750-3841.13586.
86. Benhur, D. R., Kumar, V., & Sinha, N. K. (2015). Development and standardization of sorghum pasta using extrusion technology. *Journal of Food Science and Technology, 52*(10), 6828–6833. https://doi.org/10.1007/s11483-015-0924-1.
87. Hagenimana, A., Ding, X., & Fang, T. (2006). Evaluation of rice flour modified by extrusion cooking. *Journal of Cereal Science, 43*(1), 38–46. https://doi.org/10.1016/j.jcs.2005.06.001.
88. Dlamini, N. S., & Khan, S. W. K. (2016). Effect of ingredients ratio on physicochemical and sensory properties of sorghum, maize, and soya protein concentrate blend extrudates. *International Journal of Food Science and Nutrition Engineering, 6*(5), 112–120.
89. Sahu, C., & Patel, S. (2020). Optimization of maize-millet based soy fortified composite flour for preparation of RTE extruded products using D-optimal mixture design. *Journal of Food Science and Technology, 58*(7), 2651–2660. https://doi.org/10.1007/s11483-020-09772-8.
90. Kaushal, P., Kumar, V., & Sharma, H. K. (2012). Comparative study of physicochemical, functional, antinutritional and pasting properties of taro (*Colocasia esculenta*), rice (*Oryza sativa*) flour, pigeonpea (*Cajanus cajan*) flour and their blends. *LWT–Food Science and Technology, 48*(1), 59–68. https://doi.org/10.1016/j.lwt.2012.03.018.
91. Barbiroli, A., Dalla Rosa, M., & Bortolomeazzi, R. (2013). Process conditions affect starch structure and its interactions with proteins in rice pasta. *Carbohydrate Polymers, 92*(2), 1865–1872. https://doi.org/10.1016/j.carbpol.2012.11.066.
92. Marti, A., Ferrer, E., & de la Cueva, C. (2017). *In vitro* starch digestibility of commercial gluten-free pasta: The role of ingredients and origin. *Journal of Food Science, 82*(4), 1012–1019. https://doi.org/10.1111/1750-3841.13688.
93. Schoch, T. J., & Mears, M. (1968). Preparation and properties of various legume starches. *Cereal Chemistry, 45*, 564–573.
94. Kaur, L., Singh, N., & Singh, J. (2007). Physico-chemical, rheological and structural properties of fractionated potato starches. *Journal of Food Engineering, 82*(3), 383–394. https://doi.org/10.1016/j.jfoodeng.2007.02.004.
95. Aluwi, N. A., Murphy, K. M., & Ganjyal, G. M. (2017). Physicochemical characterization of different varieties of quinoa. *Cereal Chemistry, 94*(5), 847–856. https://doi.org/10.1094/CCHEM-11-16-0252-R.
96. Campanella, O. H., Li, P. X., Ross, K. A., & Okos, M. R. (2002). The role of rheology in extrusion. In J. Welti-Chanes, G. V. Barbosa-Canovas, & J. M. Aguilera (Eds.), *Engineering and Food for the 21st Century* (pp. 393–413). Technomic Publishing.
97. Ragaee, S., & Abdel-Aal, E. S. M. (2006). Pasting properties of starch and protein in selected cereals and quality of their food products. *Food Chemistry, 95*(1), 9–18. https://doi.org/10.1016/j.foodchem.2005.01.024.
98. Sumathi, A., Ushakumari, S. R., & Malleshi, N. G. (2009). Physico-chemical characteristics, nutritional quality and shelf-life of pearl millet-based extrusion cooked supplementary foods. *International Journal of Food Sciences and Nutrition, 58*(5), 350–362. https://doi.org/10.1080/09637480802109816.
99. Onyeoziri, I. O., Onyenekwe, P. C., & Okoro, R. (2021). Descriptive sensory analysis of instant porridge from stored wholegrain and decorticated pearl millet flour cooked,

stabilized, and improved by using a low-cost extruder. *Journal of Food Science, 86*(9), 3824–3838. https://doi.org/10.1111/1750-3841.16321.

100. Camire, M. E. (1998). Chemical changes during extrusion cooking. In J. Welti-Chanes, G. V. Barbosa-Canovas, & J. M. Aguilera (Eds.), *Process-Induced Chemical Changes in Food* (pp. 109–121). Springer.

CHAPTER 9

3D PRINTING TECHNOLOGY FOR PRODUCTION OF LEGUMES AND VEGETABLES-BASED READY-TO-EAT SNACKS

DIBYA RANJAN DASH, SUSHIL KUMAR SINGH, and POONAM SINGHA

Department of Food Process Engineering, National Institute of Technology, Rourkela, Odisha, India

ABSTRACT

3D food printing technology is a digitally operated automatic technique in which complicated 3D designed food products layer on a layer can be prepared. The increasing demand for 3D printing in the food industry is due to the production of more attractively customized and modified textured products that contain functional ingredients and food additives having higher nutritional value. This chapter discusses the 3D printing application to produce legume and vegetable based ready to eat snacks containing good quality proteins, bioactive compounds, and other health-forming nutrients. An elaborative study is performed to develop revolutionary shape and texture of legume and vegetable based printed snacks in terms of distinctive appearance, layered design, recipe reformulation and automated nutrition control. The printing material and the process variables are deeply associated in evaluating the quality of the end products. Thus, selecting the proper combination of legumes and vegetable ingredients and modifying them to make more compatible for printing is described here. The impact of printing parameters, post-printing factors and formulation ingredients on

Ready-to-Eat Snacks: Emerging Technologies for Production and Safety.
Ravi Pandiselvam and Gülşah Çalışkan Koç

the attributes of the 3D printed ready to eat snacks is also explored in this chapter. Legumes can also be used as a substitute to the gluten sources like wheat, rye, oat, and barley for making gluten-free, ready-to-eat snacks.

9.1 INTRODUCTION

Recently, 3D food printing (3DFP) is counted as an innovative technique in the research and commercial field to develop functional, personalized and nutrient-rich food products [1–3]. 3D printing (3DP) technology can convert multi-step processes into a single step and is applied to prepare ready-to-eat (RTE) and ready-to-cook (RTC) food materials with personalized shapes and appearances along with the application of a wide variety of raw materials [4, 5]. The most used 3DP process is the extrusion-based 3D printing which comes under fused deposition modeling technology [6–9]. Many reviews discussed the recent work on 3 DP and its potential utilization in food sectors [4, 10, 11]. The extrusion-based 3DP process of food products is explained in Figure 9.1. Layer on layer structure of the objects formed by forcing the food ink (ingredient paste) such as puree, jelly, cheese, etc., through the nozzle in 3D printing extrusion process. In selective sintering printing, the 3D structure of the product is formed due to the sintering of the fat or sugar present in printing material such as sugar or

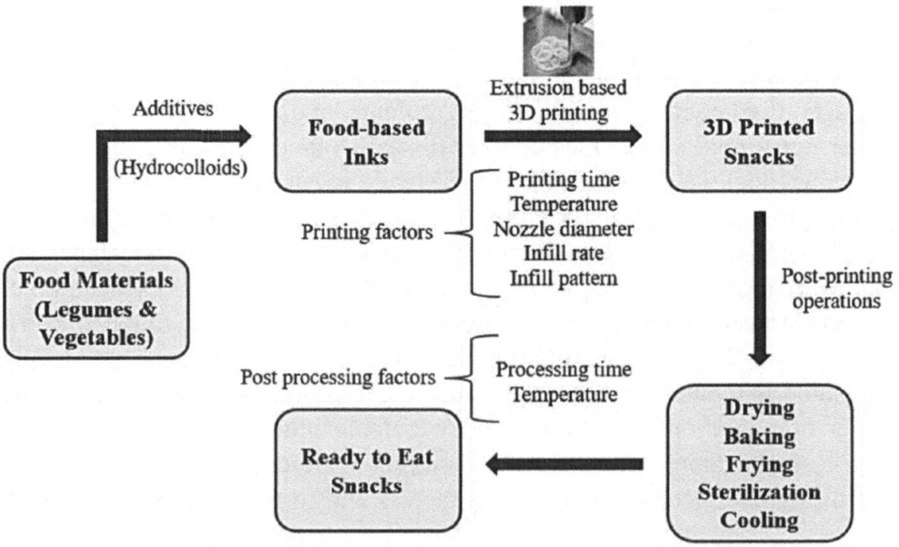

FIGURE 9.1 Extrusion-based 3D printing process of ready-to-eat snacks.

chocolate powder by heating the powder material using a laser source. In the same way, binding jetting produces layer upon layer 3D structure by binding the powder material with the use of a liquid binder through the nozzle. At last, inkjet printing is applied for the surface filling of food materials like cakes and cookies by dropping low viscosity through gravity fall [12].

RTE snacks products are now getting more popularized in the food industry due to their convenience in consumption, demand from the young populations and functional food nature [13]. The great interest of people in RTE snacks paves the way for 3D food printing technology for making legume and vegetable-based snacks with improved nutritional, functional and sensory characteristics. Legumes such as soybean and pulses can be used in place of wheat, barley, rye, and other grains for making 3D printed snacks due to the presence of gluten protein which may cause the occurrence of Coeliac disease, an auto-immune disorder [14]. Around 1.4% of the global population are affected by this disease, which can cause malnourishment, lactose intolerance, reduction in bone density and intestinal lymphoma cancer in the worst case [15]. There are various examples of snacks products in which gluten-free flour is used instead of the conventional flour (wheat, rye and other grains) containing gluten. Some of the studies include gluten-free (GF) bread by using soy, quinoa, or amaranth flour [16], okra, and soy flour used to improve the moistness of GF bread [17], firmer GF cookies prepared from sweet potato [18] and GF pasta prepared from chickpea replacing wheat flour [19]. There are also various instances of using lupin flour replacing wheat flour for making baked products like white bread, sourdough bread and cakes [20–22]. Severini, Derossi [8] developed a 3D printed fruits and vegetable-based smoothie taking carrots, kiwi fruits, pears, broccoli Raab leaves and avocado as the ingredients.

A report published by Martin, Schmidt [23] explained the development of snacks containing both legume protein and pseudo cereals or alone legume materials like chickpea, faba bean and lupin flour with the help of a low moisture extrusion process. This could be a motivation to produce functionalized ready to eat snacks by taking legumes and vegetables using 3D food printing technology with attractive geometry, customized ingredients and unique texture. Lille, Nurmela [24] found the utilization of protein, fiber and starch rich paste for making functional and customized snacks by using 3D printing extrusion technology where faba bean protein concentrate was used as a base material. The characteristics of protein based (pea protein isolate as one component) combined vegetable and fruit ink system was studied by Chen, Zhang [25] using a 3D printer to check the feasibility of the ink to develop customized ready to eat snacks. Hence, 3D printing technology is used to develop individually personalized and functionalized ready to eat snacks from legume and vegetable materials with health-forming benefits taking sustainable and balanced diet system into consideration. Apart from the advantages of 3DFP, there are some disadvantages associated with it, which is mentioned briefly in Figure 9.2.

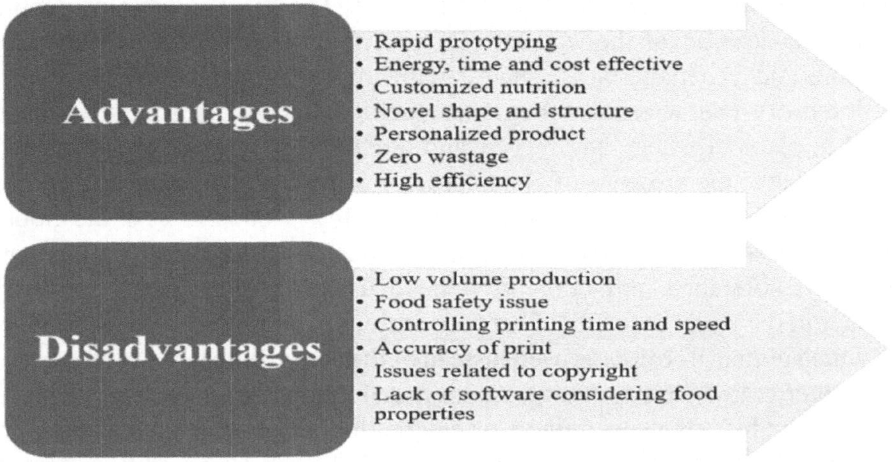

FIGURE 9.2 Advantages and disadvantages of 3D food printing.

This chapter discusses the application of 3D printing technique in the development of legume and vegetable based ready to eat snacks containing good quality proteins, bioactive compounds and other health forming nutrients. Also, the influence of printing parameters, post-printing factors and formulation ingredients on the characteristics of the 3D printed ready to eat snacks is explored in this chapter.

9.2 PRINTABILITY OF LEGUME AND VEGETABLE BASED INGREDIENTS

Printability of any 3D printing formulation defines the ability to hold up the weight of the material by giving proper dimensions [26]. Printability is governed by various factors like temperature, formulation ingredients, viscosity and additives. Factors like nozzle size, extrusion rate, speed and nozzle height may influence the printability of the formulation [27].

According to the printing ability, the 3D printing ingredients are divided into three groups like naturally printable, non-printable and alternative materials. The printable ingredients are self-efficient to be extruded easily from the syringe. They include materials like hydrogel, cheese, chocolate, dough, hummus, pasta, mashed potato, and butter etc., or powder like materials such as starch, sugar, etc. [4, 28–31]. They can support their own shape and structure and do not need extra post-printing operations. Nonprintable materials like vegetable, fruit, meat and rice are not naturally extrudable [32]. Some hydrocolloids like additives such as gelatin, alginate, starch, xanthan gum, pectin, agar, etc., are mixed with the non-printable materials to improve their viscosity and flow ability and provide support for their extrudability and structural stability [33–35]. Alternative ingredients constitute the functional and biologically active components like protein, fiber from food and agricultural residue, algae, microorganisms, etc. [36].

Printability depends on the rheology of the food formulation. The application of hydrocolloids with legumes and vegetable ingredients improves printability by interacting and affecting the molecular weight with the rheological characteristics of the formulation. Chen, Mu [37] observed the improvement in rheological characteristics of the mixture of gelatin, soy protein isolate, and sodium alginate, thereby increasing the printability of the formulation and enhancing the textural characteristics of the final 3D geometry. A 3D printing extrusion test was performed by Lille, Nurmela [24], considering a paste that contains a mixture of ingredients such as starch, faba bean protein concentrate, rye bran, milk powder and cellulose nanofiber where starch acts a thickening agent. The low viscous formulation was observed to be suitable for passing through a small nozzle at a higher shear rate; however, the viscosity must be higher at rest to obtain good structural stability of the printed samples. Chuanxing, Qi [38] reported that pea protein was a suitable ingredient to increase the overall properties and printability of potato starch-based material. Derossi, Caporizzi [39] noticed the impact of speed and flow level of printing on the deposition rate of the paste forming from banana, white beans, dried mushroom, and nonfat milk for making functionalized customized snacks. Fiber rich ready to eat snacks are good for health, especially they have positive effects on the digestive system. But processing and post-processing of fiber rich formulation create problems due to the bad printability. Negative impact on structural stability, printability and textural quality is noticed by printing large size fiber ingredients because of the problems like nozzle clogging and agglomeration due to shearing. Hence, the addition of starch or protein to the high fiber ingredients provides plasticizing impact during the printing [24, 40]. Krishnaraj, Anukiruthika [41] developed fiber and protein rich nutritious snacks by taking flours from a combination of sources like green grams, ajwain (carom) seeds, fried grams and barnyard millet. Optimization of the process parameters like motor speed of the extruder, height and diameter of nozzle, X, Y and Z axis movement speed and printing speed was conducted to obtain the best stability of the products. The printing performance is influenced by the shape and extrusion stability of the food inks which depends on their printability.

9.3 CHARACTERIZATION OF 3D PRINTED LEGUME AND VEGETABLE BASED SNACKS

Characterization of 3D printed ready to eat snacks depends on the printing and post-processing parameters. Printing factors include nozzle diameter, layer height, printing speed, printing temperature, infill level, filling pattern, etc., whereas baking time and baking temperature are considered as the post-processing/post-printing parameters. Various characteristics of the 3D printed ready to eat snacks affected by the above factors are physicochemical properties, color, dimensional stability, textural properties and microstructure. Table 9.1 shows the development of vegetable and legume-based 3D printed snacks, including the post-processing operations and factors involved.

TABLE 9.1 Summary of 3D Printed Developed Ready-to-Eat Snacks Containing Legumes and Vegetables

Major Formulation İngredients	Post-Processing Operation	Factors İnvolved	Products	References
Yam powder: potato processing by-product (0:10, 1: 9, 2: 8 and 3: 7)	Air frying	Infill level (%), infill pattern	Yam snacks	[42]
Soy protein isolate: Sodium alginate: gelatin	Baking	Printing temperature and ingredient composition	Soy protein isolate paste snacks	[43]
Lupin flour, pea protein isolates and chickpea flour	Baking (Forced convection oven)	Ingredient composition and baking time	Gluten free snacks	[44]

TABLE 9.1 *(Continued)*

Major Formulation İngredients	Post-Processing Operation	Factors İnvolved	Products	References
Celery root, carrot, onion, chicken broth, boneless chicken thigh, whole wheat flour, gelatin (240.5 Bloom), kappa-carrageenan	Baking	Nozzle height, gelatin concentration, and feed rate	Snacks	[45]
Potato flakes, sodium alginate, sodium bicarbonate, citric acid	N/A	Potato flakes content, pH	Sweet potato puree	[46]
Carrot puree, wheat flour	Baking	Nozzle diameter, filling rate, baking temperature and time	Carrot appetizer cake	[47]
Whole milk powder, whole grain rye flour	Baking	Ingredient composition, baking process	Protein and fiber rich snacks	[48]
Soy protein isolate, jackfruit seed, finger millet, unsalted butter, xanthan gum, baking powder	Baking	Fill density, water butter ratio, printing temperature	Cookie	[49]
Green gram, fried gram, barnyard millet, ajwain (carom) seeds	Deep frying, hot air drying-deep frying and microwave drying	Nozzle diameter, printing rate, layer height, printing speed	Fiber rich snacks	[41]
Potato flakes, wheat flour, bovine fat	Air frying	Infill levels, infill pattern	Potato snacks	[50]

*N/A: Not applicable

9.3.1 PHYSICOCHEMICAL PROPERTIES AND COLOR CHARACTERISTICS

Moisture, lipid, carbohydrate, protein, crude fiber, and ash content are the parameters used to estimate the proximate composition of the printed snacks before and after the printing operation [51]. The proximate composition data is used to calculate the total energy and recommended dietary allowance (RDA) of the developed snacks. Water activity of the printed snacks is measured to know the shelf-life of the snacks which is also associated with their crispiness [52]. Bulut and Candoğan [45] reported the effect of post-processing operation and addition of gelatin on the proximate composition, water activity, pH, and color of the developed functionalized snacks. The ash content and water activity of the 1.97% gelatin added sample did not show any changes before and after baking. The protein concentration of the gelatin added sample was enhanced as compared to the control one (without gelatin) owing to the greater protein content of gelatin. The melting of lipids due to temperature rise followed by releasing from the matrix caused the lipid loss due to baking. Also, the moisture loss occurred due to baking influenced the texture of the product. Baking reduced the pH of the printed sample because of the formation of free fatty acids by the degradation of lipid. The water activity value of both control and gelatin added snacks were reduced to 0.92 and 0.93 respectively after baking from a value of 0.99 without baking operation. Color characteristics of the printed snacks can affect their consumer preference and acceptability. From the above study, the color of the final product was affected due to the combined effect of baking and gelatin addition by imparting changes in the L*, a* and b* values. Krishnaraj, Anukiruthika [41] conducted proximate analysis and color study of the fiber rich snacks and found that microwave post-treated samples showed good nutritional profile as compared to the deep fried (DF) and hot air dried- deep fried (HAD-DF) samples. Microwave treatment retained nearly the same amount of protein and fiber in the snacks as compared to other post-processing operations like DF and HAD-DF. High temperature (165°C) deep frying decreased the amount of protein in the snacks in contrast to the raw flours used for dough formulation.

9.3.2 DIMENSIONAL STABILITY

Various shapes and geometries are printed by a 3D printing mechanism where the post-processing operation provides proper shapes and dimensions to the product. The product must resist the changes due to the post-processing operation that can produce minimal shape deformations and dimensional variability. Jagadiswaran, Alagarasan [53] observed the influence of the baking operation on the dimensional variability and collapse of the structure of functionalized cookies by measuring the dimensions like height and diameter before and after baking of the printed snacks. The shape fidelity (%) and dimensional variation (%) were also calculated. It was noticed that the dimensional variability was seen more in the baked sample with respect to the unbaked sample. Mixing of grapes pomace powder in the dough showed dimensional variability between the control (without grapes pomace powder) and functional cookies because of the difference in viscoelastic property of the dough. The dimensional coherence between the printed cookie and the 3D model was observed because of the variation of shape fidelity from 104–107% between functional and control cookies.

The impact of factors like infill rate, nozzle diameter, baking time and temperature on the change of maximum height and width of the printed carrot appetizer cake was studied by [47] where the height and width denoted to the swelling and deformation of the product after baking. Infill rate controlled the shape of the cake by changing the width and height of the cake. Swelling and deformation of the cake were increased by increasing the height and width respectively at a higher infill rate. At a higher infill rate, the selection of proper nozzle diameter could be able to control the deformation and swelling of the cake. The appearance of the cake baked at 145°C and in 3.4 mm diameter nozzle was similar irrespective of the baking time (15 or 25 min) and infill rate (55 or 85%). However, baking at different temperatures such as 145°C and 195°C, made more difference in the appearance of the cakes and baking time is the crucial parameter to be taken care of at higher baking temperatures like 195°C. The optimum process parameters to achieve maximum swelling and minimum deformation of the appetizer cake were found to be 3.4 mm of nozzle diameter, 71% of infill rate, 25 min of baking time and 177°C of baking temperature.

Feng, Zhang [42] reported the effect of air frying, infill level and pattern on the dimensional properties of the snacks prepared from potato processing by-products and yam powder by using 3DP based extrusion process. The size of the printed snacks was increased as compared to the 3D model due to the swelling phenomena during extrusion which was attributed to the release of elastic potential energy at nozzle that was previously stored in the shrinkage section of the nozzle diameter causing tensile deformation. This swelling phenomenon is governed by the swelling ratio (ratio of printed sample diameter to nozzle diameter) which depends on inlet flow volume and viscosity of the material. Some degree of shrinkage and deformation occurred by the air frying of the snacks due to loss of water; however, the thickness of the products remained uniform all the time (3.1 ± 0.3 mm). At the same infill level and pattern, with an increase in the content of potato by-products, the bending height of air fried snacks reduced showing less degree of deformation which was mainly due to the stronger structure formed with the addition of extra fiber by potato processing by-products. The bending heights of the cross-structured snacks were higher than the parallel-structured filled snacks at a 50% infill level. This was due to the generation of two shrinkage directions of the cross-structured sample during dehydration because of the double direction extrusion line process while forming the cross structure.

9.3.3 TEXTURE

Textural properties determine the springiness, hardness, gumminess, chewiness, and fracturability of the printed snacks developed before and after post-processing. The texture is the most crucial property among the other characteristics of the 3D printed snacks to impact their acceptability [54]. In 3D printing of snacks, baking parameters, printing parameters and the ingredient used in the formulation can affect the textural properties. Krishnaraj, Anukiruthika [41] observed the influence of different post-printing operations like HAD-DF, microwave drying and deep frying on textural properties of printed snacks developed from a composite flour containing ajwain(carom) seeds, green gram, millet and fried gram. In deep frying of the snacks, the sample became harder due to maximum loss of water occurred at higher frying temperature (165°C) but a less hardened product was formed by HAD-DF because the moisture was evaporated by hot air drying before the frying. With the increase in power during microwave drying, the product was formed having more hardness owing to rapid evaporation of water which also increased the fracturability of the snacks. Bulut and Candoğan [45] prepared chicken-based snacks by taking functional ingredients like carrots, white onions, carrot fiber, turmeric and garlic powder and observed the effect gelatin percentage and post-processing baking operation on the quality of the snacks. The combined gelation effect of *kappa*-carrageenan and gelatin made textural changes to the snacks after baking [55]. The addition of 1.79% of gelatin enhanced the hardness, chewiness and gumminess and reduced the springiness. A robust gel was formed due to the strong interaction between the gelatin and the *kappa*-carrageenan which brought distinctive changes in the physiochemical properties of the snacks [56]. Also, the combined effect of baking and 1.79% gelatin addition made the microstructure compact and uniform.

A protein and fiber rich snack were prepared by Lille, Nurmela [24] by taking faba bean and oat protein, starch, nano-cellulose fiber, rye bran and milk powder as the ingredients. Faba bean and oat protein contained snacks were dried by oven drying method for 30 min. Faba bean protein-based snacks were softer than oat protein sample due to less water removal from the former one. Also, a softer product was formed when a part of the milk powder was replaced by cellulose nanofiber which was attributed to the less dry matter content of the cellulose contained product. The semi skim milk powder produced the hardest snacks after oven drying. Hence, it was proved that the ingredient composition of the printing paste affects the hardness of the snacks. He, Zhang [46] studied the textural attributes of the 3DP mashed potato (MP) puree where the pH and potato flake content affected the gumminess and hardness of the product. Sodium alginate and sweet potato powder were also added to the MP puree. Citric acid and sodium bicarbonate were mixed to maintain the acidic and alkalinity of the mashed potato puree. The hardness and gumminess of the mashed potato-citric acid were increased when the potato flake rose from 15% to 27%. Adhesion and hardness of mashed potato-sodium bicarbonate sample were enhanced due to the change in the amount of potato flake from 19% to 27%. Overall, the increasing trend of hardness of the samples was reported to be in the order of alkaline-mashed potato, neutral-mashed potato, and acidic mashed potato, respectively.

Guénard-Lampron, Masson [47] reported the influence of the interaction of filling rate, nozzle diameter, time, and temperature of baking on the textural quality of 3DP carrot appetizer cake. The combined impact of nozzle diameter and post-printing factors (temperature and time of baking) showed a significant impact on crust strength as compared to the interaction between post-processing factors. From the experiment, it was concluded that the formation of high crust strength of the cake could be prevented by printing in a higher diameter nozzle (5 mm) when a longer baking time (25 min) or higher baking temperature (195°C) was used for baking. The interaction effect of filling rate with nozzle diameter and post-processing factors on crust strength was found insignificant. Agarwal, Wallace [44] developed gluten free 3D printed snacks by considering chickpea and lupin flour with oil and without oil as the formulation ingredients. Figure 9.3 shows the difference in

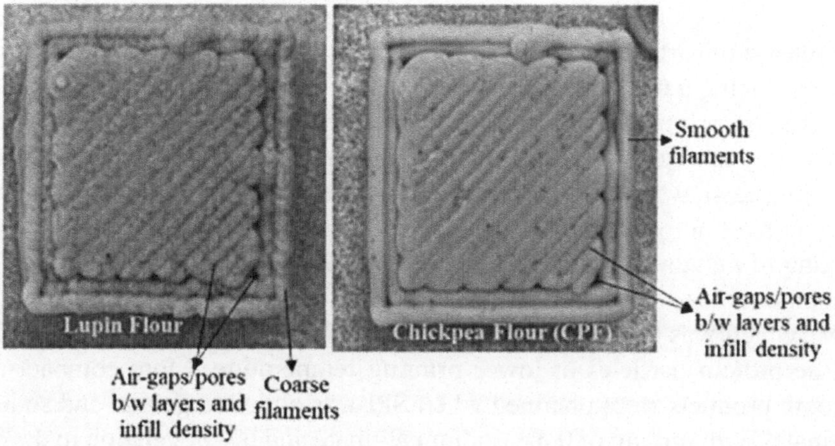

FIGURE 9.3 Images of lupin and chickpea flour-based 3D printed formulation before baking.

Source: Reprinted from Ref. [43]. Copyright Elsevier 2022. https://creativecommons.org/licenses/by/4.0/

the textural characteristics of the post-processing snacks which was noticed by the filament quality. The bigger particle size of the lupin flour produced coarse filament as compared to chickpea flour. Same kind of observations were also noticed in lupin-based breads [57] and chickpea-based biscuits [58].

9.3.4 MICROSTRUCTURE

The microstructure of the 3DP snacks is changed by the formulation ingredients, printing, and post-processing parameters. Microstructure of the snacks was observed by various researchers using Micro-CT imager analysis, X-ray micro tomography (XRT) and scanning electron microscopy (SEM) [43–45, 49].

Varghese, Wolodko [49] observed the formation of pores in the microstructure of the snacks made from soy protein isolate, finger millet and jackfruit seed as the main ingredients by using Micro-CT technology. The image analysis showed the formation of closed pores (bubbles) and open pores (connecting networks) which were influenced by filling density. The open pore space created at 50% filling density was higher than at 100% filling density because at higher filling density, more material was required to pack the structure to reduce pore space. The closed pores of the cookie were increased with an increase in filling density from 50% to 100%. At 100% filling density, expansion of the cookie occurred due to the formation of carbon dioxide with the mixing of baking powder in the dough which caused closed pore structure. At higher filling densities like 90% and 100%, the release of carbon dioxide was very less as compared to other infill densities producing a puffy texture cookie.

The 3D printing process conducted by Chen, Sun [43] to develop the product by taking gelatin and sodium alginate along with soy protein isolate (SPI) paste showed that the printing temperature and matrix had a greater role on the microstructure of the product. The control product (without the mixing of gelatin and sodium alginate) printed at 35°C and 45°C showed a smoother cross-sectional structure as compared to the product developed at 25°C which was attributed to the weak interaction and easier dispersibility of the protein particles at lower printing temperature. More compact and smooth products were obtained when SPI was added to gelatin and sodium alginate. With mixing of 0.5% sodium alginate and 6% of gelation in the SPI paste, the product formed at printing temperatures of 35°C and 45°C showed small pores in the microstructure as compared to the products at 25°C.

This happened owing to the higher mechanical strength and viscosity of the formulation at 25°C that caused loose bonding between the layers of the products. The addition of gelatin in the paste helped in binding the protein particles tightly. In paste having 0.5% sodium alginate and 2% gelatin, it was noticed that a thin film of gelatin was found surrounding the interconnected protein particles. Hence, gelatin supported a stable 3D structure by entrapping the protein particles in the sample.

Agarwal, Wallace [44] mentioned the notable impact of ingredient and 3DP process on the microstructure of snacks developed from lupin and chickpea flour and pea protein isolate. The chickpea flour-based 3D printed snacks showed dense protein-starch network in contrast to the conventionally processed snacks whereas the reverse happened to the lupin flour-based snacks showing more dense protein-starch interaction by the conventional process. When comparison was made in between only 3D printed snacks of chickpea flour and lupin flour, larger air gaps were found in lupin flour-based snacks as compared to the chickpea flour. This might be happened due to the variation of particle size in both the flours causing more micropore generation in the microstructure because of which more water loss occurred forming larger air gaps. Agarwal, Wallace [44] compared the microstructure of 3D printed and conventionally produced lupin and chickpeas flour-based snacks by taking the images using non-invasive XRT. The sizes of the air gaps were clearly noticeable in the XRT images shown in Figure 9.4. There

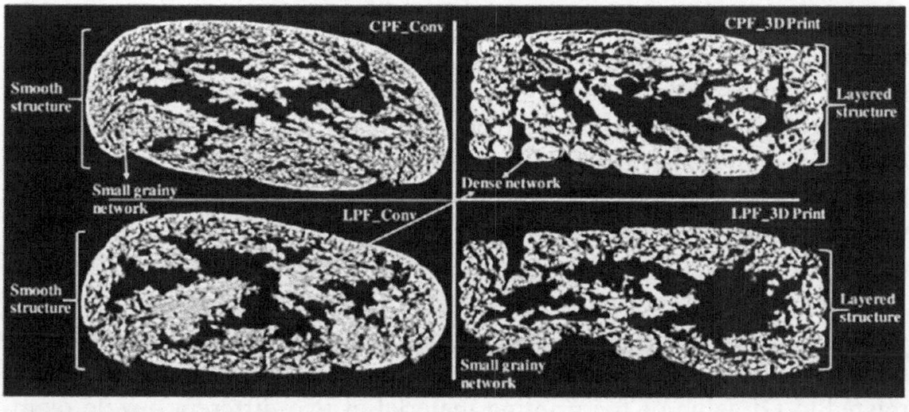

FIGURE 9.4 X-ray micro tomography images of lupin and chickpea flour based conventionally and 3D printed snacks after baking.

Source: Reprinted from Ref. [43]. Copyright Elsevier 2022. https://creativecommons.org/licenses/by/4.0/

*Conv: Conventional, CPF: Chickpea flour, LPF: Lupin flour.

were larger air gaps found in the 3D printed snacks as compared to the conventional one. Lupin flour-based snacks produced larger air pore in the microstructure as compared to the chickpea flour for both processing. The larger air gap might be attributed to the higher particle size of lupin flour resulting in higher moisture loss during baking due to the formation of larger micropores.

9.3.5 RHEOLOGICAL CHARACTERISTICS OF PRINTING FORMULATION

3D printing-based extrusion technology for legume and vegetable based ready to eat snacks depends on supplying of printing formulation through an orifice at higher pressure which can be accomplished by using screw, piston or pneumatics [59]. The flowability of the printing formulation depends on the shear stress applied on the formulation and determines the suitability of the printing materials for the printing operation [60, 61]. The flowability of the printing material can be studied by using rheological equipment. Rheological behavior also depends on the force exerted on the material at the time of printing. In a extrusion-based 3D printer, the printing material is undergone pressure drop while pushing from a larger orifice to smaller orifice through the feed syringe barrel. The shear rate and shear stress correlation study of the printed material can be included in the rheological characterization [62]. In addition, stress sweep test is conducted under oscillatory measurement at varying shear rate to get the maximum storage and loss modulus corresponding to linear viscoelastic region of the printing material [63]. Keerthana, Anukiruthika [64] reported the effect of addition of mushroom powder on the viscosity and shear profile of the printing material to develop fiber rich snacks. The shear thinning behavior of the formulation was noticed by decreasing the viscosity with the rise in the shear rate. It was observed that 20% mushroom powder in the formulation provided better flowability and printability for the process. Zhu, Stieger [65] reported the correlation between shear at collapse of the printed snacks and flow stress, storage modulus (G') and loss modulus (G") of the printing materials. The stress value at rupture was directly related to G' value of the printing material at two different nozzle diameters, i.e., 0.8 mm and 1.2 mm. The storage modulus defines the rigidity, i.e., the elastic property of the printing materials whereas the G" shows the viscous nature of the materials. The rheological characteristics of the food ink for preparing of air fried potato snacks was reported by Liu, Dick

[50]. The food ink showed shear thinning nature which was needed for the printing process. The higher storage modulus showed high mechanical strength of the food ink proving self-support to printing material for better shape stability. Liu, Zhang [63] observed that the addition of 2% massed potato along with potato starch showed good printability and extrudability having smooth shape and better stability. The yield stress and consistency index were found to be 312.16 Pa and 118.44 $Pa.s^n$ respectively, for the snacks printing. In another study conducted by Agarwal, Wallace [44], the Visco-elastic gel nature of the formulation prepared by lupin flour and chickpea flour with oil and without oil was reported by plotting the value of G' and G" of the 3D printing formulation against shear strain. Figure 9.5 showed the plot of G' and G" against the shear strain where the G' value was noticed to be more as compared to G" indicating the predominance of elastic characteristic of formulation over fluid character. Hence, rheological characteristics have crucial role in the printing process of the 3DP snacks. These properties also affect the internal structure, shape, texture, and strength of the developed snacks.

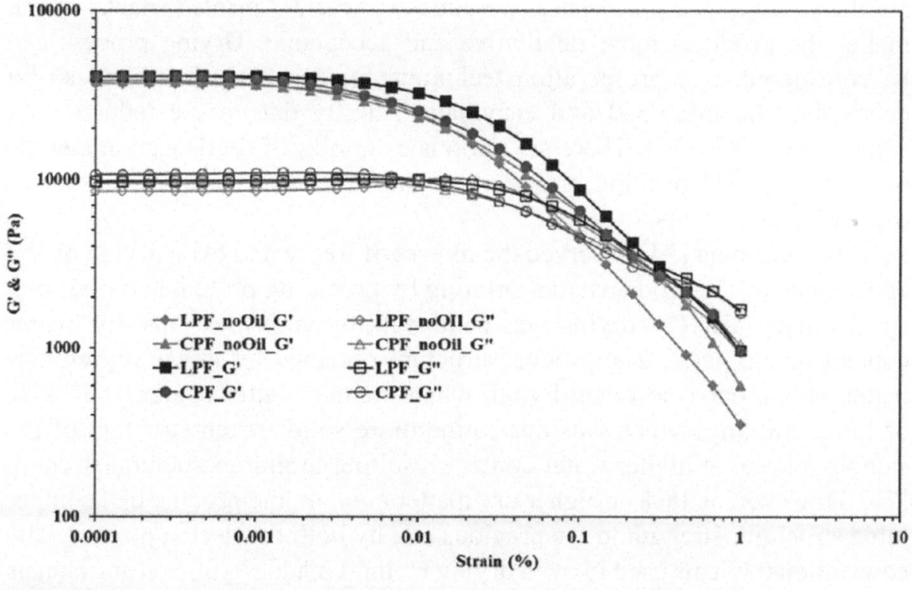

FIGURE 9.5 Plot of storage modulus (G') and loss modulus (G") against shear strain of lupin and chickpea-based 3D printed formulation with and without oil.

Source: Reprinted from Ref. [43]. Copyright Elsevier 2022. https://creativecommons.org/licenses/by/4.0/

*CPF: Chickpea flour, LPF: Lupin flour, no Oil: Without addition of oil.

9.4 POST-PRINTING OPERATIONS FOR DEVELOPMENT OF SNACKS

Very less percentage of the 3D printed snacks can be consumed directly, however majority of them are treated with various post-deposition treatment like drying, cooling, frying and baking, etc., which results in the change of textural and structural characteristics of the products due to the penetration of heat at various level [66, 67]. Protein denaturation, change in nutritional value, color, volume, and moisture reduction of the printed snacks are some of the phenomena that are noticed after the post-printing operations. The stability of the snacks is improved by the post-printing operations and the product characteristics obtained by 3D food printing is superior to the other conventional manufacturing process.

9.4.1 *DRYING*

Freeze drying, vacuum drying, oven drying, and vacuum microwave drying are the value-added processing operations for the 3D printed snacks which makes the products more qualitative and acceptable. Drying process can be considered as a preservation technique for the 3D printed snacks by decreasing the microbial and enzymatic activity due to the reduction of water content [68–70]. There are various examples of food snack materials processed by 3D printing and then, drying operation is applied as post-deposition treatments.

Lille, Nurmela [24] observed the impact of freeze and oven drying on the appearance of the products after printing by preparing protein and fiber rich snacks materials. The drying was conducted by taking various dry matter content of the paste. The product structure obtained by freeze drying was better as compared to oven drying, when the dry matter content was 35% or lower initially, which was due to the more solid frozen structure of the sample formed at higher water content resulting to more structural strength [71]. However, at 45% or higher dry matter content, the product had comparable structural strength to the previous one by both the drying process. The conventional hot air used in oven drying method made significant shrinkages in the products [72]. The experiment performed by Yang, Zhang [69] for the preparation of 3D printed mango gel by taking potato starch and concentrated mango juice as the formulation ingredients where microwave vacuum drying was used for the post processing treatment. As compared to the treatment of 0 min and 2 min, microwave vacuum drying of 4 min resulted in smooth surface and very less deformed products with higher shape accuracy. The combination of 3D printing and post-printing drying process develops customized probiotic-based products. Yoha, Anukiruthika [73] developed 3D printing manufacturing of probiotics encapsulated by different methods like spray drying, freeze drying and spray-freeze drying to develop functionalized products where post-printing operations like microwave drying, refractance window drying, freeze drying and hot air oven drying were considered to assess the viability of the probiotic strain in the product. It was found from the report that 3D printing had no bad effect on probiotic viability and the survival rate obtained for all the encapsulated probiotics was found to be 98–99% after the printing operation. Freeze drying was the best post-printing operation in terms of the survival rate of probiotic incorporated in the 3D printed products. Kuo, Clark [74] studied an integrated manufacturing process of probiotic encapsulation, 3DP process and post-printing operation like freeze drying by developing stable, functionalized snacks or supplement materials rich in probiotics like Lactobacillus and Bifidobacterium. The post-processing freeze-drying operation modified the printed product by enhancing the hardness and reducing the water activity due to the change of the semi-solid state of the product to solid like structure. The printing process can change the textural property of the post-printing dried snacks by controlling their internal structure. Liu and Zhang [75] studied the design of infill structure of 3DP potato-based snacks. Microwave vacuum drying was the best post-printing method to obtain promising textural properties by modifying the infill structure. More research data is still required to verify the accuracy of the process. However, drying is proved as the promising post-printing technique to provide more acceptable 3D printed ready to eat snacks with best structural and textural attributes.

9.4.2 FRYING

Deep fried snacks with more oil absorption have bad impact on human health like high cholesterol, obesity and overweight. Novel frying technology has come out to solve the problem of high oil absorption of the deep-fried food [76]. Now a days, vacuum frying technology has the advancement of frying food without the presence of air which can be accomplished with low temperature in comparison with the conventional deep frying [77]. Recently, the fried snacks quality has been improved by the use of vacuum frying in combination with microwave and ultrasound technologies [78]. According to the various heat transfer sources, some emerging frying techniques have been developed. They include electric field frying [79], radiation frying [80], spray frying [81], oil-water mixture frying [82], air frying [83], etc. Each frying process gives different product characteristics depending on the application principle where frying time, temperature and product microstructure are the main process parameter [84]. Many researchers used to fry as the post-processing technique after 3D printing of various snacks and interpreted the impact of frying on appearance, textural and structural attributes of the food.

Many works have been conducted to develop 3D printed snacks by considering the impact of infill levels and structures on the texture of the products [85–87]. However, more studies should be needed on the modifications of the printing factors on the texture of fried snacks. Feng, Zhang [42] prepared air fried yam snacks enriched with fiber using 3DP by adding potato processing byproduct at various proportions considering the filling level and infill structures as the printing variables. Printing variables influenced the textural attributes of air fried snacks. With rise in the infill level, the porosity of the snacks decreased causing the product structure to become denser which resulted in the requirement for more breaking force to break the fried product. The change in hardness and porosity of the fried snacks were affected by different filling structures. The higher the hardness of the fried snacks, the lower the porosity of the snacks, which gives the idea for the utilization of printing technology to develop air fried snacks with modified texture. Liu, Dick [50] applied air frying technique to dry potato snacks and observed the impact on the textural attributes of snacks by its internal structure. The decrease in the infill level reduced the hardness and fracturability of the snacks which was also influenced by the infill pattern. The change in the internal structure of the 3D printed air fried snacks could modify the textural attributes of the snacks. Hence, air frying application produces low oil absorbed snacks with personalized internal structure when combines with 3D printing retaining the oil soluble vitamins with best quality product [88]. Also, plant proteins were used to make steak like snacks by using 3D printers where textured soybean protein or drawing soy protein and hydrocolloids (guar gum, xanthangum, hydroxyethyl cellulose and sodium alginate) were used as the ingredient materials [89]. Different hydrocolloids showed different gelling characteristics after the application of frying as a post-processing method. Among them, xanthan gum in combination with textured soybean protein showed the best product characteristics after frying and product shape was maintained as the shape before frying. Xanthan and guar gum also lowered the absorbed oil by the fried snacks [90, 91]. The textural characteristics of the fried snacks were affected by various infill rates and pattern. The fried sample had the closest chewiness, gumminess, and hardness with respect to the control sample when the triangular infill pattern was used at a 60% infill rate.

9.4.3 BAKING

Among the post-processing methods, the biochemical and physiochemical reactions happening in the baking process make the 3D printed ready-to-eat snacks more palatable and digestible. Baking makes the 3D printed products more uniform and retains the bioactive compounds intact resulting in better nutrient profile of the snacks as compared to other cooking methods, especially deep oil frying [92–95]. Many researchers reported that baking is the most preferable post-printing operation for making ready to eat snacks despite other methods like steaming, air frying, deep oil frying, broiling and infrared cooking are used for cooking of various cookies formulations [47, 96–98]. Lille, Kortekangas [99] prepared a dietary fiber and protein rich paste to prepare 3D printed snacks by taking milk powder and whole grain rye flour. The proportion of each ingredient in the paste depends on the smoothness of extrusion, stability, and accuracy of the printing pattern. The post-printing baking operation was carried out by using oven at 150°C. Baking caused the total collapse of the multiple layers of the milk powder basked snacks due to vigorous expansion, however, adding rye flour to milk powder helped in maintaining the shape and appearance of the snacks. With the increase in the milk powder proportion in the paste, the sensory properties of snacks were also improved. It was observed that the fracturability of single layered product made from the mixture of both rye flour and milk powder was higher than the product made only from one ingredient, i.e., rye flour/milk powder. The texture of the snacks was influenced by the synergetic effect of various factors like composition of solid phage, intermediate porosity, etc. Some researchers also studied the baking parameter such as baking temperature and time on the quality of the 3DP products. Guénard-Lampron, Masson [47] reported the impact of nozzle diameter, filling rate and baking parameters (baking temperature, time) on the printing process of paste made from carrot puree to produce carrot appetizer cake. The optimized values obtained for the above parameters were 3.4 mm nozzle diameter, 71% filling rate and baking temperature of 177°C for a duration of 25 min. Prolonged baking process or baking at higher temperature in association with the nozzle diameter caused influence in the textural properties of the developed snacks. For instance, the use of higher diameter nozzle (5 mm) reduced from forming harder crust of the snacks due to higher temperature of baking at 195°C or for longer time (25 min) of baking. The decrease in water content of the cake due to baking influenced the textural properties. The excessive reduction in water content due to longer baking time and higher baking temperature resulted in a pore percentage, harder crust, and firmness inside the cake. Swelling and deformation of the developed cake were enhanced after baking due to arise in filling rate, which is associated with the improvement in mass after printing. Deep insights into the factors like filling percentage and filling pattern affecting the instability in the shape and appearance of 3D printed snacks after baking are needed to design superior quality of ready to eat snacks products. Pulatsu, Su [100] conducted research on the interaction of baking condition, mechanical properties, and structural changes for restructuring the cookies dough by 3D food printing. The deformation in shape at different baking factors and geometrical parameters were studied. Baking at higher temperatures caused breaking and cracking among the layers of the structure which might be attributed to the enhanced drying rate at the surface of the product. Various geometrical shapes of the snacks like cubic, cylindrical, or triangular at an infill percentage of 30% or 50% showed better structural accuracy without the formation of collapse. Hence, a better understanding of the ingredient behavior and post-processing variables are the prerequisites for the development of best quality 3DP RTE snacks.

9.4.4 COOLING

The cooling process is applied in 3D food printing for making cold foods like 3D printed desserts, ice-cream, cake decoration and many others. Besides, the distinctive aesthetic design of the cake, 3D printing can resolve the issues of variability in the design of individual decorators, enhance the effectiveness of the cake decoration and, therefore, lower the processing and production cost of pastry, cake and cream [101]. The freeze-thaw dripping loss is a great problem for vegetables and fruits due to their increased moisture level which is mainly based on the position, configuration and size of ice crystals generated at the time of freezing and thawing [102]. Therefore, high moisture content is the biggest problem for 3D printing systems having vegetable-based materials. Huang, Zhang [103] developed vegetable-based gel as a food ink for printing comprising of carrot powder, cabbage puree and xanthan gum as the ingredients where the freeze-thaw dripping loss was improved due to the printing process by controlling the sample porosity. There was a significant ($p<0.05$) correlation between freeze-thaw loss and various percentage of porosity. The percentage of drip loss was increased due to the increase in porosity of the printed snacks and the snacks porosity could be reduced by increasing the percentage of infill without requiring any post-printing operation. Hence, the frozen storage of high moisture printed products with minimum drip loss was only possible by decreasing the porosity of the printed materials with the increase in percentage of infill by using 3D printing process. However, a greater amount of research needs to be conducted to get concrete ideas on the impact of freezing and cooling process on 3D printed snacks and other products.

9.5 CONCLUSION

3D food printing, a novel technique, is used to produce novel food products like functionalized and customized ready to eat snacks by considering a range of ingredients such as legumes, vegetables, cereals, dairy materials maintaining sustainable environmental practices. The 3DP process makes unique modifications to structural, textural, and nutritional characteristics of the ready to eat snacks which may not be possible in the conventional manufacturing process. Legume and vegetable based ready to eat snacks would benefit more because of their higher protein, fiber, and antioxidant level. The impact of individual factors like printing temperature, printing time, nozzle diameter and their interaction on flowability, rheology and ingredient properties of the printing paste plays a crucial role in obtaining better structured and textured products. The by-products of legume crops and vegetables can also be used in the production of functionalized snacks which not only helps in the proper utilization of the food waste but also reduces their adverse effect on the environment. The most recent applications of printing technology are the encapsulation of probiotics, bioactive compounds, and other functional component. Apart from the above technological advancement, the regulatory requirement of individual country about microbial safety is of major concern for using 3DP technology to produce RTE snacks. The utilization of non-thermal or hurdle technologies as pretreatment or post-printing operation of 3D printed snacks can enhance their microbial safety and security. However, an extensive study should be conducted to obtain a concrete idea of the impact of these technologies on the microbial safety of the 3DP snacks.

KEYWORDS

- **3D printing technology**
- **food inks**
- **printability**
- **printing parameters**
- **ready-to-eat snacks**

REFERENCES

1. Nachal, N., Moses, J. A., Karthik, P., & Anandharamakrishnan, C. (2019). Applications of 3D printing in food processing. *Food Engineering Reviews, 11*(3), 123–141. https://doi.org/10.1007/s12393-019-09198-7.
2. Nida, S., Anukiruthika, T., Moses, J. A., & Anandharamakrishnan, C. (2020). 3D printing of grinding and milling fractions of rice husk. *Waste and Biomass Valorization, 12*(1), 81–90. https://doi.org/10.1007/s12649-020-00986-9.
3. Anukiruthika, T., Moses, J. A., & Anandharamakrishnan, C. (2020). 3D printing of egg yolk and white with rice flour blends. *Journal of Food Engineering, 265*, 109691. https://doi.org/10.1016/j.jfoodeng.2019.109691.
4. Liu, Z., Zhang, M., Bhandari, B., & Wang, Y. (2017). 3D printing: Printing precision and application in food sector. *Trends in Food Science & Technology, 69*, 83–94. https://doi.org/10.1016/j.tifs.2017.08.018.
5. Shubham, K., Anukiruthika, T., Dutta, S., Kashyap, A. V., Moses, J. A., & Anandharamakrishnan, C. (2020). Iron deficiency anemia: A comprehensive review on iron absorption, bioavailability and emerging food fortification approaches. *Trends in Food Science & Technology, 99*, 58–75. https://doi.org/10.1016/j.tifs.2020.02.021.
6. Liu, J., Sun, L., Xu, W., Wang, Q., Yu, S., & Sun, J. (2019). Current advances and future perspectives of 3D printing natural-derived biopolymers. *Carbohydrate Polymers, 207*, 297–316. https://doi.org/10.1016/j.carbpol.2018.11.091.
7. Schutyser, M. A. I., Houlder, S., de Wit, M., Buijsse, C. A. P., & Alting, A. C. (2018). Fused deposition modelling of sodium caseinate dispersions. *Journal of Food Engineering, 220*, 49–55. https://doi.org/10.1016/j.jfoodeng.2017.09.015.
8. Severini, C., Derossi, A., Ricci, I., Caporizzi, R., & Fiore, A. (2018). Printing a blend of fruit and vegetables. New advances on critical variables and shelf life of 3D edible objects. *Journal of Food Engineering, 220*, 89–100. https://doi.org/10.1016/j.jfoodeng.2017.04.026.

9. Autho. (2017). Simulation of 3D food printing extrusion and deposition. In *Nanosensors, Biosensors, Info-Tech Sensors and 3D Systems* (pp. 1016717).
10. Yang, F., Zhang, M., & Bhandari, B. (2015). Recent development in 3D food printing. *Critical Reviews in Food Science and Nutrition, 57*(14), 3145–3153. https://doi.org/10.1080/10408398.2015.1095154.
11. Wegrzyn, T. F., Golding, M., & Archer, R. H. (2012). Food layered manufacture: A new process for constructing solid foods. *Trends in Food Science & Technology, 27*(2), 66–72. https://doi.org/10.1016/j.tifs.2012.04.006.
12. Tomašević, I., Putnik, P., Valjak, F., Pavlić, B., Šojić, B., Bebek Markovinović, A., & Bursać Kovačević, D. (2021). 3D printing as a novel tool for fruit-based functional food production. *Current Opinion in Food Science, 41*, 138–145. https://doi.org/10.1016/j.cofs.2021.04.004.
13. Singha, P., Singh, S. K., Muthukumarappan, K., & Krishnan, P. (2018). Physicochemical and nutritional properties of extrudates from food grade distiller's dried grains, garbanzo flour, and corn grits. *Food Science & Nutrition, 6*(7), 1914–1926. https://doi.org/10.1002/fsn3.743.
14. Singh, P., Arora, A., Strand, T. A., Leffler, D. A., Catassi, C., Green, P. H., Kelly, C. P., Ahuja, V., & Makharia, G. K. (2018). Global prevalence of celiac disease: Systematic review and meta-analysis. *Clinical Gastroenterology and Hepatology, 16*(6), 823–836.e2. https://doi.org/10.1016/j.cgh.2017.06.037.
15. Nardo, G. D., Villa, M. P., Conti, L., Ranucci, G., Pacchiarotti, C., Principessa, L., Raucci, U., & Parisi, P. (2019). Nutritional deficiencies in children with celiac disease resulting from a gluten-free diet: A systematic review. *Nutrients, 11*(7), 1588. https://doi.org/10.3390/nu11071588.
16. Rybicka, I., Doba, K., & Bińczak, O. (2019). Improving the sensory and nutritional value of gluten-free bread. *International Journal of Food Science & Technology, 54*(9), 2661–2667. https://doi.org/10.1111/ijfs.14211.
17. Lian, H., Luo, K., Gong, Y., Zhang, S., & Serventi, L. (2019). Okara flours from chickpea and soy are thickeners: Increased dough viscosity and moisture content in gluten-free bread. *International Journal of Food Science & Technology, 55*(2), 805–812. https://doi.org/10.1111/ijfs.14347.
18. Giri, N. A., & Sakhale, B. K. (2019). Development of sweet potato flour based high protein and low-calorie gluten-free cookies. *Current Research in Nutrition and Food Science Journal, 7*(2), 427–435. https://doi.org/10.12944/CRNFSJ.7.2.21.
19. de Lima, B. R., Botelho, R. B. A., & Zandonadi, R. P. (2017). Gluten-free pasta: Replacing wheat with chickpea. *Journal of Culinary Science & Technology, 17*(1), 1–8. https://doi.org/10.1080/15428052.2017.1333936.
20. Bartkiene, E., Juodeikiene, G., Vidmantiene, D., Viskelis, P., & Urbonaviciene, D. (2011). Nutritional and quality aspects of wheat sourdough bread using *L. luteus* and *L. angustifolius* flours fermented by *Pediococcus acidilactici*. *International Journal of Food Science & Technology, 46*(8), 1724–1733. https://doi.org/10.1111/j.1365-2621.2011.02679.x.
21. Paraskevopoulou, A., Provatidou, E., Tsotsiou, D., & Kiosseoglou, V. (2010). Dough rheology and baking performance of wheat flour–Lupin protein isolate blends. *Food Research International, 43*(4), 1009–1016. https://doi.org/10.1016/j.foodres.2010.01.013.

22. Levent, H., & Bilgiçli, N. (2011). Enrichment of gluten-free cakes with lupin (*Lupinus albus* L.) or buckwheat (*Fagopyrum esculentum* M.) flours. *International Journal of Food Sciences and Nutrition, 62*(7), 725–728. https://doi.org/10.3109/09637486.2011.582460.
23. Martin, A., Schmidt, V., Osen, R., Bez, J., Ortner, E., & Mittermaier, S. (2021). Texture, sensory properties and functionality of extruded snacks from pulses and pseudocereal proteins. *Journal of the Science of Food and Agriculture, 101*(12), 5011–5021. https://doi.org/10.1002/jsfa.11273.
24. Lille, M., Nurmela, A., Nordlund, E., Metsä-Kortelainen, S., & Sozer, N. (2018). Applicability of protein and fiber-rich food materials in extrusion-based 3D printing. *Journal of Food Engineering, 220*, 20–27. https://doi.org/10.1016/j.jfoodeng.2017.02.007.
25. Chen, Y., Zhang, M., & Phuhongsung, P. (2021). 3D printing of protein-based composite fruit and vegetable gel system. *LWT, 141*, 110978. https://doi.org/10.1016/j.lwt.2021.110978.
26. Godoi, F. C., Prakash, S., & Bhandari, B. R. (2016). 3D printing technologies applied for food design: Status and prospects. *Journal of Food Engineering, 179*, 44–54. https://doi.org/10.1016/j.jfoodeng.2016.01.025.
27. Hao, L., Mellor, S., Seaman, O., Henderson, J., Sewell, N., & Sloan, M. (2010). Material characterisation and process development for chocolate additive layer manufacturing. *Virtual and Physical Prototyping, 5*(2), 57–64. https://doi.org/10.1080/17452751003753212.
28. Lanaro, M., Forrestal, D. P., Scheurer, S., Slinger, D. J., Liao, S., Powell, S. K., & Woodruff, M. A. (2017). 3D printing complex chocolate objects: Platform design, optimization and evaluation. *Journal of Food Engineering, 215*, 13–22. https://doi.org/10.1016/j.jfoodeng.2017.06.017.
29. Le Tohic, C., O'Sullivan, J. J., Drapala, K. P., Chartrin, V., Chan, T., Morrison, A. P., Kerry, J. P., & Kelly, A. L. (2018). Effect of 3D printing on the structure and textural properties of processed cheese. *Journal of Food Engineering, 220*, 56–64. https://doi.org/10.1016/j.jfoodeng.2017.06.008.
30. Li, P., Mellor, S., Griffin, J., Waelde, C., Hao, L., & Everson, R. (2014). Intellectual property and 3D printing: A case study on 3D chocolate printing. *Journal of Intellectual Property Law & Practice, 9*(4), 322–332. https://doi.org/10.1093/jiplp/jpu048.
31. Wang, L., Zhang, M., Bhandari, B., & Yang, C. (2018). Investigation on fish surimi gel as a promising food material for 3D printing. *Journal of Food Engineering, 220*, 101–108. https://doi.org/10.1016/j.jfoodeng.2017.06.017.
32. Sun, J., Peng, Z., Zhou, W., Fuh, J. Y. H., Hong, G. S., & Chiu, A. (2015). A review on 3D printing for customized food fabrication. *Procedia Manufacturing, 1*, 308–319. https://doi.org/10.1016/j.promfg.2015.09.057.
33. Gong, J., Shitara, M., Serizawa, R., Makino, M., Kabir, M. H., & Furukawa, H. (2014). 3D printing of meso-decorated gels and foods. *Materials Science Forum, 783–786*, 1250–1254. https://doi.org/10.4028/www.scientific.net/MSF.783-786.1250.
34. Kim, H. W., Bae, H., & Park, H. J. (2017). Classification of the printability of selected food for 3D printing: Development of an assessment method using hydrocolloids as reference material. *Journal of Food Engineering, 215*, 23–32. https://doi.org/10.1016/j.jfoodeng.2017.07.017.

35. Baiano, A. (2020). 3D printed foods: A comprehensive review on technologies, nutritional value, safety, consumer attitude, regulatory framework, and economic and sustainability issues. *Food Reviews International, 38*(5), 986–1016. https://doi.org/10.1080/87559129.2020.1762091.
36. Sun, J., Zhou, W., Huang, D., Fuh, J. Y. H., & Hong, G. S. (2015). An overview of 3D printing technologies for food fabrication. *Food and Bioprocess Technology, 8*(8), 1605–1615. https://doi.org/10.1007/s11947-015-1528-6.
37. Chen, J., Mu, T., Goffin, D., Blecker, C., Richard, G., Richel, A., & Haubruge, E. (2019). Application of soy protein isolate and hydrocolloids-based mixtures as promising food material in 3D food printing. *Journal of Food Engineering, 261*, 76–86. https://doi.org/10.1016/j.jfoodeng.2019.05.018.
38. Chuanxing, F., Qi, W., Hui, L., Quancheng, Z., & Wang, M. (2018). Effects of pea protein on the properties of potato starch-based 3D printing materials. *International Journal of Food Engineering, 14*(3), 20170297. https://doi.org/10.1515/ijfe-2017-0297.
39. Derossi, A., Caporizzi, R., Azzollini, D., & Severini, C. (2018). Application of 3D printing for customized food. A case on the development of a fruit-based snack for children. *Journal of Food Engineering, 220*, 65–75. https://doi.org/10.1016/j.jfoodeng.2017.04.034.
40. Lee, J. H., Won, D. J., Kim, H. W., & Park, H. J. (2019). Effect of particle size on 3D printing performance of the food-ink system with cellular food materials. *Journal of Food Engineering, 256*, 1–8. https://doi.org/10.1016/j.jfoodeng.2019.01.021.
41. Krishnaraj, P., Anukiruthika, T., Choudhary, P., Moses, J. A., & Anandharamakrishnan, C. (2019). 3D extrusion printing and post-processing of fiber-rich snack from indigenous composite flour. *Food and Bioprocess Technology, 12*(10), 1776–1786. https://doi.org/10.1007/s11947-019-02368-6.
42. Feng, C., Zhang, M., Bhandari, B., & Ye, Y. (2020). Use of potato processing by-product: Effects on the 3D printing characteristics of the yam and the texture of air-fried yam snacks. *LWT, 125*, 109265. https://doi.org/10.1016/j.lwt.2020.109265.
43. Chen, J., Sun, H., Mu, T., Blecker, C., Richel, A., Richard, G., Jacquet, N., Haubruge, E., & Goffin, D. (2022). Effect of temperature on rheological, structural, and textural properties of soy protein isolate pastes for 3D food printing. *Journal of Food Engineering, 323*, 110917. https://doi.org/10.1016/j.jfoodeng.2022.110917.
44. Agarwal, D., Wallace, A., Kim, E. H. J., Wadamori, Y., Feng, L., Hedderley, D., & Morgenstern, M. P. (2022). Rheological, structural and textural characteristics of 3D-printed and conventionally-produced gluten-free snack made with chickpea and lupin flour. *Future Foods, 5*, 100134. https://doi.org/10.1016/j.fufo.2022.100134.
45. Bulut, E. G., & Candoğan, K. (2022). Development and characterization of a 3D printed functional chicken meat-based snack: Optimization of process parameters and gelatin level. *LWT, 154*, 112768. https://doi.org/10.1016/j.lwt.2021.112768.
46. He, C., Zhang, M., & Guo, C. (2020). 4D printing of mashed potato/purple sweet potato puree with spontaneous color change. *Innovative Food Science & Emerging Technologies, 59*, 102250. https://doi.org/10.1016/j.ifset.2019.102250.
47. Guénard-Lampron, V., Masson, M., Leichtnam, O., & Blumenthal, D. (2021). Impact of 3D printing and post-processing parameters on shape, texture and microstructure of carrot appetizer cake. *Innovative Food Science & Emerging Technologies, 72*. https://doi.org/10.1016/j.ifset.2021.102744.

48. Lille, M., Kortekangas, A., Heiniö, R. L., & Sozer, N. (2020). Structural and textural characteristics of 3D-printed protein- and dietary fibre-rich snacks made of milk powder and wholegrain rye flour. *Foods, 9*(11), 1527. https://doi.org/10.3390/foods9111527.
49. Varghese, C., Wolodko, J., Chen, L., Doschak, M., Srivastav, P. P., & Roopesh, M. S. (2020). Influence of selected product and process parameters on microstructure, rheological, and textural properties of 3D printed cookies. *Foods, 9*(7), 917. https://doi.org/10.3390/foods9070917.
50. Liu, Z., Dick, A., Prakash, S., Bhandari, B., & Zhang, M. (2020). Texture modification of 3D printed air-fried potato snack by varying its internal structure with the potential to reduce oil content. *Food and Bioprocess Technology, 13*(3), 564–576. https://doi.org/10.1007/s11947-020-02411-z.
51. Dash, D. R., Pathak, S. S., & Pradhan, R. C. (2021). Extraction of oil from *Terminalia chebula* kernel by using ultrasound technology: Influence of process parameters on extraction kinetics. *Industrial Crops and Products, 171*, 113893. https://doi.org/10.1016/j.indcrop.2021.113893.
52. McClements, D. J. (2019). Food architecture: Building better foods. In *Future Foods* (pp. 27–60). Elsevier. https://doi.org/10.1016/B978-0-12-816793-4.00002-9.
53. Jagadiswaran, B., Alagarasan, V., Palanivelu, P., Theagarajan, R., Moses, J. A., & Anandharamakrishnan, C. (2021). Valorization of food industry waste and by-products using 3D printing: A study on the development of value-added functional cookies. *Future Foods, 4*, 100036. https://doi.org/10.1016/j.fufo.2021.100036.
54. Kumar, R., & Chambers, E. (2019). Understanding the terminology for snack foods and their texture by consumers in four languages: A qualitative study. *Foods, 8*(10), 484. https://doi.org/10.3390/foods8100484.
55. Derkach, S. R., Ilyin, S. O., Maklakova, A. A., Kulichikhin, V. G., & Malkin, A. Y. (2015). The rheology of gelatin hydrogels modified by κ-carrageenan. *LWT–Food Science and Technology, 63*(1), 612–619. https://doi.org/10.1016/j.lwt.2015.03.062.
56. Derkach, S. R., Voron'ko, N. G., Kuchina, Y. A., Kolotova, D. S., Gordeeva, A. M., Faizullin, D. A., Gusev, Y. A., Zuev, Y. F., & Makshakova, O. N. (2018). Molecular structure and properties of κ-carrageenan-gelatin gels. *Carbohydrate Polymers, 197*, 66–74. https://doi.org/10.1016/j.carbpol.2018.05.063.
57. Villarino, C. B. J., Jayasena, V., Coorey, R., Chakrabarti-Bell, S., & Johnson, S. K. (2015). The effects of Australian sweet lupin (ASL) variety on physical properties of flours and breads. *LWT – Food Science and Technology, 60*(1), 435–443. https://doi.org/10.1016/j.lwt.2014.06.038.
58. Yadav, R. B., Yadav, B. S., & Dhull, N. (2011). Effect of incorporation of plantain and chickpea flours on the quality characteristics of biscuits. *Journal of Food Science and Technology, 49*(2), 207–213. https://doi.org/10.1007/s13197-011-0280-3.
59. Guo, C., Zhang, M., & Bhandari, B. (2019). Model building and slicing in food 3D printing processes: A review. *Comprehensive Reviews in Food Science and Food Safety, 18*(4), 1052–1069. https://doi.org/10.1111/1541-4337.12443.
60. Liu, Y., Liang, X., Saeed, A., Lan, W., & Qin, W. (2019). Properties of 3D printed dough and optimization of printing parameters. *Innovative Food Science & Emerging Technologies, 54*, 9–18. https://doi.org/10.1016/j.ifset.2018.05.015.
61. Liu, Y., Yu, Y., Liu, C., Regenstein, J. M., Liu, X., & Zhou, P. (2019). Rheological and mechanical behavior of milk protein composite gel for extrusion-based 3D food printing. *LWT, 102*, 338–346. https://doi.org/10.1016/j.lwt.2018.12.052.

62. Coogan, T. J., & Kazmer, D. O. (2019). In-line rheological monitoring of fused deposition modeling. *Journal of Rheology, 63*(1), 141–155. https://doi.org/10.1122/1.5054642.
63. Liu, Z., Zhang, M., Bhandari, B., & Yang, C. (2018). Impact of rheological properties of mashed potatoes on 3D printing. *Journal of Food Engineering, 220*, 76–82. https://doi.org/10.1016/j.jfoodeng.2017.03.021.
64. Keerthana, K., Anukiruthika, T., Moses, J. A., & Anandharamakrishnan, C. (2020). Development of fiber-enriched 3D printed snacks from alternative foods: A study on button mushroom. *Journal of Food Engineering, 287*, 110116. https://doi.org/10.1016/j.jfoodeng.2020.110116.
65. Zhu, S., Stieger, M. A., van der Goot, A. J., & Schutyser, M. A. I. (2019). Extrusion-based 3D printing of food pastes: Correlating rheological properties with printing behavior. *Innovative Food Science & Emerging Technologies, 58*, 102214. https://doi.org/10.1016/j.ifset.2019.102214.
66. He, C., Zhang, M., & Fang, Z. (2019). 3D printing of food: Pretreatment and post-treatment of materials. *Critical Reviews in Food Science and Nutrition, 60*(14), 2379–2392. https://doi.org/10.1080/10408398.2019.1641065.
67. Sun, J., Zhou, W., Yan, L., Huang, D., & Lin, L. Y. (2018). Extrusion-based food printing for digitalized food design and nutrition control. *Journal of Food Engineering, 220*, 1–11. https://doi.org/10.1016/j.jfoodeng.2017.02.028.
68. Chong, C. H., Figiel, A., Law, C. L., & Wojdyło, A. (2013). Combined drying of apple cubes by using heat pump, vacuum-microwave, and intermittent techniques. *Food and Bioprocess Technology, 7*(4), 975–989. https://doi.org/10.1007/s11947-013-1094-2.
69. Yang, F., Zhang, M., & Liu, Y. (2018). Effect of post-treatment microwave vacuum drying on the quality of 3D-printed mango juice gel. *Drying Technology, 37*(14), 1757–1765. https://doi.org/10.1080/07373937.2018.1537365.
70. Nowacka, M., & Wedzik, M. (2016). Effect of ultrasound treatment on microstructure, color and carotenoid content in fresh and dried carrot tissue. *Applied Acoustics, 103*, 163–171. https://doi.org/10.1016/j.apacoust.2015.06.014.
71. Ratti, C. (2001). Hot air and freeze-drying of high-value foods: A review. *Journal of Food Engineering, 49*(4), 311–319. https://doi.org/10.1016/S0260-8774(01)00098-3.
72. Mayor, L., & Sereno, A. M. (2004). Modelling shrinkage during convective drying of food materials: A review. *Journal of Food Engineering, 61*(3), 373–386. https://doi.org/10.1016/S0260-8774(03)00144-X.
73. Yoha, K. S., Anukiruthika, T., Anila, W., Moses, J. A., & Anandharamakrishnan, C. (2021). 3D printing of encapsulated probiotics: Effect of different post-processing methods on the stability of *Lactiplantibacillus plantarum* (NCIM 2083) under static *in vitro* digestion conditions and during storage. *LWT, 146*, 111461. https://doi.org/10.1016/j.lwt.2021.111461.
74. Kuo, C. C., Clark, S., Qin, H., & Shi, X. (2022). Development of a shelf-stable, gel-based delivery system for probiotics by encapsulation, 3D printing, and freeze-drying. *LWT, 157*, 113075. https://doi.org/10.1016/j.lwt.2021.113075.
75. Liu, Z., & Zhang, M. (2021). Texture properties of microwave post-processed 3D printed potato snack with different ingredients and infill structure. *Future Foods, 3*, 100017. https://doi.org/10.1016/j.fufo.2021.100017.
76. Pankaj, S. K., & Keener, K. M. (2017). A review and research trends in alternate frying technologies. *Current Opinion in Food Science, 16*, 74–79. https://doi.org/10.1016/j.cofs.2017.06.004.

77. Fan, L. P., Zhang, M., & Mujumdar, A. (2005). Vacuum frying of carrot chips. *Drying Technology, 23*(3), 645–656. https://doi.org/10.1081/DRT-200054200.
78. Devi, S., Zhang, M., & Mujumdar, A. S. (2019). Influence of ultrasound and microwave-assisted vacuum frying on quality parameters of fried product and the stability of frying oil. *Drying Technology, 39*(5), 655–668. https://doi.org/10.1080/07373937.2019.1700961.
79. Yang, D., Wu, G., Lu, Y., Li, P., Qi, X., Zhang, H., Wang, X., & Jin, Q. (2021). Comparative analysis of the effects of novel electric field frying and conventional frying on the quality of frying oil and oil absorption of fried shrimps. *Food Control, 128*, 108195. https://doi.org/10.1016/j.foodcont.2021.108195.
80. Lloyd, B. J., Farkas, B. E., & Keener, K. M. (2004). Quality comparison of French fry style potatoes produced by oven heating, immersion frying and controlled dynamic radiant heating. *Journal of Food Processing and Preservation, 28*(6), 460–472. https://doi.org/10.1111/j.1745-4549.2004.tb00761.x.
81. Udomkun, P., Tangsanthatkun, J., & Innawong, B. (2020). Influence of process parameters on the physico-chemical and microstructural properties of rice crackers: A case study of novel spray-frying technique. *Innovative Food Science & Emerging Technologies, 59*, 102271. https://doi.org/10.1016/j.ifset.2019.102271.
82. Ma, R., Gao, T., Song, L., Zhang, L., Jiang, Y., Li, J., Zhang, X., Gao, F., & Zhou, G. (2016). Effects of oil-water mixed frying and pure-oil frying on the quality characteristics of soybean oil and chicken chop. *Food Science and Technology, 36*(2), 329–336. https://doi.org/10.1590/1678-457X.0004.
83. Zaghi, A. N., Barbalho, S. M., Guiguer, E. L., & Otoboni, A. M. (2019). Frying process: From conventional to air frying technology. *Food Reviews International, 35*(8), 763–777. https://doi.org/10.1080/87559129.2019.1588294.
84. Salamatullah, A. M., Ahmed, M. A., Alkaltham, M. S., Hayat, K., Aloumi, N. S., Al-Dossari, A. M., Al-Harbi, L. N., & Arzoo, S. (2021). Effect of air-frying on the bioactive properties of eggplant (*Solanum melongena* L.). *Processes, 9*(3), 435. https://doi.org/10.3390/pr9030435.
85. Liu, Z., Bhandari, B., Prakash, S., & Zhang, M. (2018). Creation of internal structure of mashed potato construct by 3D printing and its textural properties. *Food Research International, 111*, 534–543. https://doi.org/10.1016/j.foodres.2018.05.073.
86. Mantihal, S., Prakash, S., & Bhandari, B. (2019). Textural modification of 3D printed dark chocolate by varying internal infill structure. *Food Research International, 121*, 648–657. https://doi.org/10.1016/j.foodres.2018.12.053.
87. Severini, C., Derossi, A., & Azzollini, D. (2016). Variables affecting the printability of foods: Preliminary tests on cereal-based products. *Innovative Food Science & Emerging Technologies, 38*, 281–291. https://doi.org/10.1016/j.ifset.2016.10.001.
88. Heredia, A., Castelló, M. L., Argüelles, A., & Andrés, A. (2014). Evolution of mechanical and optical properties of French fries obtained by hot air-frying. *LWT–Food Science and Technology, 57*(2), 755–760. https://doi.org/10.1016/j.lwt.2014.01.044.
89. Chen, Y., Zhang, M., & Bhandari, B. (2021). 3D printing of steak-like foods based on textured soybean protein. *Foods, 10*(9), 2011. https://doi.org/10.3390/foods10092011.
90. Akdeniz, N., Sahin, S., & Sumnu, G. (2006). Functionality of batters containing different gums for deep-fat frying of carrot slices. *Journal of Food Engineering, 75*(4), 522–526. https://doi.org/10.1016/j.jfoodeng.2005.04.039.

91. Sahin, S., Sumnu, G., & Altunakar, B. (2005). Effects of batters containing different gum types on the quality of deep-fat fried chicken nuggets. *Journal of the Science of Food and Agriculture, 85*(14), 2375–2379. https://doi.org/10.1002/jsfa.2262.
92. Bainy, E. M., Bertan, L. C., Corazza, M. L., & Lenzi, M. K. (2014). Effect of grilling and baking on physicochemical and textural properties of tilapia (*Oreochromis niloticus*) fish burger. *Journal of Food Science and Technology, 52*(8), 5111–5119. https://doi.org/10.1007/s13197-014-1628-4.
93. De Pilli, T., & Alessandrino, O. (2018). Effects of different cooking technologies on biopolymers modifications of cereal-based foods: Impact on nutritional and quality characteristics review. *Critical Reviews in Food Science and Nutrition, 60*(4), 556–565. https://doi.org/10.1080/10408398.2018.1540078.
94. Escobedo, A., & Mojica, L. (2021). Pulse-based snacks as functional foods: Processing challenges and biological potential. *Comprehensive Reviews in Food Science and Food Safety, 20*(5), 4678–4702. https://doi.org/10.1111/1541-4337.12789.
95. Rico, D., González-Paramás, A. M., Brezmes, C., & Martín-Diana, A. B. (2020). Baking optimization as a strategy to extend shelf-life through the enhanced quality and bioactive properties of pulse-based snacks. *Molecules, 25*(16), 3716. https://doi.org/10.3390/molecules25163716.
96. Severini, C., Azzollini, D., Albenzio, M., & Derossi, A. (2018). On printability, quality and nutritional properties of 3D printed cereal-based snacks enriched with edible insects. *Food Research International, 106*, 666–676. https://doi.org/10.1016/j.foodres.2018.01.034.
97. Zhang, L., Lou, Y., & Schutyser, M. A. I. (2018). 3D printing of cereal-based food structures containing probiotics. *Food Structure, 18*, 14–22. https://doi.org/10.1016/j.foostr.2018.01.002.
98. Pulatsu, E., Su, J. W., & Lin, J. (2020). Factors affecting 3D printing and post-processing capacity of cookie dough. *Innovative Food Science & Emerging Technologies, 61*, 102316. https://doi.org/10.1016/j.ifset.2020.102316.
99. Lille, M., Kortekangas, A., Heiniö, R. L., & Sozer, N. (2020). Structural and textural characteristics of 3D-printed protein- and dietary fiber-rich snacks made of milk powder and wholegrain rye flour. *Foods, 9*(11), 1527. https://doi.org/10.3390/foods9111527.
100. Pulatsu, E., Su, J. W., Kenderes, S. M., Lin, J., Vardhanabhuti, B., & Lin, M. (2022). Restructuring cookie dough with 3D printing: Relationships between the mechanical properties, baking conditions, and structural changes. *Journal of Food Engineering, 319*, 110911. https://doi.org/10.1016/j.jfoodeng.2022.110911.
101. Demei, K., Zhang, M., Phuhongsung, P., & Mujumdar, A. S. (2022). 3D food printing: Controlling characteristics and improving technological effect during food processing. *Food Research International, 156*, 111120. https://doi.org/10.1016/j.foodres.2022.111120.
102. Jha, P. K., Chevallier, S., Xanthakis, E., Jury, V., & Le-Bail, A. (2020). Effect of innovative microwave assisted freezing (MAF) on the quality attributes of apples and potatoes. *Food Chemistry, 309*, 125594. https://doi.org/10.1016/j.foodchem.2019.125594.
103. Huang, Y., Zhang, M., & Pattarapon, P. (2022). Reducing freeze-thaw drip loss of mixed vegetable gel by 3D printing porosity. *Innovative Food Science & Emerging Technologies, 75*, 102893. https://doi.org/10.1016/j.ifset.2021.102893.

CHAPTER 10

VACUUM IMPREGNATION TECHNOLOGY FOR THE PRODUCTION OF READY-TO-EAT SNACKS

TARUNA VARGHESE[1] and N. VENKATACHALAPATHY[2]

[1]*PhD Scholar, Food Process Engineering, Department of Food Engineering, National Institute of Food Technology, Entrepreneurship, and Management, Thanjavur, Tamil Nadu, India*

[2]*Professor and Head, Department of Food Engineering, National Institute of Food Technology, Entrepreneurship, and Management, Thanjavur, Tamil Nadu, India*

ABSTRACT

The growing interest of scientists to exploit the microstructural changes of a food product since its physical and chemical characteristics are closely linked to its food matrix has led to the development of the concept of Food Matrix Engineering. Vacuum Impregnation is a vital operation in the food matrix study where the void properties of the food matrix are altered for incorporation or fortification of the required compound. This process is essentially used to fortify foods with preservatives, firming agents, cryoprotectants, anti-browning agents, probiotics, phenolic compounds, etc., that confers a wide range in the production of functional foods. It could also be used to produce fortified and functional ready-to-eat snacks that are healthy and shelf-stable without adding chemical or unhealthy preservatives.

Ready-to-Eat Snacks: Emerging Technologies for Production and Safety.
Ravi Pandiselvam and Gülşah Çalışkan Koç

10.1 INTRODUCTION

Ready-to-eat foods are a category of food products that greatly gained prime importance in today's growing fast-food culture. The increasingly hectic lifestyle has changed people's food preferences from unprocessed whole foods to processed ready-to-eat foods. These food products also expand the convenience of storage and improve sensory attributes, thus introducing newer varieties into the food services with good sensorial and textural characteristics. As per the US-FDA, Ready-to-eat food (RTE food) means any food that is normally eaten in its raw state or any other food, including processed food, for which it is reasonably foreseeable that the food will be eaten without further processing that would significantly minimize biological hazards [1]. Further, RTE foods are intended to perform basic preparative steps such as light washing, slicing, chopping, portioning, marinating or preservation. Snacks are portions of food smaller than a regular meal, generally eaten between meals. There is no standard definition for a snack, but it is defined based on the amount consumed and time consumed. Traditionally, snacks included savory chips and crisps produced by baking, frying and extrusion. The concern of society toward healthy snacks have included veggies and fruits in this category that could inculcate healthy food pattern for better living [2].

Vacuum or the use of sub-atmospheric pressure holds a significant role in the processing and preservation of food products. It is used in various applications, from drying, evaporation, filtration, and impregnation. Vacuum Impregnation of a porous product essentially involves an exchange of the internal gases or liquid occluded in the open pores and intercellular spaces with an external liquid phase, essentially by the action of hydrodynamic mechanisms (HDM) promoted by pressure changes [3, 4]. This process essentially involves the application of the vacuum phase followed by restoration of atmospheric pressure. The detailed mechanism of vacuum impregnation is explained in this chapter. Initially, this method was used for vacuum osmotic drying (VOD) to improve water loss and increase solid gain during drying. Later, this technology was extended to incorporate fortificants and other functional components such as cryoprotectants, osmotic agents, firming agents, acidulants, etc., onto the food. These techniques could also be used to apply edible coatings over food products [5–7]. In this manner, vacuum impregnation confers various benefits such as preservation, nutritional enrichment or fortification, and modification of sensory and textural properties [8]. This technology is mostly used for fruit and vegetable matrices and applied in the fish, meat and cheese industry [9]. This technology could also be appropriate for improved pretreatments for traditional preservation methods such as canning, freezing and drying.

This chapter explains in detail the method of vacuum impregnation and the various factors affecting vacuum impregnation. In subsequent sections, it explains the types of equipment and the modes of application of vacuum impregnation. It further explains the application of vacuum impregnation (VI) in producing ready-to-eat snacks. The advantages and disadvantages of the process and future opportunities and challenges are also addressed in this chapter.

10.2 MECHANISM OF VACUUM IMPREGNATION (VI)

As explained in the previous section, vacuum impregnation is the process by which a vacuum is applied to a porous food product that produces internal pressure or volume changes in its structure, succeeded by restoration of atmospheric pressure. The vacuum creates a suction pressure inside its structure together with the expulsion of liquid and internal gases from the food matrix. Atmospheric pressure restoration promotes absorption or impregnation of liquid solution in this porous structure. Thus, vacuum impregnation could be mainly divided into two processes: vacuum application and atmospheric pressure restoration. The process is well explained in Figure 10.1. The various operations of VI include:

- Immersion of food matrix to the impregnating solution.
- Application of vacuum to a particular level.
- The product is maintained for the specified time at that particular vacuum level.
- Atmospheric pressure restoration followed by immersion for a specific time
- Draining of the impregnating solution followed by stabilization techniques for fixing the desired compound onto the food matrix.

10.2.1 IMMERSION OF FOOD MATRIX TO THE IMPREGNATING SOLUTION

The food product is initially immersed in the impregnating solution. These are mostly immersed in an isotonic solution that prevents the movement of water molecules from and to the cell structure due to osmotic pressure caused by the concentration difference between the cell and the impregnating solution. Prevention of water movement to and from the cell structure is essential to maintain the cell integrity, which in turn preserves the texture and structure of the cellular matrix [10]. In the case where vacuum treatment is done as an aid for osmotic dehydration, a hypertonic solution is used that allows water outflow from the cell matrix. However, a slight amount of water ingress into the cell happens due to capillary action. Capillary flow is negligible and essentially happens through the narrow capillaries of the food matrix.

10.2.2 APPLICATION OF VACUUM TO A PARTICULAR LEVEL

Then vacuum pressure is applied to the food immersed in the impregnating solution. This creates a compression force acting on the food matrix such that all the internal gases and liquid present in the voids and capillaries of the food matrix get compressed. This causes considerable changes in the cell structure volume, although the food matrix's total integrity is maintained. This results in the formation of porous structures that increase the effective porosity of the food matrix. Thus, in this case, impregnation of the solution occurs due to hydrodynamic flow, where the pressure difference created between the impregnation fluid and the food matrix causes the flow. In this case, an increase in porosity creates channels for uptake of the impregnating solution into the food matrix [11, 12].

10.2.3 THE PRODUCT IS MAINTAINED FOR THE SPECIFIED TIME AT THAT PARTICULAR VACUUM LEVEL

The absorption or movement of the impregnating solution continues as per the vacuum pressure applied up to equilibrium pressure is maintained between the outside impregnating pressure and inside food matrix pressure.

10.2.4 ATMOSPHERIC PRESSURE RESTORATION FOLLOWED BY IMMERSION FOR A SPECIFIC TIME

Restoration of atmospheric pressure after the process of vacuum treatment produces a suction pressure inside the food matrix. This suction pressure promotes absorption or impregnation of the desired impregnation solution into the matrix. As absorption increases, the food product will be restored to its original dimensions with an impregnation solution entrapped in this food matrix. In such a manner, liquid impregnation occurs through the deformation-relaxation phenomenon [13, 14].

10.2.5 DRAINING OF THE IMPREGNATING SOLUTION FOLLOWED BY STABILIZATION TECHNIQUES FOR FIXING THE DESIRED COMPOUND ONTO THE FOOD MATRIX

The impregnation solution is finally drained by taking the food out of the impregnating solution so that the surface moisture and impregnating solution are drained out. It was followed by a suitable stabilization step such as drying that reduces the excess moisture present in the food matrix. Drying is usually done by vacuum drying or by use of convective drying at lower temperatures of 40ºC. This assures lesser structural and nutritional changes in the food matrix.

Thus, the major forces that come into play during the vacuum impregnation process include hydrodynamic force. The detailed method of vacuum impregnation is detailed in Figure 10.1. As shown in the figure, at food matrix remains intact before treatment. As the product is immersed in the impregnating solution, capillary action occurs. The solution impregnates into the food matrix due to the pressure difference between the inside of the food matrix and the impregnating solution (P_{liquid}>P_{inside}). As vacuum pressure is applied to the system, it creates a pressure difference between the sample and the impregnating solution. The liquid pressure intensifies the impregnation of the liquid to the food matrix through the hydrodynamic pressure. The application of vacuum also promotes compression or deformation of the food matrix. After process time t, atmospheric pressure is restored. Restoration of atmospheric pressure causes expansion of the deformed cell structure that creates a suction force causing the impregnation solution to adhere or absorb to the food matrix. It usually continues until an equilibrium pressure is maintained between the impregnating solution and the porous food structure is attained.

10.3 FACTORS AFFECTING VACUUM IMPREGNATION

Various factors affect the impregnation of the impregnating liquid onto the matrix. It mainly depends on the pressure levels applied to the food-solution

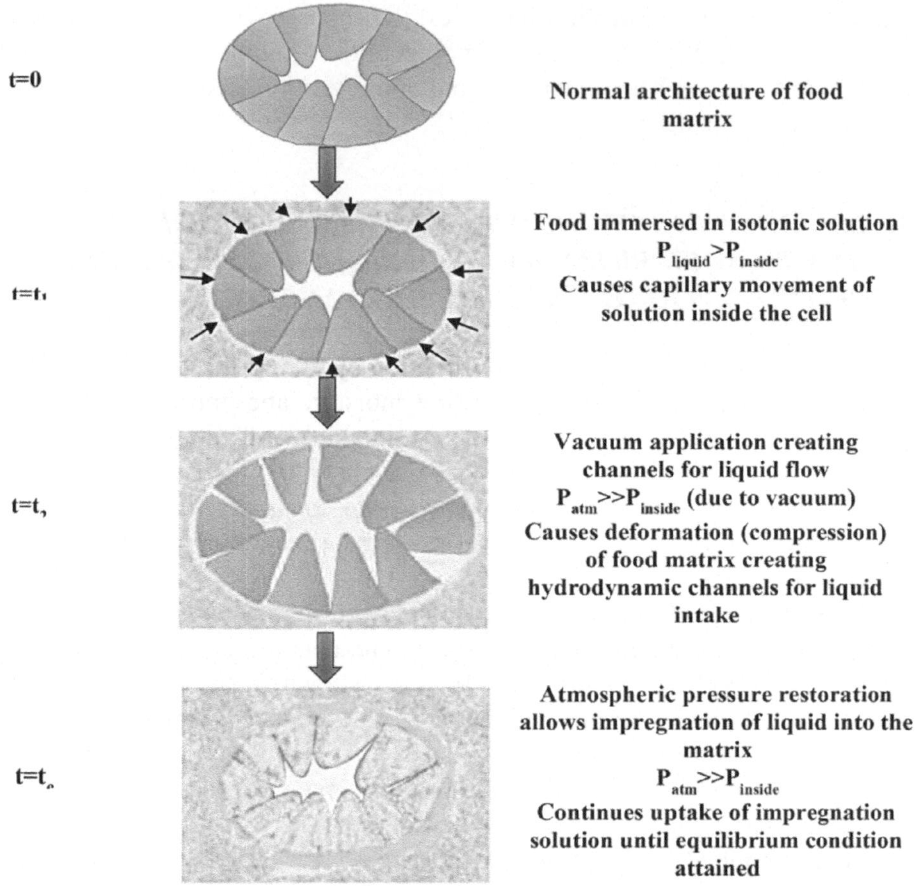

FIGURE 10.1 Detailed process of vacuum impregnation.

mixture, the impregnation solution's properties and the food matrix's properties. These factors have been explained below.

10.3.1 VACUUM CONDITIONS

Vacuum pressure is an important parameter as it represents the pressure gradient between the void phase of food and the atmosphere surrounding the external liquid that forces impregnation. Practical vacuum applied ranged between 50 and 600 mbar. An increase in vacuum pressure causes an increase in impregnation level (X) as a higher release of native liquids and gases occurs with increased Hydrodynamic force and Deformation-Relaxation force.

The vacuum period is defined as the time during which food microstructure tends to reach an equilibrium situation during which expulsion of gases and native liquids from the pores and the partial impregnation of pores occur. Relaxation time is the period during which food structure moves toward an equilibrium situation after the restoration of atmospheric pressure where decompression of pores occurs by taking up impregnation solution. Mostly, an increase in vacuum and relaxation time causes an increase in the impregnation rate. Beyond a particular time, no significant change could be measured [13]. However, too high vacuum levels produce greater cell structure damage.

10.3.2 PROPERTIES OF IMPREGNATION SOLUTIONS

The size of particles and viscosity of the impregnating solution have a greater influence on the impregnation rate. Impregnation causes the movement of smaller micro-molecules such as small protein fractions, salts and sugars. However, it does not allow the passage of large proteins and macromolecules into the food matrix [11]. The desired particles must impregnate the food structure through its porous structure due to capillary movement. Viscosity is another parameter affecting impregnation where high viscosity reduces surface tension and capillary force and decelerates liquid uptake into the food matrix. It could also cause deformation of the food matrix due to the increased stress produced on them [12].

Similarly, low viscosity liquids could easily penetrate the food structure, causing impregnation [14]. Since high viscous solutions cannot impregnate the food matrix, they form a coating over them. Viscosity also affects the retention of the impregnating compound on the food matrix after the pressure restoration phase, where the high viscosity could stick to the porous structure solution after the vacuum treatment. Similarly, surface tension also influences the extensibility and wettability of the applied liquid onto the product surface. The osmolarity of the impregnating solution also has a great impact on the impregnation of the solution. The hyper- or hypotonic solution overpowers the hydrodynamic force by using the hyper- or hypotonic solution. The increase in osmotic pressure causes changes in cell potential that results in the production of permanent deformation or shrinkage in the plant matrix resulting in an alteration of rigidity of the cell structure [15]. The use of the hypertonic solution is however used for osmo-vacuum drying. The incorporation of the hypotonic solution was lower initially but kept on decreasing with the increase in time due to dilution of the pores with the liquid phase [14]. For fortification and enrichment with functional ingredients, isotonic solutions are used that account for the absorption of functional ingredients mainly by hydrodynamic force and not merely by osmotic pressure [16]. Thus, the combination of high osmolarity liquids with higher viscous liquids has proven to affect the impregnation of liquid into the matrix negatively. Turbidity of the impregnating solution also has a greater effect that blocks the juice flow inside the porous matrix. A compound higher in turbidity usually possesses a greater amount of proteinaceous suspension that is difficult to be separated [17].

The property of the impregnation solution also has a greater influence on the physical characteristics of the food product. Impregnation of water and other solutions decreases the crispiness of the food matrix even when the product is dried afterward due to its structural changes, especially when the impregnation solution used exerts any osmotic pressure onto the product [15]. The use of firming agents such as calcium increases hardness. There may also be changes in color attributes according to the color of the impregnation solution used. There would be an improvement in the functional properties of the fruits mainly due to the exclusion of oxygen and other gas molecules from the fruit matrix, which prevents deteriorative oxidative reactions inside the fruit matrix [18]. Higher deformations caused by the vacuum treatment could also cause structural changes that would ultimately cause reduced shelf life, as in watermelons.

Vacuum impregnation could be coupled with novel processing techniques such as Ultrasound and Ohmic heating that could improve cell permeability without causing cell rupture. Ultrasound treatment up to 130 kPa does not cause any cell rupture substances into apple tissue without any increase in vacuum pressure [19]. Similarly, ohmic heating produces a localized temperature increase by 50–55ºC that subsequently increases molecular agitation and impregnation rates [20].

10.3.3 FOOD MATRIX CHARACTERISTICS

The porosity of the food matrix is an important parameter that tells about the extent to which the liquid can impregnate its structure. The amount of liquid impregnated is directly proportional to the porosity value of the fruit [21]. It also depends on the architecture, i.e., the shape and size of the pores inside the food matrix that determine the tortuosity of the food matrix [22]. Tortuosity is dependent on the structure and texture of the food matrix. As hardness decreases, the fruit becomes soft and deformation relaxation flow is dominant, helping to hold up more impregnation fluid in its structure. Accumulation of yeast cells is observed in the narrowest zones of the Intercellular spaces and *L. casei* viable cells in the dendritic regions of the apple matrix as observed by cryo-SEM [23].

Thus, the properties of the food matrix, impregnating solution and vacuum conditions have a great effect on the vacuum impregnation solution, as summarized below:

- **Food Matrix Properties:** Porosity, shape, size and distribution of pores, turtuosity, texture and structure of the food matrix.
- **Impregnation Solution Properties:** Size of particles, viscosity, surface tension, osmolarity, turbidity, properties of micro molecules present in them.
- **Process Parameters:** Vacuum pressure, vacuum attainment time, relaxation time.

10.4 MATHEMATICAL MODELING OF VACUUM IMPREGNATION PROCESS

As discussed in the above sections, the major forces that cause the liquid flow through the cell structure are hydrodynamic and deformation-relaxation forces. Fito and Pastor, 1994 was the pioneer in modeling of vacuum osmo-dehydration process, who found out that the major force that comes into play while osmotic drying under vacuum conditions was the hydrodynamic force, i.e., as a result of pressure gradients created as a result of vacuum process [24]. The other forces that come into play include the capillary pressure (Pc) due to the porous structure of the plant cell. He defined volume fraction and porosity are the important factors that cause this impregnation. The volume fraction of the food matrix is defined as the ratio of capillary force to the force given by the external fluid.

$$X_v = \frac{P_c}{P_c + P_2} = 1 - \frac{1}{r} \tag{1}$$

$$r = R + P_r = \left(\frac{P_2}{P_1}\right) + \left(\frac{P_c}{P_1}\right) \tag{2}$$

$$P_c = \sigma\left[\frac{1}{r_1} + \frac{1}{r_2}\right]\text{€os} \tag{3}$$

where: P_c is the capillary force; P_2 is the force exerted by the external fluid during the restoration of atmospheric pressure; P_1 is the force exerted during the application of vacuum; and r is the actual compression ratio, which is defined as the sum of apparent compression ratio and reduced capillary pressure as in Eqn. (2).

Capillary force is the result of the pore characteristics of the plant cell. The Young–Laplace equation defines it as in Eqn. (3), where σ is the interfacial tension (N/m), r_1 and r_2 are the principal radii of the pores in (m), and θ is the contact angle (rad). This equation holds good only if the distribution of two immiscible fluids, such as liquid and air, inside a porous medium has been extensively studied using resin micromodels [25]. This holds good by considering the microstructure of the cell structure that is usually characterized by fixed channel and pore cross-section geometry, as well as channel depth, the mean curvature of the liquid-air interface, and liquid flow rate. This could be identified by a low Bond number (Bo≪1):

$$B_o = \frac{gr^2\sigma}{\rho} \tag{4}$$

where: g is the acceleration due to gravity; r is the pore radius which is 5×10^{-6} and 4×10^{-5} m identified spinach and apple tissue respectively; and σ and ρ are the interfacial tension and the liquid density (7.2×10^{-2} N/m and 10^3 kg/ m^3).

Similarly, effective porosity was also found to be an effective parameter for impregnation as per the HDM model. It is defined as the total volume occupied by the gas molecules to the total volume of the food sample. The volumetric fraction of the liquid transferred (x) by the HDM will be the product of the volume fraction and the porosity of the product.

$$E_e = \frac{x}{x_v} \tag{5}$$

$$E_e = \frac{x}{1-\frac{1}{r}} \tag{6}$$

HDM model was effective in modeling vacuum impregnation flow but did not consider the deformations created in the food product by applying the vacuum conditions. This was further included by Fito et al. [37] who proposed an HDM-DRM by incorporating a deformation characteristic term γ that caused a change in the volume fraction and the effective porosity as:

$$E_e = \frac{x-\gamma}{1-\frac{1}{r}} \tag{7}$$

$$\gamma = \frac{V_{s2} - V_{so}}{V_{so}} \tag{8}$$

where: γ represents the structural deformation that happened in the food matrix, considering food as a viscoelastic material. The structural deformation factor is the ratio of the difference between the final and initial sample volume against the initial sample volume as in Eqn. (8), where V_{so} and V_{s2} are the initial and final sample volume before and after treatment [26].

The HDM and HDM-DRM models describe the vacuum impregnation process under initial and final equilibrium conditions and do not consider the dynamics of the process. To study the dynamics of the process, it is required to monitor the continuous changes in the mass of the food matrix as studied by Laurindo et al. [27, 65]. The changes in the volume of the fluid could be correlated to the mass change of the food matrix by the Volume-of-Fluid (VOF) model. VOF is a simulation model that considers the displacement of two or more immiscible fluids in a control volume and is applied in biphasic immiscible liquid flow in porous foods. Carciofi et al. [29] used this model to simulate flow during vacuum impregnation of foods [28]. The volume-of-Fluid (VOF) model uses the Eulerian approach, which uses physics and mathematics equations to describe the impregnation step. It uses the phase-indicating function and the conservation of mass and momentum in the liquid and gas phases to evaluate the system [29]. This model allows for evaluating the fraction of a gas phase, the interface position and the fraction of the porous medium occupied by liquid (X_L) by the Eqn. (9)

$$X_L = E_e\left(1 - \frac{\int V_s \alpha_G dV}{V_s}\right) \tag{9}$$

Where; E_e is the porosity; V_s is the volume of the solid matrix; α_G is the fraction of the gasphase; and dV is the total change in volume during the process.

10.5 MODES OF VACUUM IMPREGNATION

Vacuum impregnation is essentially done in two ways, namely pre-immersion and post-immersion impregnation modes. In post-immersion mode, the desired food sample is kept in a vacuum chamber and reduced, or sub-atmospheric pressure is introduced into the pressure vessel. After subjecting the food product to the required vacuum treatment time, the desired impregnation liquid is slowly poured over the food product until the atmospheric pressure is restored, followed by immersion in the impregnation solution for the defined time. In this method, impregnation fluid can be introduced onto the vacuum chamber using a breaker valve [30]. Pre-immersion impregnation is the most commonly used mode of vacuum impregnation. This method introduces the food product immersed in the desired impregnation solution to the pressure vessel. The desired vacuum level is applied and maintained in the pressure vessel for the defined time. Then, atmospheric pressure is restored in the pressure vessel by the pressure releasing valve or any other mechanism. Vacuum impregnation could also be performed in a vacuum dryer where the product is dipped in a particular impregnation fluid and kept in the trays of a vacuum dryer [10].

Vacuum impregnation can be done in single or pulsed mode. In a single mode, the desired vacuum is applied to the food product, and the impregnation fluid and atmospheric pressure are restored inside the system. The pulsed mode of impregnation is done with the intermittent change of pressure levels inside the vacuum system. It promotes greater cellular stress inside the food matrix and helps to take up better absorption of impregnation solution. Thus, a pulsed vacuum is essentially performed as an add-on to the normal vacuum impregnation method to loosen the cell structure and improve the permeability of the food matrix. It is especially done for rigid and robust food structures where a single vacuum is found insufficient to create the required permeability and thus impregnation.

The vacuum impregnation unit could be coupled with ultrasound for improved impregnation as per [67]. The custom-made equipment consists of stainless steel cylindrical chamber (a total of 38.5 L internal volume), piezoelectric crystals (ultrasonic transducers), a vacuum pump, air and solution drainage valves, and pressure-temperature controllers. The equipment provides six different–mass-dependent -ultrasound powers (72–840 W) at 35 kHz for distilled water within the chamber that promotes both ultrasound and vacuum treatment to be performed within the same system [19].

10.6 APPLICATIONS OF VACUUM IMPREGNATION IN PREPARATION OF READY-TO-EAT FOODS

In today's world, the food consumption trend is changing from just 'food for nutrition to the concept of 'functional food.' Food with a specific functional property is a growing topic that alleviates specific deficiencies and improves nutritional properties, especially bioactive compounds present in the food product. In addition, these also possess some specific functions that greatly impact food storage shelf life and acceptability. Vacuum impregnation is an important operation that could be effectively used to produce these functional food products. Vacuum impregnation could be used for several functions, such as enrichment of functional and nutritional ingredients into the food structure. This method could incorporate functional ingredients such as firming agents, color pigments, flavorants and other structure improvement components into the system. This method can also infuse nutrient enrichment such as probiotics and other bioactive compounds into the food system. This method has also been proven to improve the osmotic dehydration process by increasing water loss and reducing solid gain. Vacuum impregnation is usually combined with other stabilization operations such as drying and refrigeration storage that improve the stability and shelf life of the impregnated food product [10, 11, 23, 31, 32]. Without stabilization operations, the impregnated product would be overfilled with the liquid phase, which negatively impacts sensorial and shelf-life stability. Vacuum drying is the most preferred method for drying vacuum-impregnated food products with the least effect on quality aspects; however, convective drying up to 40°C could also produce nutritionally stable food products [17].

10.6.1 FORTIFICATION

Fortification, as per Codex Alimentarius Commission, is defined as the addition of one or more essential nutrients to a food system, whether or not it is normally contained in the food, for the purpose of prevention or correction of a demonstrated deficiency of one or more nutrients in the population or specific population groups [3]. In the conventional method of food fortification, specific fortificants such as vitamins or minerals are added to the food according to RDA recommendations in its formulations. The major disadvantage of this conventional fortification is the difficulty in maintaining the added micronutrient homogenously in the food system since the amount used is very less. The concept of fortification has changed over time from conventional to food-to-food fortification. Food-to-food fortification (FtF) is defined as adding micronutrient-dense foods to a recipe or food formulation rather than adding a meager amount of micronutrients into the food product [33]. It ensures uniform dispersion of micronutrients in the food product. Vacuum impregnation can be used for this FtF method where the food product with desired micronutrients could be used as the impregnating solution that could be successfully introduced into the food matrix system. Traditional methods of incorporation of fortificants of a food system include food formulation and blending. However, these produce changes in natural structure, texture and thus the fresh-like appearance of the food product [23]. The vacuum impregnation of probiotics, prebiotics, bioactives and other micronutrients into fruits and RTE snacks has been summarized in Table 10.1.

Infusion of prebiotics and probiotics into the food system is very much important for maintaining the healthy microbial bio flora in the intestine that confers various digestive and gastrointestinal benefits [30]. Probiotics are living microorganisms that confer great health benefits to the host when administered in adequate amounts. The functionality of these probiotics is due to the competitive exclusion of pathogens and bacteriocins from the intestinal tract, modulation of enzymatic activities related to metabolization of several carcinogens and other toxic substances and also aid in the production of volatile fatty acids, namely, Short Chain Fatty Acids in the intestine. Thus, they reduce digestive disorders such as diarrhea, irritable bowel syndrome and other allergic disorders in the intestine [34]. They are mostly present in dairy products, but people suffering from lactose intolerance and cholesterol control may be deprived of their benefits [23]. So, probiotics need to be consumed with non-dairy food products to maintain their viability. Incorporating these probiotics into high fiber-rich food has demonstrated protection against harmful temperature effects of drying operation. This, in turn, resulted in higher retention of phenolic and other volatile components in the food after drying [35]. The viability of microbe in a food product should be in the range of 10^7–10^8 CFU/g for its health benefits [16]. However, after the process of VI, stabilization by drying is used for fixing or maintaining probiotics in the food matrix. Drying after VI treatment has resulted in a very slight decrease in the viability of the probiotics based on the drying method employed. Vacuum drying is a supreme drying method for maintaining the viability of probiotics. However, air drying at lower temperatures, less than 40°C, has proven to preserve the viability of the probiotics to an acceptable level [23, 36]. Vacuum impregnation of probiotics successfully maintained the viable count of 10^7 CFU/ml in the food product that confers human health benefits. It was also found to be effective against the incidence of harmful microorganisms *H. pylori* when incorporated into food [20]. Probiotics could also be applied as emulsions or mixed with fruit juices making them more stable in the gastrointestinal tract [37]. Prebiotics are compounds that

TABLE 10.1 Use of Vacuum Impregnation for Fortification

Sl. No.	Product	Product Impregnation	Conditions	Result	References
1	Murtaberries	Probiotic addition: *Lactobacillus casei* var. *rhamnosus* inoculated to apple juice followed by drying	V: 150 mbarfor 15 min APR: 10 min	Shape, color and appearance are maintained due to lesser structural damage. Higher rehydration and water holding capacity The probiotic viability was better for the convective dried sample at 40ºC than the vacuum dried samples.	[45]
2	Apple slices	Probiotic addition: *Lactobacillus rhamnosus* microcapsule made as w/o/w emulsion	V: 114 mbar for 20 min APR: 150 min Followed by dehydration	Increased solid content and vacuum enhanced impregnation. Number of viable cells in the apple slices was in the range of 10^6–10^8 CFU/g db. Microcapsules of 1–2 μm could be incorporated into apple tissue.	[30]
3	Apple snack	Probiotic addition: Incubated *L. rhamnosus* into apple juice water (1:1 w/v, pH 5–5.2) incubated for 6 h at 37ºC.	V: 150 mm Hg, 15 min. APR: 15 min Followed by radiant dehydration	Air drying at 40ºC with radiant energy drying produces probiotic-enriched dried apple with more bacterial stability for 30 days at 25°C and 180 days at 4ºC	[46]
4	Apple fruit	Probiotic addition:10^{10} CFU/g of *B. coagulans* in isotonic sucrose followed by draining	V: 22 mm Hg, 5 min APR: >120 min	10^7 CFU/g of the probiotic was observed	[35]
5	Apple snack	Probiotic addition: Mandarin and pineapple/grape were infused with 10^9 CFU/ml *L.salivarius spp. Salivarius* followed by air drying	V: 50 mbar for 10 min APR: 10 min. Dehydration: 30°C for 24 hrs	Produce stable snack (moisture content 0.14 g water/ g sample) Drying affects the viability of microbe, but the concentration in the final product was 10^7 CFU/g	[28]
6	Minimally processed melon	Probiotic addition:10^{10} CFU/mL *L. acidophilus LA–3*.	V: 500 mm Hg for 2 min. APR: 2 min	3.6 times more incorporation was done by the VI technique than atmospheric immersion. Lesser changes in color and firmness. Incorporation by VI was 0.6 Log CFU/g more than the soaking technique	[34]

TABLE 10.1 *(Continued)*

Sl. No.	Product	Product Impregnation	Conditions	Result	References
7	Peeled pears	Prebiotic addition: FOS, $CaCl_2$ and citric acid followed by canning	V: 70 mbar Solution introduced through breaking vacuum valve until APR is restored.	Vacuum treatment enhanced absorption by 14 times that of osmotic treatment.	[33]
8	Semi-dried organic strawberries	Antioxidant property: Bilberry juice-based solution followed by drying	V: 5 min at 200 mbar APR: 4 min Followed by dehydration	Increased antioxidant properties, especially phenolic content for VI treated sample, which reduced slightly upon drying	[15]
9	Minimally processed apples	Antioxidant property: Green tea extract with ascorbic acid and sucrose combinations	V: 200 mbar for 5 min, APR: 5 min.	Decreased weight loss and respiration loss No change in texture and color Greater retention of phenolic activity even after 7 days.	[18]
10	Dried Courgette slices	Antioxidant property: Freshly squeezed onion and kale (50:50) with 3% NaCl solution. Followed by drying	V: 0.06 MPa for 2 min APR: 20 min.	Higher dried matter and water activity by freeze-drying as compared to convective drying Logistic model described dehydration. Final moisture content decreased by NaCl incorporation Use of NaCl alone showed a loss of quercetin content.	[44]
11	Apple	Antioxidant and firmness property: Calcium lactate and black carrot concentrate (BCC), Mannitol	Ultrasound: 130 kPaV: 211 mmHg, 3 min APR: 20 min	No cell rupture up to 130 kPa. Lower psychrophilic and mesophilic growth and greater stability due to incorporating phenolics onto its structure.	[19]
12	Apple slices	Bioactive addition: Apple juice (11.1°Brix) enriched with apple peel extract.	V: 100–800 mbar for 5 min. APR: 10 min. Apple slices were dripped out for 10 min.	Higher vacuum and lower pectin promoted more quercetin incorporation	[14]
13	Apple	Antioxidant property: Blueberry juice Followed by air drying	V: 5 kPa for 10 minAPR: 10 min. Followed by air drying	Air-drying at 40°C maintains around 50% of the initial anthocyanin and antioxidant activities.	[16]

TABLE 10.1 *(Continued)*

Sl. No.	Product	Product Impregnation	Conditions	Result	References
14	Apple snacks	Antioxidant property: Mandarin juice followed by air drying	V: 50 mbar for 10 min APR: 10 min. Air-dried at 40°C	40 g VI final product provide same quantity of hesperidin as 250 mL of fresh mandarin juice	[28]
15	Low-fat sweet potato chips	Anti-oxidant properties: Polyphenol extract (95% proanthocyanidins). Followed by low-pressure cooking (sous vide)	V: 0.8 bar, 25 min Sous vide: Samples vacuum-packed inside polyamide-polyethylene cooked in a bath at 80°C for 30 min.	VI increases phenolic content by 473% Color and texture remained stable	[40]
16	Potato Chips	Antioxidant properties: Followed by vacuum frying	V: 600 mm Hg for 10 min APR: 60 min.	Moisture loss was faster at higher temperatures Chips vacuum fried have higher TPC and increased with frying temperature. 27% higher TPC at 120°C vacuum frying where traditional frying produced a 9% decrease in TPC Color change was also less for vacuum frying	[25]
17	Low-fat potato chips	Fortification: Calcium and vitamins C and E, followed by microwave vacuum drying	V: 77.3 KPa, for 3 min. APR: 4 min Microwave power density: 1.7 W/g and absolute pressure of 4 kPa.	Higher than 90% of retention of calcium, 53% of vitamin C and 72% of vitamin E by VI and MVD	[24]
18	Apple snack	Fortification: Folic acid by use of Ohmic heating	V: 5 kPa for 5 min OH-ACdone at 60 Hz, 100 V, 105 min, 50ºC, followed by 60°C drying	VI at 40°C and VI/OH sample followed by drying presented high folic acid content.	[42]

TABLE 10.1 *(Continued)*

Sl. No.	Product	Product Impregnation	Conditions	Result	References
19	Pineapple snacks	Fortification: Calcium chloride	V: 88 kPa for 10 min APR: 10 min. Followed by drying	91% higher concentration of Ca^{2+} than impregnated at atmospheric pressure Increased drying rates and better product quality Produced porous-and-crunchy, with high glass transition temperature favoring its preservation during storage.	[36]
20	Fresh-cut apples (Fuji)	Edible coating for fortification: hydroxypropyl methylcellulose, calcium caseinate, tocopherol acetate, gluconal and zinc lactate (ZL)	V: 100 mm Hg APR: 30 min	Hydroxypropyl methylcellulose does not impregnate into the food matrix and thus forms a coating. Incorporation of other minerals along with Vit E may decrease the impregnation of Vit E into the food matrix	[47]

act as fuel to the growth of probiotics inside the small intestine. These work synergistically with probiotics to maintain their viability and promote the controlled release of probiotics into the food [38]. Fructo-oligosaccharide as prebiotics was found to be successfully infused into the pears to prepare functional RTE- pears [30].

Oxidative stress is the major cause of various lifestyle diseases like cardiovascular diseases, cancer and neurological disorders in human beings. So, it is very important to include foods with antioxidant properties into the daily diet for healthy living [39]. Fruits and vegetables, especially some exotic berries, citrus juice, and green tea, are the special source of some of the bioactive compounds that can be incorporated into the food system. Besides providing greater health benefits, it also brings about varieties to the food. So, it is very important to incorporate these antioxidant compounds into common foods to reap their health benefits. The vacuum impregnation technique has successfully incorporated various bioactive compounds such as phenolic content into the commonly consumed and minimally processed fruits and vegetables such as apples, potatoes, etc. [8, 12, 14]. Like probiotic impregnation, air drying up to 40°C is the most common stabilization method after impregnation treatment. Although it could cause some loss of phenolic compounds, the choice of novel methods like vacuum frying and low-pressure cooking confer more retention of phenolics. Vacuum frying at 120ºC caused maximum total phenolic retention of 27% higher as compared to 9% in traditional frying methods [39]. Other novel processing methods, such as low-pressure cooking and canning, also confer higher phenolic and other nutritional content retention in their food structure [39, 40]. These novel methods also aid in putting a wide variety of healthy snacks on the consumer's table. Similarly, novel pretreatments such as ultrasound confers a greater degree of impregnation [41].

Fortification with essential minerals and vitamins are crucial for promoting food with health benefits that combat malnutrition and help lead a healthy and balanced lifestyle. Fortification with vacuum impregnation was earlier done by forming fortificants enriched edible coating over the product [42]. Later, the apple was successfully fortified with folic acid by VI, followed by drying at 60°C [43]. Similarly, fortified low-fat potato chips were produced with vitamins C and E with VI coupled with microwave vacuum drying [9]. It caused the highest retention of over 90% of calcium, 53% of vitamin C and 72% of vitamins in the apple. In pineapple fortification with Calcium, vacuum impregnation has been found to increase 91% fortificant retention after freeze-drying compared to atmospheric immersion [44].

The novel method of ohmic heating coupled with VI has also been employed for increased absorption of fortificant to the food matrix [43].

10.6.2 FOOD PRESERVATION

Food preservation includes the techniques applied to a food product to improve its shelf life by slowing down or inhibiting microbial and other chemical reactions in the food. Drying is the most traditional food preservation method that prevents microbial deterioration by reducing the water content or water activity of the food product. Traditional drying methods such as convective drying uses higher temperature that may cause huge losses of nutritional and sensory attributes. The osmotic dehydration method is employed mainly as a substitute for the high-temperature processes, which uses hypertonic (high concentration) osmotic solutions such as salt and sugar to remove moisture from the food matrix. In the osmotic drying method, moisture and solute migration takes place by the concentration difference between the product and osmotic solution causing the flow of solute to the product and moisture from the product until the attainment of equilibrium on both sides of the semi-permeable membrane. Infusion of salt and sugar into the product has the advantage of preserving by reducing water activity. Various fruit snacks could be produced from the osmotic drying of apples, guavas, pears, gooseberry, and others. The vacuum impregnation technique was coupled with osmotic drying to improve process efficiency and mass transfer kinetics. Various studies used vacuum osmotic drying from the 1990s.

Vacuum dehydration produced higher water loss and solid gain, especially in the initial drying stages where the hydrodynamics mechanism (HDM) is predominant and later drying rate is reduced [48]. As modeling of dehydration process is considered, vacuum osmo-dehydration process could not be explained just by diffusion kinetics alone as in osmotic drying. The hydrodynamic model better fitted the experimental data than the diffusion model. It was better fitted by hydrodynamic models such as Peleg, followed by Weibull [49]. The hydrodynamic mechanism coupled with Fick's diffusion law better explained the mass transfer phenomenon of the process [50]. Barbosa Junior et al. model fitted water loss, solid gain and water activity kinetics in dehydrated guava. Due to increasing concern about the harmful effects of sodium salts, osmotic dehydration was done with substituted salts such as Calcium and Potassium. These salts produced increased preservative retention and ion absorption without much change in the color and texture attributes of guava [51].

Enzymatic browning is a major chemical deteriorative reaction in fresh fruits and vegetables during their cutting and other processing operations that hinder their sale in the form of minimally processed fruits. This type of reaction happens by the oxidation action of the innate enzyme polyphenol oxidase when it is cut and exposed to the atmosphere. Chemical inhibition of PPO activity includes acidification or reduction using antioxidants, chelating agents, or natural extracts such as honey. Acidulants decrease the pH of the fruit, thus inhibiting enzyme activity. Vacuum impregnation has been successfully used to incorporate natural and chemical anti-browning agents such as honey, lemon juice, ascorbates, etc., in fruits like apple, litchi and pear. It helps to increase the shelf life of cut fruits to 24 days in litchi and 20 days in pear, and 14 days in apple [2, 4, 10, 46, 47]. Another important method of food preservation is the reduction of pH, which causes a reduction in the microbial population. pH value lower than 4.5 inhibits the growth of Clostridium botulinum and the related toxin production [13]. Vacuum acidification has been applied to eggplant but resulted in non-uniform acidification and decreased pH by 20% [22]. Model fitting of pH with the pulsed vacuum of zucchini slices revealed the Peleg model better-fitted RpH values [13] (Table 10.2).

Vacuum impregnation can also be used to form edible coatings that confer various functions, such as inhibition of respiration and other physiological changes that greatly affect food's shelf life. It is especially of great importance in minimally processed fruits and vegetables where their physiological mechanism is impaired by cutting, slicing and other operations. Edible coating by vacuum impregnation has reduced the condition of white blush, especially in carrots [7]. The composition of edible coating has to be selected judiciously for good edible coating where chitosan reduces surface tension and promotes uniform edible coat [53].

10.6.3 MODIFICATION OF SENSORY AND TEXTURAL PROPERTIES

Food texture is an important property that determines consumer acceptability. Besides this, VI also greatly influences the preservation of quality during transportation, storage and freezing. During transportation and storage, delicate fruits such as raspberries may be subjected to mechanical injuries, which also influence their microbial proliferation. In such cases, it is important to add a firming agent to the fruit product to maintain its

TABLE 10.2 Use of Vacuum Impregnation for Food Preservation

Sl. No.	Product	Product Impregnation	Conditions	Result	References
1	Strawberries	Osmotic drying: 25 and 50% sucrose solution at 40°C	V: 50 mbar for 10 min. APR: 170 min.	Peleg's model fitted the experimental water loss and solid gain kinetics	[31]
			US frequency: 40 kHz and the power of 150 W	Highest solid gain and less firmness occurred with vacuum causing it to be more suitable than ultrasound	
2	Guavas	Osmotic drying: sucrose solutions at 40, 50 and 60ºBrix	V: 100 mbar for 10/15 min	Higher sucrose solution concentrations and greater vacuum pulse increases water loss and reduce the solid gain.	[24]
			APR: 15, 30, 60, 120, 180 and 300 min	Hydrodynamic model better explained the process than the diffusional model.	
3	Eggplant slices	Osmotic drying: KCl (2.5, 4%), $CaCl_2$ (0.5, 1%), NaCl (5, 7.5, 10%). Entire impregnation solution contain 10% salts in combinations. slices were immersed for 3 min in 1% ascorbic acid and 2% citric acid prior	V: 145 mbar for 10 min	Barbosa Junior et al. model fitted water loss, solid gain and water activity kinetics Vacuum and calcium increased ascorbic acid retention by 29.33 and 85.06%. Vacuum increased water loss up to 53% and ions potassium ion incorporation by 648%. Color and texture of VOD sample comparable to OD product.	[52]
4	Fresh-cut apple slices	Anti-browning agent: Honey, juice of lemon and pineapple	V: 400 mbar for 10 min APR: 10 min	Honey and lemon showed more inhibition against browning, and preservation effects. Mass gain was lesser for honey than juice due to lower viscosity. Vacuum was more effective than atmospheric immersion at 30 min. Shelf life increased to 14 days at 30ºC and 90% RH	[53]
5	Litchi	Anti-browning agent: Sucrose syrup with anti-browning agents such as cysteine, ascorbic acid, 4-hexylresorcinol, and calcium lactate.	V: 570 mmHg for 10 min APR: 10 min Followed by vacuum packaging at –355 mmHg	Increases shelf life to 24 days. Increased mass gain, TSS content and reduced drip loss. Increased firmness	[41]

TABLE 10.2 *(Continued)*

Sl. No.	Product	Product Impregnation	Conditions	Result	References
6	Pear	Anti-browning agent: Ascorbate, 4-HR, EDTA, citrate with 1 g/100 g calcium lactate (firming agent)	V: 5 kPa for 5 min APR: 10 min Storage: Room temp.	Hardness and crispiness decreased with VI, but lactate stiffness could be maintained with the addition of calcium. Shelf life was increased by 20 days. 4-HR was most effective as an anti-browning agent.	[54]
7	Egg plant	Acidification: Lactic acid soln. of pH of 3	V: 100 mbar for 3 min APR: 5 min	Decrease pH by 20%. Non-uniform acidification	[26]
8	Zucchini slices	Acidification: Lactic acid solution of pH_3	V: 400 and 200 mbar applied for 2 min. APR done at 20, 40, 60, 120, 240, 300, 360 and 720th min	Pulsed vacuum improves pH reduction Peleg model well fits R_{pH} values. Increase in vacuum from 400 to 200 mbar caused lesser pH. No significant volume changes at both vacuum levels.	[55]
9	Minimally processed pumpkin	Edible coating: Chitosan (1.5% w/v) dispersed in acetic acid	V: 500 mmHg for 4 min and stored for 16 days	Thicker coating formed over the surface. No change in water content and carotenoid content. Lesser microbial load	[6]
10	Fresh-cut carrot	Edible coating: Chitosan (1%, w/v) dispersed in acetic acid (1%, w/v)	V: 5 kPa for 4 min. APR: 2 min.	Diminished white blush during storage. Improved water vapor transmission Better preservation of the color and mechanical response during cold storage.	[7]
12	Apple products	Inhibit browning and preserve fluid loss: Isotonic carbohydrate solutions	V: 738 mbar for 10s	Maximum penetration with minimal effects on fruit composition and quality. Treatment with sorbitol produced slightly higher firmness than a fresh sample.	[42]

structure and texture. Calcium is the major compound used for improving the firmness of the tissue owing to its property to structurally bind with the food matrix and maintain its texture and integrity. Firming agents have been found to preserve the structure of the product even after giving drying operations. This also results in the formation of various snacks, such as crunchy pineapple snacks [44]. Even though calcium impregnation improved texture, the vacuum has reduced the shelf life of minimally processed melons that used 60 MPa pressure [56]. Thus, it is required to maintain the vacuum pressure limit that causes fewer deformations in the structure. Fruits are highly perishable and require to be stored for a longer period. Freezing is the most effective method for arresting the microbial growth in fruits and vegetables. However, they cause structural changes in them, changing the texture mostly to a soft and pliable state. Cryoprotectants are the substances used in food products to preserve their structural integrity during long-term frozen storage or continuous temperature abuse. Cryoprotectant infusion works based on the reduction of freezable water in the food matrix that makes the product stable. Cryoprotectant increases the liquid phase's glass transition temperature (Tg'), thus conferring textural and structural changes in the fruit matrix. In this way, it reduces drip loss and increases the firmness of the fruit in frozen apple snacks, marionberry, etc. [6, 55].

Use of VI technology could also be used to infuse color and flavoring agents onto the food matrix with little change to the sensory attributes [14, 44]. Bellary and Rastogi [20] used Ultrasound together with vacuum and blanching treatments to incorporate curcuminoids for banana slices. It was concluded that using ultrasound and vacuum increased the impregnation rate. In contrast, vacuum along with blanching treatment produced a greater impregnation amount that might be due to structural deformation in its matrix by this treatment [4] (Table 10.3).

10.7 ADVANTAGES AND DISADVANTAGES OF VACUUM IMPREGNATION

10.7.1 ADVANTAGES

Vacuum impregnation is the extension of the immersion impregnation method to improve the rate of impregnation. Application of vacuum causes higher impregnation as it increases calcium impregnation by 91% in pineapple [44], 10% more β-carotene and lutein in minimally processed fruit salad [2] and 3.6 times more incorporation of *L. acidophilus* in minimally processed melon as against atmospheric immersion method of impregnation

TABLE 10.3 Use of Vacuum Impregnation for Modification of Sensory and Textural Properties

Sl. No.	Product	Product Impregnation	Conditions	Result	References
1	Minimally Processed Melon	Firming agent: 5% Calcium lactate	V: 60 MPa for 5 min. APR: 5 min.	Reduced shelf life due to higher deformation of the fruit matrix. Darker appearance and textural improvement	[57]
2	Pineapple snacks	Firming agent:0.01% $CaCl_2$	V: 88 kPa for 10 min APR: 10 min.	VI presented 91% higher Ca than atmospheric impregnation. Increased convective drying rates Produced porous-and-crunchy snacks High glass transition temperature of snacks	[36]
3	Zucchini	Firming agent: Maltodextrins, salt and $CaCl_2$	V: 25 mbar for 10 min APR: 30 min	NaCl (alone or in combination) led to significant water loss but not observed in solutions with MD and $CaCl_2$, Caled to more interconnected structures that limited water and solids exchange related to the deformation phenomena MD prevented damage due to shrinkage In mixed solutions, presence of MD limited damages induced by NaCl and $CaCl_2$	[58]
4	Red raspberries	Firming agent: Low Methoxy Pectin (LMP-Ca)/ Pectinmethylesterase (PME)-calcium	V: 50.8 kPa for 7 min. APR: 5 min.	LMP-Ca was more effective as a firming agent than PME-Ca Increase in solution temperature from 20 to 40°C affected the firmness of the fruit	[59]
3	Frozen Marionberry	Cryoprotectant: High fructose corn syrup (HFCS) and high methoxyl pectin (HMP) as, Calcium gluconal (CG) and zinc lactate (ZL) as nutraceutical ingredients.	V: 50 mmHg for 15 min. APR: 30 min.	VI increased maximum compression force from 45% to 137%, Reduced drip loss from 28% to 48%. Ca improved texture quality, Zn improved color stability.	[60]

TABLE 10.3 *(Continued)*

Sl. No.	Product	Product Impregnation	Conditions	Result	References
4	Apple snacks	Cryoprotectant: Concentrated grape musts and pectin solutions	V: 50 mbars for 5 min APR: 25 min	Pectin improved frozen product stability. Peleg and Maxwell models properly fit the relaxation curves	[61]
5	Apple sticks	Flavoring agent: Isotonic fructose solution with ascorbic acid and green apple flavoring	V: 280 mbar for 5 min and APR: 1.5 min US: 35-kHz	V and V+US treatments gave the highest aroma enrichment at 5.0 min, but the US and atmospheric produced lesser aroma.	[50]
6	Banana slice	Color: Curcuminoids	V: 20 kPa for 30 min APR: 5 hr US: frequency 35 kHz. V+US combinations	US+blanching produced the fastest infusion (in less time) but US+V produced the highest infusion. The use of NaCl solution for infusion resulted in increased curcuminoid absorption with higher moisture loss from the sample	[20]
7	Minimally processed fruit salad	Color enhancement: β-carotene and lutein	V: 500 mmHg for 7 min.	VI improved color incorporation at 10% as against 4% by the soaking method. No change in pH, acidity, and total soluble solids	[62]

[63]. The use of this technology has also been shown to increase the retention of phenolic, fortificants or probiotics even after the process of drying operations. Probiotic viability after the vacuum impregnation treatment was maintained above 10^7 CFU/ml even after vacuum drying or convective air drying up to 40°C [35, 37, 54]. The phenolic compound was retained in the fruit even after even after 7 days of storage, as seen in minimally processed apples [8]. So, it is proved to be an effective method of food fortification with maximum retention of impregnating solution activity.

This technology could cause minimal changes in the textural, color and sensorial properties of the food product. Lesser change in textural properties could be conferred to the lesser structural variations caused in the food matrix. These could also be used to improve these sensorial properties, as explained in Section 10.6.3. Zura-Bravo et al. [45] studied vacuum impregnation of probiotics into murta berry at 150 mbar for 15 min and found no change in shape, color and appearance, which was further confirmed by the microstructural studies [35]. Similar results were found for the impregnation of probiotics in melons [63], green tea polyphenols in apples [8] and several others. The vacuum impregnation method produced a better product than the ultrasound method. VI caused better results in terms of the amount of solution impregnated than that of ultrasound and blanching treatments in the impregnation of curcuminoids in banana slices [64]. It also caused the highest solid gain and less firmness when compared with vacuum impregnation studies [48]. Vacuum impregnation could be combined with methods such as microwave vacuum drying, low-pressure cooking, vacuum frying, and other methods that help prepare healthy fortified snacks. These methods were used to produce low-fat, phenolic content fortified potato chips [9, 40, 65].

The vacuum impregnation equipment is less costly than other novel impregnation methods such as ultrasound-assisted and high-pressure impregnation methods. Also, a vacuum impregnation system could be coupled with ultrasound and ohmic heating systems, showing its usage flexibility [43, 52].

Thus, the various advantages of vacuum impregnation in foods could be summarized as follows:

- Improved impregnation as compared to the atmospheric immersion method.
- Higher retention of phenolic compounds and fortificants in the food system.
- Greater retention of phenolic compounds during drying and storage.
- Greater retention and viability of probiotics after their impregnation.

- Lesser structural changes are given by lesser changes in textural and other properties.
- The method could be easily coupled with other methods, such as drying and frying, that help in healthier snack production.
- It also increases the variety of fortified high-quality snack products.
- Equipment used for VI could be coupled with systems of Ultrasound and Ohmic heating systems.
- Equipment is less costly and easier to handle.

10.7.2 DISADVANTAGES

Plant cells mostly account for their anisotropic and heterogenous making it more complicated to model and study in detail. Cell structure, especially pores size and architecture, changes with each food matrix and cannot be generalized. There can be hydrophobic porous structures in the food matrix that do not cause the wetting of the plant tissues and hinders the impregnation of the solvent to its structure, as found out by Gas in Scattering Media Absorption Spectroscopy in apples [62]. As explained in the previous section, vacuum impregnation is essentially the function of the pressure differences between the vacuum pressure applied and the atmospheric pressure restored in the system. However, in some food products, such as spinach leaves, incipient bubbling or dissipation of air from the voids takes place at relatively lower vacuum pressure levels due to the narrow pore size [25]. Also, the structure of the food matrix may not be uniform, resulting in heterogenous impregnation intake. In the study of quercetin in apple matrix, isotonic solution uptake in the inner cortex was higher than in the outer part of the apple owing to its smaller cells and lower intercellular spaces [10].

Even though vacuum impregnation is an effective technique for incorporating desired compounds into its structure, the use of pressures has been found to affect the cell structure and cause deformations negatively. It may also influence the reduced shelf life of the food product [48]. It happens especially in soft-structured or high moisture content foods such as melons. The change in structure is attributed to the deformations created in the structure. So, the use of this method is limited to more elastic bodies at a particular vacuum pressure level. This method has also been found to cause an increase in the time of impregnation compared to the Ultrasound method. Also, the method may not work with a rigid structure that makes its impregnation difficult and thus have to be coupled with other methods such as ultrasound, ohmic heating, etc.

Thus, the disadvantages of VI could be summarized as follows:

- Non-uniform absorption of the impregnating solution.
- May cause deformations and structural changes based on the vacuum pressure applied and the structure of the plant cell.
- Non-uniform plant cell structure makes it difficult to model the process.
- Technology may not work with a firm, rigid cell structure.
- It may be slower than other novel methods.

10.8 OPPORTUNITIES AND CHALLENGES

As mentioned in the above sections, VI could be used to prepare healthy, fortified snack foods. This technology could be extended to fortify various components such as proteins, vitamins and other food components. This technology could be coupled with other novel pretreatment methods such as Ultrasound, high-pressure processing, and electric field-mediated techniques that increase its impregnation efficiency in the food system. This technology could also be used as a pretreatment for oil and other bioactive extraction where the solvent for extraction could be incorporated into the food matrix by VI technology. VI was successfully done with other processes like vacuum frying, drying and other methods, but the continuous system that couples these processes was not well developed. Continuous industrial equipment and process lines could be developed to process healthy snacks on an industrial scale continuously. Similarly, pretreatment setups could also be incorporated into the process line for industrial implementation of this technology. Also, it is very important to optimize the vacuum conditions, time for vacuum treatment, and atmospheric pressure restoration time since the change in these process parameters could pose structural and textural changes in the food.

The important challenge in applying VI treatment is the control of pressure conditions and the pressure seal required to maintain vacuum conditions. Vacuum treatment would result in water loss and volatile compound loss to the impregnating solution, which may reduce its usefulness. The loss of structure, shape and integrity during its treatment is another factor that reduces its usefulness, especially in high moisture and semi-solid foods.

10.9 CONCLUSION

Functional snack food production with nutritional and other functional components is crucial to combat malnutrition and other micronutrient deficiencies in today's society. Vacuum impregnation is proven to be an important method that could be used as a tool for fortification and improvement of textural properties, thus aiding in the production of functional snack products. The choice of optimum process parameters aid in maintaining the texture and structure of the food matrix. It is also required to commercialize these technologies on an industrial scale and couple them with other unit operations such as drying and frying for inline production of snack foods. This technology is cost-effective and easy to implement and find its applications in small-scale and large-scale industries.

KEYWORDS

- **fortification**
- **preservation**
- **ready-to-eat foods**
- **texture modification**
- **vacuum impregnation**

REFERENCES

1. U.S. Food and Drug Administration (US-FDA). (2022). Part 117–Current good manufacturing practice, hazard analysis, and risk-based preventive controls for human food.
2. Ostermeier, R., Hill, K., Toäpfl, S., & Jaäger, H. (2020). Pulsed electric field as a sustainable tool for the production of healthy snacks. In *Pulsed Electric Fields to Obtain Healthier and Sustainable Food for Tomorrow* (pp. 103–128).
3. Chiralt, A., Fito, P., Barat, J. M., Andres, A., González-Martínez, C., Escriche, I., & Camacho, M. M. (2001). Use of vacuum impregnation in food salting process. *Journal of Food Engineering, 49*(2–3), 141–151. https://doi.org/10.1016/S0260-8774(00)00215-0.
4. Fito, P., Chiralt, A., Barat, J. M., Andrés, A., Martínez-Monzó, J., & Martínez-Navarrete, N. (2001). Vacuum impregnation for the development of new dehydrated products. *Journal of Food Engineering, 49*(4), 297–302. https://doi.org/10.1016/S0260-8774(00)00222-8.

5. Batpho, K., Boonsupthip, W., & Rachtanapun, C. (2017). Antimicrobial activity of collagen casing impregnated with nisin against foodborne microorganisms associated with ready-to-eat sausage. *Food Control, 73*, 1342–1352. https://doi.org/10.1016/j.foodcont.2016.11.009.
6. Soares, C., Gomes, E., Dahlke, F., De Rolt, C., Plentz, P., Dantas, M., & Scussel, V. (2020). Use of IoT for real-time monitoring of storage silos and ozone gas fungal decontamination strategy. *International Journal of Computers and Applications, 175*, 1–7.
7. Vargas, M., Chiralt, A., Albors, A., & González-Martínez, C. (2009). Effect of chitosan-based edible coatings applied by vacuum impregnation on quality preservation of fresh-cut carrot. *Postharvest Biology and Technology, 51*(2), 263–271. https://doi.org/10.1016/j.postharvbio.2008.07.019.
8. Radziejewska-Kubzdela, E., Biegańska-Marecik, R., & Kidoń, M. (2014). Applicability of vacuum impregnation to modify physico-chemical, sensory and nutritive characteristics of plant-origin products—A review. *International Journal of Molecular Sciences, 15*(9), 16577–16610. https://doi.org/10.3390/ijms150916577.
9. Derossi, A., Francavilla, M., Monteleone, M., Caporizzi, R., & Severini, C. (2021). From biorefinery of microalgal biomass to vacuum impregnation of fruit: A multidisciplinary strategy to develop innovative food with increased nutritional properties. *Innovative Food Science & Emerging Technologies, 70*, 102677. https://doi.org/10.1016/j.ifset.2021.102677.
10. Panarese, V., Dejmek, P., Rocculi, P., & Gómez Galindo, F. (2013). Microscopic studies providing insight into the mechanisms of mass transfer in vacuum impregnation. *Innovative Food Science & Emerging Technologies, 18*, 169–176. https://doi.org/10.1016/j.ifset.2013.01.008.
11. Alzamora, S. M., Salvatori, D., Tapia, M. S., López-Malo, A., Welti-Chanes, J., & Fito, P. (2005). Novel functional foods from vegetable matrices impregnated with biologically active compounds. *Journal of Food Engineering, 67*(1–2), 205–214. https://doi.org/10.1016/j.jfoodeng.2004.05.067.
12. Guillemin, A., Degraeve, P., Noel, C., & Saurel, R. (2008). Influence of impregnation solution viscosity and osmolarity on solute uptake during vacuum impregnation of apple cubes (Var. Granny Smith). *Journal of Food Engineering, 86*(4), 475–483. https://doi.org/10.1016/j.jfoodeng.2007.11.021.
13. Derossi, A., De, T., & Severini, C. (2012). The application of vacuum impregnation techniques in the food industry. In *Scientific, Health and Social Aspects of the Food Industry* (pp. 25–56).
14. Schulze, B., Peth, S., Hubbermann, E. M., & Schwarz, K. (2012). The influence of vacuum impregnation on the fortification of apple parenchyma with quercetin derivatives in combination with pore structures X-ray analysis. *Journal of Food Engineering, 109*(3), 380–387. https://doi.org/10.1016/j.jfoodeng.2011.11.015.
15. Tylewicz, U., Mannozzi, C., Romani, S., Castagnini, J. M., Samborska, K., Rocculi, P., & Dalla Rosa, M. (2019). Chemical and physicochemical properties of semi-dried organic strawberries enriched with bilberry juice-based solution. *LWT, 114*, 108377. https://doi.org/10.1016/j.lwt.2019.108377.
16. Castagnini, J. M., Betoret, N., Betoret, E., & Fito, P. (2015). Vacuum impregnation and air-drying temperature effect on individual anthocyanins and antiradical capacity

of blueberry juice included into an apple matrix. *LWT–Food Science and Technology, 64*(2), 1289–1296. https://doi.org/10.1016/j.lwt.2015.07.034.

17. Carciofi, B. A. M., Prat, M., & Laurindo, J. B. (2011). Homogeneous volume-of-fluid (vof) model for simulating the imbibition in porous media saturated by gas. *Energy and Fuels, 25*(5), 2267–2273. https://doi.org/10.1021/ef200126m.
18. Tappi, S., Tylewicz, U., Romani, S., Dalla Rosa, M., Rizzi, F., & Rocculi, P. (2017). Study on the quality and stability of minimally processed apples impregnated with green tea polyphenols during storage. *Innovative Food Science & Emerging Technologies, 39*, 148–155. https://doi.org/10.1016/j.ifset.2016.12.002.
19. Yilmaz, P., Demirhan, E., & Özbek, B. (2021). Microwave drying effect on drying characteristics and energy consumption of *Ficus carica* Linn leaves. *Journal of Food Process Engineering, 44*(10), e13869. https://doi.org/10.1111/jfpe.13869.
20. Bellary, A. N., & Rastogi, N. K. (2014). Effect of selected pretreatments on impregnation of curcuminoids and their influence on physico-chemical properties of raw banana slices. *Food and Bioprocess Technology, 7*(10), 2803–2812. https://doi.org/10.1007/s11947-014-1311-9.
21. Ajav, E. A., & Yinusa, B. (2015). Factors affecting forage chopper performance. *International Agricultural Engineering Journal, 17*(3), 112–116.
22. Deng, Y., & Zhao, Y. (2008). Effect of pulsed vacuum and ultrasound osmopretreatments on glass transition temperature, texture, microstructure, and calcium penetration of dried apples (Fuji). *LWT–Food Science and Technology, 41*(9), 1575–1585. https://doi.org/10.1016/j.lwt.2007.10.018.
23. Betoret, E., Betoret, N., Arilla, A., Bennár, M., Barrera, C., Codoner, P., & Fito, P. (2012). No invasive methodology to produce a probiotic low humid apple snack with potential effect against Helicobacter pylori. *Journal of Food Engineering, 110*(2), 289–293. https://doi.org/10.1016/j.jfoodeng.2011.11.008.
24. Duarte-Correa, Y., Díaz-Osorio, A., Osorio-Arias, J., Sobral, P. J. A., & Vega-Castro, O. (2020). Development of fortified low-fat potato chips through vacuum impregnation and microwave vacuum drying. *Innovative Food Science & Emerging Technologies, 64*, 102437. https://doi.org/10.1016/j.ifset.2020.102437.
25. Moreira, R. G., & Almohaimeed, S. (2018). Technology for processing of potato chips impregnated with red root beet phenolic compounds. *Journal of Food Engineering, 228*, 57–68. https://doi.org/10.1016/j.jfoodeng.2018.02.014.
26. Derossi, A., Iliceto, A., De Pilli, T., & Severini, C. (2015). Application of vacuum impregnation with anti-freezing proteins to improve the quality of truffles. *Journal of Food Science and Technology, 52*(11), 7200–7208. https://doi.org/10.1007/s13197-015-1872-1.
27. Iwabuchi, K., & Kurata, K. (2003). Short-term and long-term effects of low total pressure on gas exchange rates of spinach. *Advances in Space Research, 31*(1), 241–244. https://doi.org/10.1016/S0273-1177(02)00749-2.
28. Betoret, E., Sentandreu, E., Betoret, N., Codoner-Franch, P., Valls-Bellés, V., & Fito, P. (2012). Technological development and functional properties of an apple snack rich in flavonoid from mandarin juice. *Innovative Food Science & Emerging Technologies, 16*, 298–304. https://doi.org/10.1016/j.ifset.2012.07.008.
29. Carciofi, B. A. M., Prat, M., & Laurindo, J. B. (2012). Dynamics of vacuum impregnation of apples: Experimental data and simulation results using a VOF model. *Journal of Food Engineering, 113*(2), 337–343. https://doi.org/10.1016/j.jfoodeng.2012.06.013.

30. Flores-Andrade, E., Pascual-Pineda, L. A., Alarcón-Elvira, F. G., Rascón-Díaz, M. P., Pimentel-González, D. J., & Beristain, C. I. (2017). Effect of vacuum on the impregnation of Lactobacillus rhamnosus microcapsules in apple slices using double emulsion. *Journal of Food Engineering, 202*, 18–24. https://doi.org/10.1016/j.jfoodeng.2017.01.019.
31. Cheng, X. F., Zhang, M., Adhikari, B., & Islam, M. N. (2014). Effect of power ultrasound and pulsed vacuum treatments on the dehydration kinetics, distribution, and status of water in osmotically dehydrated strawberry: A combined NMR and DSC study. *Food and Bioprocess Technology, 7*(10), 2782–2792. https://doi.org/10.1007/s11947-014-1297-3.
32. Fito, P., & Pastor, R. (1994). Non-diffusional mechanisms occurring during vacuum osmotic dehydration. *Journal of Food Engineering, 21*(4), 513–519. https://doi.org/10.1016/0260-8774(94)90050-7.
33. Hidalgo, I. G., Borras, M. T. G., & García, J. M. R. (2019). Comparison of two technological procedures to enrich peeled pears in halves with fructooligosaccharides for later canning as a new functional food. *Emirates Journal of Food and Agriculture, 31*(5), 328–334. https://doi.org/10.9755/ejfa.2019.v31.i5.1947.
34. de Oliveira, P. M., Ramos, A. M., Martins, E. M. F., Vieira, É. N. R., Soares, A. de S., & de Noronha, M. C. (2017). Comparison of vacuum impregnation and soaking techniques for addition of the probiotic *Lactobacillus acidophilus* to minimally processed melon. *International Journal of Food Science & Technology, 52*(12), 2547–2554. https://doi.org/10.1111/ijfs.13533.
35. Soto-Caballero, M. C., Acosta-Muñiz, C. H., Chávez-Leal, V., González-Aguilar, G., Soria-Hernández, C. G., & Avila-Quezada, G. D. (2021). Enrichment of sliced apple fruit with *Bacillus coagulans*. *Emirates Journal of Food and Agriculture, 33*(1), 12–19. https://doi.org/10.9755/ejfa.2021.v33.i1.2262
36. de Lima, M. M., Tribuzi, G., de Souza, J. A. R., de Souza, I. G., Laurindo, J. B., & Carciofi, B. A. M. (2016). Vacuum impregnation and drying of calcium-fortified pineapple snacks. *LWT–Food Science and Technology, 72*, 501–509. https://doi.org/10.1016/j.lwt.2016.05.026.
37. Fito, P., Andrés, A., Chiralt, A., & Pardo, P. (1996). Coupling of hydrodynamic mechanism and deformation-relaxation phenomena during vacuum treatments in solid porous food-liquid systems. *Journal of Food Engineering, 27*(3), 229–240. https://doi.org/10.1016/0260-8774(94)00073-5.
38. Corrêa, J. L. G., Pereira, L. M., Vieira, G. S., & Hubinger, M. D. (2010). Mass transfer kinetics of pulsed vacuum osmotic dehydration of guavas. *Journal of Food Engineering, 96*(4), 498–504. https://doi.org/10.1016/j.jfoodeng.2009.08.031.
39. Kruger, J., Taylor, J. R. N., Ferruzzi, M. G., & Debelo, H. (2020). What is food-to-food fortification? A working definition and framework for evaluation of efficiency and implementation of best practices. *Comprehensive Reviews in Food Science and Food Safety, 19*(6), 3618–3658. https://doi.org/10.1111/1541-4337.12615.
40. Abalos, R. A., Naef, E. F., Aviles, M. V., & Gómez, M. B. (2020). Vacuum impregnation: A methodology for the preparation of a ready-to-eat sweet potato enriched in polyphenols. *LWT, 131*, 109773. https://doi.org/10.1016/j.lwt.2020.109773.
41. Shah, N. S., & Nath, N. (2008). Changes in qualities of minimally processed litchis: Effect of antibrowning agents, osmo-vacuum drying and moderate vacuum packaging.

LWT–Food Science and Technology, 41(4), 660–668. https://doi.org/10.1016/j.lwt.2007.05.020.

42. Moreno, J., Espinoza, C., Simpson, R., Petzold, G., Nunez, H., & Gianelli, M. P. (2016). Application of ohmic heating/vacuum impregnation treatments and air drying to develop an apple snack enriched in folic acid. *Innovative Food Science & Emerging Technologies, 33*, 381–386. https://doi.org/10.1016/j.ifset.2016.01.001.
43. Li, Y. H., Qi, Y. R., Wu, Z. F., Wang, Y. Q., Wang, X. C., Wang, F., & Yang, M. (2017). Comparative study of microwave-vacuum and vacuum drying on the drying characteristics, dissolution, physicochemical properties, and antioxidant capacity of Scutellaria extract powder. *Powder Technology, 317*, 430–437. https://doi.org/10.1016/j.powtec.2017.04.034.
44. Kręcisz, M., Stępien, B., Pasławska, M., Popłonski, J., & Dulak, K. (2021). Physicochemical and quality properties of dried courgette slices: Impact of vacuum impregnation and drying methods. *Molecules, 26*(15), 4597. https://doi.org/10.3390/molecules26154597.
45. Zura-Bravo, L., Rodríguez, A., Stucken, K., Castillo, L., Ah-Hen, K. S., García-Segovia, P., & Vega-Galvez, A. (2019). Nutritional and organoleptic properties of murta (*Ugni molinae* Turcz) berries impregnated with *Lactobacillus casei* var. rhamnosus and dehydrated by different methods. *Food Chemistry, 299*, 125104. https://doi.org/10.1016/j.foodchem.2019.125104.
46. Noorbakhsh, R., Yaghmaee, P., & Durance, T. (2013). Radiant energy under vacuum (REV) technology: A novel approach for producing probiotic enriched apple snacks. *Journal of Functional Foods, 5*(3), 1049–1056. https://doi.org/10.1016/j.jff.2013.03.013.
47. Park, S., Kodihalli, I., & Zhao, Y. (2005). Nutritional, sensory, and physicochemical properties of vitamin E- and mineral-fortified fresh-cut apples by use of vacuum impregnation. *Journal of Food Science, 70*(9), S593–S599. https://doi.org/10.1111/j.1365-2621.2005.tb09992.x.
48. Champagne, C. P., Ross, R. P., Saarela, M., Hansen, K. F., & Charalampopoulos, D. (2011). Recommendations for the viability assessment of probiotics as concentrated cultures and in food matrices. *International Journal of Food Microbiology, 149*(3), 185–193. https://doi.org/10.1016/j.ijfoodmicro.2011.07.005.
49. Cunningham, M., Azcarate-Peril, M. A., Barnard, A., Benoit, V., Grimaldi, R., Guyonnet, D., Holscher, H. D., Hunter, K., Manurung, S., Obis, D., & Petrova, M. I. (2021). Shaping the future of probiotics and prebiotics. *Trends in Microbiology, 29*(8), 667–685. https://doi.org/10.1016/j.tim.2021.03.007.
50. Comandini, P., Blanda, G., Mújica Paz, H., Valdez Fragoso, A., & Gallina Toschi, T. (2010). Impregnation techniques for aroma enrichment of apple sticks: A preliminary study. *Food and Bioprocess Technology, 3*(6), 861–866. https://doi.org/10.1007/s11947-009-0317-5.
51. Gras, M. L., Vidal, D., Betoret, N., Chiralt, A., & Fito, P. (2003). Calcium fortification of vegetables by vacuum impregnation: Interactions with cellular matrix. *Journal of Food Engineering, 56*(2–3), 279–284. https://doi.org/10.1016/S0260-8774(02)00267-2.
52. de Jesus Junqueira, J. R., Correa, J. L. G., de Mendonça, K. S., Resende, N. S., & de Barros Vilas Boas, E. V. (2017). Influence of sodium replacement and vacuum pulse on the osmotic dehydration of eggplant slices. *Innovative Food Science & Emerging Technologies, 41*, 10–18. https://doi.org/10.1016/j.ifset.2017.01.011.

53. Gavrilaş, S., Ursachi, C. Ş., Dochia, M., & Perţa-Crişan, S. (2020). The influence of different vacuum impregnation solutions on the color of fresh-cut apple slices. *Scientific Study & Research. Chemistry & Chemical Engineering, Biotechnology, Food Industry, 21*(4), 523–531. https://doi.org/10.3402/fnr.v59.25971.
54. Perez-Cabrera, L., Chafer, M., Chiralt, A., & Gonzalez-Martinez, C. (2011). Effectiveness of antibrowning agents applied by vacuum impregnation on minimally processed pear. *LWT–Food Science and Technology, 44*(10), 2273–2280. https://doi.org/10.1016/j.lwt.2011.07.006.
55. Derossi, A., Fiore, A. G., de Pilli, T., & Severini, C. (2011). A review on acidifying treatments for vegetable canned food. *Critical Reviews in Food Science and Nutrition, 51*(10), 955–964. https://doi.org/10.1080/10408398.2010.490883.
56. Plaza-Diaz, J., Ruiz-Ojeda, F. J., Gil-Campos, M., & Gil, A. (2019). Mechanisms of action of probiotics. *Advances in Nutrition, 10*(suppl_1), S49–S66. https://doi.org/10.1093/advances/nmy063.
57. Tappi, S., Tylewicz, U., Romani, S., Siroli, L., Patrignani, F., Dalla Rosa, M., & Rocculi, P. (2016). Optimization of vacuum impregnation with calcium lactate of minimally processed melon and shelf-life study in real storage conditions. *Journal of Food Science, 81*(11), E2734–E2742. https://doi.org/10.1111/1750-3841.13520.
58. Occhino, E., Hernando, I., Llorca, E., Neri, L., & Pittia, P. (2011). Effect of vacuum impregnation treatments to improve quality and texture of zucchini (*Cucurbita pepo* L.). *Procedia Food Science, 1*, 829–835. https://doi.org/10.1016/j.profoo.2011.09.124.
59. Quintanilla, A., Mencía, A., Powers, J., Rasco, B., Tang, J., & Sablani, S. S. (2018). Vacuum impregnation of firming agents in red raspberries. *Journal of the Science of Food and Agriculture, 98*(10), 3706–3714. https://doi.org/10.1002/jsfa.8888.
60. Zhao, Y., & Xie, J. (2004). Practical applications of vacuum impregnation in fruit and vegetable processing. *Trends in Food Science & Technology, 15*(9), 434–451. https://doi.org/10.1016/j.tifs.2004.01.007.
61. Joshi, A. P. K., Rupasinghe, H. P. V., & Pitts, N. L. (2010). Sensory and nutritional quality of the apple snacks prepared by vacuum impregnation process. *Journal of Food Quality, 33*(6), 758–767. https://doi.org/10.1111/j.1745-4557.2010.00338.x.
62. Santana Moreira, M., de Almeida Paula, D., Maurício Furtado Martins, E., Nascif-Rufino Vieira, É., Mota Ramos, A., & Stringheta, P. C. (2018). Vacuum impregnation of β-carotene and lutein in minimally processed fruit salad. *Journal of Food Processing and Preservation, 42*(3). https://doi.org/10.1111/jfpp.13756.
63. Magalhães da Veiga Moreira, I., de Figueiredo Vilela, L., da Cruz Pedroso Miguel, M. G., Santos, C., Lima, N., & Freitas Schwan, R. (2017). Impact of a microbial cocktail used as a starter culture on cocoa fermentation and chocolate flavor. *Molecules, 22*(5), 812. https://doi.org/10.3390/molecules22050812.
64. Batpho, K., Boonsupthip, W., & Rachtanapun, C. (2017). Antimicrobial activity of collagen casing impregnated with nisin against foodborne microorganisms associated with ready-to-eat sausage. *Food Control, 73*, 1342–1352. https://doi.org/10.1016/j.foodcont.2016.09.027.
65. Laurindo, J. B., Stringari, G. B., Paes, S. S., & Carciofi, B. A. M. (2007). Experimental determination of the dynamics of vacuum impregnation of apples. *Journal of Food Science, 72*(8), E470–E475. https://doi.org/10.1111/j.1750-3841.2007.00585.x.
66. Tylewicz, U., Lundin, P., Cocola, L., Dymek, K., Rocculi, P., Svanberg, S., Dejmek, P., & Gomez Galindo, F. (2012). Gas in scattering media absorption spectroscopy

(GASMAS) detected persistent vacuum in apple tissue after vacuum impregnation. *Food Biophysics, 7*(1), 28–34. https://doi.org/10.1007/s11483-011-9242-6.

67. Yılmaz, F. M., & Bilek, S. E. (2018). Ultrasound-assisted vacuum impregnation on the fortification of fresh-cut apple with calcium and black carrot phenolics. *Ultrasonics Sonochemistry*, *48*, 509–516.

CHAPTER 11

MICROWAVE APPLICATIONS FOR THE PRODUCTION OF READY-TO-EAT SNACKS

B. POOJA, KRISHNA H. MENON, and G. NAGAMANIAMMAI

Department of Food Process Engineering, School of Bioengineering, College of Engineering and Technology, SRM IST, Kattankulathur, Tamil Nadu, India

ABSTRACT

Microwave heating has been extensively used in food processing. The major applications are heating, frying, extraction, and drying. However, an increasing trend in the usage of microwave processing for the dehydration of ready-to-eat food products and extraction of value-added components has been observed. The microwave heating process creates an electric field that polarizes the charges in the food by two mechanisms such as polarization and conduction, upon the application of heat. The frequency ranges from 300 MHz to 300 GHz, while 915 MHz is used in most food industries. The main advantage of microwave processing is the retention of significant nutrients compared to conventional heating techniques. The application of microwaves in ready-to-eat snacks includes fruit bars, low-fat snacks, potato chips, noodles, etc. This processing can be applied to first, second, and third-generation ready-to-eat snacks. Microwave processed snacks are widely accepted by consumers in terms of palatability and are safe for consumption as per the recommended standards.

Ready-to-Eat Snacks: Emerging Technologies for Production and Safety.
Ravi Pandiselvam and Gülşah Çalışkan Koç

11.1 INTRODUCTION

The present-day food items produced in the market are processed using conventional thermal techniques. It is found that the foods processed using microwaves have a greater shelf life and good qualitative traits. Microwave processing has been used in industrial processing for many years but has been in action for almost 40 years [1, 2]. Microwaves are part of electromagnetic waves, and the frequency ranges between 300 MHz and 300 GHz and wavelengths of 1 mm to 1 m [3]. Microwave technology has emerged as a promising processing tool for the production of ready-to-eat snacks due to its ability to rapidly heat and cook food products. Ready-to-eat snacks are convenient food products that require minimal preparation and can be consumed directly without further cooking. Examples of such snacks include popcorn, chips, nuts, and other similar products. The food sector has utilized microwaves for a variety of purposes, including pasteurization, cooking, sterilization, frying, drying, and blanching. The microwave application has a significant advantage in that the heat penetrates the food in a short time and has a greater energy efficiency [4]. Microwave processing is getting popular in a number of industries, such as food processing, pulp & paper, medicine, and materials science [5]. A wide range of foods, such as vegetables, have been exposed to MW treatments [6]. The action of microwaves takes place between the water molecules and the charged particles present in the food, causing uniform and rapid heat generation in the food material. The dipole rotation of the water molecules present in the food can lead to the generation of heat hence water is one of the most important components that interact with the microwave and causes the microwave [7]. The microwave system consists of the following components: magnetron, transformer, oven cavity, waveguide, and sample. The frequency of domestic microwaves is 2.54 GHz. The benefits of microwave heating include quick processing, consistent heating, energy efficiency, increased product quality, and avoiding huge temperature gradients. As the name suggests, ready-to-eat snacks can be directly consumed without additional cooking or preparations [8].

Ready-to-eat snacks are pre-cooked or processed foods that require minimal preparation before consumption, making them convenient and time-saving for consumers. Ready-to-eat snacks are classified into three generations [9]. Snack foods from the first generation consist of snacks such as roasted almonds, potato chips, popcorn, etc. The majority of snacks, single-ingredient snacks; instantaneously enlarged snacks; and simple-shaped products like maize crisps, puffed snack foods, papads, and fryums are examples of second-generation snack foods [10, 11]. Third-generation snack foods (half-products or pellets): This category includes pellets and snacks manufactured from numerous components and cooked by extrusion. Overall, microwave technology is a versatile tool that has a number of applications in the production of ready-to-eat snacks. It allows for faster cooking times, improved quality, and increased safety, making it a valuable asset in the food industry.

11.2 MICROWAVE HEATING

11.2.1 PRINCIPLE

Microwave heating is a type of electromagnetic heating that uses microwave radiation to cook or heat food (Figure 11.1). Microwave heating works by creating an electromagnetic field within the microwave oven, which causes water molecules in the food to vibrate and generate heat through friction. This results in rapid and uniform heating of the food, particularly in the outer layers, which can help to reduce cooking times and improve the quality of ready-to-eat snacks.

The principle of microwave heating is based on the fact that water molecules have a positive and negative end, which causes them to align themselves with an alternating electromagnetic field. When the field changes direction, the water molecules vibrate and collide with each other, which generates heat. This heat is transferred to the food, causing it to cook or heat up [12, 13]. The penetration depth of microwaves is determined by the frequency of the electromagnetic radiation used and the dielectric properties of the food Figure 11.2. Ready-to-eat snacks, which are typically low-moisture foods, have lower dielectric constants and loss factors compared to high-moisture foods. As a result, microwave heating in these snacks primarily occurs in the outer layers, which can lead to a crispy and browned texture. The use of microwave technology for the production of ready-to-eat snacks can also lead to improved quality and reduced cooking times, as it can help to preserve the flavor, texture, and nutritional content of the food. The heating of the food results from the electromagnetic field in the chamber of the microwavable device; the microwave converts the electromagnetic energy into thermal energy. The main components present in microwave include:

1. A high voltage transformer: a microwave oven requires high voltage to generate power; hence a high voltage transformer is used.

2. Cavity magnetron: the cavity magnetron is a vacuum chamber that converts electrical energy into microwaves.
3. Microcontroller- a microcontroller helps to communicate between the user and the device. It consists of processing cores along with programmable inputs and outputs.
4. Waveguide–guides the waves produced at the magnetron to the cavity for output.
5. Cooling fan- a cooling fan reduces the temperature produced by the magnetron [14] (Figure 11.1).

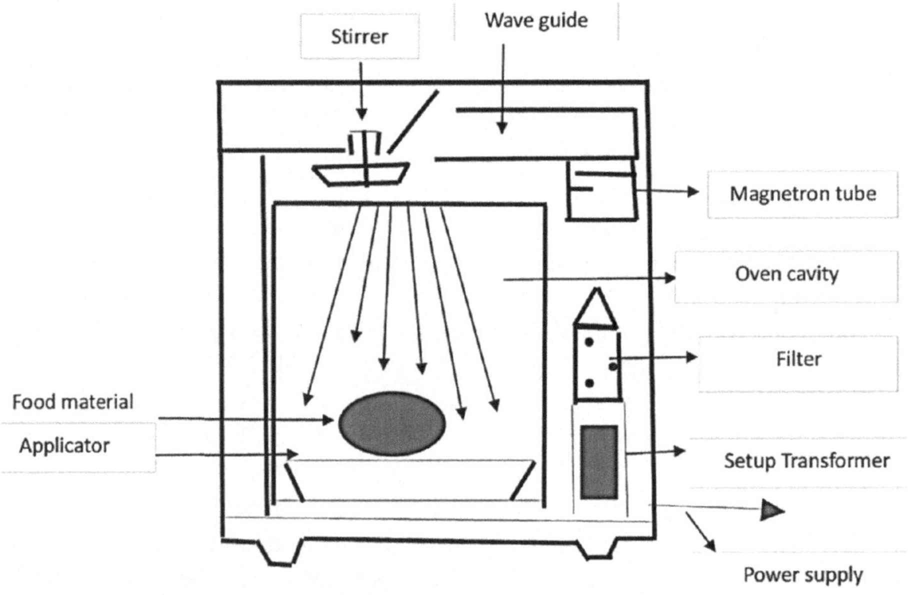

FIGURE 11.1 Components of microwave.

11.3 DIELECTRIC PROPERTIES OF RTE SNACK FOOD UPON OHMIC HEATING

The dielectric properties of food play an important role in the interaction between microwaves and food. Dielectric properties are the electrical properties of a material that determine its response to an electric field. The food's dielectric property can be viewed as the key variable in calculating the electromagnetic coupling and coupling stated in the matter of relative permittivity:

$$e_r = e_r^{''} - je_r^{''} \qquad (1)$$

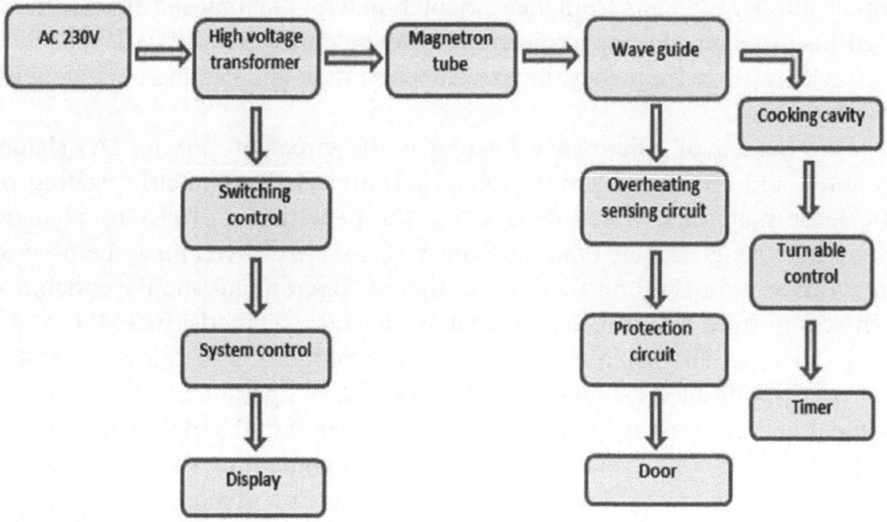

FIGURE 11.2 Block diagram of microwave working.

Where; $j = \sqrt{-1}, \varepsilon_r'$ is dielectric constant, and $j\varepsilon_r''$ is the loss factor (as per Eqn. (1)) [15]. The dielectric property depends on food composition, temperature, salt content, ingredients, and microwave frequency. The dielectric property of food decreases as the moisture content decreases; specifically, the imaginary part ε_r'' diminishes as the water level declines [16]. Numerous factors influences the dielectric characteristics of most materials with moisture content being a crucial component of electricultural products. Additionally, the absorption of microwaves (MW) is determined by these properties.

11.3.1 FOOD MATERIALS ENERGY AND SUBSEQUENT HEATING BEHAVIOR IN DIELECTRIC HEATING

Physical features of raw materials, surface area, components present in food, and volume, are some factors that influence the microwave heating process. Other properties such as moisture content, temperature, dielectric frequency, and density affect the food upon microwave heating [17]. Different dielectric frequencies can also give uniform heating and permeability to various food products. The dielectric characteristics of meals would vary depending on the dielectric frequencies used. The polarization of molecules having permanent dipole moments results from their orientation with the imposed electric field, significantly contributing to dielectric characteristics frequency dependence [18] with a lower frequency, improved penetration and heating homogeneity was achieved.

One benefit of Microwave heating is the effect of varying (A) shape, volume, and (B) surface on power absorption [19] volumetric heating of dielectric materials, which depends on the penetration of electromagnetic waves. It was guaranteed that microwave (2,450/915 MHz) may be used to treat different in-shell meals (e.g., almonds). Operational and dependability can be improved by using low-frequency dielectric methods (915 MHz MW) [20]. There is an increase in the dielectric properties of the ready-to-eat extruded foods such as noodles and pasta with temperature as the moisture content in these food items is meagre. The increase in the dielectric constant is due to water-bound relaxation. The products containing sugar, e.g., cakes, biscuits, etc., showed that as the concentration of glucose increases, the dielectric constant decreased as the water content was less to react with the electric field [21]. The inclusion of salt to starch solutions increased their dielectric loss factor.

11.4 MICROWAVE PROCESSING ON DIFFERENT READY-TO-EAT FOODS

11.4.1 TEMPERING AND THAWING

The lower thermal conductivity of water limits the traditional thawing of frozen goods compared to ice. The thawing process slows as the water layer thickens [22]. Microwave ovens can also be used for thawing frozen snacks such as frozen pizzas and chicken nuggets. The microwaves quickly and evenly thaw the food, allowing it to be cooked or heated up in a shorter amount of time.

Microwaves are used to swiftly thaw small quantities of food and dissolve lipids, for example, food items such as butter, chocolate, fondant cream, etc. The water heats up quickly as the ice melts because it has a higher loss factor than ice [23]. Because thawing may not proceed evenly throughout the massive blocks, certain parts of food may cook while others remain frozen. Products such as frozen chicken that can be defrosted in the microwave are examples of food goods [24]. Due to water's weaker heat conductivity than ice, traditional methods of defrosting frozen objects are constrained. The thawing process slows as the water layer thickens.

11.4.2 HEATING AND BAKING

Microwave ovens are commonly used for cooking and heating ready-to-eat snacks. The microwaves penetrate the food and excite the water molecules, causing them to heat up quickly and evenly. This allows snacks like popcorn, frozen meals, and potato chips to be cooked or heated up in a matter of minutes. Microwave baking conserves nutrition while saving time, energy, and space [25]. Baking involves heat transfer as well as mass transfer that induces both biological, chemical, and physical modifications, for example, gelatinization of starch, emission of carbon dioxide by leavening agents, evaporation of water, volume expansion, browning reaction, and crust development [26]. The surface temperature of dishes cooked in the microwave can exceed 212°F. The combination of halogen lamp with microwave baking is a revolutionary technique in the bakery industry. Traditional baked items offer advantages that microwave baked goods do not. Cakes, bread, doughnuts, muffins, and biscuits include edible forms [27].

11.4.3 DRYING

The microwaves create an electromagnetic field inside the substance during the drying process, and the internal pressure differential forces the water to the surface. Chips and cookies drying to perfection is a common occurrence. During microwave drying, pressure develops inside the food casing, causing evaporation to speed up. Pasta noodles are made using this method. Microwave technology can be used for drying snacks such as fruits and vegetables. The microwaves remove the moisture from the food quickly and efficiently, resulting in a crispy and crunchy texture. This is particularly useful for snacks that need to have a long shelf life. Microwaves and other systems are used to prepare various foods. Pasta, egg yolk powder, tomato paste, chocolate powder, and snack food can all be made using microwave-hot air drying. Drying potato chips and other snack foods in the microwave is a common practice. To package products in flat containers, the bulk of contemporary options use thermoform/vacuum/gas flush seal technology. Packages are heated at 158–194°F using a combination of hot air and microwaves in up to 25-meter-long tunnel conveyors. During the equilibration phase, the packages' slowest heating regions reach 176–185°F in 10 minutes. These packages are then cooled to 33.8–35.6°F and have a storage period of about 40 days at 46.4°F [28–30].

11.5 INTERACTION OF MICROWAVES ON RTE SNACK FOOD COMPONENTS

Extrusion systems followed by microwave processing for producing 3rd generation snacks are widely used nowadays [31–33]. In extrusion cooking, the pellets acquire a stable minimum moisture content after cooking and later undergo puffing during microwave cooking. Microwave processing can be performed in pre-packed puffing and popping, where the waves interfere with the internal properties of the food to cause desirable structural and sensory changes [33]. The sequence changes in the ready-to-eat puffed and popped snack product occurs due to the expansion of internal moisture due to the heat developed by the microwave, resulting in the popping of build-up moisture through the internal shell of the kernel. In puffing, the thermal gradient causes the sudden release of water vapor, thereby expanding the starchy endosperm to gain a higher puff yield. The degree of puffing and popping depends on internal moisture content, degree of gelatinization, residence time, amylose content, microwave power density, and the variety of the grain [34]. The studies conducted based on the above evidence concluded that the microwave puffing of a paddy variety was maximum at a moisture content of 14% and a microwave power density of 800 Watt [35]. Again, these variables depend upon the diversification of coatings applied to the food product, such as butter, salt, hydrogenated oil, RTE flavors, as well as the addition of other components [36]. The moisture content of the kernel is a critical process parameter that determines the expansion behavior of pellets. The presence of high moisture leads to high vapor pressure and uneven heat transmission, while the presence of less moisture leads to uneven expansion of the pellets [37]. The optimum moisture content of 10% is preferable along with half gelatinization of starch to acquire uniform shape and air cell distribution in the pellets [38]. The vibrational energy of microwave heating leads to the routing and friction of water molecules, which produce heat due to dipole rotation and ion polarization. The alignment of polar water molecules in the food will produce collective potential energy, producing heat. In ion polarization, the acceleration of ions in food happens due to applying an electric field, giving rise to a consecutive rise in heat. The degree of acceleration is again determined by the amount of moisture content in the food. Moisture produces superheated steam in the nuclei of the food's kernel, a glassy matrix due to the pre-gelatinization of starch, to create a localized pressure [39]. The whole expansion behavior of the food matrix occurred in two transition phases: first, the transition of the glassy matrix to the amorphous state and again to the glassy matrix (Figure 11.3). The superheated heat is responsible for the first transition, and the reversal occurs when the former is released at the end of the microwave heating process. This whole process is also known as volumetric heating within the sample [40, 41]. In addition to chemical composition and physical content, there are certain supplementary factors that influence the expansion behavior of food. This includes the presence of carboxymethyl cellulose (CMC), fat, salt, sucrose, and xanthan gum (XG). CMC, sugar, and salt have the ability to mimic the dielectric properties of food, thereby influencing the whole process [43, 44].

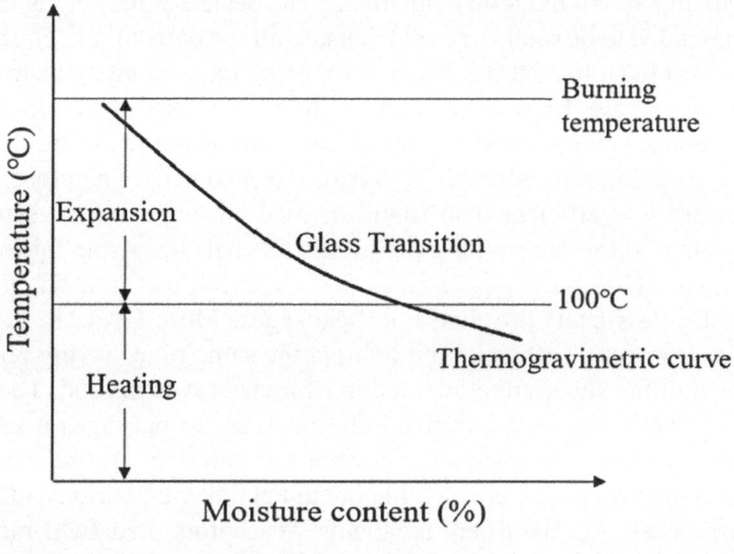

FIGURE 11.3 Microwave expansion of cereal matrix.
Source: Adapted from Ref. [41].

11.6 IN-PACKAGE MICROWAVE PROCESSING OF RTE SNACKS

The microwavable packages should provide optimum microbial safety, preserve the quality of the product, prevent interaction between the packaging material and food, and provide uniform microwave heating of the product. The packaging material should not interfere with the chemical composition of the food and should be convenient for processing. Folded or molded paperboard containers are usually used for microwave treatment, along with ethylene-vinyl alcohol providing oxygen barrier properties. They are preferable because of their solid structure and compatibility [45].

Other materials such as aluminum, soda-lime, or borosilicate glass, metals cannot be used because they are prone to leaching, breakage, and corrosion. Aluminum foils are widely used for food wrapping and packaging, but they prevent microwave interaction with food as it reflects the waves into the source. The same action in metallic packaging also momentously affects the food quality. Glass containers are susceptible to breakage during microwave treatment due to the heating of food [46]. Also, these bottles/jars are not transportation convenient, restricting its application as a packaging material [47, 48]. The packaging material should withstand the applied power and temperature of the microwave treatment and ensure an unvarying heat transfer to all parts of the food component. The design of packaging material is recommended to be round or oval to ensure all the parts of the food receive the same heat [47]. In contrast, square or rectangular packaging restricts this uniform microwave heating of food, with temperature in the core lesser than temperature along the edges [49]. Similar temperature differences were observed in studies performed by various researchers. Therefore, edged packages are less efficient than round or oval packages due to symmetry [50]. Another factor determining the product's shelf life is the thickness of the packages. Thin packages ensure a symmetrical heat transfer but have considerably less shelf life than thick packages. More layers in the latter will have a protective effect on the food, at the same time, having possibilities to restrict the superlative interaction of microwaves in food. Therefore, optimizing the microwave power on the thickness of packages is essential to consider the layers of packages for each food product. Active containers are the modification of microwavable containers, having various additional properties. They are classified as shields, susceptors, and field modifiers [51]. Shields avert the absorption of heat by the microwave, while susceptors use the method of conduction to transfer the heat onto the surface layer of the food. As the name suggests, the field modifier modifies the distribution of heat within the food product.

11.7 EFFECTS OF MICROWAVE PROCESSING ON THE NUTRITIVE VALUE OF RTE SNACKS

Pulse-based snacks have a wider reception in the food industry due to their nutritional supplementation than any other food. Pulse-based RTE snacks have significantly increased in number compared to other RTE snacks. In addition to the nutrient content, other factors such as minimum moisture content, physical structure, and the complementation of various functional components in the pulses have led to its wide acceptance in this area. Studies performed have observed that the microwaved RTE snack based on fish had enhanced flavor and protein aggregation, which accounts for the product quality [53–55].

11.7.1 EFFECT ON CARBOHYDRATES

The predominant storage carbohydrates found in most kernels are starch, constituting 30–50% of the whole kernel dry mass. Starch has wider applications in the food and chemical industries. Human dietary consumption of starch includes cassava, cereals, rice, grains, bread, etc. The major components of starch are amylose and amylopectin, surrounded by a protein matrix. Various studies have shown that microwave processing changes the structure of carbohydrates, along with some changes in their physicochemical properties. The addition of sucrose contributes to a volumetric expansion of the product [37]. The optimum starch content for uniform expansion was observed to be 60%. The above experiment indicated that the MW elevates the gelatinization temperatures. In contrast, other works suggest that the molecular weight of these wet starches was significantly decreased along with the reduction in viscosity after the processing [56]. Solubility and crystallinity were also diminished to a lesser value after MW. Microwave treatment also catalyzes reactions such as the synthesis and hydrolysis of starch components. Investigations have also concluded that microwave processing increases the activity of the enzyme glucose isomerases [57]. However, the retrogradation of starch occurs during MW heating at 750 W with an exposure time of 40 seconds. In-situ heat generation during microwave processing leads to expansion and causes the release of certain active components in food and the inactivation of beta-glucan lytic enzymes [58]. This property can be utilized to produce gluten-free products as specific undesirable enzymes are eliminated during this process.

11.7.2 EFFECT ON PROTEIN

Proteins are one of the most important components in food. RTE snacks with sufficient protein content have wide acceptability among consumers as a recent trend has been seen inclined towards healthy snacks. Proteins are essential for most biological activities in the body, from cell building to DNA replication [60]. Proteins exist in diverse structures in each food, but the majority are of three-dimensional or tertiary structures. Pulse RTE snacks have more protein content than most cereal-based RTE snacks [61]. MW treatment of proteins is associated with various physical and chemical changes [62]. In addition, proteins are subjected to degradation, denaturation, unfolding in their structure, and acceleration of reactions upon the various MW treatment power density administered [63]. Amino acids are formed due to the hydro degradation of proteins during the MW treatment. The MW treatment could lead to bioavailable free amino acids in the food product, but the stability of those amino acids is questionable. The experiment conducted on the in vitro protein digestibility and antinutrient factors such as trypsin inhibitors of soymilk proteins upon MW treatment at 185°F for a time period of 600 seconds concluded that an increase of 93% and a decrease of 52% were observed for trypsin inhibition activity and protein digestibility respectively [64]. Also, an increase in the secondary structure of proteins was also observed, which confirms its digestibility. Enhancement of surface hydrophobicity due to the exposure of hydrophobic residues due to protein denaturation due to MW treatment was observed in certain research [65] in the study of characteristics of quinoa protein aggregates upon MW irradiation. Polymerization and intracellular disulfide bond formation of proteins have also been observed. This experiment noted a significant reduction in the protein gliadin during polymerization, which reduced the cross-reactivity reaction with the administered antibody (R5 antibody). Other studies have shown that gluten content decreases significantly with different microwave times and power densities. Various studies have been performed to analyze the effect of MW treatment on total polyphenolic content (TPC) and antioxidant activity. These works have also demonstrated that the TPC of the phenolic extracts was highest (2.64 mg GAE g^{-1}) in microwaved black quinoa compared to conventionally treated samples [68–71]]. Similarly, high antioxidant and α-glucosidase inhibitory activity were also observed in the former, which confers the application of microwave processing in various cereal and pulse-based RTE snacks. MW treatment has also been found to reduce allergens in shrimp containing RTE snacks. The allergenicity reduction of the protein tropomyosin in shrimp has reduced to 75% upon MW administration at 275°F for 900 seconds [72]. Furthermore, the study also concluded that the increase in β sheets had affected the total amino acid content as well as the assimilability of shrimp protein.

11.7.3 EFFECT ON LIPIDS

The lipids in food components could be fatty acids, monoglycerides, triglycerols, and emulsifiers. They could be free lipids associated with hexane or petroleum ether extraction and bound lipids, which can be extracted with chloroform-methanol mixtures. Most of the non-polar lipids are associated with bound lipids [73]. Lipids form a complex with starch (amylose) and exist as an outer protective covering in many RTE snacks, especially extruded products. They, at times, act as lubricants in extruded foods, facilitating easier passage through the equipment. However, this lubricating effect is also observed to contribute to the product's volumetric expansion [74]. As lipids melt during treatment, it facilitates smooth transportation of the air bubbles formed inside the matrix, preventing the bubbles' breakage and accounting for uniform expansion. During MW treatment of food, lipids undergo oxidation, binding with amylose, antioxidant and free radicle formation, and cis-trans isomerization of unsaturated fatty acids, which significantly affect the nutritional characteristics of the snacks [75].

Solid fats account for most expansion in RTE food, whereas liquid fats have the reverse effect, as they do not facilitate bubble transportation from one matrix to the other. [76] suggested that 6% solid fat content contributes to maximum expansion during MW treatment, and generally, the fat content greater than 2% accounts for uniform expansion. In addition, it also influences the texture and microstructure of the expanded product, resulting in high moisture loss. Nevertheless, microwave treatment is usually preferable to conventional treatments such as deep fat frying as it will limit the fat content of RTE products.

11.8 MICROBIAL SAFETY OF MICROWAVE PROCESSED RTE SNACKS

The term microwave pasteurization itself denotes the action of microwaves on microorganisms. The thermal effect of microwave heating was inhibitory to many possible microorganisms in food. Microwave irradiation has been extensively used to inactivate pathogens in various industries [77]. The advantage of microwave treatment over conventional heat treatments is that the loss of nutrients due to exposure to high temperatures can be avoided. Microwave heating inactivates many bacteria, such as *Escherichia coli*, *Streptococcus faecalis*, *Clostridium perfringens*,

Staphylococcus aureus, *Salmonella*, and *Listeria* spp; actinomycetes, fungi, and bacteriophages (bacteriophage PL-1) [78]. Investigations on the microbiological stability of punnets of pear with a target temperature of 194°F have observed a 2-log reduction in the standard plate count of bacteria, yeast, and mold. The count number stayed the same throughout the cold storage of the pears at 39.2°F [79]. Another study indicated that the modified atmospheric packaging system (MAPS) of the microwave-assisted RTE snacks had controlled most of the pathogens, with an aerobic plate count (APC) of 105 to 106 CFU/g upon the storage of 6 weeks at 45.14–35.6°F. A minimum reduction of *Clostridium botulinum* spores and *Listeria monocytogenes* by 6 logs with an APC of not greater than 107 CFU/g is preferred for optimum microbiological safety and stability. Other studies demonstrated the MW treatment at 1,300 W for 80 seconds and 100 seconds, respectively, for the microbial inactivation of white shrimp, along with induction heating microwave-assisted induction heating (MAIH) for the comparison. The work concluded that higher inactivation was observed in microwave-assisted induction heating contrary to MW treatment [80]. But another research has contrasted all the previously performed works and remarked that microwave treatment could not be relied on as the final word of pasteurization. Due to different product sizes and shapes, the non-uniform heat transmission led to the failed inactivation of a few microorganisms in the sample taken. The experimental study showed the unsatisfactory inactivation of Listeria monocytogenes at temperatures ranging from 46.4–60.8°F [77]. The complete microbial inactivation during the microwave treatment occurs only if the sample contains water. One experiment administered a microwave power of 2,450 MHz on dry bacteria and viruses and inferred that the microorganisms remained intact. It is primarily due to the inadequate energy absorbed in the microbial cells. Moreover, most of the inactivated microorganisms were not lysed in the food. The thermal effect of microwaves induces the aggregation of proteins in their cells, which results in denaturation [81].

11.9 CHALLENGES ASSOCIATED WITH MICROWAVE PROCESSING

Like any other food processing technique, microwave processing is also associated with its pros and cons. From the safety and quality point of view, one of the significant drawbacks is the irregular transfer of heat throughout the product, creating hot spots and cold spots, leading to failure in complete microbial inactivation [82]. This mainly depends upon the thickness of the food product and the type of packaging material used. The ultimate aim of this processing technique is "heat and eat;" therefore, the product's microbiological safety cannot be compromised. Hence, the optimization of microwave power density for food products with specific volume, weight, and active packaging material has to be performed and is considered a future research area [83]. Several researchers have performed experiments to reduce uneven heat transfer and microwave treatment, but most of the correction methods are in pipelines. Many models have also been developed to combat this complication but are not commercially applied. Several researchers have conducted studies on the uniformity of temperature distribution on RTE snacks. Their study, which consisted of microwave heating of a few food products, observed that the temperature difference between the hottest and coldest spots was more than 158°F, even at a processing time of 10 seconds. Also, in cylindrical-shaped food packages, the hottest spots were in the center of the axis line, while the coldest spots were in the area above the axis line. Similarly, the temperature difference between spaghetti and rice was found to be 65°C [83, 84]. To achieve uniform temperature dispersal, few methods have been employed commercially. One such method is using metal bands to shield the food materials during microwave processing. Another method is keeping the food products side by side rather than stacking them. The design of the oven, ingredients in food, and a few other variables can also be controlled regarding the same. Moisture accumulation in the product is another problem that needs to be addressed (Table 11.1) [85]. Such events are unpleasant and could lead to lesser 149°F consumer acceptance of RTE snacks. Retention of moisture after microwaving is avoidable, yet research is being carried out to reduce the former.

TABLE 11.1 Temperature of Ring, Oval, and Oblong-Shaped Beef-Pork Loaves

Power	Time of Cooking, Seconds	Ring Shaped Loaves Temperature, °F	Time of Cooking, Seconds	Oval Shaped Loaves Temperature, °F	Time of Cooking, Seconds	Oblong Shaped Loaves Temperature, °F
Intense	660	136.04–152.96	780	118.04–156.02	1,020	125.96–152.06
	780	149.0–157.82	1,020	145.94–168.98	1,320	145.94–156.02
Moderate	1,020	129.056–138.92	1,200	136.94–141.08	1,560	133.88–140.0
	1,200	149.0–156.92	1,440	150.98–167.0	2,040	161.96–183.92

11.10 CONCLUSION

Numerous food processes have been effectively enforced using microwave treatment. The non-uniform temperature and hotspot caused by the microwave oven heating have led to nutritive deterioration in the food quality. Advancements in specific areas were the focus of exploration. Specifically, some efficient methods for obtaining finished foods. It is necessary to investigate foods that have improved sensory and nutritional properties. It is important to know about microwaved processed food's nutritional and sensory evaluation. These are some of the areas:

1. The interaction of a variety of dietary ingredients. It is debatable whether the heat effect is the only element influencing food component properties during MW treatment, particularly nutritional loss. The protein, lipids, and carbohydrates are sensitive to microwave heat treatment, but the mechanism that modifies the food components is unknown.
2. Improving the pilot-scale level and carrying out detailed optimized research.
3. An efficient equipment design and operation. A mixture of heating methods was adopted to address the downside of single MW treatment and unite the excellence of several heating techniques. When traditional and MW treatment approaches are compared, it becomes clear that MW-related technology has immense potential.
4. Geometrical homogenization and food features.

The measurement of electromagnetic absorption of distinct sample parts pointed out that the distribution of treatment was not nearly uniform owing to the diverse components in the materials [64]. The hybrid microwave methods, including enhanced sprout bed dehydration, vacuum drying, freeze-drying, and microwave-assisted air drying, have been successfully optimized [65–67, 90–109]. In order to overcome the inconsistencies, homogeneity is a good way. Pre-treatment of ohmic heating can be applied to the foods to increase the microwave heating rate. Homogeneity is an excellent way to eliminate these inconsistencies, for this packaging sample with a surrounding media can be used during microwave heating. Low e" materials can be employed, although e' of medium or packed materials can also be used.

Despite the widespread usage of microwaves in the food processing industry, several areas require further future research. With the increasing demand for personalized and customized snacks, microwave technology could be used to create snacks that are tailored to individual preferences. For example, a microwave oven equipped with sensors and artificial intelligence could detect a person's taste preferences and adjust the cooking time and temperature accordingly, resulting in a snack that is perfectly cooked and seasoned to their liking.

- **Sustainable Packaging:** Microwavable packaging has become increasingly popular in the food industry, and this trend is expected to continue in the future. Microwave technology could be used to create sustainable packaging materials that are biodegradable or recyclable, reducing the environmental impact of ready-to-eat snacks [110–115].
- **Health and Nutrition:** Microwave technology could be used to create snacks that are healthier and more nutritious. For example, microwave-assisted drying could be used to preserve the nutrients in fruits and vegetables, resulting in snacks that are rich in vitamins and minerals. Microwave-assisted cooking could also be used to reduce the fat and salt content of snacks, making them healthier and more appealing to health-conscious consumers [88, 116].
- **Automation:** The food industry is increasingly adopting automation and robotics to improve efficiency and reduce costs. Microwave technology could be integrated into automated snack production lines to speed up cooking and packaging processes, allowing for faster production and lower costs. Currently, there are a few potential future prospects for microwave application in ready-to-eat snacks. As technology continues to evolve, there will likely be many more innovative uses for microwave technology in the food industry [117–124].

KEYWORDS

- **aerobic plate count**
- **food safety**
- **frying**
- **microwave heating**
- **modified atmospheric packaging system**
- **ready-to-eat snacks**

REFERENCES

1. Soni, A., Smith, J., Thompson, A., & Brightwell, G. (2020). Microwave-induced thermal sterilization: A review on history, technical progress, advantages, and challenges as compared to the conventional methods. *Trends in Food Science & Technology, 97*, 433–442. https://doi.org/10.1016/j.tifs.2020.01.013.
2. Viji, P., Debbarma, J., & Ravishankar, C. N. (2022). Research developments in the applications of microwave energy in fish processing: A review. *Trends in Food Science & Technology, 123*, 222–232. https://doi.org/10.1016/j.tifs.2022.05.011.
3. Agrawal, D. (2013). Microwave sintering of metal powders. In *Advances in Powder Metallurgy* (pp. 361–379). Woodhead Publishing.
4. Sakai, N. (2010). Dielectric properties of food and microwave heating. *Japan Journal of Food Engineering, 11*, 19–30.
5. Schiffmann, R. (2017). Packaging for microwave foods. In *The Microwave Processing of Foods* (pp. 273–299). Woodhead Publishing.
6. Ruan, G., Chen, Z., Wei, M., Liu, Y., Li, H., & Du, F. (2013). The study on microwave-assisted enzymatic digestion of ginkgo protein. *Journal of Molecular Catalysis B: Enzymatic, 94*, 23–28. https://doi.org/10.1016/j.molcatb.2013.04.001.
7. Yüksel Aydar, A., & Alev, M. (2019). Emerging extraction technologies in olive oil production. In *Technological Innovation in the Olive Oil Production Chain* (pp. 51–69). IntechOpen.
8. Yang, R., & Chen, J. (2021). Mechanistic and machine learning modeling of microwave heating process in domestic ovens: A review. *Foods, 10*(9), 2029. https://doi.org/10.3390/foods10092029.
9. Wongsa, J., Uttapap, D., Lamsal, B. P., & Rungsardthong, V. (2016). Effect of puffing conditions on physical properties and rehydration characteristics of instant rice product. *International Journal of Food Science & Technology, 51*(3), 672–680. https://doi.org/10.1111/ijfs.12967.
10. Lamacchia, C., Landriscina, L., & D'Agnello, P. (2016). Changes in wheat kernel proteins induced by microwave treatment. *Food Chemistry, 197*, 634–640. https://doi.org/10.1016/j.foodchem.2015.10.027.
11. Utrilla-Coello, R. G., Bello-Perez, L. A., Vernon-Carter, E. J., Rodriguez, E., & Alvarez-Ramirez, J. (2013). Microstructure of retrograded starch: Quantification from lacunarity analysis of SEM micrographs. *Journal of Food Engineering, 116*(4), 775–781. https://doi.org/10.1016/j.jfoodeng.2013.01.024.
12. Guzik, P., Kulawik, P., Zając, M., & Migdał, W. (2022). Microwave applications in the food industry: An overview of recent developments. *Critical Reviews in Food Science and Nutrition, 62*(29), 7989–8008. https://doi.org/10.1080/10408398.2021.1949678.
13. Ohlsson, T., & Bengtsson, N. (2001). Microwave technology and foods. In *Advances in Food and Nutrition Research* (Vol. 43, pp. 65–140). https://doi.org/10.1016/S1043-4526(01)43003-0.
14. Wray, D., & Ramaswamy, H. S. (2015). Novel concepts in microwave drying of foods. *Drying Technology, 33*(7), 769–783. https://doi.org/10.1080/07373937.2015.1018288.
15. Buffler, C. R. (1995). Advances in dielectric measurement of foods. In *Characterization of Food* (pp. 213–232).
16. Gaukel, V. T., Siebert, U., & Erle, U. (2017). Microwave-assisted drying. In *The Microwave Processing of Foods* (pp. 152–178). Woodhead Publishing.

17. Venkatesh, M. S., & Raghavan, G. S. V. (2004). An overview of microwave processing and dielectric properties of agri-food materials. *Biosystems Engineering, 88*(1), 1–18. https://doi.org/10.1016/j.biosystemseng.2004.01.013.
18. Gutiérrez, J. D., Catalá-Civera, J. M., Bows, J., & Peñaranda-Foix, F. L. (2017). Dynamic measurement of dielectric properties of food snack pellets during microwave expansion. *Journal of Food Engineering, 202*, 1–8. https://doi.org/10.1016/j.jfoodeng.2017.01.002.
19. Ryynänen, S., Tuorila, H., & Hyvönen, L. (2001). Perceived temperature effects on microwave heated meals and meal components. *Food Service Technology, 1*(3), 141–148. https://doi.org/10.1046/j.1471-5731.2001.00015.x.
20. Xin, Y., Zhang, M., Xu, B., Adhikari, B., & Sun, J. (2015). Research trends in selected blanching pre-treatments and quick-freezing technologies as applied in fruits and vegetables: A review. *International Journal of Refrigeration, 57*, 11–25. https://doi.org/10.1016/j.ijrefrig.2015.04.007.
21. Meda, V., Orsat, V., & Raghavan, V. (2017). Microwave heating and the dielectric properties of foods. In *The Microwave Processing of Foods* (pp. 23–43). Woodhead Publishing.
22. Li, R., Zhang, S., Kou, X., Ling, B., & Wang, S. (2017). Dielectric properties of almond kernels associated with radio frequency and microwave pasteurization. *Scientific Reports, 7*(1), 1–10. https://doi.org/10.1038/s41598-017-14687-x.
23. Virtanen, A. J., Goedeken, D. L., & Tong, C. H. (1997). Microwave assisted thawing of model frozen foods using feed-back temperature control and surface cooling. *Journal of Food Science, 62*(1), 150–154. https://doi.org/10.1111/j.1365-2621.1997.tb10958.x.
24. Yang, R., & Chen, J. (2022). Dynamic solid-state microwave defrosting strategy with shifting frequency and adaptive power improves thawing performance. *Innovative Food Science & Emerging Technologies, 81*, 103157. https://doi.org/10.1016/j.ifset.2022.103157.
25. Sumnu, G. (2001). A review on microwave baking of foods. *International Journal of Food Science & Technology, 36*(2), 117–127. https://doi.org/10.1046/j.1365-2621.2001.00446.x.
26. Chavan, R. S., & Chavan, S. R. (2010). Microwave baking in food industry: A review. *International Journal of Dairy Science, 5*(3), 113–127. https://doi.org/10.3923/ijds.2010.113.127.
27. Atuonwu, J. C., & Tassou, S. A. (2019). Energy issues in microwave food processing: A review of developments and the enabling potentials of solid-state power delivery. *Critical Reviews in Food Science and Nutrition, 59*(9), 1392–1407. https://doi.org/10.1080/10408398.2018.1474160.
28. Ashraf, S., Saeed, S. M. G., Sayeed, S. A., & Ali, R. (2012). Impact of microwave treatment on the functionality of cereals and legumes. *International Journal of Agriculture & Biology, 14*(3).
29. Awuah, G., Tony Koral, & Dongsheng Guan. (2014). Radio-frequency baking and roasting of food products. In *Radio-Frequency Heating in Food Processing: Principles and Applications* (pp. 231–243). CRC Press.
30. Chandrasekaran, S., Ramanathan, S., & Basak, T. (2014). Microwave food processing—A review. *Food Research International, 52*(1), 243–261. https://doi.org/10.1016/j.foodres.2013.07.024.
31. Delgado-Nieblas, C., Aguilar Palazuelos, E., Gallegos-Infante, A., Rocha-Guzmán, N., Zazueta-Moral, J., & Caro-Corrales, J. (2012). Characterization and optimization of

extrusion cooking for the manufacture of third-generation snacks with winter squash (*Cucurbita moschata* D.) flour. *Cereal Chemistry, 89*(1), 65–72. https://doi.org/10.1094/CCHEM-01-12-0003-R.

32. Dhumal, C. V., Pardeshi, I. L., Sutar, P. P., & Babar, O. A. (2014). Optimization of process parameters for development of microwave puffed product. *Journal of Ready to Eat Food, 1*(3), 111–119. https://doi.org/10.5958/2321-5265.2014.00016.6.
33. Panakbalentić, J. E. L. E. N. A., Babić, J., Jozinović, A., Ačkar, Đ., Milicevic, B., Muhamedbegović, B., & Šubarić, D. (2018). Production of third-generation snacks. *Croatian Journal of Food Science and Technology, 10*(1), 98–105. https://doi.org/10.17508/CJFST.2018.10.1.11.
34. Quitain, A. T., Daimon, H., Fujie, K., Katoh, S., & Moriyoshi, T. (2006). Microwave-assisted hydrothermal degradation of silk protein to amino acids. *Industrial & Engineering Chemistry Research, 45*(13), 4471–4474. https://doi.org/10.1021/ie0514602.
35. Rahman, H. A., Salleh, M., & Salim, N. M. (2019). Effect of moisture content and microwave power on puffed yield and expansion volume of Malaysian paddy variety MR297. *Malaysian Applied Biology, 48*(1), 139–143.
36. Kutlu, N., Pandiselvam, R., Saka, I., Kamiloglu, A., Sahni, P., & Kothakota, A. (2022). Impact of different microwave treatments on food texture. *Journal of Texture Studies, 53*(6), 709–736. https://doi.org/10.1111/jtxs.12783.
37. Kraus, S., Enke, N., Gaukel, V., & Schuchmann, H. P. (2014). Influence of degree of gelatinization on expansion of extruded, starch-based pellets during microwave vacuum processing. *Journal of Food Process Engineering, 37*(3), 220–228. https://doi.org/10.1111/jfpe.12095.
38. Kraus, S., Enke, N., Schuchmann, H. P., & Gaukel, V. (2014). Influence of sucrose content on expansion of extruded, starch-based pellets during microwave vacuum processing. *Journal of Food Process Engineering, 37*(6), 628–634. https://doi.org/10.1111/jfpe.12097.
39. Verma, D. K., Mahanti, N. K., Thakur, M., Chakraborty, S. K., & Srivastav, P. P. (2020). Microwave heating: Alternative thermal process technology for food application. In *Emerging Thermal and Nonthermal Technologies in Food Processing* (pp. 25–67). CRC Press.
40. Kraus, S., Schuchmann, H. P., & Gaukel, V. (2014). Factors influencing the microwave-induced expansion of starch-based extruded pellets under vacuum. *Journal of Food Process Engineering, 37*(3), 264–272. https://doi.org/10.1111/jfpe.12093.
41. Boischot, C., Moraru, C. I., & Kokini, J. L. (2003). Factors that influence the microwave expansion of glassy amylopectin extrudates. *Cereal Chemistry, 80*(1), 56–61. https://doi.org/10.1094/CCHEM.2003.80.1.56.
42. Delgado-Nieblas, C. I., Zazueta-Morales, J. J., Aguilar-Palazuelos, E., Jacobo-Valenzuela, N., Aguirre-Tostado, F. S., Carrillo-López, A., & Telis-Romero, J. (2018). Physical, microstructural and sensory characteristics of extruded and microwave-expanded snacks added with dehydrated squash. *Revista Mexicana de Ingeniería Química, 17*(3), 805–821. https://doi.org/10.24275/rmkiq/ingenieriaquimica/v17n3a12.
43. Tiwari, A., & Jha, S. K. (2017). Extrusion cooking technology: Principal mechanism and effect on direct expanded snacks–An overview. *International Journal of Food Studies, 6*(1). https://doi.org/10.7455/ijfs/6.1.2017.a7.
44. Yolacaner, E. T., Sumnu, G., & Sahin, S. (2017). Microwave-assisted baking. In *The Microwave Processing of Foods* (pp. 117–141). Woodhead Publishing.

45. Tang, J., Feng, H., & Lau, M. (2002). Microwave heating in food processing. *Advances in Bioprocessing Engineering, 1*, 1–43.
46. Regier, M. (2014). Microwavable food packaging. In *Innovations in Food Packaging* (pp. 495–514). Academic Press.
47. Muthukumarappan, K. C., Marella, V., Sunkesula, V., & Kutz, M. (2019). *Handbook of Farm, Dairy and Food Machinery Engineering* (pp. 389–415). Elsevier.
48. Auksornsri, T., & Songsermpong, S. (2017). Lethality and quality evaluation of in-packaged ready-to-eat cooked jasmine rice subjected to industrial continuous microwave pasteurization. *International Journal of Food Properties, 20*(8), 1856–1865. https://doi.org/10.1080/10942912.2016.1209496.
49. Kanishka, B., Shyam, S., Sablani, J. T., & Barbara, R. (2013). Migration of chemical compounds from packaging polymers during microwave, conventional heat treatment and storage. *Comprehensive Reviews in Food Science and Food Safety, 12*, 523–545. https://doi.org/10.1111/1541-4337.12020.
50. Nguyen, T. P., & Songsermpong, S. (2022). Microwave processing technology for food safety and quality: A review. *Agriculture and Natural Resources, 56*(1), 57–72. https://doi.org/10.1016/j.anres.2021.08.005.
51. Bhattacharya, M., & Basak, T. (2016). Review on the susceptor assisted microwave processing of materials. *Energy, 97*, 306–338. https://doi.org/10.1016/j.energy.2015.11.068.
52. Steed, L. E., Truong, V. D., Simunovic, J. O. S. I. P., Sandeep, K. P., Kumar, P. R. A. B. H. A. T., Cartwright, G. D., & Swartzel, K. R. (2008). Continuous flow microwave-assisted processing and aseptic packaging of purple-fleshed sweet potato purees. *Journal of Food Science, 73*(9), 455–462. https://doi.org/10.1111/j.1750-3841.2008.00845.x.
53. Cai, L., Feng, J., Cao, A., Zhang, Y., Lv, Y., & Li, J. (2018). Denaturation kinetics and aggregation mechanism of the sarcoplasmic and myofibril proteins from grass carp during microwave processing. *Food and Bioprocess Technology, 11*(2), 417–426. https://doi.org/10.1007/s11947-017-2031-3.
54. Staroszczyk, H., Fiedorowicz, M., Opalińska-Piskorz, J., & Tylingo, R. (2013). Rheology of potato starch chemically modified with microwave-assisted reactions. *LWT-Food Science and Technology, 53*(1), 249–254. https://doi.org/10.1016/j.lwt.2013.02.013.
55. Duan, Z. H., Jiang, L. N., Wang, J. L., Yu, X. Y., & Wang, T. (2011). Drying and quality characteristics of tilapia fish fillets dried with hot air-microwave heating. *Food and Bioproducts Processing, 89*(4), 472–476. https://doi.org/10.1016/j.fbp.2010.11.006.
56. Lisiecka, K., Wójtowicz, A., & Gancarz, M. (2021). Characteristics of newly developed extruded products supplemented with plants in a form of microwave-expanded snacks. *Materials, 14*(11), 2791. https://doi.org/10.3390/ma14112791.
57. Pérez-Quirce, S., Ronda, F., Melendre, C., Lazaridou, A., & Biliaderis, C. G. (2016). Inactivation of endogenous rice flour β-glucanase by microwave radiation and impact on physico-chemical properties of the treated flour. *Food and Bioprocess Technology, 9*(9), 1562–1573. https://doi.org/10.1007/s11947-016-1828-1.
58. Harasym, J., & Olędzki, R. (2018). The mutual correlation of glucose, starch, and beta-glucan release during microwave heating and antioxidant activity of oat water extracts. *Food and Bioprocess Technology, 11*(4), 874–884. https://doi.org/10.1007/s11947-018-2173-7.
59. Villanueva, M., Harasym, J., Muñoz, J. M., & Ronda, F. (2019). Rice flour physically modified by microwave radiation improves viscoelastic behavior of doughs and its

bread-making performance. *Food Hydrocolloids, 90*, 472–481. https://doi.org/10.1016/j.foodhyd.2019.01.016.

60. Xiang, S., Zou, H., Liu, Y., & Ruan, R. (2020). Effects of microwave heating on the protein structure, digestion properties and Maillard products of gluten. *Journal of Food Science and Technology, 57*, 2139–2149. https://doi.org/10.1007/s11483-020-09637-4.
61. Cork, S. D., Blanchard, C., Mawson, A. J., & Farahnaky, A. (2022). Pulse flaking: Opportunities and challenges, a review. *Comprehensive Reviews in Food Science and Food Safety, 21*(3), 2873–2897. https://doi.org/10.1111/1541-4337.12883.
62. Plagemann, R., Von Langermann, J., & Kragl, U. (2014). Microwave-assisted covalent immobilization of enzymes on inorganic surfaces. *Engineering in Life Sciences, 14*(5), 493–499. https://doi.org/10.1002/elsc.201300142.
63. Jiang, H., Liu, Z., & Wang, S. (2018). Microwave processing: Effects and impacts on food components. *Critical Reviews in Food Science and Nutrition, 58*(14), 2476–2489. https://doi.org/10.1080/10408398.2017.1350350.
64. Vanga, S. K., Wang, J., & Raghavan, V. (2020). Effect of ultrasound and microwave processing on the structure, in-vitro digestibility and trypsin inhibitor activity of soymilk proteins. *LWT, 131*, 109708. https://doi.org/10.1016/j.lwt.2020.109708.
65. Zhang, Y., Yan, Y., Li, W., Huang, K., Li, S., Cao, H., & Guan, X. (2022). Microwaving released more polyphenols from black quinoa grains with hypoglycemic effects compared with traditional cooking methods. *Journal of the Science of Food and Agriculture, 102*(13), 5948–5956. https://doi.org/10.1002/jsfa.11812.
66. Huang, K., Shi, J., Li, M., Sun, R., Guan, W., Cao, H., Guan, X., & Zhang, Y. (2022). Intervention of microwave irradiation on structure and quality characteristics of quinoa protein aggregates. *Food Hydrocolloids, 130*, 107677. https://doi.org/10.1016/j.foodhyd.2022.107677.
67. Jian, F., Jayas, D. S., White, N. D. G., Fields, P. G., & Howe, N. (2015). An evaluation of insect expulsion from wheat samples by microwave treatment for disinfestation. *Biosystems Engineering, 130*, 1–12. https://doi.org/10.1016/j.biosystemseng.2014.11.005.
68. Padalino, L., Del Nobile, M. A., la Gatta, B., Rutigliano, M., Di Luccia, A., & Conte, A. (2019). Effects of microwave treatment of durum wheat kernels on quality characteristics of flour and pasta. *Food Chemistry, 283*, 454–461. https://doi.org/10.1016/j.foodchem.2019.01.113.
69. Ruiz-Armenta, X. A., Zazueta-Morales, J. D. J., Delgado-Nieblas, C. I., Carrillo-López, A., Aguilar-Palazuelos, E., & Camacho-Hernández, I. L. (2019). Effect of the extrusion process and expansion by microwave heating on physicochemical, phytochemical, and antioxidant properties during the production of indirectly expanded snack foods. *Journal of Food Processing and Preservation, 43*(12), e14261. https://doi.org/10.1111/jfpp.14261.
70. Marra, F., Zhan, L., & Lyng, J. G. (2009). Radio frequency treatment of foods: Review of recent advances. *Journal of Food Engineering, 91*(4), 497–508. https://doi.org/10.1016/j.jfoodeng.2008.11.007.
71. Yang, H. W., & Gunasekaran, S. (2004). Comparison of temperature distribution in model food cylinders based on Maxwell's equations and Lambert's law during pulsed microwave heating. *Journal of Food Engineering, 64*(4), 445–453. https://doi.org/10.1016/j.jfoodeng.2004.01.009.

72. Dong, X., Wang, J., & Raghavan, V. (2021). Impact of microwave processing on the secondary structure, in-vitro protein digestibility and allergenicity of shrimp (*Litopenaeus vannamei*) proteins. *Food Chemistry, 337*, 127811. https://doi.org/10.1016/j.foodchem.2020.127811.
73. Inchingolo, R., Cardenia, V., & Rodriguez-Estrada, M. T. (2013). The effects of microwave heating on edible oils and lipid-containing food. *Lipid Technology, 25*(3), 59–61. https://doi.org/10.1002/lite.201300008.
74. Deng, X., Huang, H., Huang, S., Yang, M., Wu, J., Ci, Z., & Zhang, D. (2022). Insight into the incredible effects of microwave heating: Driving changes in the structure, properties and functions of macromolecular nutrients in novel food. *Frontiers in Nutrition, 9*. https://doi.org/10.3389/fnut.2022.1016463.
75. Mishra, G., Joshi, D. C., & Panda, B. K. (2014). Popping and puffing of cereal grains: A review. *Journal of Grain Processing and Storage, 1*(2), 34–46.
76. Abo Bakr, T. M. (2020). Microwave applications in food processing: An overview. *Alexandria Journal of Food Science and Technology, 17*(2), 11–22. https://doi.org/10.21608/ajfst.2020.23596.1056.
77. Matera, A., Altieri, G., Ricciardi, A., Zotta, T., Condelli, N., Galgano, F., & Di Renzo, G. C. (2022). Microbiological stability and overall quality of ready-to-heat meals based on traditional recipes of the Basilicata region. *Foods, 9*(4), 406. https://doi.org/10.3390/foods9040406.
78. Devi, A. F., Au, X. N., Weerakkody, R., Sanguansri, P., Swiergon, P., Singh, T., Ng, S., & Gamage, T. (2021). Microwave pasteurised pear snack: Quality and microbiological stability. *Food and Bioprocess Technology, 14*(9), 1615–1630. https://doi.org/10.1007/s11947-021-02743-3.
79. Montero, M. L., Sablani, S., Tang, J., & Ross, C. F. (2020). Characterization of the sensory, chemical, and microbial quality of microwave-assisted, thermally pasteurized fried rice during storage. *Journal of Food Science, 85*(9), 2711–2719. https://doi.org/10.1111/1750-3841.15336.
80. Zhang, L. G., Lyng, J. G., & Brunton, N. P. (2007). The effect of fat, water and salt on the thermal and dielectric properties of meat batter and its temperature following microwave or radio frequency heating. *Journal of Food Engineering, 80*(1), 142–151. https://doi.org/10.1016/j.jfoodeng.2006.01.015.
81. Yalcin, E., Sakiyan, O., Sumnu, G., Celik, S., & Koksel, H. (2008). Functional properties of microwave-treated wheat gluten. *European Food Research and Technology, 227*, 1411–1417. https://doi.org/10.1007/s11483-008-0075-2.
82. Vadivambal, R., & Jayas, D. S. (2010). Non-uniform temperature distribution during microwave heating of food materials—A review. *Food and Bioprocess Technology, 3*(2), 161–171. https://doi.org/10.1007/s11947-008-0135-7.
83. Ozkoc, S. O., Sumnu, G., & Sahin, S. (2014). Recent developments in microwave heating. In *Emerging Technologies for Food Processing* (pp. 361–383). https://doi.org/10.1016/B978-0-12-411516-1.00019-3.
84. Coutinho, L. S., Batista, J. E. R., Caliari, M., & Soares Júnior, M. S. (2013). Optimization of extrusion variables for the production of snacks from by-products of rice and soybean. *Food Science and Technology, 33*, 705–712. https://doi.org/10.1590/S0101-20612013005000093.

85. Campanone, L. A., & Zaritzky, N. E. (2005). Mathematical analysis of microwave heating process. *Journal of Food Engineering, 69*(3), 359–368. https://doi.org/10.1016/j.jfoodeng.2004.07.014.
86. Wang, A., Zhou, C., Du, Z., Liu, M., Zhu, S., Shen, S., & Ouyang, P. (2009). Enhancement of microwave-assisted covalent immobilization of penicillin acylase using macromolecular crowding and glycine quenching. *Journal of Bioscience and Bioengineering, 107*(3), 219–224. https://doi.org/10.1016/j.jbiosc.2008.10.012.
87. Wang, Y., Wig, T. D., Tang, J., & Hallberg, L. M. (2003). Dielectric properties of foods relevant to RF and microwave pasteurization and sterilization. *Journal of Food Engineering, 57*(3), 257–268. https://doi.org/10.1016/S0260-8774(02)00256-7.
88. Ozcelik, M., Heigl, A., Kulozik, U., & Ambros, S. (2019). Effect of hydrocolloid addition and microwave-assisted freeze drying on the characteristics of foamed raspberry puree. *Innovative Food Science & Emerging Technologies, 56*, 102183. https://doi.org/10.1016/j.ifset.2019.102183.
89. Ekezie, F. G. C., Sun, D. W., Han, Z., & Cheng, J. H. (2017). Microwave-assisted food processing technologies for enhancing product quality and process efficiency: A review of recent developments. *Trends in Food Science & Technology, 67*, 58–69. https://doi.org/10.1016/j.tifs.2017.06.007.
90. Bassey, E. J., Cheng, J. H., & Sun, D. W. (2021). Novel nonthermal and thermal pretreatments for enhancing drying performance and improving quality of fruits and vegetables. *Trends in Food Science & Technology, 112*, 137–148. https://doi.org/10.1016/j.tifs.2021.03.003.
91. Zhang, X., Zhang, M., & Adhikari, B. (2020). Recent developments in frying technologies applied to fresh foods. *Trends in Food Science & Technology, 98*, 68–81. https://doi.org/10.1016/j.tifs.2020.01.003.
92. Orsat, V., Raghavan, V., & Meda, V. (2005). Microwave technology for food processing: An overview. In *The Microwave Processing of Foods* (pp. 105–118). https://doi.org/10.1016/B978-1-85617-457-4.50009-8.
93. Hassoun, A., Siddiqui, S. A., Smaoui, S., Ucak, İ., Arshad, R. N., Garcia-Oliveira, P., & Bono, G. (2022). Seafood processing, preservation, and analytical techniques in the age of Industry 4.0. *Applied Sciences, 12*(3), 1703. https://doi.org/10.3390/app12031703.
94. Ghaffari-Bohlouli, P., Jafari, H., Taebnia, N., Abedi, A., Amirsadeghi, A., Niknezhad, S. V., & Shavandi, A. (2022). Protein by-products: Composition, extraction, and biomedical applications. *Critical Reviews in Food Science and Nutrition*, 1–46. https://doi.org/10.1080/10408398.2022.2144188.
95. Chen, B., Zhang, M., Wang, Y., Devahastin, S., & Yu, D. (2022). Comparative study of conventional and novel combined modes of microwave and infrared-assisted thawing on quality of frozen green pepper, carrot, and cantaloupe. *LWT, 154*, 112842. https://doi.org/10.1016/j.lwt.2021.112842.
96. Manyatsi, T. S., Al-Hilphy, A. R., Majzoobi, M., Farahnaky, A., & Gavahian, M. (2022). Effects of infrared heating as an emerging thermal technology on physicochemical properties of foods. *Critical Reviews in Food Science and Nutrition*, 1–20. https://doi.org/10.1080/10408398.2022.2156388.
97. Gonzalez, J. J., McCarthy, K. L., & McCarthy, M. J. (2000). Textural and structural changes in lasagna after cooking. *Journal of Texture Studies, 31*(1), 93–108. https://doi.org/10.1111/j.1745-4603.2000.tb00280.x.

98. Kar, S., & Sutar, P. P. (2022). Enhancing the efficacy of microwave blanching-cum-black mold inactivation of whole garlic (*Allium sativum* L.) bulbs using ultrasound: Higher inactivation of peroxidase, polyphenol oxidase, and *Aspergillus niger* at lower processing temperatures. *Food and Bioprocess Technology, 15*(3), 635–655. https://doi.org/10.1007/s11947-022-02768-5.
99. Huang, R., Pan, X., Lv, J., Zhong, W., Yan, F., Duan, F., & Jia, L. (2018). Effects of explosion puffing on the nutritional composition and digestibility of grains. *International Journal of Food Properties, 21*(1), 2193–2204. https://doi.org/10.1080/10942912.2018.1488797.
100. Yao, Y., Zhang, B., Zhou, L., Wang, Y., Fu, H., Chen, X., & Wang, Y. (2022). Steam-assisted radio frequency blanching to improve heating uniformity and quality characteristics of stem lettuce cuboids. *Food and Bioprocess Technology, 15*(8), 1907–1917. https://doi.org/10.1007/s11947-022-02790-7.
101. Liu, J., Liu, Y., Li, X., Zhu, J., Wang, X., & Ma, L. (2023). Drying characteristics, quality changes, parameters optimization and flavor analysis for microwave vacuum drying of garlic (*Allium sativum* L.) slices. *LWT, 173*, 114372. https://doi.org/10.1016/j.lwt.2023.114372.
102. Urbizo-Reyes, U., San Martin-González, M. F., Garcia-Bravo, J., Vigil, A. L. M., & Liceaga, A. M. (2019). Physicochemical characteristics of chia seed (*Salvia hispanica*) protein hydrolysates produced using ultrasonication followed by microwave-assisted hydrolysis. *Food Hydrocolloids, 97*, 105187. https://doi.org/10.1016/j.foodhyd.2019.105187.
103. Sun, Y., Zhang, M., & Chen, H. (2020). LF-NMR intelligent evaluation of rheology and printability for 3D printing of cookie dough pretreated by microwave. *LWT, 132*, 109752. https://doi.org/10.1016/j.lwt.2020.109752.
104. Solaesa, Á. G., Villanueva, M., Muñoz, J. M., & Ronda, F. (2021). Dry-heat treatment vs. heat-moisture treatment assisted by microwave radiation: Techno-functional and rheological modifications of rice flour. *LWT, 141*, 110851. https://doi.org/10.1016/j.lwt.2021.110851.
105. Yi, C., Zhu, H., Bao, J., Quan, K., & Yang, R. (2020). The texture of fresh rice noodles as affected by the physicochemical properties and starch fine structure of aged paddy. *LWT, 130*, 109610. https://doi.org/10.1016/j.lwt.2020.109610.
106. Nasrabadi, M. N., Doost, A. S., & Mezzenga, R. (2021). Modification approaches of plant-based proteins to improve their techno-functionality and use in food products. *Food Hydrocolloids, 118*, 106789. https://doi.org/10.1016/j.foodhyd.2021.106789.
107. Habinshuti, I., Mu, T. H., & Zhang, M. (2020). Ultrasound microwave-assisted enzymatic production and characterization of antioxidant peptides from sweet potato protein. *Ultrasonics Sonochemistry, 69*, 105262. https://doi.org/10.1016/j.ultsonch.2020.105262.
108. Venkateswara Rao, M., CK, S., Rawson, A., & DV, C. (2021). Modifying the plant proteins techno-functionalities by novel physical processing technologies: A review. *Critical Reviews in Food Science and Nutrition*, 1–22. https://doi.org/10.1080/10408398.2021.1904594.
109. Guzik, P., Szymkowiak, A., Kulawik, P., Zając, M., & Migdał, W. (2022). The confrontation of consumer beliefs about the impact of microwave-processing on food and human health with existing research. *Trends in Food Science & Technology, 119*, 110–121. https://doi.org/10.1016/j.tifs.2022.02.010.

110. Li, S., Zhang, R., Lei, D., Huang, Y., Cheng, S., Zhu, Z., & Cravotto, G. (2021). Impact of ultrasound, microwaves and high-pressure processing on food components and their interactions. *Trends in Food Science & Technology, 109*, 1–15. https://doi.org/10.1016/j.tifs.2021.03.005.
111. Rifna, E. J., Singh, S. K., Chakraborty, S., & Dwivedi, M. (2019). Effect of thermal and non-thermal techniques for microbial safety in food powder: Recent advances. *Food Research International, 126*, 108654. https://doi.org/10.1016/j.foodres.2019.108654.
112. Wang, X., Muhoza, B., Wang, X., Feng, T., Xia, S., & Zhang, X. (2019). Comparison between microwave and traditional water bath cooking on saltiness perception, water distribution and microstructure of grass carp meat. *Food Research International, 125*, 108521. https://doi.org/10.1016/j.foodres.2019.108521.
113. Sattar, S., Ahmad, T., Nisa, M. U., Imran, M., Holmes, M., Maycock, J., & Khan, M. K. (2019). Microwave processing impact on physicochemical and bioactive attributes of optimized peach functional beverage. *Journal of Food Processing and Preservation, 43*(7), e13952. https://doi.org/10.1111/jfpp.13952.
114. Sompong, R., & Songsermpong, S. (2022). Production method and optimization of puffing tofu skin snack using home microwave oven. *Agriculture and Natural Resources, 56*(3), 525–536. https://doi.org/10.1016/j.anres.2022.02.001.
115. Yang, R., & Chen, J. (2022). Recent application of artificial neural network in microwave drying of foods: A mini-review. *Journal of the Science of Food and Agriculture, 102*(14), 6202–6210. https://doi.org/10.1002/jsfa.11923.
116. Lara, N., Portilla, A., Osorio, F., & Ruales, J. (2022). Modeling of the microwave heating-toasting time-related variables and characterization of non-isothermal rheological properties of floury and sweet specialty maize kernels. *Applied Food Research, 2*(1), 100119. https://doi.org/10.1016/j.afres.2022.100119.
117. Ritudomphol, O., & Luangsakul, N. (2019). Optimization of processing condition of instant rice to lower the glycemic index. *Journal of Food Science, 84*(1), 101–110. https://doi.org/10.1111/1750-3841.14446.
118. Moraru, C. I., & Kokini, J. L. (2003). Nucleation and expansion during extrusion and microwave heating of cereal foods. *Comprehensive Reviews in Food Science and Food Safety, 2*(4), 147–165. https://doi.org/10.1111/j.1541-4337.2003.tb00025.x.
119. Tsai, Y. H., Hwang, C. C., Lin, C. S., Lin, C. Y., Ou, T. Y., Chang, T. H., & Lee, Y. C. (2021). Comparison of microwave-assisted induction heating system (MAIH) and individual heating methods on the quality of pre-packaged white shrimp. *Innovative Food Science & Emerging Technologies, 73*, 102787. https://doi.org/10.1016/j.ifset.2021.102787.
120. Zhang, Y., Soldatov, S., Papachristou, I., Nazarova, N., Link, G., Frey, W., & Silve, A. (2022). Pulsed microwave pretreatment of fresh microalgae for enhanced lipid extraction. *Energy, 248*, 123555. https://doi.org/10.1016/j.energy.2022.123555.
121. Liu, Z., & Zhang, M. (2021). Texture properties of microwave post-processed 3D printed potato snack with different ingredients and infill structure. *Future Foods, 3*, 100017. https://doi.org/10.1016/j.fufo.2021.100017.
122. Martins, C. P., Ramos, G. L., Pimentel, T. C., Freitas, M. Q., Duarte, M. C. K., Azeredo, D. P., & Cruz, A. G. (2022). How microwave technology is perceived? A food safety cross-cultural study between Brazil and Portugal. *Food Control, 134*, 108763. https://doi.org/10.1016/j.foodcont.2022.108763.

123. Qin, Y., Duan, Z., Zhou, S., & Wei, Z. (2022). Effect of intermittent microwave drying on nutritional quality and drying characteristics of persimmon slices. *Food Science and Technology, 42*. https://doi.org/10.1590/fst.18421.

124. Wu, F., Chi, B., Xu, R., Liao, H., Xu, X., & Tan, X. (2022). Changes in structures and digestibility of amylose-oleic acid complexes following microwave heat-moisture treatment. *International Journal of Biological Macromolecules, 214*, 439–445. https://doi.org/10.1016/j.ijbiomac.2022.02.030.

CHAPTER 12

PACKAGING MATERIALS AND TECHNOLOGIES FOR READY-TO-EAT SNACKS

M. SELIM ŞILBIR and A. BURCU ATALAY

Department of Food Engineering, Faculty of Engineering, Iğdır University, Iğdır, Turkey

ABSTRACT

This book chapter provides a comprehensive review of packaging materials and technologies for "ready to eat" snacks. When looking for quick and wholesome snack options, consumers are increasingly turning to ready-to-eat (RTE) foods. However, the packaging of these snacks is crucial in ensuring their safety, freshness, and quality. The first point made in the chapter is how crucial package components and technology are for maintaining the quality and safety of RTE snacks, extending their shelf lives, minimizing food waste, and improving consumer convenience. Different packaging materials are covered, including flexible and rigid packaging such as plastics, paper, glass, and metals. Additionally, the chapter discusses innovative technologies like edible coatings and modified atmosphere, vacuum, and active packaging. The review is based on current research published within the last decade (2013 to 2023) and includes references to the most recent studies in the field. Moreover, the chapter highlights emerging nanotechnology applications in food packaging. The importance of sustainability in packaging is also underlined, and opportunities and challenges related to using green materials and technology are discussed. The chapter includes real-world examples of innovative packaging solutions that have been developed to address these challenges. This comprehensive guide to packaging materials and technologies

Ready-to-Eat Snacks: Emerging Technologies for Production and Safety.
Ravi Pandiselvam and Gülşah Çalışkan Koç

for RTE snacks will be of interest to food scientists, packaging engineers, and all stakeholders involved in the production and marketing of RTE snacks. Overall, the chapter provides insights into many of the latest developments in packaging techniques and materials, placing particular emphasis on the significance of sustainable and innovative solutions.

12.1 INTRODUCTION

Ready-to-eat snacks (RTE) snacks are getting more and more well-liked by consumers due to their accessibility and convenience. These snacks must be packaged properly to guarantee their quality, freshness, and safety. Packaging materials and technologies have advanced in recent years, providing various options for food manufacturers to choose from. An overview of the various packaging materials and methods used for ready-to-eat snacks is given in this chapter.

Most recently, the demand for RTE has increased significantly, driven by the growing preference for convenient, healthy, and tasty snack options among consumers. As a result, the RTE snack market has experienced steady growth, with demographic and lifestyle changes, including the rise of urbanization and the popularity of snacking, playing a major role in driving this trend. Figure 12.1 exhibits distinct layers of conventional flexible packaging material.

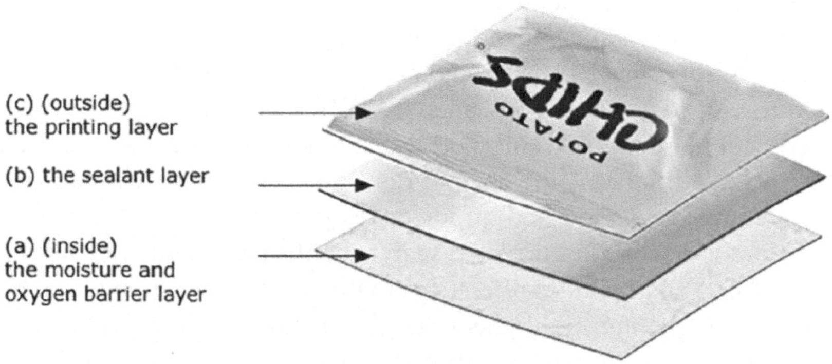

FIGURE 12.1 Different layers of typical flexible packaging materials used for ready-to-eat snacks, (a) the moisture and oxygen barrier layer, (b) the sealant layer for sealing the package, and (c) the printing layer for branding and product information.

However, in order to ensure the quality and safety of these products, there are increasing demands for effective and efficient packaging materials and technologies. In the RTE snack market, packaging is essential for keeping food safe from infection, prolonging shelf life, and improving consumer convenience. In order to meet the growing need for efficient, environmentally friendly, and sustainable packaging options, it is now crucial to develop novel packaging solutions. Various packaging technologies and snack production steps are demonstrated in the flowchart depicted in Figure 12.2.

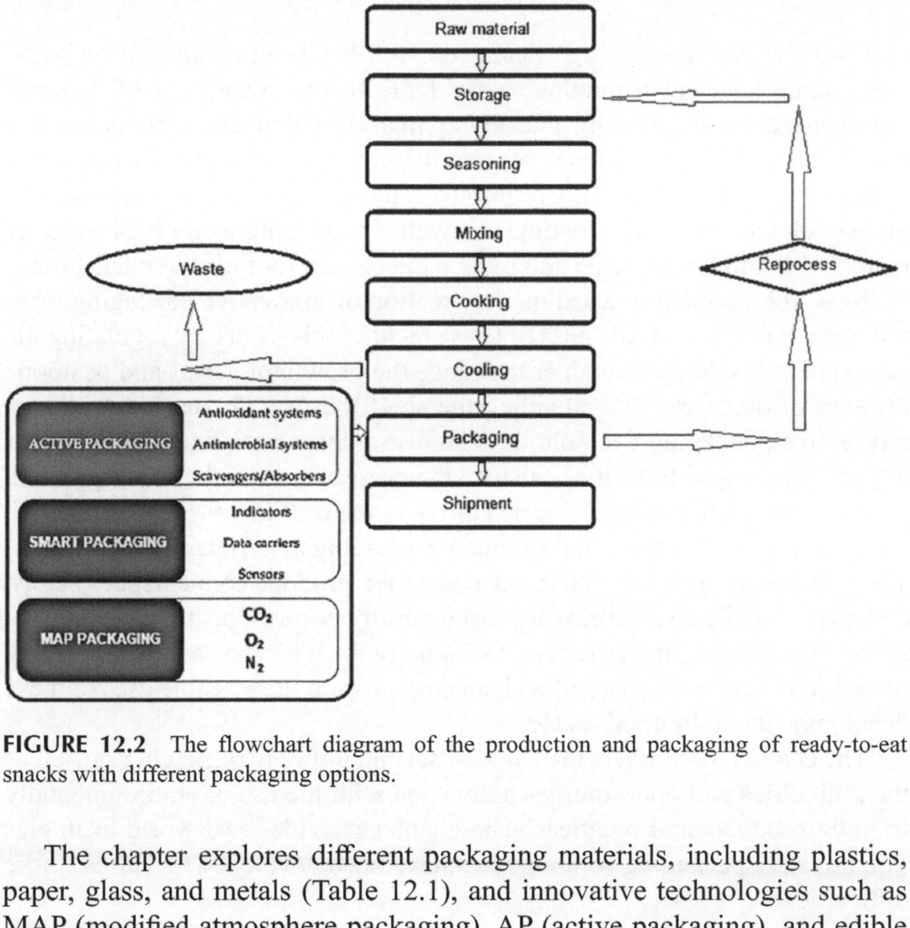

FIGURE 12.2 The flowchart diagram of the production and packaging of ready-to-eat snacks with different packaging options.

The chapter explores different packaging materials, including plastics, paper, glass, and metals (Table 12.1), and innovative technologies such as MAP (modified atmosphere packaging), AP (active packaging), and edible coatings. The chapter emphasizes the value of packaging in maintaining snack goods' quality and safety, cutting down on food waste, and enhancing consumer convenience.

TABLE 12.1 Commonly Used Packaging Materials for Ready-to-Eat Snacks

Material	Characteristics
Plastic	Versatile, moisture barrier, transparent
Paper	Lightweight, economical
Aluminum foil	Moisture and gas barrier, opaque
Glass	Transparent, inert, can be reused
Metal cans	Rigid, durable, airtight
Flexible pouch	Lightweight, cost-effective, customizable shape

Over the last decade, significant research has been conducted on packaging materials and technologies for RTE snacks. One area of research has focused on developing packaging materials that are sustainable and eco-friendly, as a consequence of growing consumer demand for items that respect the environment. This prompted the creation of novel materials like biodegradable films and coatings as well as the optimization of existing materials to minimize waste and reduce the carbon footprint of packaging.

Research has also focused on the creation of innovative packaging technologies, including MAP and AP. These technologies work by regulating the atmosphere inside the container to reduce the growth of germs and postpone the start of deterioration, extending the shelf life of RTE snacks. Additionally, active packaging technologies can also release antimicrobial agents or oxygen scavengers to further enhance the preservation of the snack product.

In summary, the research carried out over the past 10 years has resulted in substantial advancements in RTE snack packaging materials and technology. These innovations have enhanced consumer product convenience, safety, and quality while also addressing sustainability issues. The newest improvements in packaging materials and technology for RTE snacks are thoroughly reviewed in this book chapter, with an emphasis on the possible uses of these developments in the food sector.

The chapter also covers the value of sustainability in packaging as well as the difficulties and opportunities associated with the use of environmentally friendly products and practices. The chapter provides real-world examples of innovative packaging solutions that have been developed to address these difficulties.

Overall, this book chapter serves as a comprehensive guide to packaging materials and technologies for RTE snacks, providing insights into many recent advancements in the industry. It will be of interest to food scientists, packaging engineers, and all stakeholders involved in the production and marketing of RTE snacks.

12.2 PACKAGING MATERIALS FOR READY-TO-EAT SNACKS

The types of foodstuffs consumed and consumption habits have changed as a direct result of lifestyle changes, necessitating the development of new packaging and containers by the food industry [1]. RTE is a result of consumers' altering attitudes and busy lifestyles [2]. Regardless of their lipid content, these RTE snacks are particularly prone to lipid oxidation, which reduces sensory perception, nutritional quality, and food safety during storage while also rendering the item unpleasant. Preserving and extending the shelf life of RTE snacks is, therefore, crucial [3].

Due to its role as a barrier against oxygen and moisture loss or gain, food packaging has a big impact on how long food goods can be kept on the shelf. Packaging prevents exposure to germs and pollutants while protecting against physical damage, unpleasant odors, and the outside environment. It should be inexpensive, nontoxic, transparent, simple to open, and tamper-resistant in addition to offering these advantages [4]. Laminated packaging is mostly utilized for the packing of snacks to maintain the product's textural characteristics, moisture content, grease content, and air proofing [5]. Different RTE items need varied packaging qualities and materials due to their varying shelf lives [2]. There are a variety of packaging options for RTE snacks, which are chosen based on the material's compatibility with the processing method, ease of use for consumers, and other technical characteristics that ensure the product's safety. Packaging materials and layers for RTE meals include plastic, aluminum foil, and paper [6]. For many different liquid and semi-solid food products, metal and glass are also frequently employed.

The packaging's overall environmental impact depends on several factors, including the packaged food, distances, and distribution logistics; renewable energy sources; easily accessible waste disposal systems; re-use rate; biodegradability and/or recyclability; the packaging material's end-of-life scenario; use; disposal; and the sum total of production costs. As a result, research and development in food packaging are expanding quickly. Modern advances in food packaging components and designs have been studied and invented to improve the sustainability of food chains and decrease food waste. Significant attempts have been made to develop biodegradable and/or bio-based materials, such as cardboard, paper, and bioplastics to produce more environmentally friendly food packaging [1]. Due to customers' increased need for goods that are safer, of higher quality, and have a longer shelf life, as well as their preference for natural and biodegradable materials over synthetic and non-biodegradable ones, there has been an increase in interest in edible packaging for food applications over the past few years. The global market's interest in edible packaging solutions has expanded with the usage of edible, biodegradable, and renewable materials [7]. RTE snacks are produced using a wide range of technologies since they are simple to handle, don't need additional cooking or preparation, and concentrate only on appealing packaging. The term "RTE food" refers to a variety of snacking goods, such as biscuits, bread, sandwiches, rolls, soups, chips, fruit and vegetable salads, breakfast cereals, dairy products, etc. [5]. Table 12.2 provides an itemized comparison between packaging materials that can be classified as either rigid or flexible.

TABLE 12.2 Advantages and Disadvantages of Rigid and Flexible Packaging Materials Used for Ready-to-Eat Snacks [8, 9]

Material	Properties	Advantages	Disadvantages
Flexible packaging	Lightweight, cost-effective, easy to transport, versatile	Good moisture and oxygen barrier, custom printing options	Limited shelf life, not suitable for heavy products or high-temperature applications
Rigid packaging	Durable, stackable, and good for heavier products	Good barrier properties, excellent protection from external damage	Not flexible, may be less cost-effective, difficult to transport and store

12.2.1 PLASTICS

Plastic materials interact with food because of their virtually opposite characteristics, such as imperfect barriers and non-inertness. Yet, the usage of plastic materials in packaging has increased dramatically during the past few decades. Polymers are more affordable, lighter, and less energizing than other materials while still being comparably durable [10]. Due to their adaptability, affordability, and durability, plastics are also extensively employed in the field of food packaging.

Many plastic materials, including polyethylene terephthalate (PET), polypropylene (PP), polystyrene (PS), and polyethylene (PE) are used to make RTE snack packaging. PET and PS are frequently used for hard packagings, such as trays and containers, whereas PE and PP are frequently used for flexible packaging, such as pouches and bags. Gas can pass through polystyrene's (PS) barrier characteristics. Low-density polyethylene has minimal water vapor permeability and is gas permeable. Grease resistance and minimal water vapor penetration characterize high-density polyethylene. The dual oven ability of crystallized polyethylene terephthalate

(CPET) trays is a particular benefit. Its other benefits include adaptability in design, resilience to grease and oil, and little to no impact on food flavor. Polyethylene terephthalate (PET): which typically has a translucent look, is particularly popular for transporting goods that are kept in freezers, refrigerators, or shelves, such as ready-to-eat salads and vegetables. Polypropylene (PP): Often used for microwave-safe packaging, this material can withstand temperatures of up to 125°C. Grease doesn't damage it.

Possible layers include foil, various plastics, paper, and glue. Polyethylene (PE) is the second-most produced plastic in the world after polyvinyl chloride, and polypropylene (PP) is the third-most used bulk plastic [10]. The process of making the trays, filling the containers, and getting the containers topped with lids are covered by the principal packaging for RTE meals. The capacity of plastics to form a barrier against moisture, oxygen, and other gases contributes to the shelf life of RTE snacks. Polymers also offer other benefits. Polymers are also lightweight and are simple to modify to meet the unique requirements of various RTE snacks. Plastics have recently been looked into as a food packaging material, nevertheless, due to worries about their potential negative impacts on the environment and human health.

The two pieces that make up commonly used MAP materials are upper films and tray portions. In general, PS foam, molded plastics, injection-molded plastics, pre-molded pulps that are coated in plastic materials, or laminations of cartoon and plastic films are used to create the trays that serve as the base of MAP [6]. Moreover, pasteurized ready-to-eat meals may be vacuum-packed as long as the packing material has low oxygen permeability and is light-impermeable. For this kind of food product, pouches made of PA/PE (polyamide/polyethylene), PET/PE (polyethylene terephthalate/ polyethylene), and PA/PP (polyamide/polypropylene) laminations can be utilized. When using a single plastic material in plastic packaging, there are instances when a combination of materials must be used. Lamination, coextrusion, and extrusion coating are some of these techniques. Further, convenience foods, including soups, ready meals, baby food, and snacks, can be packaged in hard retort food bowls. Deep-drawing plastic plates made from lidding foils and PS or copolymerized PS materials are used to package desserts like flan and pudding [11].

12.2.2 PAPER AND PAPERBOARD

Because they are recyclable and biodegradable, paper and paperboard are better choices for the environment than plastics. They can also be easily modified, inexpensive, and lightweight. Aluminum foil's barrier qualities include low water vapor permeability and grease resistance. Paper and paperboard packaging is among the oldest and most commonly used for foodstuffs such as bakery goods, beverages, dry powders, milk, and milk-based products, confectionary, etc., because of its reputation for being environmentally friendly.

In composite cans, paper, particularly kraft-type paper, is the main layer, with layers of high-density polyethylene, aluminum, and polypropylene, added to improve the characteristics. Containers for snacks, chilled dough, frozen foods, nuts, dried fruits, chips, powdered foods, dried meats, solid or liquid shortening, salt and spices, cookies and crackers, chilled dough, and confectionery items are just a few examples of food applications. Aluminum foil, paperboard, and polymer composite cans are usually used to package milk powder in place of metal cans [12]. The restricted barrier qualities of paper and paperboard, however, may prevent their usage in several RTE foods. RTE snack packaging frequently includes paper and paperboard, especially cereal bars and other baked products. Further packaging for cooked meals can be made of paper-based shields to shield them from light-induced oxidation.

12.2.3 METAL

In addition to being recyclable, easy to form into a variety of shapes, resistant to high heating temperatures, having a rigid construction, long-distance transport, and unique decorating options, metal-based packaging offers good protection against light, gas, and moisture. The drawbacks include resource depletion, the leaching of dangerous compounds into food from containers, and global warming caused by carbon dioxide emissions from metal manufacturing facilities. Metals do not have a neutral relationship with food goods; they are covered in protective lacquers to stop metal components from migrating and interacting with the food [13].

They are used as closures for several packaging types, including glass bottles and composite cans. Metal packaging is made of coated steel, tin plate, and free steel; stainless and polymer-coated steel, and aluminum. Metal is used in the packaging of dairy goods, drinks, fruits-vegetables, seafood meat, and poultry as well as bakery goods and sweets, coffee, and tea.

12.2.4 ALUMINUM FOIL

Aluminum is flexible and light in weight. It keeps the food fresh for a prolonged period by protecting it from UV radiation, oxygen, oils and fats, light, bacteria, and water vapor. Instant beverages, ready meals, snacks, biscuits, peanuts, sausage, and dried meat are among the products best packaged in aluminum foil and laminates.

Indirect contact between acidic or salty foods and aluminum is not recommended, although composite sheets with an aluminum layer can be employed.

In RTE snack packaging, aluminum foil is frequently used as a barrier material, especially for potato chips and other snack foods that are delicate to oxygen and moisture. RTE snacks' shelf life is increased by the efficient barrier that aluminum foil offers against these elements. Aluminum foil, however, is not eco-friendly and might be challenging to recycle.

Packaging options for ready meals include paper boxes with aluminum foil layers, pouches with aluminum foil layers, and curved metal trays [11].

12.2.5 GLASS

Glass packaging is a style of container that is frequently used in meals, including heat-treated or pressure-packed solid, powdered, and liquid items. RTE foods like salsa and dips are frequently packaged in glasses. RTE foods' freshness and flavor are maintained by the great barrier that glass offers against moisture, oxygen, and other gases. Glass is an environmentally friendly material since it can be recycled and reused. Glass is a popular choice among consumers for food packaging due to its many advantages.

It doesn't need any additional special coating because it is a strong, resilient, and chemically inert substance. It works with the vacuum, filling, and closing techniques [7]. Glass can only be used in a certain sort of RTE snack due to its weight and ease of breaking. The negative aspects of glass, which some favor because they believe it to be healthy, exist as well. The food becomes discolored as a result of the way it allows light to pass through. Its tendency to crack when subjected to forces including impact, heat, shock, and extreme internal pressure could cause issues with its manufacture, shipping, storage, and sale. Beverage bottles, soda bottles, natural mineral water bottles, and fruit juice bottles must all be packaged in glass.

12.3 PACKAGING TECHNOLOGIES FOR READY-TO-EAT SNACKS

The lifestyle of consumers, value, profit, food biosafety, safety, packaging legislation, and environmental concerns are some of the key socioeconomic needs that support advances in food packaging, and these needs are the main driving force behind the food packaging industry. Innovations in food packaging have improved the quality and safety of the food as well as the system's overall sustainability [14]. Technologies used in packaging create a delicate synergy between the product and packaging that suits the needs of the distributor and customer as well as the producer. Thus, packaging technology is crucial for consumer convenience as well as food protection [2]. Convenient packings, such as microwave susceptors and easy-open packaging, draws customers' attention because of its efficiency and ease of use. RTE snacks, which make up the majority of handy foods, are now the answer to these issues.

RTE foods are divided into three categories: products for direct consumption, products to be cooked, and precooked products [6]. Nevertheless, two factors should be considered when examining RTE foods and their packing needs: (i) horticultural products that breathe, as fresh vegetables/fruits and that have just been mildly processed; and (ii) non-respiratory goods, such as RTE or use items like dairy, meat or other items [15]. With some processing and packaging methods, the variety of these items increases. Packaging for RTE meals is a significant issue because it has a direct impact on the shelf life of these products. For RTE that has just been freshly cooked, such as main dishes for takeout, lunch boxes, and fast food (such as burgers and tortillas), the packaging and container must be heat- and oil-resistant. Examples of sterilized RTE products include salsa, kimchi, guacamole, and salad dressing. These products often need high-integrity packaging with acceptable flavor and a gas barrier [2]. With packaging methods like vacuum packing and modified atmosphere packaging, RTE foods preserve their natural characteristics without the need for chemical preservatives [6]. Skin packaging, a new technology, provides an additional packaging choice for RTE meals. AP, such as MAP and antimicrobial packaging, has drawn a lot of interest because of its capacity to increase the safety and shelf life of food [16]. The use of intelligent packaging technologies in deep-freeze and chilled food cold chains is increasing, despite the RTE industry's extensive use of novel food packaging techniques like active packaging technology (antimicrobials, oxygen scavengers, respiration regulators, and aroma/odor absorbers) [2, 17]. With the requirements and capacities of consumers in mind, new packaging designs have recently been created, including easy-open packaging, microwaveable packaging with self-ventilation, and biodegradable packaging. The RTE industry has seen the entry of many cutting-edge packaging materials, including microwave protection, self-heating packaging, active, and smart.

With regard to the packaging techniques of RTE products, this review intends to inform readers and present innovative research. With RTE packaging technology, it's critical to comprehend the demands of industry and consumers. The production of next-generation packaging technology needs to take this into account and include further related developments. Future developments in active and intelligent packaging, edible coating, packaging with nanotechnologies for food, antimicrobial, self-cooling/self-heating, flavor/odor absorbers, water-soluble, sustainable, and other unique food packaging technologies are included [17]. Additionally, future evaluation and research should focus on the influence of external factors on packaging patterns, including the impact of modern consumer behavior, contemporary global issues like pandemics, marine pollution, climate change, the circular economy, and emerging technology. Modern customers desire to minimize their overall environmental impact, fueling the packaging trend towards the use of green, recyclable, sustainable, and biodegradable green materials. A few research have specifically addressed RTE foods and their package designs when taking into account newly released material. This might be because RTE foods are still relatively new. Their complexity in terms of ingredients is another factor that could make the examination more difficult [18]. Table 12.3 exhibits a summary of several packaging technologies, materials, and their corresponding advantages.

12.3.1 MODIFIED ATMOSPHERE PACKAGING (MAP)

The term refers to a method of adding air to a package before it is sealed using a prepared gas combination. Once the package is sealed, no more effort is made to regulate the atmosphere inside [28]. O_2, CO_2, and N_2 are the three gases used in MAP the most frequently. During storage, product respiration uses up O_2 and releases CO_2. An inert gas called nitrogen is utilized in MAP as a "filler" gas to balance off the volume loss caused by CO_2 absorption and prevent package collapse [29]. However, nitrogen simply serves to prevent the oxidation of lipids by acting as an oxygen substitute [30, 28]. Carbonic acid lowers the pH of the meal as a result of carbon dioxide and water. It hinders the development of several aerobic bacteria and the majority of molds [31].

Generally, the greater the concentration, of the gas, the more powerful its inhibiting effect is. By reducing the rate of chemical and biological deteriorative processes and the growth of spoilage organisms, food storage in a modified gaseous environment can preserve quality and lengthen product shelf life. The fundamental idea behind MAP is that a modified environment can be produced either actively by adding a certain gas combination in conjunction

TABLE 12.3 A Summary of Various Packaging Technologies and Materials Used for Ready-to-eat Snacks, along with Their Potential Benefits

Type of Packaging Technologies/Packaging Materials	Type of RTE Snacks	Potential Benefit	References
Innovative bio-based packaging system (Active packaging)	Organic ready-to-eat iceberg lettuce	-Improvement in overall freshness and odor -Demonstration of antimicrobial activity Demonstrating a positive impact in comparison to a traditional plastic film -Confirmation that it is an environmentally friendly solution potentially providing added value related to quality	[19]
Oxygen scavenger /Active scavenging systems (absorber)	Seeds, nuts, and oils; instant powders with added fat; fried snacks; and goods made from dried meat	The avoidance of rancidity	[20]
Antimicrobial packaging systems / Active releasing systems (emitter)	Dairy goods, fresh and processed fruits and vegetables, meat, fresh and smoked fish, fresh seafood, ready-to-eat meals, grains, cereals	The prevention or slowing of bacterial development and an increase in shelf life	[20]
Skin and vacuum packaging	Fish	Skin-packaged filets had a larger drip loss than filets that were normally vacuum-packed and were also paler, less reddish, and less yellowish. The conclusion is that the filet parts' packaging extends their microbiological shelf life and exhibits high-quality characteristics. Due to its more appealing packaging, the use of super chilled storage along with vacuum-skin packaging may be an attractive choice for customer preferences.	[21]

TABLE 12.3 *(Continued)*

Type of Packaging Technologies/Packaging Materials	Type of RTE Snacks	Potential Benefit	References
Packaging materials (Metallic foil laminated low-density polyethylene (MFL), metallic foil laminated low-density polyethylene (ML), aluminum foil laminated low-density polyethylene (AFL))	Ready-to-eat high-energy food paste	The RTE-TF paste can be packaged in AFL and MFL pouches to maintain a more acceptable form and a longer shelf life than ML pouches.	[22]
Vacuum-Packaging	Beef Meat	According to this investigation, VP meat that had been frozen after being chilled during storage had better microbial indicators.	[23]
Packaging materials (LDPE, Aluminum foil, and Aluminum foil pouches with Nitrogen gas)	Aonla and Cereal-Based Extruded RTE Snack	In comparison to aluminum foil and LDPE pouches, the physical and nutritional changes are significantly less in aluminum foil with nitrogen gas.	[24]
Passive modified atmosphere packaging	Ready-to-eat cherry tomatoes	MAP had no impact on nutritional quality, as seen by the 11% antioxidant capacity and 40% lycopene losses compared to control tomatoes, respectively. Ready-to-eat cherry tomatoes wrapped in PE showed shelf life when sensory quality was taken into account.	[25]
PE, PP and PET packages, passive and active atmospheres (Modified Atmosphere Packaging (MAP))	Ready–to–eat whole strawberries	With extending shelf-life, PE, PP, and PET produced sufficient packaging solutions that were only constrained by sensory characteristics or microbial development.	[33]

TABLE 12.3 *(Continued)*

Type of Packaging Technologies/Packaging Materials	Type of RTE Snacks	Potential Benefit	References
Aseptic packaging	Juices, iced tea, dairy-based beverages, low-acid foods, wine, small-particle foods like soups and salsas, spring water, vitamins, and energy drinks, and customized functional waters	Food safety, convenience, affordability, cost-effectiveness in transit and storage with low environmental impact, and the absence of a cold chain are among the significant benefits of aseptic packing.	[26]
Intelligent packaging	A wide range of applications, from meat products to liquids and fruits and vegetables.	Intelligent packaging is a system that performs smart tasks like detecting, sensing, recording, tracking, transmitting, and using scientific logic to increase quality and safety, extend shelf life, and make decisions easier. It also warns customers about potential issues. To reduce the prevalence of foodborne diseases caused by pathogenic microorganisms.	[27]

with packaging materials' permeability or passively by employing correctly permeable packaging materials [2]. Fruits and vegetables are examples of products that introduce passive modification since they continue to breathe while in their packaging, producing CO_2 while using oxygen. Additionally, packing materials are typically O_2-impermeable. Active MAP is a focused adjustment of gas composition that does not change over time [6].

MAP, the most popular method for "ready to eat" meals, involves altering the atmosphere's makeup inside the package. Using MAP, a product's shelf life is extended and its quality is improved. Ready to eat food items either have a short time on the shelf or a long duration on the shelf. With varying shelf lives, different RTE items require different packaging features and materials. To extend shelf life and minimize surface bruises, fresh-cut vegetables often require modified-environment, antibacterial and extremely durable packaging. Perforated-mediated MAP technology is currently being considered a substitute for a conventional MAP for fresh-cut vegetables with high respiration, such as mushrooms, kale and lettuce [2]. Using MAP, vacuum skin packaging, and microwaveable containers, fresh-cut asparagus's ability to remain edible as a RTE product was increased while maintaining quality of product and meeting consumer demand for a convenient food. Broccoli, artichokes, cabbage, cantaloupes, mangos, and pineapple are just a few of the fresh fruits and vegetables that have been shown to benefit from the use of MAP in preserving their quality and shelf life [32]. Different gas compositions for different RTE food types contribute to the complexity of this packing method. For some foods, there are some unique limitations on gas usage. The collapse of packages of meat with high water and lipid content is a limiting issue for excessive CO_2 consumption in package atmospheres because CO_2 is extremely soluble in water and oil. The fact that different types of fruits and vegetables have varying minimum and maximum O_2 and CO_2 tolerance levels is another fascinating element of gas concentration [6].

Passive MAP is a suitable method for preserving the quality of whole, RTE strawberries. In comparison to passive MAP, active MAP did not increase the strawberry cultivar's shelf life under the evaluated packing conditions [33]. MAP preserved the red color and firmness of the tomato while reducing weight loss and sensory quality loss. When compared to cherry tomatoes that had not been packed, MAP had no impact on the antioxidant capacity or lycopene concentration [25]. In comparison to the control, postharvest preservation of cherry tomatoes utilizing MAP technology resulted in less weight loss and sensory degradation. Utilizing atmosphere packing (MAP) in conjunction with refrigeration has allowed for the preservation of safety and shelf life for whole, minimally processed fruits and vegetables [34]. To improve food safety, a further barrier (hurdle technology) is required. The microbiological and sensory quality of the food may be improved by hurdle technology as a result of the synergistic effects of numerous preservation treatments [35]. The initial physiological state of fresh vegetables, suitable handling and sanitation procedures, the packing system and storage conditions can all extend the duration of the product by impacting both microbial load and chemical deterioration [32]. Figure 12.3 illustrates the fundamental principle of MAP alongside a food product that has been MAP-packaged.

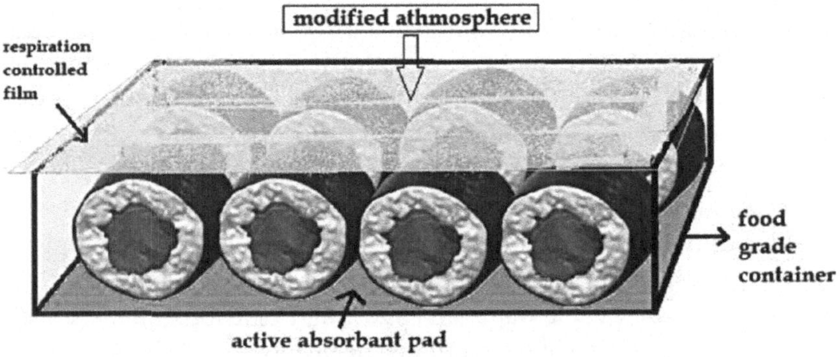

FIGURE 12.3 The MAP principle and MAP packaged sushi slices.

12.3.2 VACUUM PACKAGING

The process of vacuum packing (VP) involves eliminating gases from the packaging environment, with a focus on removing oxygen to induce an anaerobic condition that increases the product's shelf life and terminates the growth of pathogenic bacteria. So as to make sure product containment, meat products are frequently vacuum-packed in plastic bags that are heat-shrinkable and have low gas permeability [36]. Preventing oxidation reactions such as lipid oxidation, vitamin loss, oxidative browning, pigment loss, etc., is the main goal of VP. Other benefits of VP include volume reduction and the increased rigidity of flexible packages. For cans, jars, trays, and pouches, vacuum-pump devices are available before the item is sealed [37].

The first MAP likely to be created for commercial use was VP. The cheapest and most popular method of packaging, particularly among small seafood producers, is vacuum packaging [38]. In order to reduce the deteriorative effects of microbiological and endogenous enzymatic activity during storage, MA and VP in combination with low temperatures have been employed for decades to prolong seafood's shelf life [39]. VP is a technology that works well with fatty fish because it prevents the unpleasant odor that causes lipid oxidation of fat [40].

The preservation of dried ready-to-eat grasshoppers can be accomplished using either chilled storage or vacuum packaging, depending on the resources at hand [41]. This method prolongs the life span of boneless meat, primal, and subprimal meat pieces, and processed meat products, including sausages, patties, nuggets, and also hard cheese [42]. The quality of the meat product was maintained over the course of storage, following physicochemical and microbiological information related to the processing of a long-life vacuum-packed "RTE meat product based on traditional Portuguese seasoned meat" [43]. Vacuum packing aims to reduce the amount of oxygen that comes into contact with the product, halt or stop all oxidative reactions, and prevent microbial growth. The product's natural flavor is maintained. by vacuum packaging, which also has a number of other benefits like reducing weight loss, saving space and energy during storage, transit, and distribution, improving keeping characteristics, and improving product presentation [44]. Fresh produce's shelf life is extended by VP, which also prevents aerobic decomposition and lowers quality degradation [4]. Anaerobic deterioration by microorganisms and oxidative spoilage are the concerns associated with vacuum packaging. Vacuum packaging combined with chilling or freezing storage may resolve this issue [23, 45].

12.3.3 SKIN PACKAGING

In a more recent development known as vacuum skin packaging (VSP), food is placed on a tray and then securely sealed with a thin, flexible sheet [21]. VSP, also known as "skin packaging," refers to the process of totally eliminating the pack's interior air and improving food products under a vacuum, which reduces the amount of oxygen used for microbial growth and oxidation [40]. Modern preservative meat packaging that does not rely on changed atmospheres or traditional vacuum sealing is known as VSP. The purge loss during storage is higher in conventional vacuum packaging; the lowest values are seen in VSP. Vacuum packaging allowed lactic acid bacteria to proliferate and stabilize the microflora of packaged meat [46].

In comparison to conventional forms, skin packaging has a greater production cost, a longer shelf life, and less purge loss. But in other ways, skin packaging gives the meat better qualities than MAP or vacuum packaging [46]. Microbial growth decrease, oxidation prevention, improved protection for shipping and distribution, appearance, and display are just a few of the numerous advantages that VSP packaging offers. For sliced meat products such as ham, pastrami, chicken, cheese, seafood, smoked fish filets, and RTE meals, VSP is frequently the preferred packaging option. VSP is unique in that it covers trays made of robust, resilient, leak-proof materials with a thin, transparent plastic film, also known as skin film. A major portion of skin film is made of polyethylene or polyvinyl chloride (PVC). Figure 12.4 depicts a sushi-tray packaged with a skin-packaging method.

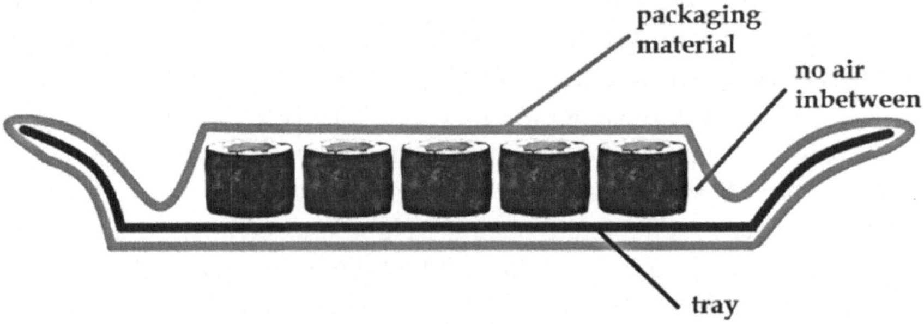

FIGURE 12.4 A skin-packaged sushi-tray.

12.3.4 ASEPTIC PACKAGING

To prevent reinfection, sterile food products are placed in sterile containers under sterile conditions, and the containers are then sealed. This technique is known as aseptic packing [47]. Aseptic packaging has probably been utilized in food service applications all over the world as a safe and superior packaging option [26]. Aseptic packing, sterilization, and other industrial methods for food processing and microbial deactivation all rely on increased temperatures and chemical-physical principles. These methods also have advantages and disadvantages [48]. Although pasteurization, sterilization, and aseptic packaging have all been successful at inactivating bacteria and spoilage enzymes during thermal preservation and processing [49], aseptic processing has several advantages over traditional package processing, including better nutritional and sensory maintenance functions, a larger range of container sizes and materials, and a longer shelf life [50].

Packaging for customer or consumer use must have the following qualities: modified atmosphere packaging, aseptic and antimicrobial packaging for RTE packed salad with fresh fruit, vegetables, and salad dressing. For fresh juice, acid-resistant, high-integrity packaging is also an option [51].

Several aseptic processing designs are utilized for distinct foods. For aseptic processing, foods with a high viscosity, low acidity (pH>4.6), and discrete particles are in demand [48]. Aseptic techniques have recently been used to package wine, and small-particle meals including soups and sauces, natural water, vitamin and vitality waters, and special functioned waters. The current product uses for aseptic packaging are particularly common in the conventional core industries of low-acid meals, juices, dairy-based beverages, and iced teas [26]. Paper and plastic are common components of traditional aseptic packaging. Aseptic packing can also be done with foil, metal cans, and other metal containers. Our variables in the form of packaging materials improve output and decrease process costs [52]. Figure 12.5 illustrates an aseptic packaging material composed of five distinct layers.

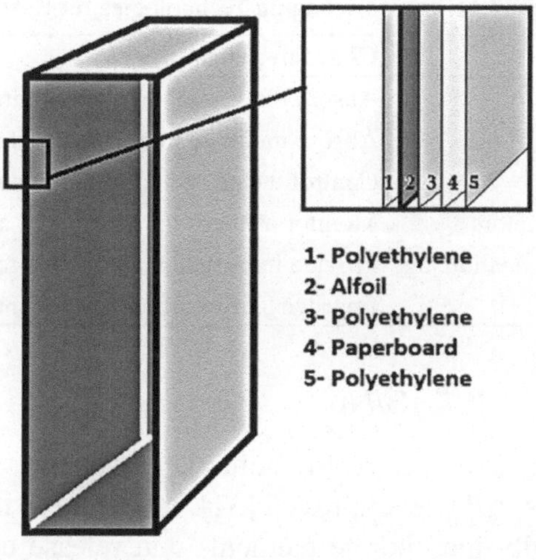

FIGURE 12.5 The layers of aseptic packaging material.

12.3.5 ACTIVE (AP) AND INTELLIGENT (IP) PACKAGING

Numerous new food items have been released on the market, necessitating regular consumer evaluation of their quality up until consumption as well as the creation of a number of cutting-edge packaging technologies like IP and AP to avoid spoilage of food while being stored [53]. AP and IP have been produced to extend shelf life. Specific kinds of packaging have the ability to alter the atmosphere inside the box, share knowledge, maintain records of the product supply chain, and prevent counterfeiting. This will properly improve food quality and safety, which will raise the product's value and satisfaction ratings from consumers as well as increase convenience [54].

The packaging materials known as "active packaging" and "intelligent packaging" not only enhance a product's fundamental functions but also react to external stimuli. These systems are new-generation packaging systems that offer "containment, preservation, communication, and facilitation" services in addition to the essential transportation and protection functions [55]. The need for RTE and minimally processed food goods led to the development of AP and IP, often known as "innovative packaging" [17]. The development of research in the quickly growing field of AP and IP materials is tightly associated with the pursuit of ecologically sustainable packaging solutions [56]. Table 12.4 displays an assortment of modern active and intelligent packaging technologies [57].

TABLE 12.4 Active and Intelligent Packaging Technologies for Ready-to-eat Snacks [57]

Technology	Characteristics
Oxygen scavengers	Absorb oxygen to maintain product freshness
Moisture absorbers	Absorb moisture to prevent product spoilage
Antimicrobial agents	Control the growth of microorganisms in the product
Time-temperature indicators	Monitor changes in temperature and time
Radio-frequency identification tags	Provide information about the product to the consumer
Nanocomposites	Improve barrier properties and increase shelf life

12.3.5.1 ACTIVE PACKAGING

In response to consumer desire for natural, recyclable, and biodegradable packaging materials, AP was improved [31]. AP refers to materials and products that purposefully incorporate elements that release components into or through the environment or the packaged food around them to prolong the life span on the shelf, and maintain, or enhance the condition of food packages [58]. The future invention of novel food packaging is headed in a direction that looks promising: active packaging. Various approaches are used to add antimicrobial agents to packaging. This involves combining the addition of pads or sachets with antimicrobials, antioxidants, or absorbent materials, direct coating of the packaging material's surface, direct integration of antimicrobial substances inside the packaging, or being absorbed or immobilized in the material used to form the films [59]. Food interacts with active packaging, which releases/absorbs substances from the environment/food and prolongs shelf life. Active packaging, smart or intelligent packaging, indicators of time-temperature (TTIs), microwave, gas, and other advancements in packaging technology were some of the advances that occurred in the 20th century (AP; including antimicrobials, oxygen scavengers, and moisture absorbers) [60].

Active scavenging (absorbers) and releasing (emitters) systems are the two types of AP systems. Two main types of AP are active scavenging packaging, which includes carbon dioxide, moisture absorbers, and scavengers, and active releasing and antioxidant packaging, which include packaging that emits or generates carbon dioxide and antimicrobial packaging [61]. Figure 12.6 presents an example of active packaging method.

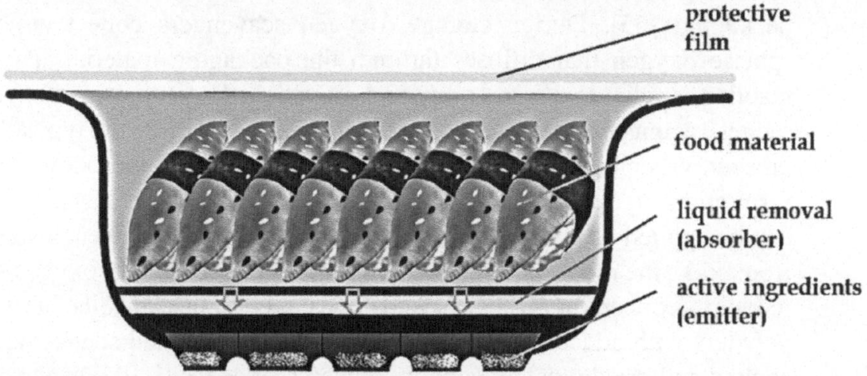

FIGURE 12.6 A schematic model of active packaging.

- **Ethylene Scavengers:** Ethylene scavengers that take ethylene from the environment generated by fresh vegetables were created to manage the ethylene (C_2H_4) gas in the packaging and increase the time span of the product on the shelf [53]. Fruits and vegetables' physiological characteristics (growth, ripening, and senescence) are greatly affected by ethylene [62]. Potassium permanganate in silica is the most well-known, least expensive, and extensively utilized ethylene absorption technique. Another method of ethylene adsorption involves covering zeolite with a quaternary ammonium cation after it has been treated with potassium permanganate. This system can absorb additional organic substances like benzene, toluene, and xylene in addition to ethylene from the atmosphere [55]. Thus, it is crucial for post-harvest preservation to use ethylene production scavengers or inhibitors to delay the ripening of vegetables and fruits [63]. It applies to fruits like apples, avocados, apricots, mangoes, and tomatoes, as well as vegetables like potatoes, carrots [64].

- **Oxygen Scavengers:** Oxygen Scavengers are a class of active packaging components that have been developed to prevent or delay the oxidative degradation of food products by scavenging any residual oxygen present in the packaging environment. Because oxygen in packaging causes oxidation of the product or encourages the growth of aerobic microbes, which results in color modifications, sensory changes, or nutritional losses, and also quality and shelf life are negatively impacted [61]. In addition to improving product quality and shelf life, oxygen-scavenging devices can replace vacuum or gas-flushing packaging [65]. During storage, oxygen scavengers collect atmospheric oxygen that diffuses through the packaging material. Their reaction is reliant on iron compound oxidation. In oxygen-permeable sachets, agents that capture oxygen using elements of iron are typically present. When oxygen absorbers are present, carbon dioxide slows the respiration of fresh items and prevents package collapse [31]. They can be applied as labels, sachets, or directly to the package enclosure. Iron oxide, the most common oxygen scavenger, reacts with oxygen to reduce its concentration [17]. Baked goods, seeds, nuts, and oils; instant powders with added fat; fruit and vegetable juices; fried snacks; and cooked and dried meat all contain oxygen scavengers [20]. Displayed in Figure 12.7 is a model of an oxygen scavenger sachet.

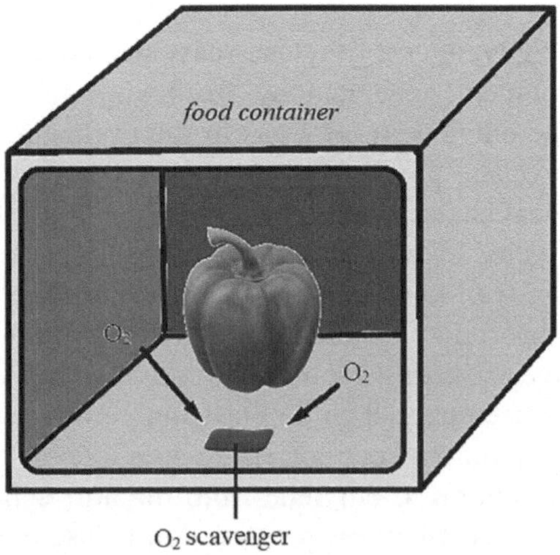

FIGURE 12.7 A working model of oxygen scavenger sachet used in food packaging.

- **Moisture Scavengers:** Moisture scavengers maintain low moisture levels in food products and extend shelf-life by selectively removing excess water vapor from the packaging environment. By preserving moisture content, lowering moisture condensation in packaging, enhancing appearance, and minimizing browning or discoloration, the scavenger has the effect of increasing shelf life [55]. Water molecules from the environment are physically absorbed and held onto by moisture scavengers. Moisture absorbers work by physically adsorbing a moisture-absorbing substance and are frequently made as pads, sachets, films, or trays [31]. It is applicable to meat, fresh fish, grains, mushrooms, strawberries, maize, tomatoes, and seeds [55].
- **Carbon Dioxide Emitters:** Carbon dioxide emitters release controlled amounts of carbon dioxide to maintain MAP environments. When oxygen absorbers are present, carbon dioxide slows the respiration of fresh items and prevents package collapse [31]. Absorbers for carbon dioxide are designed to take carbon dioxide out of the air to maintain pressure during storage and prevent a package from bursting. It occurs in roasted coffee, fresh fish, beef jerky, dried poultry products, and meat [56].
- **Antimicrobial Packaging Systems:** Antimicrobial packaging systems are solutions that contain active ingredients used to perform antimicrobial packaging. Antimicrobial agents are used during plastic film production procedures. Antimicrobial packaging can use substances that have built-in antimicrobial qualities that can be triggered chemically or physically, such as enzymes and bacteriocins [66]. The creation of antimicrobial packaging is essential for actively limiting the growth of bacteria. These technologies promote the quality and safety of food while also extending the shelf life of food products by lowering the development phase and prolonging the usage of microorganisms' lagged phase [67]. This system is applied in processed/fresh vegetables and fruits, dairy products, fresh/smoked fish, fresh seafood, cereals and baked goods, and RTE meals. Meat products such as RTE meat and cattle, ham, turkey breast, chicken, and fish filets are frequently covered with edible antimicrobial agents, including extracts from plants and essential oils, etc. [2].

- **Antioxidant Packaging:** Antioxidant packaging is an active packaging technology that incorporates antioxidants into the packaging material to prevent or delay oxidation. By preventing lipid, pigment, and protein oxidation, antioxidant chemicals increase the foods' shelf life after being applied, retaining qualities including flavor, texture, fragrance, color, and overall product quality [67]. When creating an antioxidant packaging material, the polymer of the packaging material is intimately mixed with the antioxidant agent. This can be done in one of three ways: (i). utilizing coating technology to apply the solution to a substrate after dissolving both in a suitable solvent, (ii). extrusion technologies are used to melt the polymer, incorporate, and mix it into the melting agent, or (iii). by immobilizing the antioxidant [54]. Antioxidant packaging is used for fried foods, instantaneous powders with fat, seeds, nuts, oils, and fresh, fatty fish and meat [68]. Antioxidant-active packaging is commonly utilized to improve product quality and extend the duration of storage of meat and its products by reducing the amount of oxygen that the food item is exposed to. To extend the shelf life of food, especially beef or processed meat, antioxidants in light- and oxygen-barrier packaging are essential factors [69]. The substitution of synthetic chemicals with natural antioxidants is a significant trend in packaging meat with antioxidant activity. The most well-known natural antioxidants are tocopherols, extracts from herbs like oregano [54].

12.3.5.2 INTELLIGENT PACKAGING (IP) TECHNOLOGIES

IP is a system that performs intelligent actions, including detecting, recording, tracking, and sending, by using scientific reasoning with the aim of enhancing quality and safety, prolonging shelf life, and assisting in decision-making. Additionally, it alerts clients to potential problems [27]. It is generally acknowledged that three primary technologies can be used to develop intelligent packaging systems: (i). indicators used to increase customer convenience and/or notify them of the food quality; (ii). Barcodes and radiofrequency identification tags RFID are two examples of data carriers created specifically for storing, disseminating, and traceability, and (iii). sensors that make it possible to quickly and accurately measure the analytes in food [70]. The creation of mobile applications and smart refrigerators opens up new opportunities for contact with customers via sensors, RFID, or "near-field communication" (NFC) tags used as smart packaging.

Despite the fact that this technology is still in development and has a limited presence in the RTE market right now, it is possible to obtain a commercially viable colorimetric indication with fresh products, including meat and seafood, as well as chilled and deep-frozen options. From the time of production, until the product is in the consumer's hands, the smart label keeps track of the product's quality [2]. The true meaning of intelligent packaging is to assess the capability and stability of AP systems [17].

Intelligent packaging that is capable of more than merely protecting and containing contents is where the future of food packaging lies. IP is a type of communicative container that effectively closes the evaluation gap that now exists for a variety of food goods, including liquid foods, meat, fish, dairy, fruits, and vegetables [71]. It is possible that intelligent and active packaging may enter the market and even overtake it in a few years due to the advantages of adopting smart packaging in terms of logistics, safety, and marketing. Intelligent packaging monitors the product's conditions, doesn't affect food, and also gives information [72].

- **Freshness Indicators:** Freshness Indicators can be applied to foods that are packaged and perishable, including milk, fruits, vegetables, cakes, and other cold items [31]. Two different types of changes can occur in fresh food products: (i) pH changes, which can lead to the production of harmful substances, unpleasant aromas, and other microbiological changes; and (ii) oxidation of lipids and pigments, which can cause undesirable flavors and scents, color changes, and other adverse chemical and biological reactions.
- **Pathogen Indicators:** Pathogen Indicators are biosensors, which are small analytical devices that can detect, record, and communicate data about biochemical processes caused by pathogens [54]. This helps determine whether hazardous bacteria are present by looking for immobilized antibodies [31]. Moreover, pathogenic microorganism-contaminated meat and meat products contamination indicators have been developed for more precise detection. It applies to items that spoil quickly, like meat, fish, and poultry.
- **Time-Temperature Indicators (TTI):** Time-Temperature Indicators (TTI) give details regarding storage conditions while continuously tracking and reflecting the temperature history across the food storage facility's cold chain (Figure 12.8). These indicators' operating theory is based on the diverse reactions that occur when two or more materials are combined (chemical, mechanical, enzymatic, etc.) [53].

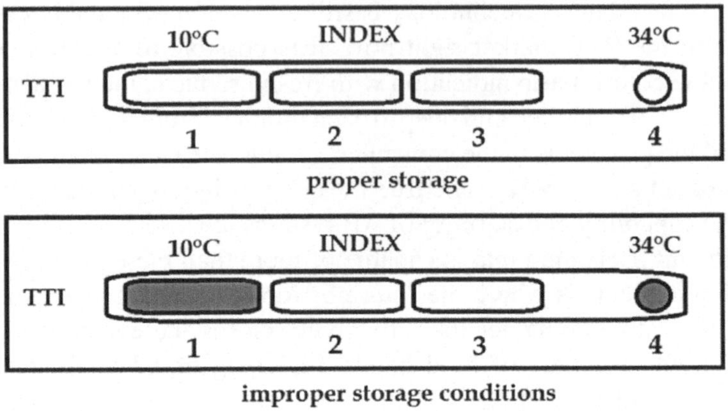

FIGURE 12.8 A schematic example of time-temperature indicators used in snack packaging.

- **Gas İndicator:** Gas indicator offers information on the condition of the food's quality based on the inside environment, checks the effectiveness and integrity of packed food depending on the particular mixture of gases within MAP, and spots package leaks [53]. Foods packaged with the necessary gas content [54].
- **Barcode:** Barcode is applied to the products using the barcoding process, also known as automated identification (Auto-ID), which was developed to help large supermarkets and retail businesses manage their inventory. A central system receives the information contained in barcodes after they are scanned by an optical barcode scanner [73]. Inventory control is made easier by helping with product identification and product authenticity. If harmful germs develop inside the package while it is being stored, the barcode can detect them and experience color changes, rendering the barcode unreadable [71].
- **Radiofrequency İdentification (RFID):** Radiofrequency İdentification (RFID) is a data collection and archiving technique that produces precise, real-time data without the need for human intervention (Figure 12.9). Many benefits of RFID technology in supply chain management are predicted, including reduced costs for storage, handling, and distribution; greater sales through a drop in stock-outs; and improved cash flow through an increase in inventory rotations. RFID technology is also thought to play a vital role in preventing product recalls [74].

FIGURE 12.9 An RFID example in snack shrimp packaging.

- **Sensors:** Sensors are objects or systems that have software, a network of connections, and electronics for control and processing [70]. Sensors determine the presence, quantity, and makeup of specific physical or chemical analytes. A novel kind of food packaging uses edible sensors, which are a safe way to check for food deterioration [55].
- **Biosensors:** Biosensors can detect and monitor specific indicators of food quality and safety, such as microbial contamination, spoilage, and chemical changes. The recognition layer is where chemical sensors and biosensors differ most from one another. Chemical sensors use a chemical substance as its receptor, whereas biological elements including nucleic acids, enzymes, and antigens are used in the recognition stage of biosensors [70]. For quality control and observing color changes during the period of time when chicken and beef filets deteriorate, biological analyte-interactive biosensors are examples of biosensors. Biosensors include electrochemical systems that use nylon-6 nanofiber membranes as the base and glucose oxidase as the operational enzyme to detect the presence of glucose in a range of beverages [55].

12.4 ADVANCEMENTS AND FUTURE TRENDS

New materials and technologies are constantly being developed in the packaging industry for RTE snacks to meet the evolving needs of consumers and the food industry. For ready-to-eat snacks in particular, the sector is very dependent on the packaging. Manufacturers of food are now capable of extending the duration of storage, security, and convenience of snacks for consumers thanks to recent improvements in packaging technology. Since the packaging industry is aware of the damage that packaging waste causes to the environment, biodegradable and compostable materials have been established. Active packaging that makes use of nanotechnology and antimicrobial compounds is also gaining popularity as a way to enhance food safety and extend shelf life. Additionally, smart packaging that makes use of sensors and other technologies can give consumers essential information like nutritional information and expiration dates.

The latest developments and future trends of biodegradable and compostable materials, active packaging, nanotechnology, and smart packaging for ready-to-eat snacks will all be addressed in this subtitle, along with any possible implications for the food industry and consumer behavior.

12.4.1 BIODEGRADABLE AND COMPOSTABLE MATERIALS

It is becoming more popular to employ biodegradable and compostable materials for RTE snack packaging in order to reduce the adverse effects of packaging waste. These substances are made to degrade naturally over time without releasing pollutants or dangerous chemicals into the environment.

The demand for sustainable packaging solutions is rising as customers become more ecologically concerned. For ready-to-eat snacks, biodegradable and compostable packaging materials have emerged as a possible replacement for conventional plastic packaging. The amount of waste that gets up in landfills and oceans is decreased as a result of these materials' natural decomposition in the environment [75].

The most recent developments in ready-to-eat snack packaging that is biodegradable and compostable have been concentrated on making the packaging more useful and durable while yet being environmentally friendly. In contrast to conventional plastic packaging, some packaging companies have created compostable packaging films that are transparent, waterproof, and heat-sealable [76].

In addition to their environmental benefits, biodegradable and compostable packaging materials also appeal to consumers who prioritize sustainability. Nearly 70% of consumers, according to a new McKinsey & Company survey, are willing to spend more for environmentally friendly items [77].

It is projected that in the future, the utilization of compostable and biodegradable packages would expand. Even now, some businesses are looking into using food waste, including potato peels and cornstarch, as a source for packaging raw materials.

It's crucial to remember that biodegradable and compostable packaging is not a universally effective solution. To ensure that these materials can effectively decompose in the environment, proper disposal is essential. Additionally, some detractors have expressed worry about the possibility that packaging made of biodegradable and compostable materials could increase the amount of food wasted since customers could think they are less detrimental to the environment and more inclined to throw them away carelessly. Biodegradable and compostable packaging materials have a bright future in the food business as the demand for environmentally friendly packaging keeps growing [78].

Packaging materials for ready-to-eat snacks that are compostable and biodegradable are not only good for the environment but also have several advantages. These substances have minimal to no negative effects on the environment because they come from natural sources and can decompose into non-toxic components. Biodegradable and compostable packaging materials are lightweight in addition to being environmentally friendly, which lowers transportation costs and the carbon footprint of packaging materials [79].

In the upcoming years, it is projected that the market for compostable and biodegradable packaging materials would expand dramatically. According to a recent report, the worldwide revenues for biodegradable packaging will rise at a CAGR of 14.3% between 2021 and 2026, reaching $25.5 billion [80].

One of the most recent advances in biodegradable and compostable packaging materials for ready-to-eat snacks is the use of plant-based materials like bamboo, wheat straw, and sugarcane to produce packaging that is both ecologically friendly and useful. Additionally, edible packaging is growing in popularity since it provides customers with a distinctive and sustainable experience. Edible packaging is manufactured from substances like seaweed and starch [81–83].

Several companies are also producing packaging materials that are compostable and biodegradable utilizing cutting-edge methods. For instance, a company has invented seaweed-based packaging materials that dissolve in hot water, while another company has produced seaweed-based edible water containers [83].

It's important to remember that compostable and biodegradable packaging materials have significant drawbacks as well. To ensure proper decomposition, these materials need specific disposal techniques, and the absence of uniform certification and labeling procedures may cause misunderstanding among consumers. Additionally, the production costs of biodegradable and compostable packaging materials are higher than those of conventional packaging materials, which may prevent their widespread use [84–85].

In conclusion, biodegradable and compostable packaging materials are an increasingly popular solution for environmentally conscious consumers who want to reduce their carbon footprint. With continued innovation and advancements in technology, these materials have the potential to revolutionize the packaging industry and help reduce waste in the environment.

12.4.2 ACTIVE PACKAGING

The phrase "active packaging" refers to the use of substances or materials in packaging systems that engage actively both the packaged food and its surroundings. The goals of AP are to raise the food item's quality and safety, lengthen its shelf life, and enhance its sensory attributes. AP systems can be designed to carry out a variety of diverse functions, such as flavor release, oxygen scavenging, moisture control, and antibacterial qualities. The active components used in these systems may be integrated into the packaging materials or used separately from the packaging.

The field of active packaging is developing, and new materials and technologies are being created to fulfill the requirements of the food business, including the rising need for environmentally friendly and sustainable packaging options. Here are some of the potential future developments in active packaging:

12.4.2.1 USE OF NANOTECHNOLOGY

The advent of intelligent and responsive packaging thanks to nanotechnology has the ability to completely transform the active packaging industry. In addition, the application of nanosensors in active packaging can assist in detecting and signaling changes in moisture, temperature, and oxygen levels, allowing for increased food safety and quality.

Active packaging systems with nanosensor integration have the ability to offer real-time monitoring of numerous parameters that may have an impact on food quality and safety. For instance, nanosensors can be utilized to track and detect changes in food packaging's bacterial contamination, temperature, humidity, and gas composition [86].

Recent research has demonstrated that the use of nanosensors in packing films can aid in the early detection of hazardous bacteria and toxins, hence preventing foodborne diseases [87]. Additionally, the application of nanotechnology to active packaging can facilitate the creation of packaging elements that adapt to environmental changes. For instance, nanoparticles can be utilized to create packaging materials that react to temperature changes by dispensing antimicrobial chemicals to stop the growth of bacteria [88].

Overall, by enhancing food quality and safety, extending shelf life, and lowering food waste, the application of nanotechnology in active packaging holds an opportunity to completely transform the food packaging sector. However, there are still certain issues that must be resolved, especially the probable relocation of nanoparticles from packaging materials to food, which could have harmful impacts on health [89, 90].

12.4.2.2 INCREASED USE OF NATURAL ANTIMICROBIAL AGENTS

Active packaging that uses natural antibacterial agents instead of synthetic preservatives is becoming more and more popular. Bacteriocins, plant extracts, and essential oils are a few examples. Natural antimicrobial chemicals are being used more frequently, which benefits the food business by reducing the need for synthetic preservatives in active packaging. In this method, organic substances such as bacteriocins, extracts from plants, and essential oils are used to prolong the life on the shelf of food products. In order to help prevent the spread of harmful microorganisms and maintain the quality of food, these chemicals can be applied to packaging elements where they will slowly release. According to research, this concept may help decrease microbial development and increase the safety of food items, making it desirable to use in the food industry [67, 91–93].

12.4.2.3 INTEGRATION OF İNTELLIGENT SYSTEMS

Active packaging that incorporates intelligent technology can deliver live data on the nutritional value and security of packed food. By monitoring a food product's temperature and humidity during storage and transportation, for instance, intelligent devices could warn buyers if the item has been subjected to undesirable conditions.

The food sector is paying more and more attention to the integration of intelligent technology into active packaging since it can deliver real-time data on the quality as well as safety of food in packaging. During the packaging, transportation, and storage processes, intelligent systems can monitor a number of parameters, including humidity, temperature, gas composition, and pressure. a recent research study developed an innovative packaging system capable of monitoring the temperature, humidity, and gas composition of perishable food products during transportation. The research discovered that the smart system was efficient in determining the product's shelf life and monitoring its quality [94].

In a different study, a pH-responsive color indicator was created by adhering shikonin to cellulose paper. This indicator demonstrated outstanding stability and responsiveness to pH changes, and it was effectively employed for assessing the freshness of seafood and meat while in storage, implying its potential application in intelligent food packaging uses [95].

Intelligent systems can inform customers if a food product has been subjected to unfavorable conditions along with keeping track of the product's quality. Immobilized silver nanoparticles embellished with molybdenum disulfide on a "Split-Ring Resonator" (SRR) tag were used to create a wireless "electronic gas sensor" (EGS) for ammonia and biogenic amine detection. The developed sensor offered a wirelessly appealing response when utilized to monitor the decomposition process of various meat species at room temperatures, including mutton, beef, chicken, and hog. The system was able to monitor when the meat product had spoiled, indicating to the consumer that the item is not suitable for consumption [96].

Overall, by giving live data on the quality and security of packaged food, lowering food waste, and boosting customer confidence in the goods they consume, the incorporation of intelligent technologies into active packaging holds the power to completely transform the food industry.

12.4.2.4 DEVELOPMENT OF GREEN ACTIVE PACKAGING

There is an increasing interest in creating green active packaging for ready-to-eat snacks as the food industry grows more concerned with sustainability. The term "active packaging" describes materials that actively engage with the food item to extend and enhance both its quality and shelf life.

Environmentally responsible and sustainable materials are used in the creation of green active packaging. For example, some packaging materials may be biodegradable or compostable, which means they can be broken down naturally by microorganisms and do not contribute to pollution. Additionally, packaging made from renewable resources, such as plant-based materials, can reduce the reliance on non-renewable resources like fossil fuels.

Green active packaging has the potential to increase the food industry's sustainability while simultaneously preserving the safety and quality of food goods in ready-to-eat snacks. In addition to satisfying the demands of customers who are growing more interested in sustainable products, it can assist reduce waste and environmental effects.

Sure, some future trends in eco-friendly active packaging for ready-to-eat snacks include:

1. **Incorporating Edible Packaging:** Edible packaging is made from natural materials that are safe for consumption and can potentially reduce waste. In the future, we may see more edible packaging options for ready-to-eat snacks, such as coatings made from plant-based materials [97, 98].
2. **Using Smart Packaging Technology:** Smart packaging technology can help to increase the food products' shelf lives by monitoring and controlling environmental factors that as temperature and humidity. In the future, green active packaging for RTE snacks may incorporate smart packaging technology to further improve the nutritional value and safety of food items [99, 100].
3. **Focusing on Sustainable Sourcing:** As consumers become more interested in sustainable products, there may be a greater emphasis on using packaging materials that are sourced sustainably. This may include materials that are produced using renewable energy or materials that are recycled or biodegradable [101].
4. **Incorporating Antimicrobial Properties:** Antimicrobial properties in packaging might help to decrease the risk of contamination and spoilage of food products. In the future, green active packaging for RTE snacks may incorporate natural antimicrobials like plant extracts can help to increase the food items' safety and quality [102, 103].

Overall, the future trends in green active packaging for RTE snacks will likely focus on improving sustainability, quality, and safety while meeting the needs and preferences of consumers.

12.4.2.5 PERSONALIZED PACKAGING

With the rise of personalized nutrition, there is potential for active packaging to become more tailored to individual consumers' needs. For example, packaging could be designed to release certain nutrients or supplements at specific times of the day or in response to specific physiological conditions.

The above approach could be beneficial for individuals who have unique nutritional needs or health conditions, such as athletes, pregnant women, or people with vitamin shortages. This tailored packaging may increase the efficacy and reliability of nutrition delivery while also offering consumers greater convenience and better health outcomes. This concept, however, is still in its early phases and would necessitate further research and development.

These are just a few potential future trends in active packaging, and it will be interesting to see how this field continues to develop in the coming years.

12.4.3 NANOTECHNOLOGY

With the use of nanotechnology, it is possible to engineer materials with distinctive features by modifying matter at the nanoscale. In the food industry, nanotechnology has garnered attention for its potential to enhance food packaging technology, especially for ready-to-eat (RTE) snacks. Improved barrier characteristics against oxygen, humidity, and UV light have been shown in nano-based materials, which lowers the risk of food deterioration and lengthens the duration of storage of packed RTE snacks. Moreover, the utilization of nano sensors and other nanodevices in food packaging can enable real-time monitoring of various parameters such as temperature, gas composition, and microbial contamination, providing valuable insights into the food product's quality and safety. Research studies have shown the successful implementation of nanotechnology in packaging RTE snacks, such as antimicrobial films, nanostructured coatings, and active packaging systems [104]. Therefore, nanotechnology presents an exciting avenue for the creation of creative packaging options that can respond to the changing requirements of the RTE snack industry.

12.4.4 SMART PACKAGING

Utilizing cutting-edge technology, smart packaging entails developing materials that can communicate with consumers, offer nutritional data, and even alter in color or feel depending on the state of the RTE food within. In addition to potentially revolutionizing the way RTE snacks are packaged and consumed, smart packaging has the potential to significantly improve consumer health and reduce food waste.

The integration of technological advances into package materials to improve functioning and provide additional value to consumers is referred to as smart packaging or intelligent packaging. The way RTE snacks are packed and consumed in the food industry has the potential to change as a result of smart packaging. Smart packaging, for example, can include sensors and other devices that monitor the state of the food product, such as temperature, humidity, and gas composition. This information can be delivered to consumers via mobile apps, allowing them to make accurate decisions about the product's safety and quality. Smart packaging can also give nutritional information such as calorie counts and ingredient listings to assist consumers in making healthier food choices.

Furthermore, smart packaging can be programmed to change color, texture, or shape in response to changes in the condition of the RTE food contained beneath it. This feature might alert consumers of probable deterioration or contamination, lowering the risk of foodborne illness and waste. A recent study investigated how nanotechnology and other advanced materials can be used to create smart packaging for RTE snacks, such as bio-based sensors, colorimetric indicators, and edible films [104]. These materials hold significant promise for developing creative packaging solutions that can suit the changing needs of the RTE snack market while enhancing consumer health.

12.5 CONCLUSION

The convenience and availability of RTE foods are major contributors to their popularity. RTE snacks must be packaged properly to ensure their safety, freshness, and quality. Plastics, paper, aluminum foil, and glass are common packaging materials for RTE snacks, each with advantages and downsides. The chapter includes several packaging techniques, including MAP systems, aseptic, AP and IP systems, and vacuum packaging systems. AP and IP (smart) systems, such as antibacterial agents, moisture absorbers, oxygen scavengers, *RFID* tags, TTI, and nanocomposites are emerging technologies that can increase the security of RTE snacks.

The RTE snack packaging market is continually evolving to satisfy the changing needs of customers and the food sector. Biodegradable and compostable materials, active packaging, nanotechnology, and smart packaging are all emerging ideas having the potential to alter how to prepare and consume RTE snacks. Food manufacturers now have a variety of options to select from, according to the snack's storage circumstances and type.

To summarize, it is obvious that RTE snack packaging is critical in guaranteeing their safety and quality, and the industry is always innovating to satisfy increasing consumer and food industry needs. RTE snack packaging's future will most likely involve a blend of classic packaging technologies and new emergent trends. To fulfill these needs, the food industry must continue to adapt and evolve, and advancements in packaging materials and technologies will be crucial in guaranteeing the safety and quality of RTE snacks.

KEYWORDS

- **food safety**
- **packaging materials**
- **packaging technologies**
- **ready-to-eat snacks**
- **shelf life**

REFERENCES

1. Versino, F., Ortega, F., Monroy, Y., Rivero, S., López, O. V., & García, M. A. (2023). Sustainable and bio-based food packaging: A review on past and current design innovations. *Foods, 12*(5), 1057. https://doi.org/10.3390/foods12051057.
2. Bumbudsanpharoke, N., & Ko, S. (2022). Packaging technology for home meal replacement: Innovations and future prospective. *Food Control, 132*, 108470. https://doi.org/10.1016/j.foodcont.2021.108470.
3. Padmashree, A., Sharma, G., & Govindaraj, T. (2013). Development and evaluation of shelf stability of Flaxoat nutty bar in different packaging materials. *Food and Nutrition Sciences, 4*(5), 538–546. https://doi.org/10.4236/fns.2013.45069.
4. Sujeetha, A. P., Meenatchi, R., Paulin Patricia, & Negi, A. (2020). Effect of vacuum packaging on quality of pomegranate arils during storage. *Current Journal of Applied Science and Technology, 39*(38), 40–46.
5. Temgire, S., Borah, A., Kumthekar, S., & Idate, A. (2021). Recent trends in ready to eat/cook food products. *The Pharma Innovation Journal, 10*(5), 211–217.
6. Öztürk, K., & Nilüfer-Erdil, D. (2015). Packaging applications for ready-to-eat foods. Istanbul technical university faculty of chemical and metallurgical engineering-food engineering (Graduate Study Thesis), 53.
7. Yarış, A., & Ceyhun-Sezgin, A. (2017). Food packaging: Glass and plastic. In H. Araplıoğlu, A. Atık, R. L. Elliot, & E. Turgeon (Eds.), *Researches on Science and Art in 21st Century Turkey* (pp. 735–740). Ankara: Gece Publishing.
8. Niaounakis, M. (2019). *Recycling of Flexible Plastic Packaging*. William Andrew Applied Science Publishers, Elsevier.
9. Kumari, S. V. G., Pakshirajan, K., & Pugazhenthi, G. (2022). Recent advances and future prospects of cellulose, starch, chitosan, polylactic acid, and polyhydroxyalkanoates for sustainable food packaging applications. *International Journal of Biological Macromolecules, 221*, 163–182.

10. Awulachew, M. T. (2022). A review of food packaging materials and active packaging system. *International Journal of Health Policy Planning, 1*(1), 28–35.
11. Üçüncü, M. (2011). *Gıda Ambalajlama Teknolojisi*. Ambalaj Sanayicileri Derneği İktisadi İşletmesi.
12. Deshwal, G. K., Panjagari, N. R., & Alam, T. (2019). An overview of paper and paper-based food packaging materials: Health safety and environmental concerns. *Journal of Food Science and Technology, 56*, 4391–4403. https://doi.org/10.1007/s13197-019-03950-z.
13. Deshwal, G. K., & Panjagari, N. R. (2020). Review on metal packaging: Materials, forms, food applications, safety and recyclability. *Journal of Food Science and Technology, 57*, 2377–2392.
14. Ruippo, L., Koivula, H., Korhonen, J., Toppinen, A., & Kylkilahti, E. (2022). Innovating for sustainability: Attributes, motivations, and responsibilities in the Finnish food packaging ecosystem. *Circular Economy and Sustainability*, 1–19.
15. Ayhan, Z. (2011). Effect of packaging on the quality and shelf-life of minimally processed/ready to eat foods. *Akademik Gıda, 9*(4), 36–41.
16. Kusuma, H. S., Yugiani, P., Himana, A. I., Aziz, A., & Putra, D. A. W. (2023). Reflections on food security and smart packaging. *Polymer Bulletin*, *23*, 1–47. https://doi.org/10.1007/s00289-023-04734-4.
17. Majid, I., Ahmad, N., Nayik, G. A., Mohammad Dar, S., & Nanda, V. (2016). Novel food packaging technologies: Innovations and future prospective. *Journal Saudi Society Agricultural Sciences, 17*(4), 454–462.
18. Silberbauer, A., & Schmid, M. (2017). Packaging concepts for ready-to-eat food: Recent progress. *Journal of Packaging Technology and Research, 1*, 113–126. https://doi.org/10.1007/s41783-017-0019-9.
19. Wieczyńska, J., & Cavoski, I. (2018). Antimicrobial, antioxidant and sensory features of eugenol, carvacrol and trans-anethole in active packaging for organic ready-to-eat iceberg lettuce. *Food Chemistry, 259*, 251–260. https://doi.org/10.1016/j.foodchem.2018.03.137.
20. Kaushal, M., Vaidya, D., & Gupta, A. (2019). Novel technologies in processing of horticultural crops. *Smart Agri Post, 4*(3), 23–26.
21. Chan, S. S., Rotabakk, B. T., Løvdal, T., Lerfall, J., & Roth, B. (2021). Skin and vacuum packaging of portioned Atlantic salmon originating from refrigerated seawater or traditional ice storage. *Food Packaging and Shelf Life, 30*, 100767. https://doi.org/10.1016/j.fpsl.2021.100767.
22. Raigar, R. K., & Mishra, H. N. (2022). Ready-to-eat high-energy food paste stability and shelf-life prediction in different packaging materials. *Applied Food Research, 2*(2), 100230.
23. Rovira, P., Giannina, B., Jesica, R., Cabrera, M. C., Ali, S., Souza, G. de, Santiago, L., & Caterina, R. (2023). Microbiological changes during long-storage of beef meat under different temperature and vacuum-packaging conditions. *Foods, 12*(4), 694. https://doi.org/10.3390/foods12040694.
24. Chikkanna, G. S., Jha, S. K., Ambika, D. S., & Sadananda, G. K. (2020). Effect of packaging material and storage period on quality of aonla and cereal-based extruded ready-to-eat snack. *International Journal of Current Microbiology and Applied Sciences, 9*(1), 511–519.

25. Paulsen, E., Sofía, B., & Patricia, L. (2019). Ready-to-eat cherry tomatoes: Passive modified atmosphere packaging conditions for shelf-life extension. *Food Packaging and Shelf Life, 22*, 100407.
26. Sanjana, M. C., Hemegowda, R., & Sushma, R. E. (2019). Aseptic packaging–A novel technology to the food industry. *International Journal of Trend in Scientific Research and Development, 3*(3), 307–310. ISSN: 2456–6470.
27. Loucanova, E., Kalamárová, M., & Parobek, J. (2017). The innovative approaches to packaging–Comparison analysis of intelligent and active packaging perceptions in Slovakia. *Studia Universitatis, Vasile Goldis* Arad–*Economics Series, 27*, 33–44.
28. Wyrwa, J., & Barska, A. (2017). Innovations in the food packaging market: Active packaging–A review article. *European Food Research and Technology, 243*, 1681–1692.
29. Ijabadeniyi, O. A., Govender, A., Olagunju, O. F., & Oyedeji, A. B. (2021). The antimicrobial activity of two phenolic acids against foodborne *Escherichia coli* and *Listeria monocytogenes* and their effectiveness in a meat system. *Italian Journal of Food Science, 33*, 39–45.
30. Pal, M. (2017). Nanotechnology: A new approach in food packaging. *Journal of Food Microbiology, Safety and Hygiene, 2*, 121. https://doi.org/10.4172/2476-2059.1000121.
31. Pal, M., Devrani, M., & Hadush, A. (2019). Recent developments in food packaging technologies. *Beverage & Food World, 46*(1), 21–25.
32. Benyathiar, P., Harte, B., & Harte, J. (2020). Shelf-life extension of fresh asparagus using modified atmosphere packaging and vacuum skin packaging in microwavable tray systems. *Packaging Technology and Science, 33*(10), 407–415.
33. Paulsen, E., Sofía, B., & Patricia, L. (2021). Production of packaged ready-to-eat whole strawberries (cv. San Andreas): Packaging conditions for shelf-life extension. *Food Packaging and Shelf Life, 29*, 100696. https://doi.org/10.1016/j.fpsl.2021.100696.
34. Wiley, R. C., & Yildiz, F. (2017). Introduction to minimally processed refrigerated (MPR) fruits and vegetables. In F. Yildiz & R. Wiley (Eds.), *Minimally Processed Refrigerated Fruits and Vegetables* (pp. 3–15). Springer, Boston, MA. https://doi.org/10.1007/978-1-4939-7018-6_1.
35. Oliveira, M., Abadias, M., Usall, J., Torres, R., Teixidó, N., & Viñas, I. (2015). Application of modified atmosphere packaging as a safety approach to fresh-cut fruits and vegetables–A review. *Trends in Food Science & Technology, 46*(1), 13–26.
36. Tyuftin, A. A., & Kerry, J. P. (2023). The storage and preservation of meat: Storage and packaging. In *Lawrie's Meat Science* (pp. 315–362). Woodhead Publishing. https://doi.org/10.1016/B978-0-323-85408-5.00017-0.
37. Berk, Z. (2018). Chapter 27: Food packaging. In *Food Process Engineering and Technology* (3rd ed., pp. 625–641). Elsevier. https://doi.org/10.1016/B978-0-12-812018-7.00027-0.
38. Watts, E. (2022). Seafood handling, processing, and packaging. *Reference Module in Food Science*. https://doi.org/10.1016/B978-0-323-85125-1.00102-2.
39. Bouletis, A. D., Arvanitoyannis, I. S., & Hadjichristodoulou, C. (2017). Application of modified atmosphere packaging on aqua cultured fish and fish products: A review. *Critical Reviews in Food Science and Nutrition, 57*(11), 2263–2285. https://doi.org/10.1080/10408398.2013.862202.
40. Abd El-Hay, M. M. (2022). Processing and preparation of fish. In *Postharvest and Postmortem Processing of Raw Food Materials* (pp. 315–342). Unit Operations and Processing Equipment in the Food Industry.

41. Ssepuuya, G., Aringo, R. O., Mukisa, I. M., & Nakimbugwe, D. (2016). Effect of processing, packaging, and storage-temperature-based hurdles on the shelf stability of sautéed ready-to-eat *Ruspolia nitidula*. *Journal of Insects as Food and Feed, 2*(4), 245–253.
42. Naveena, B. M., Khansole, P. S., Shashi, Kumar, M., Krishnaiah, N., Kulkarni, V. V., & Deepak, S. J. (2017). Effect of sous vide processing on physicochemical, ultrastructural, microbial, and sensory changes in vacuum packaged chicken sausages. *Food Science and Technology International, 23*(1), 75–85. https://doi.org/10.1177/1082013216658580.
43. Laranjo, M., Gomes, A., Potes, M. E., Fernandes, M. J., Fraqueza, M. J., & Elias, M. (2016). Development of a long-life vacuum-packaged ready-to-eat meat product based on a traditional Portuguese seasoned meat. *International Journal of Food Science and Technology, 51*, 1150–1158.
44. Deuri, D., Hazarika, P., Singh, T. P., Chhangte, L., Singh, P., & Talukder, S. (2016). Effect of curing ingredients and vacuum packaging on the physico-chemical and storage quality of ready-to-eat Vawksa Rep (smoked pork product) during refrigerated storage. *Veterinary World, 9*(6), 587–594.
45. Aberoumand, A., & Baesi, F. (2020). Effects of vacuum packaging in freezer on oxidative spoilage indexes of fish *Lethrinus atkinsoni*. *Food Science & Nutrition, 14*(8), 4145–4150. https://doi.org/10.1002/fsn3.1704.
46. Kameník, J., Saláková, A., Pavlík, Z., Bořilová, G., Hulanková, R., & Steinhauserová, I. (2014). Vacuum skin packaging and its effect on selected properties of beef and pork meat. *European Food Research and Technology, 239*, 395–402. https://doi.org/10.1007/s00217-014-2233-9.
47. Voicu, G., Constantin, G. A., Stefan, E. M., Tudor, P., Munteanu, M. G., & Żelaziński, T. (2019). Aspects regarding the aseptic packaging of food products. *Acta Technica Corviniensis-Bulletin of Engineering, 12*(1), 85–90.
48. Chiozzi, V., Agriopoulou, S., & Varzakas, T. (2022). Advances, applications, and comparison of thermal (pasteurization, sterilization, and aseptic packaging) against non-thermal (ultrasounds, UV radiation, ozonation, high hydrostatic pressure) technologies in food processing. *Applied Sciences, 12*, 2202. https://doi.org/10.3390/app12042202.
49. Roobab, U., Aadil, R. M., Madni, G. M., & Bekhit, A. E. D. (2018). The impact of nonthermal technologies on the microbiological quality of juices: A review. *Comprehensive Reviews in Food Science and Food Safety, 17*, 437–457.
50. Lumby, N., & Park, J. (2021). Chapter 16: Packaging for probiotic beverages. In *Probiotic Beverages* (pp. 339–363). https://doi.org/10.1016/B978-0-12-818588-9.00014-0.
51. Park, J. A., & Jang, Y. S. (2018). Differences in consumer purchasing biases of home meal replacement by gender and involvement: Target to university students. *Culinary Science & Hospitality Research, 24*(7), 63–73.
52. Sagar, N. A., & Pareek, S. (2020). Antimicrobial assessment of polyphenolic extracts from onion (*Allium cepa* L.) skin of fifteen cultivars by sonication-assisted extraction method. *Heliyon, 6*(11), e05478. https://doi.org/10.1016/j.heliyon.2020.e05478.
53. Firouz, M. S., Mohi-alden, K., & Omid, M. (2021). A critical review on intelligent and active packaging in the food industry: Research and development. *Food Research International, 141*, 110113.
54. Fang, Z., Zhao, Y., Warner, R., & Johnson, S. (2017). Active and intelligent packaging in meat industry. *Trends in Food Science and Technology, 61*, 60–71.

55. Özcan, A. (2020). New approaches in smart packaging technologies. (Invited lecture), 21–37. https://doi.org/10.24867/GRID-2020-p1.
56. Nura, A. (2018). Advances in food packaging technology–A review. *Journal of Postharvest Technology, 6*(4), 55–64.
57. Dobrucka, R., & Przekop, R. (2019). New perspectives in active and intelligent food packaging. *Journal of Food Processing and Preservation, 43*(11), e14194.
58. Just, D. R., & Goddard, J. M. (2023). Behavioral framing and consumer acceptance of new food technologies: Factors influencing consumer demand for active packaging. *Agribusiness, 39*, 3–27. https://doi.org/10.1002/agr.21778.
59. Guo, M., Zhang, X., & Jin, Z. T. (2023). Active food packaging. In G. W. Smithers (Ed.), *Encyclopedia of Food Safety* (2nd ed.). Elsevier. https://doi.org/10.1016/B978-0-12-822521-9.00078-2.
60. Han, J., Ruiz-Garcia, L., Qian, J., & Yang, X. (2018). Food packaging: A comprehensive review and future trends. *Comprehensive Reviews in Food Science and Food Safety, 17*, 860–877. https://doi.org/10.1111/1541-4337.12343.
61. Yildirim, S., Röcker, B., Pettersen, M. K., Nilsen-Nygaard, J., Ayhan, Z., Rutkaite, R., Radusin, T., Suminska, P., Marcos, B., & Coma, V. (2018). Active packaging applications for food. *Comprehensive Reviews in Food Science and Food Safety, 17*(1), 165–199. https://doi.org/10.1111/1541-4337.12322.
62. Gaikwad, K. K., Singh, S., & Negi, Y. S. (2020). Ethylene scavengers for active packaging of fresh food produce. *Environmental Chemistry Letters, 18*(2), 269–284. https://doi.org/10.1007/s10311-019-00938-1.
63. Pathak, N. (2019). *Photocatalysis and Vacuum Ultraviolet Light Photolysis as Ethylene Removal Techniques for Potential Application in Fruit Storage* (Ph.D. thesis). Technische Universität Berlin.
64. Altaf, U., Varsha, K., & Rouf, A. (2018). Novel packaging technology for food industry. *Journal of Pharmacognosy and Phytochemistry, 7*(1), 1618–1625.
65. Sundramoorthy, A. K., Thomaati, H., Vignesh Kumar, G., & Gunasekaran, S. (2018). Graphene-based nano sensors and smart food packaging systems for food safety and quality monitoring. In *Graphene Bioelectronics; Advanced Nanomaterials* (pp. 267–306). https://doi.org/10.1016/B978-0-12-813349-1.00012-3.
66. Cierpiszewski, R. (2016). *Opakowania aktywne i inteligentne* [Active and intelligent packaging]. Wydawnictwo Uniwersytetu Ekonomicznego w Poznaniu. ISBN: 978-83-7417-872-3.
67. Alves, J., Gaspar, P. D., Lima, T. M., & Silva, P. D. (2023). What is the role of active packaging in the future of food sustainability? A systematic review. *Journal of the Science of Food and Agriculture, 103*(3), 1004–1020. https://doi.org/10.1002/jsfa.11880.
68. Rojas, A., Torres, A., Galotto, M. J., Guarda, A., & Julio, R. (2020). Supercritical impregnation for food applications: A review of the effect of the operational variables on the active compound loading. *Critical Reviews in Food Science and Nutrition, 60*(8), 1290–1301. https://doi.org/10.1080/10408398.2019.1567459.
69. Wrona, M., Silva, F., Salafranca, J., Nerín, C., Alfonso, M. J., & Caballero, M. Á. (2021). Design of new natural antioxidant active packaging: Screening flowsheet from pure essential oils and vegetable oils to ex-vivo testing in meat samples. *Food Control, 120*, 107536. https://doi.org/10.1016/j.foodcont.2020.107536.

70. Ghaani, M., Cozzolino, C. A., Castelli, G., & Farris, S. (2016). An overview of the intelligent packaging technologies in the food sector. *Trends in Food Science & Technology, 51*, 1–11. https://doi.org/10.1016/j.tifs.2016.02.008.
71. Kalpana, S., Priyadarshini, S. R., Leena, M. M., Moses, J. A., & Anandharamakrishnan, C. (2019). Intelligent packaging: Trends and applications in food systems. *Trends in Food Science & Technology, 93*, 145–157. https://doi.org/10.1016/j.tifs.2019.09.008.
72. Drago, E., Campardelli, R., Pettinato, M., & Perego, P. (2020). Innovations in smart packaging concepts for food: An extensive review. *Foods, 7*(9), 1628. https://doi.org/10.3390/foods9111628.
73. Romanello, D. (2021). *Communicative Packaging for Food Waste Reduction: Barriers and Drivers to The Adoption Along the Agri-Food Supply Chain* (Master's thesis). Politecnico Milano.
74. Janjarasskul, T., & Suppakul, P. (2018). Active and intelligent packaging: The indication of quality and safety. *Critical Reviews in Food Science and Nutrition, 58*(5), 808–831. https://doi.org/10.1080/10408398.2016.1225278.
75. Boz, Z., Korhonen, V., & Koelsch Sand, C. (2020). Consumer considerations for the implementation of sustainable packaging: A review. *Sustainability, 12*(6), 2192. https://doi.org/10.3390/su12062192.
76. Anonymous. (2023). *TIPA Corp.* Retrieved from: https://tipa-corp.com/ (accessed on 9 August 2024).
77. Anonymous. (2023). *Are Consumers Willing to Pay More for Sustainable Packaging?* McKinsey & Company. Retrieved from: https://www.mckinsey.com/business-functions/sustainability/our-insights/sustainability-blog/are-consumers-willing-to-pay-more-for-sustainable-packaging.
78. Geyer, R., Jambeck, J. R., & Law, K. L. (2017). Production, use, and fate of all plastics ever made. *Science Advances, 3*(7), e1700782. https://doi.org/10.1126/sciadv.1700782.
79. Avérous, L., & Kalia, S. (2016). *Biodegradable and Biobased Polymers for Environmental and Biomedical Applications*. John Wiley & Sons.
80. Luxton, R. (2023). Challenges and future aspects of sensor technology. In *Advanced Sensor Technology* (pp. 853–877). Elsevier.
81. Campbell, I. R., Lin, M. Y., Iyer, H., Parker, M., Fredricks, J. L., Liao, K., Jimenez, A. M., Grandgeorge, P., & Roumeli, E. (2023). Progress in sustainable polymers from biological matter. *Annual Review of Materials Research, 53*. https://doi.org/10.1146/annurev-matsci-081322-101226.
82. Hamed, I., Jakobsen, A. N., & Lerfall, J. (2022). Sustainable edible packaging systems based on active compounds from food processing byproducts: A review. *Comprehensive Reviews in Food Science and Food Safety, 21*(1), 198–226. https://doi.org/10.1111/crf3.12704.
83. Asgher, M., Qamar, S. A., Bilal, M., & Iqbal, H. M. (2020). Bio-based active food packaging materials: Sustainable alternative to conventional petrochemical-based packaging materials. *Food Research International, 137*, 109625. https://doi.org/10.1016/j.foodres.2020.109625.
84. Zhao, X., Wang, Y., Chen, X., Yu, X., Li, W., Zhang, S., Meng, X., Zhao, Z. M., Dong, T., Anderson, A., Aiyedun, A., Li, Y., Webb, E., Wu, Z., Kunc, V., Ragauskas, A., Ozcan, S., & Zhu, H. (2023). Sustainable bioplastics derived from renewable natural resources for food packaging. *Matter, 6*(1), 97–127. https://doi.org/10.1016/j.matt.2022.08.006.

85. Granato, G., Fischer, A. R., & van Trijp, H. C. (2022). The price of sustainability: How consumers trade-off conventional packaging benefits against sustainability. *Journal of Cleaner Production, 365*, 132739. https://doi.org/10.1016/j.jclepro.2022.132739.
86. Fathima, A., Chiome, T. J., Catherine, A. A., Egbuna, C., Achar, R. R., & Srinivasan, A. (2022). Smart use of nanomaterials as sensors for detection and monitoring of food spoilage. In *Application of Nanotechnology in Food Science, Processing and Packaging* (pp. 169–188). Cham: Springer International Publishing. https://doi.org/10.1007/978-3-030-73904-1_8.
87. Nile, S. H., Baskar, V., Selvaraj, D., Nile, A., Xiao, J., & Kai, G. (2020). Nanotechnologies in food science: Applications, recent trends, and future perspectives. *Nano-Micro-Letters, 12*, 1–34. https://doi.org/10.1007/s40820-020-00438-4.
88. Sánchez-López, E., Gomes, D., Esteruelas, G., Bonilla, L., Lopez-Machado, A. L., Galindo, R., Cano, A., Espina, M., Ettcheto, M., Camins, A., Silva, A. M., Durazzo, A., Santini, A., Garcia, M. L., & Souto, E. B. (2020). Metal-based nanoparticles as antimicrobial agents: An overview. *Nanomaterials, 10*(2), 292. https://doi.org/10.3390/nano10020292.
89. Ashfaq, A., Khursheed, N., Fatima, S., Anjum, Z., & Younis, K. (2022). Application of nanotechnology in food packaging: Pros and cons. *Journal of Agriculture and Food Research, 7*, 100270. https://doi.org/10.1016/j.jafr.2022.100270.
90. Kavuncuoglu, H., Yalcin, H., & Dogan, M. (2023). Development of (TiO_2-ZnO)/LDPE based active nanocomposite films and detection of migration to minced beef during storage using response surface methodology. *Food Chemistry, 402*, 134278. https://doi.org/10.1016/j.foodchem.2022.134278.
91. Ng, Z. J., Zarin, M. A., Lee, C. K., & Tan, J. S. (2020). Application of bacteriocins in food preservation and infectious disease treatment for humans and livestock: A review. *RSC Advances, 10*(64), 38937–38964. https://doi.org/10.1039/D0RA07171F.
92. Carpena, M., Nuñez-Estevez, B., Soria-Lopez, A., Garcia-Oliveira, P., & Prieto, M. A. (2021). Essential oils and their application on active packaging systems: A review. *Resources, 10*(1), 7. https://doi.org/10.3390/resources10010007.
93. Mukurumbira, A. R., Shellie, R. A., Keast, R., Palombo, E. A., & Jadhav, S. R. (2022). Encapsulation of essential oils and their application in antimicrobial active packaging. *Food Control*. https://doi.org/10.1016/j.foodcont.2022.108883.
94. Alam, A. U., Rathi, P., Beshai, H., Sarabha, G. K., & Deen, M. J. (2021). Fruit quality monitoring with smart packaging. *Sensors, 21*(4), 1509. https://doi.org/10.3390/s21041509.
95. Ezati, P., Bang, Y. J., & Rhim, J. W. (2021). Preparation of a shikonin-based pH-sensitive color indicator for monitoring the freshness of fish and pork. *Food Chemistry, 337*, 127995. https://doi.org/10.1016/j.foodchem.2020.127995.
96. Zhang, L., Yang, H., Tan, Q., Jing, L., Zhang, W., Xiong, J., & Chen, P. Y. (2021). Wireless detection of biogenic amines using a split-ring resonator with silver nanoparticles-decorated molybdenum disulfide. *Sensors and Actuators B: Chemical, 343*, 130155. https://doi.org/10.1016/j.snb.2021.130155.
97. Gaspar, M. C., & Braga, M. E. (2023). Edible films and coatings based on agrifood residues: A new trend in the food packaging research. *Current Opinion in Food Science*. https://doi.org/10.1016/j.cofs.2023.101006.
98. Díaz-Montes, E., & Castro-Muñoz, R. (2021). Edible films and coatings as food-quality preservers: An overview. *Foods, 10*(2), 249. https://doi.org/10.3390/foods10020249.

99. de Oliveira Filho, J. G., Braga, A. R. C., de Oliveira, B. R., Gomes, F. P., Moreira, V. L., Pereira, V. A. C., & Egea, M. B. (2021). The potential of anthocyanins in smart, active, and bioactive eco-friendly polymer-based films: A review. *Food Research International, 142*, 110202. https://doi.org/10.1016/j.foodres.2021.110202.
100. Moustafa, H., Youssef, A. M., Darwish, N. A., & Abou-Kandil, A. I. (2019). Eco-friendly polymer composites for green packaging: Future vision and challenges. *Composites Part B: Engineering, 172*, 16–25. https://doi.org/10.1016/j.compositesb.2019.02.039.
101. Siracusa, V., & Lotti, N. (2019). Intelligent packaging to improve shelf life. In *Food Quality and Shelf Life* (pp. 261–279). Academic Press.
102. Vodnar, D. C., & Teleky, B. E. (2023). Recent trends in antibacterial coatings and biofilm. *Coatings, 13*(2), 255. https://doi.org/10.3390/coatings13020255.
103. Taherimehr, M., YousefniaPasha, H., Tabatabaeekoloor, R., & Pesaranhajiabbas, E. (2021). Trends and challenges of biopolymer-based nanocomposites in food packaging. *Comprehensive Reviews in Food Science and Food Safety, 20*(6), 5321–5344. https://doi.org/10.1111/crf3.12657.
104. Chinchkar, A. V., Singh, A., Kamble, M. G., Prabhakar, P. K., Meghwal, M., & Sharma, A. (2023). Nanotechnology applications for quality determination of RTE and packaged food. In *Nanotechnology Applications for Food Safety and Quality Monitoring* (pp. 265–288). Academic Press.

CHAPTER 13

APPLICATION OF EMERGING TECHNOLOGIES FOR PRESERVATION OF READY-TO-EAT SNACKS

SHUBHRA SHEKHAR, SUSHIL KUMAR SINGH, and POONAM SINGHA

Department of Food Process Engineering, National Institute of Technology, Rourkela, Odisha, India

ABSTRACT

Production of ready-to-eat foods is growing daily due to its versatility in handling, no cooking requirement, and easy availability. These foods are pre-cleaned, pre-cooked, and ready to consume without any preparation. For the food safety of the product, preservation is necessary for storing the food for a longer time and can later be consumed without being spoiled. Spoiled foods are not generally safe to eat and show undesirable changes like changes in color, flavor, texture, nutritional content, etc. Many techniques like pasteurization, sterilization, cooking, drying, canning, etc., have been used for many years to preserve food from microbial spoilage. However, these techniques have some drawbacks, which will be discussed in this chapter. Non-thermal processing is considered an excellent replacer of existing conventional methods for processing food to increase shelf life, safety, reduce food poisoning and reduce economic loss. Non-thermal treatments such as pulsed electric field (PEF), high-pressure processing (HPP), cold plasma (CP), ultrasound (US), microwave heating, and oscillation magnetic field (OMF) are gaining popularity now-a-days. A combination of preservation techniques or hurdle technology may be used, which has a synergistic effect on the preservation of microorganisms. Synergistically with different techniques if

Ready-to-Eat Snacks: Emerging Technologies for Production and Safety.
Ravi Pandiselvam and Gülşah Çalışkan Koç

individual hurdles target various activities within the cell, allowing a gentler preservation treatment with less influence on product quality. The application of these emerging technologies thus becomes essential for the food processor to achieve such benefits of these technologies.

13.1 INTRODUCTION

The customers growing demand for the requirement of ready-to-eat (RTE) snacks is rising steadily and is projected to rise significantly over the next five years. Customers are looking for convenient snacks with unique sensory and textural characteristics [1–3]. Consumers prefer RTE items since they require less time and effort. As people is becoming more health-conscious, there is a more critical requirement for nutritional and high-quality RTE meals [4]. It is desirable to provide the product with the appropriate treatment so that the maximum possible amount of its nutrition can be preserved, and microorganism present can be reduced. The microorganisms that are most commonly found in RTE snacks are *Staphylococcus saprophyticus, Staphylococcus aureus,* β-*hemolytic, streptococci, Staphylococcus saprophyticus, Pseudomonas* spp., *E. coli, Shigella* spp., *Enterobacter* spp. [5]. The traditional preservation method requires high heat and additional time for preservation, resulting in the loss of nutrients; new technologies are utilized to overcome this problem. Emerging technology and techniques in the food processing area is being created to suit unique needs of consumers for food products that are healthy, safe, and very slightly processed. These technologies are environmentally friendly and need less energy and water, which allows them to circumvent some limitations of the existing food processing industry [6]. Figure 13.1 shows different technologies that are utilized for the preservation of RTE snack food.

13.2 HIGH PRESSURE PROCESSING (HPP)

HPP also named as high hydrostatic pressure (HHP) or ultra-high pressure (UHP), is a process that involves exposing solid and liquid food to 100 to 800 MPa pressure range, with or without the presence of packaging [7]. The earliest application of HPP was in the treatment of milk with the goals of lowering the microbes and increasing storage life of the produce [8] then in future of vegetables and fruits in 1914 [9]. A significant change in HPP occurred in Japan in 1992 when the first high-pressure processed product was acquainted with the market [10]. This processing method puts the food

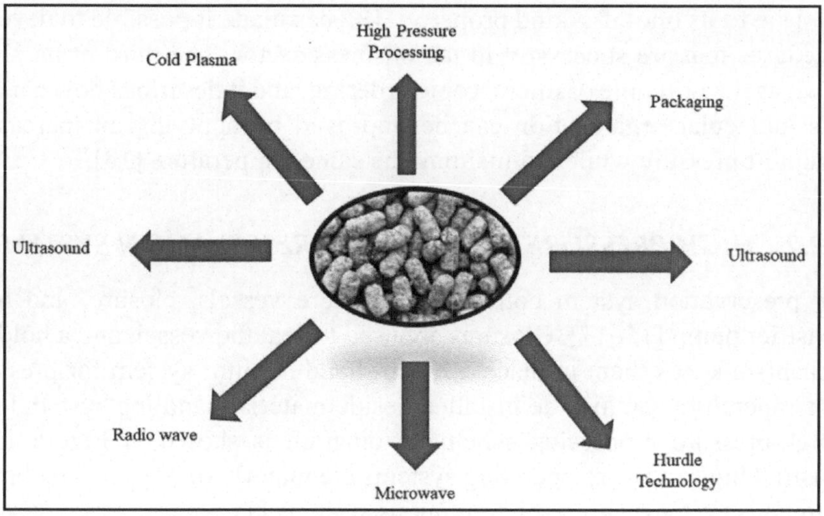

FIGURE 13.1 Application of emerging technologies for the preservation of RTE snacks.

items under both high pressure and high temperatures. To effectively transmit force, water is often utilized for a predetermined amount of time, which can last anywhere from a few seconds to many minutes. There is a wide range of possible times because it also varies depending on the food product. In addition to preservation, HPP can be utilized for variety of alterations, such as changes in texture [11, 12].

13.2.1 PRINCIPLE OF HPP

HPP is primarily guided by 3 major principles, namely, the Le Chatelier, microscopic ordering, and isostatic/isobaric principle [13]. The first principle states that at a certain temperature, variation in the pressure causes equilibrium to move closer to a condition that encourages volume reduction. This condition is known as the law of equilibrium. The way molecular ordering works, the volume of a substance will decrease proportionately with an increase in pressure. At low temperatures, HPP has a slight effect on strong covalent bonds, but it can weaken weaker bonds and induce breakage. Because of this, HPP influences the tertiary and quaternary protein structure, but not on primary and secondary structures of the protein. The next concept is called the isostatic/isobaric principle, and it asserts that regardless of the shape of the product, pressure is delivered uniformly from all directions during compression, and the product keeps its shape during decompression

[14]. Due to its one-of-a-kind property, HPP has made it possible to develop procedures that are successful in the business sector. The third principle is known as the principle of microscopic ordering, and it describes how a material's molecular organization can be improved by applying an increasing amount of pressure while maintaining the same temperature [13].

13.2.2 HIGH PRESSURE PROCESSING PRESERVATION SYSTEM

HPP preservation system consists of pressure vessels, closure, and high intensifier pump [15–17]. Closures are used to seal the vessel, and a holding mechanism keeps them in place. A control and monitor system for pressure and temperature can also be installed beside material handling system [18]. A high-pressure processing schematic diagram is shown in Figure 13.2. Industrial high-pressure operating systems are majorly of 3 types, i.e., batch, continuous systems and semi-continuous systems [7].

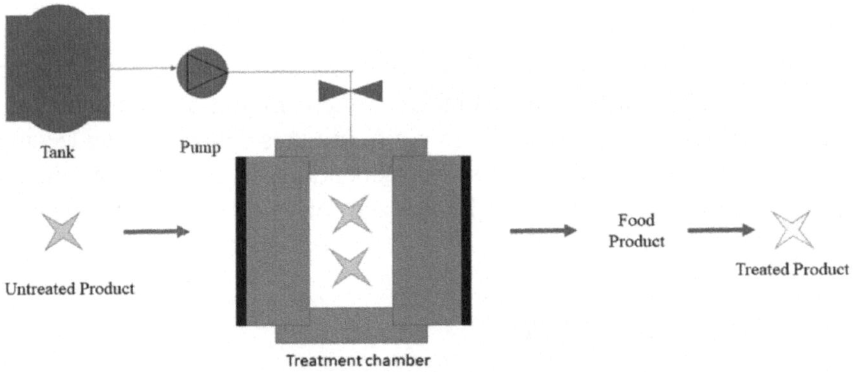

FIGURE 13.2 Schematic diagram for high pressure processing system.
Source: Adapted from Ref. [19]

Depressurization as well as off-loading of the product takes place after the holding time has been completed. As a result, one "cycle" refers to the whole amount of time spent on pressurization, holding, and depressurization [20]. As water is cheap and nontoxic, it is the most employed pressure transfer medium. The transfer of heat between the product and fluid is dependent on the liquid characteristics like specific heat, viscosity, and compressibility [21]. In this the pressure is applied equally in all directions; therefore, the structural properties are unaffected. Food's sensory and nutritional qualities are also preserved due to minimal heat impacts.

Semi-continuous HPP systems comprise of various components such as a low-pressure pump, pressure vessels for vessel filling, holding and sterilizing tanks, a high-pressure transfer pump, and control valves. A divider piston is used to isolate the product and the pressurized fluid in the pressure vessels which is used to prevent cross-contamination of treated with untreated products control valves [22]. Grapefruit juice is processed in Japan using a commercial semi-continuous HPP machine. It can process 600 liters per hour at the highest pressure of 400 MPa [23, 24].

13.2.3 APPLICATION OF HPP IN PRESERVATION AND SAFETY OF RTE SNACKS

HPP is an effective technique of preservation as it kills harmful spoilage causing microorganisms, and hence extending storage life and safety of the product [25–27]. The major impact of HPP treatment is on the cell structure and its critical function. HPP has a high lethality effect on microbial populations, which is expected to be produced by the mixture of effects, i.e., cell shape, the permeability of cell membrane, changed metabolic responses, and genetic interference in microorganism cells [13].

The efficiency of the process depends upon various parameters, including process parameters, microorganism properties as well as food matrix. This is due to the reason that some microorganisms have a higher susceptibility to high pressure than others. Bacterial spores generally tend to have higher resistance as compared to vegetative cells. In contrast, yeast and molds have more resilience to high pressure as compared to bacterial cell [25].

HPP's ability to kill food-borne microbes is influenced by both inherent and external variables. HPP causes modifications in the cell membrane, cell wall, enzyme-facilitated cellular activities, and proteins resulting in microorganism inactivation [28]. Protein denaturation [29]and accumulation of the cellular proteins [30] is the heart of HPP's microbial inactivation principles. Disrupting the cellular membrane releases internal cellular components [31]. Cell death results from the outflow of internal solutes, ribosomes, and nuclear material owing to pressurization. Insufficient pressure may cause reversible cell structural alterations that recover after a few days of storage [32]. Higher pressure causes permanent changes in cellular structure due to membrane permeability [33]. Using scanning electron micrographs correlates with fewer viable *E. coli* cells. HPP also targets proteins and enzymes to inactivate microbes [34, 35].

Bacterial spores inactivate differently than vegetative cells due to many protective layers. Inactivating bacterial spores need two procedures, unlike vegetative cells. First, spores are germinated at 600 MPa and 60°C. In the next step, germinating spores are inactivated utilizing temperatures between 60 and 120°C and pressures above 600 MPa [36].

HPP increases food safety, shelf life, and quality. Low-temperature, high pressure processing doesn't influence ascorbic acid content, but high-temperature processing does.

13.3 ULTRASOUND PRESERVATION

Sound is created when molecules in an alternating wave of air, water, or another medium vibrate. Ultrasound waves resemble sound waves in appearance but differ in frequency. The frequency, measured in cycles per second or hertz, is determined by the distance between each peak. The human ear can perceive sound waves between 16 Hz and 16–20 kHz, which is a lesser range beyond 20 kHz than the ultrasound frequencies [26, 37–39].

Ultrasound is made up of mechanical waves that move and spread via an elastic material. In an equilibrium state, ultrasound waves generate an oscillation pattern that is channeled with energy transfer through one particle to the other [40]. Such waves disperse pressure changes in the medium as they move continuously in a compression and rarefaction motion, providing various advantages in processing and preservation of the food [41].

13.3.1 PRINCIPLE OF ULTRASOUND

The fundamental idea behind ultrasound of low power is that sound waves travel over the food matrix while having mechanical properties that alternately compress and decompress them. These waves are distinguished by parameters such as velocity, wavelength, pressure, duration, and frequency [43]. When sound waves of a certain frequency strike the surface, they cause velocity to drift and sound to attenuate due to absorption and scattering mechanisms. The vapor bubbles that are formed during the process of cavitation burst during the application of low frequency and due to this, there is the generation of high pressures greater than 500 bar and the temperature below 5,000°C, causing greater shear forces, which is conclusively linked to inducing cavitation in the treated medium [26]. Figure 13.3 represent the principle of ultrasound cavitation how bubble formation occurs and implodes.

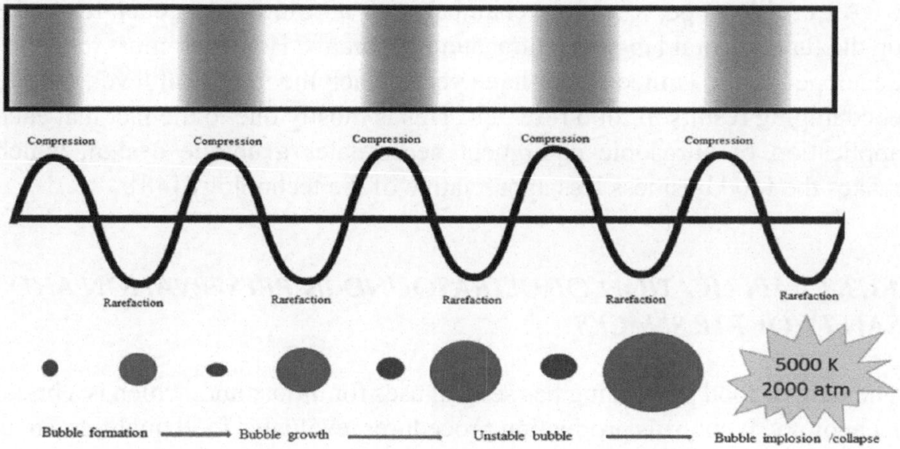

FIGURE 13.3 Principle of ultrasound cavitation.

Source: Adapted from Ref. [42]

When using higher frequency ultrasound, cavitation is also more stable (less vigorous breakup of smaller bubbles), resulting in more micro-streaming [44, 45]. Chemical, thermal, and mechanical impacts are all triggered by cavitation events. In aqueous solutions, it causes free radical production of hydrogen and oxygen ion. Second, it explains the single electron shift that occurs in the cooling phase, and there is the formation of hydrogen peroxide, which is having bactericidal properties. Heat is generated due to cavitation, which is helpful in the application of pasteurization aids. Mechanical shocks cause the destruction of cell structure and promotion of lysis and causing depolymerization that causes enzymes inactivation [46, 47].

13.3.2 ULTRASOUND PRESERVATION SYSTEM

Low-power ultrasound waves have intensities lower than 1 W/cm^2 and frequencies over 100 kHz, whereas ultrasound having 20 and 500 kHz frequency and intensities more than 1 W/cm^2 intensity are high energy ultrasound. However, for ultrasonic-technology applications, the most widely used frequency range is 20 to 500 MHz. Depending on the application, various conditions can be applied. Lower frequency ultrasound ranges are used to produce mechanical and chemical variations due to the formation of cavitation that increases the energy, with the goal of microbial inactivation, whereas ultrasound of higher frequencies are mostly used for the constitution of food [46].

An oscilloscope, a sample chamber, and an ultrasonic generator make up the fundamental lab-scale ultrasound apparatus. However, most remedies developed in the United States have yet to reach the industrial level, despite encouraging results in food research. This is mostly due to the fact that each application of ultrasonic equipment necessitates a unique design, which makes the food business less appreciative of the technology [48].

13.3.3 APPLICATION OF ULTRASOUND IN PRESERVATION AND SAFETY OF RTE SNACKS

The field of food processing has several uses for ultrasound, which is a broad technology. It controls production procedures, evaluates food qualities, spots flaws, and enhances the extraction and inactivation of microorganisms. Ultrasound is a greener alternative as it reduces the use of energy as well as chemical requirements [49].

The presence of nutrition in the food products make it susceptible to infestation of microorganisms, causing deterioration resulting in losses in both quality and quantity. Ultrasound has been successfully applied for the reduction of microorganisms without having any negative effect on the sensory attributes and nutrients of the food product as preservation of the quality is a crucial consideration when implementing any technology [50]. The acoustic cavitation process, which outcomes in the disruption of the microbial cell wall, is responsible for microbial inactivation. However, some important elements such as the amplitude, temperature, the food product composition, duration of the treatment, and microbes present to be reduced, impact the effectiveness of inactivation [51].

Ultrasound is now commonly utilized in a widespread range of food systems due to growing awareness among consumers for the demand of food items with no chemical preservatives and lesser processing yet having good quality as well as microbial safety. Numerous research investigations conducted over the years have supported the effectiveness of ultrasound as a replacement for or improvement of different traditional processing techniques. The impact of technical collaboration with ultrasound on the general food quality should be the subject of further study. In general, sound waves are thought to be safer, more environmentally friendly, and more effective than other methods. However, there are some restrictions, such as the loss of nutritional and organoleptic quality caused by high-intensity ultrasound equipment. The commercialization of approaches is also constrained by greater energy requirements and costs. Additionally, industrial-scale implementation calls for research for examining the impact of ultrasonic processing on large-scale food processing as well as optimization of various parameters and to use either alone or with combination of other technologies, e.g., mano-thermo-sonication. To assess the technology's commercial viability, commercial-scale investigations should also pay attention to how consumers feel about meals that have been treated with ultrasound.

13.4 MICROWAVE PRESERVATION

Microwaves are non-ionizing radiation that induce the motion of molecules due to ion migration and dipole rotation but do not change the molecular composition. Domestic microwave equipment normally operates at 2.45 GHz frequency, whereas industrial microwave systems typically work at 915 MHz and 2.45 GHz frequencies. Electromagnetic waves having a frequency between 300 MHz and 300 GHz are microwaves [52].

13.4.1 PRINCIPLE OF MICROWAVE

Polar molecules in food products, like water molecules, rotates in accordance with the alternating electromagnetic field as soon as microwave radiation is absorbed [53]. The water molecule is a "dipole" because it has electrically charged two ends of which one positively and one negatively charged. When the product is exposed to an electromagnetic field, these dipoles have the same orientation because water molecules behave like magnets. These water molecules rotates to produce heat for cooking [54]. So, a food's composition will have an impact on how it heats up in a microwave. Food with a higher moisture content will heat up more quickly due to the dipolar interaction. Microwave heating in non-conductors refer to dielectric heating in a particular frequency band brought on by polarization. However, inside the microwave chamber the rise in the temperature of the product is due to electrical energy. The microwave then leaves the product after dissipating which causes the temperature inside the product to rise quickly [53]. The transmission of microwave energy is due to dipole rotation and conductive migration of ions. The polarization of ions, which causes the ionic molecules to migrate backward and forth in order to arrange with the oscillating electric field, is the second primary method of heating using microwaves [55, 56].

13.4.2 MICROWAVE PRESERVATION SYSTEM

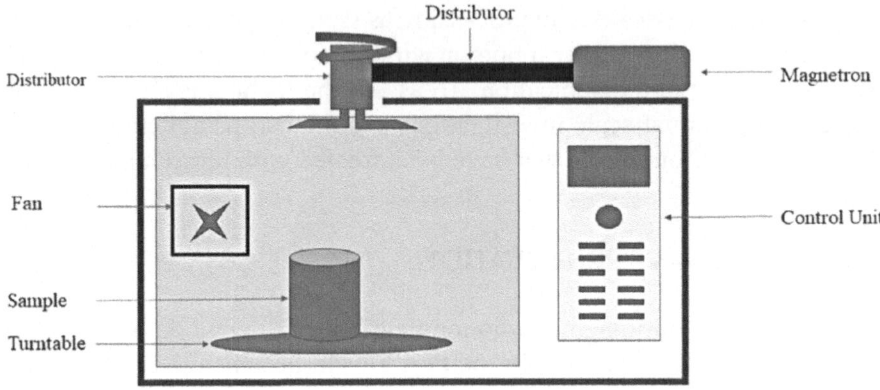

FIGURE 13.4 Schematic diagram of microwave preservation system.

Beginning with the conversion of electrical energy from low-voltage alternating current and direct high-voltage current into direct current, the process begins. After employing this direct current, the magnetron, which is also known as the heart of microwave, produces microwaves at 2.45 GHz frequency. The microwave is directed into the wave guide with the help of antenna present on the top of the magnetron. Stirrer, a device resembling a fan, disperses the microwaves inside the oven cavity after being streamlined by the waveguide. Later, the reflection of the waves is due to the internal metals present in the oven and the waves are later absorbed by food molecules. Food particles are pushed backward and forward at double frequency as that of the microwave frequency since every wave consists of positive and negative component. The brain of the microwave, i.e., the magnetron transforms the electrical energy into microwave radiation. Both direct high-voltage current and low-voltage alternating current are used by magnetrons. A capacitor and a diode work together to filter out the high voltage and convert it to direct current after a transformer lowers the input voltage to the necessary levels. Inside the magnetron, a central terminal known as a cathode emits electrons that are drawn to the positively charged anode that surrounds the cathode. Permanent magnets make the electrons move in a circular motion as opposed to a straight line. A constant pulsing magnetic field or electromagnetic radiation is produced as soon as they pass through resonant cavities. Figure 13.4 shows the image of microwave system.

13.4.3 APPLICATION OF MICROWAVE IN PRESERVATION AND SAFETY OF RTE SNACKS

The uniformity of heating and distribution of temperature of RTE meals was studied [57]. The four-component chilled ready-to-eat meal samples under examination included meat patties, sauce, mashed potatoes, and carrots. Two fiber optic measuring devices with seven sensors were used to measure temperature during microwave heating. The temperature was measured using 35 copper-constantan thermocouples following microwave heating 30 seconds after heating and 5 minutes after cooling. After 30 seconds of microwave heating, there was a temperature difference of more than 70°C between the hotter and colder parts of the mashed potatoes. 30 seconds after heating, the mean component temperatures for mashed potatoes ranged from 62.7 to 79.9°C, for beef patties from 78.8 to 87.2°C, sauce from 82.1 to 90°C, and carrots from 61.3 to 94.3°C. After a five-minute chilling interval, mean temperatures dropped, and the range shrunk as heat conduction made the temperature distribution more uniform.

13.5 IRRADIATION

Irradiation works on the principle of ionizing radiation without involving heating the food to such an extent that it causes sensory and nutritional changes in the food. Irradiation kills spoilage bacteria as well as slowing down the rate of germination, ripening, and sprouting thus increasing its storage life [27]. In this process, a low level of radiant energy is applied to the material, which has to be preserved. Gamma radiations are produced using an electron accelerator which is most commonly used in the food industries [58]. According to FAO 1984, gamma-rays emitted from Co-60 and Ce-137 and X-rays generated can be used for food irradiation (Figure 13.5) if operated at or below 5 MeV.

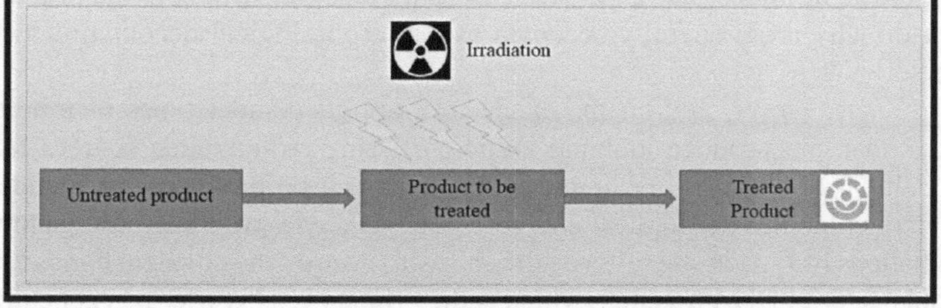

FIGURE 13.5 Schematic diagram showing the process of irradiation.

13.5.1 PRINCIPLE

As a final step in the food manufacturing process, irradiation exposes food to a specific radiation dose, which eliminates the harmful microbial load and prolongs the food storage life by preventing spoilage. Inactivation of the microorganisms is mostly by direct or indirect destruction by the ionizing radiation due to free-OH radicals produced from radiolysed water. Irradiation makes it possible for energy to travel through and strike all the food molecules. Nucleic acids are found in both insects and microorganisms. When radiation is activated, it damages DNA through ionizing particles but does not cause any significant food temperature rise. This prevents cell division by preventing DNA synthesis. Whereas indirect radiation causes the formation of active molecules like -OH and H radicals, as well as hydrated electrons, which results in lysis of the cell [16, 59, 60].

13.5.2 PROCESSING AND SYSTEM

A wide range of wavelengths is there in electromagnetic radiation, and depending on their energy, these waves can be used for a variety of things. There are 2 types of radiation depending on the energy, i.e., ionization and non-ionization energy radiation. Higher energy and frequency with shorter wavelengths are ionizing, whereas the opposite is for non-ionizing energy [61, 62].

Irradiation can be obtained from a wide range of sources, like X-rays-rays, and even beams of electron, and it always contains some amount of ionizing radiation. Gamma radiations have a maximum energy of 10 MeV and can be produced by using radionuclide sources such as Co-60 or Ce-137 in conjunction with electrons [63]. Gamma rays can destroy the covalent bonds that hold bacterial DNA together, which is the primary reason why it is generally accepted as a microbiological decontamination agent.

X-rays, γ-rays, and beams of electron are all distinct types of irradiations that can produce ionizing radiation. Using radionuclide sources like cobalt–60 or cesium–137, gamma radiations can be produced; accelerated electrons having the highest energy of 10 MeV can produce these gamma radiations [63]. Gamma rays has the ability to destroy the covalent bonds that hold bacterial DNA together, which is the primary reason why it is generally accepted as a microbiological decontamination agent [64]. Irradiation with the radioisotope cobalt–60 is frequently employed for the purpose of treating food because of its insolubility in water, which prevents environmental pollution and dangers associated with the process [65]. X-rays are created in this way. Irradiating food with X-rays that have the highest energy of 5 MeV produces results that are comparable to those produced by Co-60 in terms of their ability to penetrate.

X-rays can effectively penetrate thick material (between 30 and 40 centimeters). This technology is increasingly being utilized treatment of the food in the package to prevent recontamination [66]. Irradiation with 10 kGy dosage of X-rays is used for the reduction of microbes present [65]. The beam of electrons is driven because of high-energy electrons in an accelerator, which produces accelerated electrons at almost the speed of light. These accelerated electrons are what propel electron beams. The electron beam has less homogeneity and depth of penetration but has higher output and lower cost. Irradiation is frequently applied in the food processing items that have a thinner consistency [66].

The technologies that maintain food safety during processing have evolved with shifting customer expectations and concerns regarding the quality of food. Irradiation is a flexible technique that can be used for a diversity of applications. The technique used in the agro-food industry is extremely effective and well suited for the purpose of holding broad efficacy against a variety of microorganisms and insects. Irradiation of food goods while they are still sealed in their packaging also helps to avoid the recontamination of food [65]. As food processing sector involve a lot of different types of raw material, it invites different type of microbes. When product is subjected to radiation therapy, the energy from the radiation travels through the food and strikes each molecule that is present in the food. DNA is found in both microbes and insects. Radiant energy when it encounters DNA, causes damage, which stops the multiplication of microorganisms. The ability of microorganisms to recover from radiation harm and variances in their chemical and physical structures are some of the factors that contribute to the disparities in the dietary radiation sensitivity among these species. Irradiation has been shown in a great number of studies over the course of recent history to be an effective method for removing bacteria populations from a variety of food products.

The total aerobic bacteria of fresh and preserved mango juice decreased from $4.2 \pm 0.3 \times 10^4$ and $6.5 \pm 0.2 \times 10^5$ CFU/mL to $2.6 \pm 0.2 \times 10$ and $1.1 \pm 0.8 \times 10$ CFU/mL as a result of a irradiation dosage of 3.0 kGy [67]. There are several factors, both internal and external, that influence how various long types of food can be stored. Irradiation can increase the storage life of many food products it has shown application in field of fruits and vegetable, meat product and many more according to a number of research that have been conducted on the topic [68–70].

13.5.3 CONSUMER PERCEPTION OF FOOD IRRADIATION

Irradiation has become a technique that has been given a great deal of consideration in a variety of nation's food and agriculture sectors over the course of the last several decades. Food irradiation is universally recognized as a potential food preservation method, and this view is supported by a variety of linked organizations. The investment cost for this technology is higher for commercial use. Despite the fact that irradiation proposes a number of benefits, such as the elimination of microbes and insects thus increasing their storage life, the absence of any chemical input, without any extensive deterioration of the quality of the food, etc., it is impossible to deny that this technology also equates to some disadvantages such as it is applicable to a constrained selection of food types but not to all of them. Aside from that, consumer generally doesn't prefers to consume irradiated food due to the misconception of having radiation in it due to this perception its acceptance is very difficult [60, 71]. Labeling of irradiated goods is likewise required by the regulating connected bodies; nevertheless, only a select few nations make its application mandatory, and only a select few states allow for its optional application.10 kGy dosages for irradiation of food products have already been certified by associated institutions such as FAO and WHO for food safety. In order to increase consumer confidence and acceptance of irradiated products, the stamp "Radura," which is a green-colored sign, is affixed to irradiated food products before they are treated with radiation. Aside from that, there is a definite requirement to change the mentality of consumers and increase their level of awareness [72, 73].

13.6 COLD PLASMA

Irving Langmuir, a chemist, came up with the concept that would later be known as "plasma" in 1928 [74]. After its three unique forms—solid, liquid and gas —the fourth state of matter is plasma that scientists have discovered. When the energy level is increased, there occurs a transition state of the substance from solid to liquid then to gaseous phase. In addition, if there is a surge in the energy of the gaseous state over particular threshold limit, there is molecules ionization and formation of plasma that is the fourth state of matter [16, 75]. Ionized plasma is a type of gas that is mostly made up of reactive species. Plasma is a type of gas that is ionized [76]. One of the more recent technologies that have found use in the process of preserving and ensuring the safety of food is cold plasma. The application of an electrical current creates reactive species and generates plasma glow, which is the primary method for producing cold plasma. The active plasma components when comes in contact to the microbial cells, it damages the cell. It is possible to utilize it to efficiently inactivate microbial cells, which results in increasing shelf life of the food product [77, 78].

13.6.1 PRINCIPLE OF COLD PLASMA

The source of the plasma and the method employed to generate it can determine whether the plasma is thermal or non-thermal. In general, welding arc and spark plug generated the plasma. Another type of plasma is thermal plasma, which is produced when there is higher energy for the particles involved to be in equilibrium. The particles do not share any of their energy with one another; the heated plasma contains highly reactive species [79]. To create plasma, one of the most important steps is the ionization of the gas. Ionization can be caused by different types of fields, including electrical, thermal, magnetic, radio waves, and microwaves [80, 81]. The transfer of high energy from the outer parts of atoms that are present in the gas results in the removal of electrons present in the atomic nuclei caused the formation plasma reactive species [82]. Gaseous mixture that is employed to generate plasma influences the reactive species composition, both in terms of kind and concentration. As an illustration, if there is oxygen present in the mixture, reactive oxygen species concentration will be higher [83]. The quality characteristics of the product is not affected in any way because it do not involve increase in temperature [84]. A non-equilibrium state known as cold plasma is characterized by the presence of large particles that are, for the most part, at room temperature. Free electrons, on the other hand, possess an adequate amount of kinetic energy for ionizing collisions and the breakdown of bonds [85].

13.6.2 COLD PLASMA GENERATION: ITS EQUIPMENT AND PROCESS DESIGN

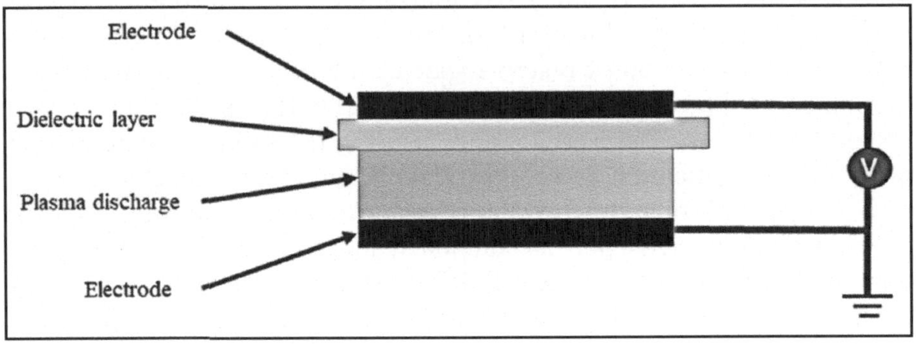

FIGURE 13.6 Schematic diagram of cold plasma dielectric barrier discharge system.
Source: Adapted from Ref. [86]

Depending on the specifics of the processing being done, the discharge systems (Figure 13.6) that are used may vary. These involve dielectric barrier, corona, glow, radio frequency, arc, and microwave discharge [85, 86]. Utilizing metal electrodes that have been coated in a dielectric substance is required to carry out a dielectric barrier discharge. The system consists of one grounded electrode and the other to high voltage source. In between the electrodes, a mixture of gases is allowed to flow. In glow plasma discharge, a gaseous mixture is subjected to high voltage, which causes an alternating current to be passed through the mixture. When pressure hits 2 Pa inside vacuum chamber plasma is generated [87]. Under atmospheric pressure, the plasma produced by a corona discharge makes use of a field of electric current that is not uniform. The resultant plasma is dimly lit and only slightly ionized after being produced using this technique. In the event of a microwave plasma discharge, the magnetron is the device that is utilized to generate a high-frequency electromagnetic field. Since heat is produced when electrons collide with gaseous atoms and molecules, it is frequently utilized in the context of processing that takes place at high temperatures. As a direct consequence of this, a significant amount of ionization takes occurs [88]. After the plasma has been formed, it can be released for use in the treatment of food [16, 89]. There are distinct application strategies for cold plasma that are accessible for the utilization in the food industry. One is by directly exposing the food product with active plasma. This method is used to sterilize food materials.

When utilizing the technique of remote exposure, the initial step is to produce the plasma. Food items are not located in the same compartment where plasma is created. This is to ensure that the treatment is effective. There is no direct touch with one another when there is generation of the plasma. Electrode contact approach is the approach in which the plasma-generating electrodes are utilized. Food is then positioned between the electrodes, which causes the electrodes to release reactive species causing the hitting of ions [90].

13.6.3 COLD PLASMA APPLICATION FOR PRESERVATION OF RTE SNACKS

Cold plasma has the potential of preservation of food without bargaining with the quality of the produce. According to research done by Phan, Phan [91], the deactivation of microbes is accomplished by reactive species on different parts of the microorganism because lipid membrane are more vulnerable to ROS [92]. The process of peroxidation consequences in the creation of radicals of fatty acids, which, upon interaction with oxygen, resulting in the formation of lipid hydroperoxide. The generation of irreversible covalent adducts by the lipid peroxides is another mechanism by which it deteriorates the proteins and DNA [93]. During plasma generation, ultraviolet photons are also produced. These photons cause genetical damage for preventing DNA imitation [94]. Photons are responsible for the oxidation of nucleotides because of ROS. Free radicals can cause changes to proteins and lipids, but they can also break down carbon-oxygen and carbon-nitrogen bonds that are exists in cell wall constituents, which can result in destruction to the wall of the cell. This is due to the free radicals can alter lipids and proteins [76]. The contents of the cell matrix get released because of cell membrane permeabilization. The reactive species are release which destructs the protein and intracellular nucleic acids, which have a bactericidal effect [95]. Scanning electron microscopy was used to determine the morphological alterations that were brought about in bacterial cells. These changes included electroporation, membrane, cell wall damage. It is toxic to both spores and vegetative cells thus a useful technique for the preservation of the product [96]. In solid foods, surface decontamination is prioritized, whereas for liquid foods, each individual particle is brought into directly in contact to plasma. Decontaminating materials used in packaging is another use for cold plasma technology.

For instance, when a radiofrequency pulse discharge was applied to the air within PET bottles, it causes a decrease of microorganisms equivalent to three log cycles and had a deodorizing impact after only a few milliseconds of exposure [97, 98]. Number of researchers has investigated cold plasma as a potential method for eliminating microbial contamination from packing materials [99]. Etching is the process that results from the volatilization of spore surface components, which is also caused by reactive plasma species [100]. Application of CP for five minutes at 200 W resulted in the population of *B. subtilis* was reduction by 3.5 log cycle [101]. For the deactivation of microorganism of indented food item more exposure time is required [102, 103]. 3-D configuration of the protein changes because of the breakage of peptide bonds [104]. Alterations to starch can also be accomplished by the utilization of cold plasma technology [105]. Crystallinity is greatly impacted when starch granules are subjected to plasma treatment [106]. On rice and cassava starch, graft polymerization with ethylene was performed using glow plasma [107]. After being applied with cold plasma, it was discovered that the germination rate of the pulse seeds enhanced between 10 and 20% [108]. Cold plasma application in food business possesses significant commercialization potential. To fulfill with the safety, optimization of the operational parameters is needed.

13.7 PULSED ELECTRIC FIELD

Pulsed electric field (PEF), is a non-thermal processing procedure that include subjecting the product to an electrical field with a higher voltage for brief amount of time. In this process electrical pulses within 10 to 80 KV most frequently utilized. PEF is a good method for inactivating enzymes and removing microbial contamination from a substance. In addition to this, it makes the process of mass transfer during drying operations better [109, 110]. When compared to conventional thermal processing methods, PEF provides a number of benefits, including improved color, flavor, and less effect on nutrients, pathogens are reduced, thus improving storage life [111]. PEF can be used to modify food proteins in order to improve their stability and usefulness which can be done in different ways [112]. Beside food industry, In the fields of genetic engineering and biotechnology PEF is also used for electrofusion and cell hybridization [113].

13.7.1 PRINCIPLE OF PULSED ELECTRIC FIELD

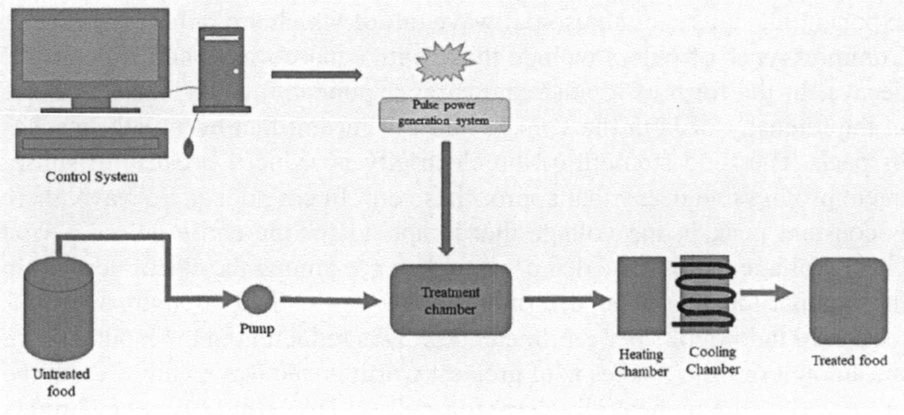

FIGURE 13.7 Pulse electric field system for the preservation of RTE snacks.

Source: Adapted from Ref. [114]

PEF uses high intensity of electric fields having pulsed voltages of 10 to 80 kV range are utilized for microseconds to milliseconds. The item to be treated is placed within the two electrodes, and then is subjected to pulse of higher intensities. Space within the electrodes is referred to as the "Treatment gap." Depending on the requirements of the processing it can be operated at different temperatures [115]. Food contains ions, which are what give food its electrical conductivity, it is possible for electrical pulses to be transmitted through the meal. This is because of the ionic composition of the product [56]. During the application of PEF number of difference phenomena occurs that includes increase in the resistive heating as well as cell membrane disruption [116]. Electroporation of microorganisms is an important process that happens during the treatment, and deactivation of microorganisms is because of electroporation. In addition to it deactivation of PEF can be utilized as a pre-processing step prior the execution of various unit operations such as drying. It doesn't have much of an effect on the nutritional or gustatory qualities of the foods it's added to.

13.7.2 DESIGN OF EQUIPMENT AND PROCEDURES

The equipment consists of a pulse generation system which converts the alternative to direct current. Switches are used to controlling the flow of electric energy into the chamber and a capacitor that stores energy. The pulse producing network also known as PFN is the formation of discharge that influence wave type that is produced or the pulse shape. It could be an exponential, square, or sinusoidal wave, all of which are rather short [117]. Common types of pulses include those with square waves and exponential decays. In the form of a pulse known as exponential decay, there is a rise in the intensity of both the voltage and the current that eventually reaches its peak. The field strength of the electricity is reduced because of subsequent progressive decay that approaches zero. In case of square wave, there is constant peak in the voltage that is upheld for the entire pulsed period [118]. Voltage that is provided per unit distance among the electrodes within the compartment is what determines the electric field strength, which is measured in kilovolts per centimeter. For 37% reduction in the input voltage the amount of time it takes is of greatest worth is one way to think about the pulse width of exponentially decaying pulses. The term "frequency" refers to the number of pulses that are carried outin certain amount of time (Hz). The specific energy of the pulse, measured in kilojoules per kilogram, is mostly dependent on the pulse width and voltage, as well as the shape and conduction of the material [118].

13.7.3 PRESERVATION OF RTE SNACKS USING PEF

The usage of high intensity electrical field cause the collapse of membrane of the cell causing formation of pore inside the cell membrane [119]. During PEF, the structural changes occurring in the membrane are governed by several different theories. In the beginning of the use with PEF, there are no changes that take place in the cellular structure. An increase in intensity that is greater than that of critical level causes irreversible cell membranes electroporation of microorganisms [120, 121]. To provide an explanation for the phenomenon of electroporation, the dielectric rupture theory was created. A trans-membrane potential rise takes place through the cellular membrane which is said to take place as a result in external field of electricity, and this theory also predicts that the membrane thickness will also decrease. The breakdown of the membrane takes place when there is a greater rise in electro compression than there is in the process of restore of the force withing the cellular membrane. Dynamic model which was developed to explain the process of pore creation, it was noted that an adequate voltage is provided to increase the size of the pore to more than the critical diameter and then increase is much more making it irreversible [122]. The membrane structure of the cell is of a lipid bilayer which possesses an overall positive electric

TABLE 13.1 Advantages and Limitations of Various Technology Used for Preservation of RTE Snacks

Technique	Advantages	Limitations
High pressure	Destruction of microbial spores and vegetative cells Absence of toxics Nutritional composition remains the same after processing. Colour and Flavour are conserved. Lesser processing time Even treatment throughout Needed texture changes occur. Processing even when the food is packed. Reduced use of chemical preservatives Attractiveness to customer	Less effect on enzymatic activity, cooling of the product is required. Few microorganisms can live even after the treatment. High cost for antimicrobial effect 40% free water must be present
Ultrasound	Deactivate enzyme and microorganism. Reduction of processing time and temperature Increase heat transfer. structure and texture of the food is changed. Can be batch or continuous. Affect enzymatic activity	Complex mode of action Penetration depth is affected by solids and air in product. Presence of free radical causing destruction Undesirable changes in the food For proper effectiveness should be used in combination Scaling problem
Microwave	Fast volumetric heating Higher drying rate Less drying time Increase in the quality of the product. Less energy consumption Reduced operation cost	Higher initial cost Partial loss of flavor and nutrients Changes in texture
Gamma radiation	Dependable Excellent food penetration Appropriate for sterilization Other application like sprout inhibition Very less quality deterioration For large scale production it is appropriate Less cost of energy Can kill insects. Appropriate for low moisture foods	Costly Localized radiation risks Less awareness in consumer Flavor changes because of oxidation

charge. Electrical field causes changes in the lipid molecules, which further increases the pores that are already there and causes new pore formation [123]. Additionally, the electric field influences the cell membranes protein channels. High intensity electric field application to protein channels, this results in the irreversible denaturation of the channels [123]. In order to induce irreversible alterations, lower electric field strength is needed in case of prokaryotes as compared to that of eukaryotic cells [124]. Bacterial spores have a far higher level of resistance to PEF when compared to that of vegetative cell. PEF in conjunction with different preservation methods enhances the technique's capacity to eradicate microorganisms. An exponential pulse of 20 kV/cm electric strength delivered at 10 pulses with a width of 30 microseconds rate causes decrease in *E. coli* but field strength of less than the critical range of 3 kV/cm was given, it was found that E. coli was not subject to any potentially fatal effects [125].

13.8 HURDLE TECHNOLOGY

The integration of several distinct preservation methods into a single preservation plan is what's known as "hurdle technology." There are over 60 possible barriers for foods that have the potential to make minimally processed items more stable or increase their overall quality. Controlling the temperature, redox potential, water activity, acidity of food, and using preservatives, changing environment, and competitive microbes, are the most critical obstacles that are typically encountered in the process of preserving food (e.g., lactic acid, bacteria). Combining obstacles allows for a relatively modest intensity of the individual preservation procedures, which helps to minimize the loss of quality, while the complete influence on microorganism development may continue to be substantial. This helps to ensure that the product is of the highest possible quality. Natural preservatives are one type of technology that is considered a hurdle in slowing down the spoiling process of food [126].

The term "hurdle technology" refers to combined methods for preservation of food, for the purpose of producing foods that are safe, stable, and nutritious. The idea behind hurdle technology is utilized in a wide variety of countries for the preservation of goods in a way that is gentle but nevertheless quite efficient. Around the world, people have been using a combination of different preservation techniques for hundreds of years, even if there isn't a solid scientific explanation behind the practice. These techniques include physical, chemical, and biological aspects. Improved understanding has allowed for the application of the concept. The idea of this technology is so compatible to consumers' trend toward foods that have seen only the barest minimum of processing, it has enjoyed a great deal of success both in the realm of practical application and in the realm of study. Although a single method of food preservation (the 'hurdle'), such as pasteurization, might be sufficient for food safety, the present inclination in the food sector is to prevent the product quality without forfeiting the safety of the food. This contrasts with the practice of using a single method of food preservation. For the effective formation of hurdle proper knowledge of the type of microorganism and the microorganism that are targeted should be well known. Figure 13.8 shows an example that provides various hurdles for preserving the food product.

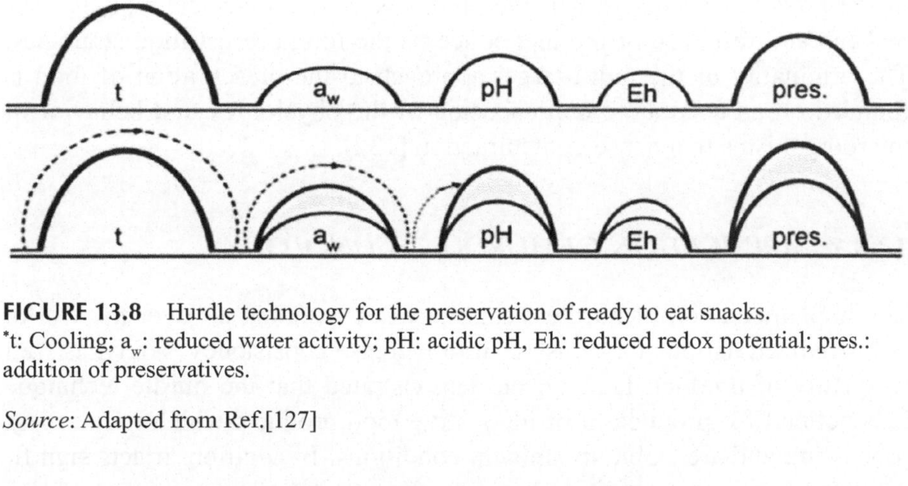

FIGURE 13.8 Hurdle technology for the preservation of ready to eat snacks.
*t: Cooling; a_w: reduced water activity; pH: acidic pH, Eh: reduced redox potential; pres.: addition of preservatives.

Source: Adapted from Ref.[127]

13.8.1 PRINCIPLE OF HURDLE TECHNOLOGY

Food preservation techniques conventionally used like thermal treatment, temperature reduction, drying and dehydration, salting, curing, syruping, brining, fermentation, smoking, etc., are utilized in order to confirm safety and consistency of the supply [128]. Several distinct challenges were cleverly merged in order to provide preservation which are easy and also exceedingly dependable [129]. Applying barriers (hurdles) simultaneously to contaminating microbes such as temperature, water activity, chemical preservatives, acidity, competing for flora, etc., is one method of controlling contamination [128]. These barriers have a synergistic effect on one another and cause damages which are severe than those caused by sole technique. The microbes must exert a particular amount of effort in order to clear each of the obstacles. When there are a greater number of obstacles, microorganisms will need to exert more effort in order to overcome them [129].

The intrinsic variables are considered to be the most crucial considerations in the obstacle technological process (aw, pH, Eh, and chemicals) elements external to the substance, such as the temperature of storage and gas environment, as well as factors inside the substance, like the processing factors (drying, heating, fermentation). In place of damaging procedures such as thermal processing, novel physical elements are now used [130]. This technique is an essential component for the manufacturing of a wide change of stable goods that are both cooked and ready-to-eat.

For the selection of the appropriate combination of obstacles that will result in injuries and fatalities, it is essential to have a solid understanding of the physiology and growth requirements of microbial cells [131]. The makeup and response of product concerns the food preservation strategies. The originality of the multi-target approach to the preservation of food is founded on an accurate comprehension of the physiology and behavior of microorganisms that cause contamination [132].

13.8.2 APPLICATION OF HURDLE TECHNOLOGY

The implementation of hurdle technology is expanding all over the world; industrialized nations make use of it to increase consistency, safety, as well as quality of product. It has been demonstrated that the hurdle technique is beneficial for production of innovative food products which require less processing and are stable in ambient conditions. In addition, it acts significantly by facilitating modification and enhancement of conventional cuisine in order to generate foods with an intermediate moisture content that are not only stable but also have improved sensory attributes [133, 134].

The term "hurdle technology-preserved foods" refers to food products that have extended shelf life, improved food safety as well as quality of the food, because of application of many preserving elements in conjunction with one another. As was mentioned, the basic techniques of food preservation were created through empirical research several centuries ago. As a result, a wide variety of foods that have been preserved in this manner are manufactured and sold. For instance, the ripening and fermentation process presents a series of challenges that must be overcome before fermented food products (such as sausages, cheeses, and vegetables) can be guaranteed to be free of harmful bacteria and maintain their freshness for extended periods of time. Salt and nitrite are key obstacles in the initial stage of the ripening because they hinder many microorganisms in the batter that are responsible for the fermentation of salami-type fermented sausages. The proliferation of other bacteria depletes the available oxygen. This results in a decrease in the product's redox potential while simultaneously increasing the Eh hurdle. This results in a decrease growth of aerobic microorganisms and a selection trend that leans more toward lactic acid causing bacteria. These are competing floras that contribute to the acidity of the product as well, which increases the pH barrier.

In the case of non-fermented foods as well, such as ready-to-eat items that contain a variety of unprocessed or minimally processed veggies, the hurdle method has been utilized to safeguard both quality and safety. The most important steps in the process of stabilizing these perishable items are refrigeration and the use of packaging in a modified environment. The condition where it is difficult to maintain lower temperature and where there are the chances of low temperature resistant microorganism number of hurdles are required to be applied. These hurdles include bio preservation, which involves the utilization of lactic acid causing bacteria which creates less acid but significant levels of anti-listerial bacteriocins. This helps to decrease the negative impact that the obstacle has on the quality of the food. Bacteriocins are tiny proteinaceous molecules that, in general, have an antibacterial action spectrum that is limited to Gram positive pathogens *L. monocytogenes* is included in this category. Natural antimicrobial compounds can be found in spices and herbs, and aromatic plants they can be utilized when broader or complementary spectrum of activity is desired [130, 135].

13.8.3 LIMITATIONS OF HURDLE TECHNOLOGY

It is generally agreed that the presence of multiple obstacles might have an additive or even a further synergic result than the presence of an only obstacle. Though, as of late, a few studies have demonstrated that hostile outcomes of grouping therapies are also probable, and consequently, collective hurdles were shown not as much of efficient at falling levels of microbes in comparison to single treatments. Most hurdle technology tactics planned can be "mild preservation" technology, they–dissimilar sterilization–do not typically kill or deactivate altogether of the microbes that are exist. Instead, they individually prevent the evolution of the microorganisms or inactivate them for short term, which might initiate an element of uncertainty with regard to the food's safety [135]. It is costly and a greater amount of humectant and preservatives are needed. Maximum storage life is generally from 3–6 months. There is high amount of antimicrobial activities [136]. For the formation of hurdle this should be taken into consideration as well as highlighting necessity for authentication should be to accomplish the result with confidence. The production of spores by certain microorganisms is both a significant limitation on the use of moderate preservation methods and an excellent illustration of how microbes adapt to their environments. The development of spores mitigates the effects of heating and prevents acids or other preservatives from penetrating the vegetative cell structure of the organism. When a cell detects stress, it immediately begins to undergo a series of adaptations that are more temporary in nature [135].

13.9 CONCLUSION

The reasons for the current popularity of RTE snacks, which are linked to improvements in supply quality and changes in value systems and lifestyles, in any case, allow us to envision a future scenario in which these items' food habits expand further, at least for some groups of people. The literature's conclusions in terms of profiling the convenience-oriented customer identify time pressure, youth, and a scarcity of cooking skills as basic traits, providing several important factors to forecast the future of these items. Young people, who today exhibit a strong interest in these items, will become the group of consumers with the most purchasing power in the future and will thus be able to exert an increasing influence on overall food demand. The development of a supply of swiftly produced foods with high quality and healthy content but at a higher cost could be an intriguing strategy to pursue.

The convenience of RTE snack items as well as the attraction for its texture and sensory qualities, customer demand for RTE snacks has been steadily increasing. The global food industry has a lot of room to make products that can give the consumer a variety of nutritional ranges, e.g., RTE snacks with a low GI and high bioactive content for people who want to eat healthily. Over the next 10 years, most of the research into how extrusion technology affects the chemical and nutritional profile of RTE snacks is likely to be conducted, which will benefit both consumers and the food industry from the most recent findings on the association between the structure of the food its nutrition profile. Emerging technologies have the potential to be used in the preservation of the ready to eat snack industry to address the need for safe and high-quality food. In the food sector, several techniques—like HPP, ultrasound, cold plasma—are already in use. On the other hand, before they can be commercialized, other uses need more study. In addition to being used as preservation methods, these processes are used to pre-process a variety of items.

KEYWORDS

- **high pressure processing**
- **microwave**
- **non-thermal preservation**
- **RTE snacks**
- **ultrasound**

REFERENCES

1. Brennan, M. A., Derbyshire, E., Tiwari, B. K., & Brennan, C. S. (2013). Ready-to-eat snack products: The role of extrusion technology in developing consumer acceptable and nutritious snacks. *International Journal of Food Science & Technology, 48*(5), 893–902.
2. Singha, P., Singh, S. K., & Muthukumarappan, K. (2019). Textural and structural characterization of extrudates containing blends of apple pomace, defatted soy flour and corn grits. *Journal of Food Process Engineering, 42*, e13046.
3. Singha, P., Singh, S. K., Muthukumarappan, K., & Krishnan, P. (2018). Physicochemical and nutritional properties of extrudates from food grade distiller's dried grains, garbanzo flour, and corn grits. *Food Science & Nutrition, 6*(7), 1914–1926.
4. Silberbauer, A., & Schmid, M. (2017). Packaging concepts for ready-to-eat food: Recent progress. *Journal of Packaging Technology and Research, 1*(3), 113–126.
5. Giwa, A. S., Memon, A. G., Ahmed Shaikh, A., Korai, R., Maitlo, G. U., Maitlo, I., Ali, S., & Ahmad, J. (2021). Microbiological survey of ready-to-eat foods and associated preparation surfaces in cafeterias of public sector universities. *Environmental Pollutants and Bioavailability, 33*(1), 11–18.
6. Toepfl, S., Mathys, A., Heinz, V., & Knorr, D. (2006). Potential of emerging technologies for energy efficient and environmentally friendly food processing. *Food Reviews International, 22*(4), 405–423.

7. Farkas, D. F., & Hoover, D. G. (2000). High pressure processing. *Journal of Food Science, 65*, 47–64.
8. Hite, B. H. (1899). The effect of pressure in the preservation of milk: A preliminary report (Vol. 58). Morgantown, WV: West Virginia agricultural experiment station.
9. Hite, B. H., & Giddings, N. J. (1914). The effect of pressure on certain microorganisms encountered in the preservation of fruits and vegetables. *Bulletin of the West Virginia University Agricultural Experiment Station, 146*, 1–67.
10. Knorr, D. W. (1993). Effects of high-hydrostatic-pressure processes on food safety and quality. *Food Technology, 47*, 156–161.
11. Ravichandran, C., Jayachandran, L. E., Kothakota, A., Pandiselvam, R., & Balasubramaniam, V. (2022). Influence of high-pressure pasteurization on nutritional, functional and rheological characteristics of fruit and vegetable juices and purees—An updated review. *Food Control, 146*, 109516.
12. Raghunathan, R., Pandiselvam, R., Kothakota, A., & Khaneghah, A. M. (2021). The application of emerging non-thermal technologies for the modification of cereal starches. *LWT, 138*, 110795.
13. Yordanov, D. G., & Angelova, G. V. (2010). High pressure processing for foods preserving. *Biotechnology & Biotechnological Equipment, 24*, 1940–1945.
14. Gomathy, K., Pandiselvam, R., Kothakota, A., & Ramesh, S. (2021). High-pressure processing: Potential applications for foods. In *Handbook of Research on Food Processing and Preservation Technologies* (pp. 3–32). Apple Academic Press.
15. Hokmollahi, F., & Ehsani, M. R. (2017). High pressure processing and its application in cheese manufacturing: A review. *Journal of Food Bioscience and Technology, 7*, 57–66.
16. Dash, D. R., Singh, S. K., & Singha, P. (2022). Recent advances on the impact of novel non-thermal technologies on structure and functionality of plant proteins: A comprehensive review. *Critical Reviews in Food Science and Nutrition, 64*(10), 3151–3166.
17. Barbhuiya, R. I., & Singha, P. (2023). Positive displacement pumps. In *Transporting Operations of Food Materials within Food Factories* (pp. 201–217).
18. Mertens, B. (1995). Hydrostatic pressure treatment of food: Equipment and processing. In G. W. Gould (Ed.), *New Methods of Food Preservation* (pp. 135–158). Springer US.
19. Nabi, B. G., Mukhtar, K., Arshad, R. N., Radicetti, E., Tedeschi, P., Shahbaz, M. U., Walayat, N., Nawaz, A., Inam-Ur-Raheem, M., & Aadil, R. M. (2021). High-pressure processing for sustainable food supply. *Sustainability, 13*(24), 13908.
20. Elamin, W. M., Endan, J. B., Yosuf, Y. A., Shamsudin, R., & Ahmedov, A. (2015). High pressure processing technology and equipment evolution: A review. *Journal of Engineering Science & Technology Review, 8*(5), 75–83.
21. Balasubramanian, S., & Balasubramaniam, V. M. (2003). Compression heating influence of pressure transmitting fluids on bacteria inactivation during high pressure processing. *Food Research International, 36*, 661–668.
22. Balci, A. T., & Wilbey, R. A. (1999). High pressure processing of milk—The first 100 years in the development of a new technology. *International Journal of Dairy Technology, 52*(4), 149–155.
23. Palou, E., Lopez-Malo, A., & Welti-Chanes, J. (2002). Innovative fruit preservation methods using high pressure. In *Engineering and Food for the 21st Century* (pp. 745–746). CRC Press.

24. Barbhuiya, R. I., & Singha, P. (2022). Pomelo wastes. In K. Muzaffar, S. Ahmad Sofi, & S. A. Mir (Eds.), *Handbook of Fruit Wastes and By-Products: Chemistry, Processing Technology, and Utilization* (pp. 19–38). CRC Press.
25. McClements, J. M. J., Patterson, M. F., & Linton, M. (2001). The effect of growth stage and growth temperature on high hydrostatic pressure inactivation of some psychotropic bacteria in milk. *Journal of Food Protection, 64*, 514–522.
26. Asaithambi, N., Singh, S. K., & Singha, P. (2021). Current status of non-thermal processing of probiotic foods: A review. *Journal of Food Engineering, 303*, 110567.
27. Barbhuiya, R. I., & Singha, P. (2021). A comprehensive review on impact of non-thermal processing on the structural changes of food components. *Food Research International, 149*.
28. Simpson, R. K., & Gilmour, A. (1997). The effect of high hydrostatic pressure on Listeria monocytogenes in phosphate-buffered saline and model food systems. *Journal of Applied Microbiology, 83*(2), 181–188.
29. Barbosa-Canovas, G., Pothakamury, U., & Swanson, B. (1995). State of the art technologies for the stabilization of foods by non-thermal processes: Physical methods. In *Food Preservation by Moisture Control: Fundamentals and Applications* (pp. 493–532).
30. Farr, D. (1990). High pressure technology in the food industry. *Trends in Food Science & Technology, 1*, 14–16.
31. Abe, F. (2007). Exploration of the effects of high hydrostatic pressure on microbial growth, physiology and survival: Perspectives from piezophysiology. *Bioscience, Biotechnology, and Biochemistry, 71*, 2347–2357.
32. Koseki, S., & Yamamoto, K. (2014). Chapter 1: High pressure microbial inactivation: Modelling and its features. In *Novel Food Preservation and Microbial Assessment Techniques* (pp. 3–24). CRC Press.
33. Yang, B., Shi, Y., & Xia, X. (2012). Inactivation of foodborne pathogens in raw milk using high hydrostatic pressure. *Food Control, 28*, 273–278.
34. Niven, G. W., Miles, C. A., & Mackey, B. M. (1999). The effects of hydrostatic pressure on ribosome conformation in Escherichia coli: An *in vivo* study using differential scanning calorimetry. *Microbiology, 145*, 419–425.
35. Ulmer, H. M., Gänzle, M. G., & Vogel, R. F. (2000). Effects of high pressure on survival and metabolic activity of *Lactobacillus plantarum* TMW1.460. *Applied and Environmental Microbiology, 66*, 3966–3973.
36. Georget, E., Sevenich, R., & Reineke, K. (2015). Inactivation of microorganisms by high isostatic pressure processing in complex matrices: A review. *Innovative Food Science & Emerging Technologies, 27*, 1–14.
37. Pandiselvam, R., Aydar, A. Y., Kutlu, N., Aslam, R., Sahni, P., Mitharwal, S., Gavahian, M., Kumar, M., Raposo, A., & Yoo, S. (2022). Individual and interactive effect of ultrasound pre-treatment on drying kinetics and biochemical qualities of food: A critical review. *Ultrasonics Sonochemistry, 106261*.
38. Asaithambi, N., Singha, P., Dwivedi, M., & Singh, S. K. (2019). Hydrodynamic cavitation and its application in food and beverage industry: A review. *Journal of Food Process Engineering, 42*(5), e13144.
39. Barbhuiya, R. I., Singha, P., & Asaithambi, N. (2022). Ultrasound-assisted rapid biological synthesis and characterization of silver nanoparticles using pomelo peel waste. *Food Chemistry, 385*, 132602.

40. Taha, A., Mehany, T., Pandiselvam, R., Anusha Siddiqui, S., Mir, N. A., Malik, M. A., Sujayasree, O., Alamuru, K. C., Khanashyam, A. C., & Casanova, F. (2022). Sono processing: Mechanisms and recent applications of power ultrasound in food. *Critical Reviews in Food Science and Nutrition*, 1–39.
41. Aslam, R., Alam, M. S., Kaur, J., Panayampadan, A. S., Dar, O. I., Kothakota, A., & Pandiselvam, R. (2022). Understanding the effects of ultrasound processing on texture and rheological properties of food. *Journal of Texture Studies, 53*(6), 775–799.
42. Astráin-Redín, L., Alejandre, M., Raso, J., Cebrián, G., & Álvarez, I. (2021). Direct contact ultrasound in food processing: Impact on food quality. *Frontiers in Nutrition, 8*, 633070.
43. Dolas, R., Saravanan, C., & Kaur, B. P. (2019). Emergence and era of ultrasonics in fruit juice preservation: A review. *Ultrasonics Sonochemistry, 58*, 104609.
44. Rastogi, N. K., & Knorr, D. (2013). *Recent Developments in High Pressure Processing Of Foods*. Springer.
45. Zhu, Z., Zhang, P., & Sun, D. W. (2020). Effects of multi-frequency ultrasound on freezing rates and quality attributes of potatoes. *Ultrasonics Sonochemistry, 60*, 104733.
46. Aydar, A. Y., Aydın, T., Karaiz, A., Alabey, F., Kothakota, A., Raposo, A., Albaridi, N. A., & Pandiselvam, R. (2023). Effect of ultrasound assisted cleaning on pesticide removal and quality characteristics of *Vitis vinifera* leaves. *Ultrasonics Sonochemistry, 92*, 106279.
47. Jacob, A., Sudagar, I., Pandiselvam, R., Rajkumar, P., & Rajavel, M. (2022). Optimization of ultrasound processing parameters for preservation of matured coconut water using a central composite design. *Quality Assurance and Safety of Crops & Foods, 14*(SP1), 33–41.
48. Awad, T. S., Moharram, H. A., & Shaltout, O. E. (2012). Applications of ultrasound in analysis, processing and quality control of food: A review. *Food Research International, 48*, 410–427.
49. Tao, Y., & Sun, D. W. (2015). Enhancement of food processes by ultrasound: A review. *Critical Reviews in Food Science and Nutrition, 55*, 570–594.
50. Saeeduddin, M., Abid, M., & Jabbar, S. (2015). Quality assessment of pear juice under ultrasound and commercial pasteurization processing conditions. *LWT–Food Science and Technology, 64*, 452–458.
51. Gallo, M., Ferrara, L., & Naviglio, D. (2018). Application of ultrasound in food science and technology: A perspective. *Foods, 7*, 164.
52. Datta, A. K., & Davidson, P. M. (2000). Microwave and radio frequency processing. *Journal of Food Science, 65*, 32–41.
53. Singha, P., & Muthukumarappan, K. (2016). Quality changes and freezing time prediction during freezing and thawing of ginger. *Food Science & Nutrition, 4*(4), 521–533.
54. Ohlsson, T. (1993). Domestic use of microwave ovens. In *Encyclopaedia of Food Science, Food Technology and Nutrition* (Vol. 2, pp. 1232–1237).
55. Kutlu, N., Pandiselvam, R., Saka, I., Kamiloglu, A., Sahni, P., & Kothakota, A. (2022). Impact of different microwave treatments on food texture. *Journal of Texture Studies, 53*(6), 709–736.
56. Lal, A. N., Prince, M., Kothakota, A., Pandiselvam, R., Thirumdas, R., Mahanti, N. K., & Sreeja, R. (2021). Pulsed electric field combined with microwave-assisted extraction of pectin polysaccharide from jackfruit waste. *Innovative Food Science & Emerging Technologies, 74*, 102844.

57. Ryynanen, S., & Ohlsson, T. (1996). Microwave heating uniformity of ready meals as affected by placement, composition, and geometry. *Journal of Food Science, 61*, 620–624.
58. Mahapatra, A. K., Muthukumarappan, K., & Julson, J. L. (2005). Applications of ozone, bacteriocins, and irradiation in food processing: A review. *Critical Reviews in Food Science and Nutrition, 45*(6), 447–461.
59. Lacroix, M. (2014). Irradiation. In *Emerging Technologies for Food Processing* (pp. 293–312). Elsevier.
60. Ravindran, R., & Jaiswal, A. K. (2019). Wholesomeness and safety aspects of irradiated foods. *Food Chemistry, 285*, 363–368.
61. Prakash, A., & Jesús Ornelas-Paz, J. (2019). Irradiation of fruits and vegetables. In *Postharvest Technology of Perishable Horticultural Commodities* (pp. 563–589). Woodhead Publishing.
62. Prakash, A. (2020). What is the benefit of irradiation compared to other methods of food preservation? In *Genetically Modified and Irradiated Food* (pp. 217–231). Academic Press.
63. Hernández-Hernández, H. M., Moreno-Vilet, L., & Villanueva-Rodríguez, S. J. (2019). Current status of emerging food processing technologies in Latin America: Novel non-thermal processing. *Innovative Food Science & Emerging Technologies, 58*, 102233.
64. Mittendorfer, J. (2016). Food irradiation facilities: Requirements and technical aspects. *Radiation Physics and Chemistry, 129*, 61–63.
65. Bisht, B., Bhatnagar, P., & Gururani, P. (2021). Food irradiation: Effect of ionizing and non-ionizing radiations on preservation of fruits and vegetables–A review. *Trends in Food Science & Technology, 114*, 372–385.
66. Ashraf, S., Sood, M., Bandral, J. D., Trilokia, M., & Manzoor, M. (2019). Food irradiation: A review. *International Journal of Chemical Studies, 7*, 131–136.
67. Naresh, K., Varakumar, S., Variyar, P. S., Sharma, A., & Reddy, O. V. S. (2014). Enhancing antioxidant activity, microbial and sensory quality of mango (*Mangifera indica* L.) juice by γ-irradiation and it's *in vitro* radioprotective potential. *Journal of Food Science and Technology, 52*(7), 4054–4065.
68. Alsager, O. A., Alnajrani, M. N., & Alhazzaa, O. (2018). Decomposition of antibiotics by gamma irradiation: Kinetics, antimicrobial activity, and real application in food matrices. *Chemical Engineering Journal, 338*, 548–556.
69. Lu, Y., Jie-Lun, H., & Jun-Yi, Y. (2020). Progress on the effect of microwave irradiation on structural characteristics of starch and its application in starch-derived food processing. *Science and Technology of Food Industry, 18*, 330.
70. Yuan, W., Xueting, L., Jinding, C., Jinfang, L., Like, M., & Yanxiang, G. (2021). Research progress of electron beam irradiation technology and its application in the food industry. *Food Industry, 7*, 257–261.
71. Lima Filho, T., Della Lucia, S. M., & Lima, R. M. (2015). Conjoint analysis as a tool to identify improvements in the packaging for irradiated strawberries. *Food Research International, 72*, 126–132.
72. Junqueira-Gonçalves, M. P., Galotto, M. J., & Valenzuela, X. (2011). Perception and view of consumers on food irradiation and the radura symbol. *Radiation Physics and Chemistry, 80*, 119–122.

73. Roberts, P. B. (2014). Food irradiation is safe: Half a century of studies. *Radiation Physics and Chemistry, 105*, 78–82.
74. Thirumdas, R., Sarangapani, C., & Annapure, U. S. (2015). Cold plasma: A novel non-thermal technology for food processing. *Food Biophysics, 10*, 1–11.
75. Luo, Q. Z., D'angelo, N., & Merlino, R. L. (1998). Shock formation in a negative ion plasma. *Physics of Plasmas, 5*, 2868–2870.
76. Misra, N. N., Pankaj, S. K., & Segat, A. (2016). Cold plasma interactions with enzymes in foods and model systems. *Trends in Food Science & Technology, 55*, 39–47.
77. Niveditha, A., Pandiselvam, R., Prasath, V. A., Singh, S. K., Gul, K., & Kothakota, A. (2021). Application of cold plasma and ozone technology for decontamination of *Escherichia coli* in foods–A review. *Food Control, 130*, 108338.
78. Sruthi, N., Josna, K., Pandiselvam, R., Kothakota, A., Gavahian, M., & Khaneghah, A. M. (2022). Impacts of cold plasma treatment on physicochemical, functional, bioactive, textural, and sensory attributes of food: A comprehensive review. *Food Chemistry, 368*, 130809.
79. Lu, X., Naidis, G. V., Laroussi, M., Reuter, S., Graves, D. B., & Ostrikov, K. (2016). Reactive species in non-equilibrium atmospheric-pressure plasmas: Generation, transport, and biological effects. *Physics Reports, 630*, 1–84.
80. Kaavya, R., Pandiselvam, R., Gavahian, M., Tamanna, R., Jain, S., Dakshayani, R., Khanashyam, A. C., Shrestha, P., Kothakota, A., Arun Prasath, V. (2022). Cold plasma: A promising technology for improving the rheological characteristics of food. *Critical Reviews in Food Science and Nutrition*, 1–15.
81. Anbarasan, R., Jaspin, S., Bhavadharini, B., Pare, A., Pandiselvam, R., & Mahendran, R. (2022). Chlorpyrifos pesticide reduction in soybean using cold plasma and ozone treatments. *LWT–Food Science and Technology, 159*, 113193.
82. Surowsky, B., Schlüter, O., & Knorr, D. (2015). Interactions of non-thermal atmospheric pressure plasma with solid and liquid food systems: A review. *Food Engineering Reviews, 7*(2), 82–108.
83. Smet, C., Baka, M., & Dickenson, A. (2018). Antimicrobial efficacy of cold atmospheric plasma for different intrinsic and extrinsic parameters. *Plasma Processes and Polymers, 15*, 1700048.
84. Mayookha, V., Pandiselvam, R., Kothakota, A., Ishwarya, S. P., Khanashyam, A. C., Kutlu, N., Rifna, E., Kumar, M., Panesar, P. S., & Abd El-Maksoud, A. A. (2022). Ozone and cold plasma: Emerging oxidation technologies for inactivation of enzymes in fruits, vegetables, and fruit juices. *Food Control*, 109399.
85. Li, H. P., Ostrikov, K., & Sun, W. (2018). The energy tree: Non-equilibrium energy transfer in collision-dominated plasmas. *Physics Reports, 770–772*, 1–45.
86. Pankaj, S. K., Wan, Z., & Keener, K. M. (2018). Effects of cold plasma on food quality: A review. *Foods, 7*(1), 4.
87. Romani, V. P., Olsen, B., & Collares, M. P. (2019). Improvement of fish protein films properties for food packaging through glow discharge plasma application. *Food Hydrocolloids, 87*, 970–976.
88. Roy, M. (2017). Protective hard coatings for tribological applications. In *Materials under Extreme Conditions* (pp. 259–292). Elsevier.
89. Sakudo, A., Misawa, T., & Yagyu, Y. (2020). Equipment design for cold plasma disinfection of food products. In *Advances in Cold Plasma Applications for Food Safety and Preservation* (pp. 289–307). Elsevier.

90. Niemira, B. A., & Gutsol, A. (2011). Nonthermal plasma as a novel food processing technology. In *Nonthermal Processing Technologies for Food* (pp. 272–288).
91. Phan, K. T. K., Phan, H. T., & Brennan, C. S. (2017). Nonthermal plasma for pesticide and microbial elimination on fruits and vegetables: An overview. *International Journal of Food Science & Technology, 52*, 2127–2137.
92. Alkawareek, M. Y., Gorman, S. P., & Graham, W. G. (2014). Potential cellular targets and antibacterial efficacy of atmospheric pressure non-thermal plasma. *International Journal of Antimicrobial Agents, 43*, 154–160.
93. Joshi, S. G., Cooper, M., & Yost, A. (2011). Nonthermal dielectric-barrier discharge plasma-induced inactivation involves oxidative DNA damage and membrane lipid peroxidation in *Escherichia coli. Antimicrobial Agents and Chemotherapy, 55*, 1053–1062.
94. Guo, J., Huang, K., & Wang, J. (2015). Bactericidal effect of various non-thermal plasma agents and the influence of experimental conditions in microbial inactivation: A review. *Food Control, 50*, 482–490.
95. Mai-Prochnow, A., Murphy, A. B., & McLean, K. M. (2014). Atmospheric pressure plasmas: Infection control and bacterial responses. *International Journal of Antimicrobial Agents, 43*, 508–517.
96. Feichtinger, J., Schulz, A., & Walker, M. (2003). Sterilization with low-pressure microwave plasmas. *Surface and Coatings Technology, 174*, 564–569.
97. Koulik, P., Begounov, S., & Goloviatinskii, S. (1999). Atmospheric plasma sterilization and deodorization of dielectric surfaces. *Plasma Chemistry and Plasma Processing, 19*, 311–326.
98. Deilmann, M., Halfmann, H., & Bibinov, N. (2008). Low-pressure microwave plasma sterilization of polyethylene terephthalate bottles. *Journal of Food Protection, 71*, 2119–2123.
99. Schneider, J., Baumgärtner, K. M., & Feichtinger, J. (2005). Investigation of the practicability of low-pressure microwave plasmas in the sterilization of food packaging materials at industrial level. *Surface and Coatings Technology, 200*, 962–966.
100. Philip, N., Saoudi, B., & Crevier, M. C. (2002). The respective roles of UV photons and oxygen atoms in plasma sterilization at reduced gas pressure: The case of N_2-O_2 mixtures. *IEEE Transactions on Plasma Science, 30*, 1429–1436.
101. Hury, S., Vidal, D. R., & Desor, F. (1998). A parametric study of the destruction efficiency of Bacillus spores in low pressure oxygen-based plasmas. *Letters in Applied Microbiology, 26*, 417–421.
102. Fernández, A., Shearer, N., & Wilson, D. R. (2012). Effect of microbial loading on the efficiency of cold atmospheric gas plasma inactivation of *Salmonella enterica* serovar Typhimurium. *International Journal of Food Microbiology, 152*, 175–180.
103. Ziuzina, D., Patil, S., & Cullen, P. J. (2014). Atmospheric cold plasma inactivation of *Escherichia coli*, *Salmonella enterica* serovar Typhimurium and *Listeria monocytogenes* inoculated on fresh produce. *Food Microbiology, 42*, 109–116.
104. Dobrynin, D., Fridman, G., & Friedman, G. (2009). Physical and biological mechanisms of direct plasma interaction with living tissue. *New Journal of Physics, 11*, 115020.
105. Rostamabadi, H., Bajer, D., Demirkesen, I., Kumar, Y., Su, C., Wang, Y., Nowacka, M., Singha, P., & Falsafi, S. R. (2023). Starch modification through its combination with other molecules: Gums, mucilage's, polyphenols and salts. *Carbohydrate Polymers, 314*.

106. Zhang, B., Xiong, S., & Li, X. (2014). Effect of oxygen glow plasma on supramolecular and molecular structures of starch and related mechanism. *Food Hydrocolloids, 37*, 69–76.
107. Lii, C. Y., Liao, C. D., & Stobinski, L. (2002). Exposure of granular starches to low-pressure glow ethylene plasma. *European Polymer Journal, 38*, 1601–1606.
108. Filatova, I., Azharonok, V., & Kadyrov, M. (2011). The effect of plasma treatment of seeds of some grain and legumes on their sowing quality and productivity. *Romanian Journal of Physics, 56*, 139–143.
109. Toepfl, S. (2006). Pulsed electric fields (PEF) for permeabilization of cell membranes in food and bioprocessing: Applications, process and equipment design, and cost analysis.
110. Onwude, D. I., Hashim, N., & Janius, R. (2017). Non-thermal hybrid drying of fruits and vegetables: A review of current technologies. *Innovative Food Science & Emerging Technologies, 43*, 223–238.
111. Stoica, M., Bahrim, G., & Cârâc, G. (2011). Factors that influence the electric field effects on fungal cells. In *Science against Microbial Pathogens: Communicating Current Research and Technological Advances* (pp. 291–302). World Scientific Publishing Co Pte Ltd.
112. Zhao, W., Yang, R., & Zhang, H. Q. (2012). Recent advances in the action of pulsed electric fields on enzymes and food component proteins. *Trends in Food Science & Technology, 27*, 83–96.
113. Ngadi, M. O., Yu, L., Amiali, M., & Ortega-Rivas, E. (2009). Food quality and safety issues during pulsed electric field processing. In *Processing Effects on Safety and Quality of Foods* (pp. 441).
114. Maged, E. A. M., & Ayman, H. A. E. (2012). Pulsed electric fields for food processing technology. In E. Ayman Amer (Ed.), *Structure and Function of Food Engineering* (pp. Ch. 11). IntechOpen.
115. Zimmermann, U., & Benz, R. (1980). Dependence of the electrical breakdown voltage on the charging time in *Valonia utricularis*. *Journal of Membrane Biology, 53*, 33–43.
116. Sitzmann, W. (1995). High-voltage pulse techniques for food preservation. In *New Methods of Food Preservation* (pp. 236–252). Springer.
117. Morris, C., Brody, A. L., & Wicker, L. (2007). Non-thermal food processing/preservation technologies: A review with packaging implications. *Packaging Technology and Science, 20*, 275–286.
118. Wouters, P. C., Alvarez, I., & Raso, J. (2001). Critical factors determining inactivation kinetics by pulsed electric field food processing. *Trends in Food Science & Technology, 12*, 112–121.
119. Simonis, P., Kersulis, S., & Stankevich, V. (2019). Pulsed electric field effects on inactivation of microorganisms in acid whey. *International Journal of Food Microbiology, 291*, 128–134.
120. Delso, C., Martínez, J. M., Cebrián, G., Condón, S., Raso, J., & Álvarez, I. (2022). Microbial inactivation by pulsed electric fields. In *Pulsed Electric Fields Technology for the Food Industry: Fundamentals and Applications* (pp. 169–207).
121. Martin-Belloso, O., & Elez-Martínez, P. (2005). Food safety aspects of pulsed electric fields. In *Emerging Technologies for Food Processing* (pp. 183–217). Elsevier.
122. Schoenbach, K. H., Joshi, R. P., & Stark, R. H. (2000). Bacterial decontamination of liquids with pulsed electric fields. *IEEE Transactions on Dielectrics and Electrical Insulation, 7*, 637–645.

123. Tsong, T. Y. (1989). Electroporation of cell membranes. In *Electroporation and Electrofusion in Cell Biology*. Springer.
124. Heinz, V., Álvarez, I., & Angersbach, A. (2001). Preservation of liquid foods by high intensity pulsed electric fields—Basic concepts for process design. *Trends in Food Science & Technology, 12*, 103–111.
125. Evrendilek, G. A., & Zhang, Q. H. (2005). Effects of pulse polarity and pulse delaying time on pulsed electric fields-induced pasteurization of *E. coli* O157: H7. *Journal of Food Engineering, 68*, 271–276.
126. Rico, D., Martín-Diana, A. B., Barat, J. M., & Barry-Ryan, C. (2007). Extending and measuring the quality of fresh-cut fruit and vegetables: A review. *Trends in Food Science & Technology, 18*(7), 373–386.
127. Jag, P., Om Pravesh Kumar, R., Sangeeta, K., & Akhilesh Kumar, S. (2021). Preservation of seafoods by hurdle technology. In R. Chhabi Lal (Ed.), *Meat and Nutrition* (pp. Ch. 8). IntechOpen.
128. Leistner, L. (1992). Food preservation by combined methods. *Food Research International, 25*(2), 151–158.
129. Leistner, L., & Gorris, L. G. M. (1995). Food preservation by hurdle technology. *Trends in Food Science & Technology, 6*(2), 41–46.
130. Guerrero, S. N., Ferrario, M., Schenk, M., & Carrillo, M. G. (2017). Hurdle technology using ultrasound for food preservation. In *Ultrasound: Advances for Food Processing and Preservation* (pp. 39–99). Elsevier.
131. Wordon, B., Mortimer, B., & McMaster, L. D. (2012). Comparative real-time analysis of *Saccharomyces cerevisiae* cell viability, injury, and death induced by ultrasound (20 kHz) and heat for the application of hurdle technology. *Food Research International, 47*(2), 134–139.
132. Leistner, L. (2000). Basic aspects of food preservation by hurdle technology. *International Journal of Food Microbiology, 55*(1–3), 181–186.
133. Shelef, L., & Seiter, J. (2005). Indirect and miscellaneous antimicrobials. In *Antimicrobials in Food* (3rd ed., pp. 573–598).
134. Abdullahi, N., & Dandago, M. A. (2021). Hurdle technology: Principles and recent applications in foods. *Indonesian Food and Nutrition Progress, 17*(1), 6–11.
135. Mukhopadhyay, S., & Gorris, L. G. M. (2014). Hurdle technology. In *Encyclopedia of Food Microbiology* (pp. 221–227).
136. Neeha, V., & Kakade, S. B. (2014). Use of hurdle technology in food preservation. *International Journal of Engineering and Management Research (IJEMR), 4*(5), 204–212.

INDEX

D

E

N

O

P

Q

R

S

T

W

X

Y

Z